The Dynamics of Partially Molten Rock

The Dynamics of Partially Molten Rock

RICHARD F. KATZ

PRINCETON UNIVERSITY PRESS
Princeton and Oxford

Published by Princeton University Press
41 William Street, Princeton, New Jersey 08540
6 Oxford Street, Woodstock, Oxfordshire OX20 1TR

press.princeton.edu

Library of Congress Control Number 2021946194
ISBN 978-0-691-17656-7
E-book ISBN 978-0-691-23264-5

British Library Cataloging-in-Publication Data is available

Editorial: Ingrid Gnerlich and Whitney Rauenhorst
Production Editorial: Jill Harris
Jacket Design: Wanda España
Production: Danielle Amatucci
Publicity: Matthew Taylor and Charlotte Coyne
Copyeditor: Gregory Zelchenko

This book has been composed in Minion Pro & Universe LT Std

Printed on acid-free paper. ∞

Printed in the United States of America

10 9 8 7 6 5 4 3 2 1

To Michael and Jake.

Contents

	Preface	xiii
	List of Symbols	xvii
CHAPTER 1	**Introduction**	1
	1.1 Motivation	1
	1.2 Basic Physical Considerations	2
	1.3 Research Questions and Applications	6
	1.4 About This Book	12
	1.4.1 Overview of the Organization and Content	12
	1.4.2 References to the Literature	14
	1.4.3 Mathematical Notation	14
	1.5 The Way Forward?	16
CHAPTER 2	**A Condensed History of Magma/Mantle Dynamics**	18
	2.1 Foundation	18
	2.2 Axial Age	21
	2.3 Exploration	24
	2.4 Generalization, Extension, and Future History	26
CHAPTER 3	**A Review of One-Phase Mantle Dynamics**	29
	3.1 Governing Equations	29
	3.2 Mantle Convection	31
	3.3 Kinematic Solutions for Corner Flow	33
	3.4 Literature Notes	35
	3.5 Exercises	36
CHAPTER 4	**Conservation of Mass and Momentum**	39
	4.1 The Representative Volume Element and Phase-Averaged Quantities	39
	4.2 Conservation of Mass	42
	4.3 Conservation of Momentum	44
	4.3.1 Stress and Pressure	45
	4.3.2 The Interphase Force	46
	4.3.3 Constitutive Equations in the Magma–Mantle Limit	48

	4.4	A Note about Disaggregation	51
	4.5	The Full Mechanical System, Assembled	52
	4.6	Special, Limiting Cases	53
		4.6.1 No Porosity, No Melting	53
		4.6.2 Partially Molten, Rigid Medium	53
		4.6.3 Constant, Uniform Solid Viscosity	53
	4.7	Literature Notes	54
	4.8	Exercises	56
CHAPTER 5		**Material Properties**	58
	5.1	Microstructure	58
		5.1.1 Grain-Size Change	59
		5.1.2 Textural Equilibration	62
	5.2	Permeability	66
	5.3	Viscosity	68
		5.3.1 Shear Viscosity of the Aggregate	68
		5.3.2 Compaction Viscosity of the Aggregate	69
		5.3.3 Shear Viscosity of the Liquid	73
	5.4	Thermodynamic Properties	73
		5.4.1 Density	73
		5.4.2 Solid–Liquid Phase Change	74
	5.5	Literature Notes	75
	5.6	Exercises	78
CHAPTER 6		**Compaction and Its Inherent Length Scale**	80
	6.1	The Compaction-Press Problem	81
	6.2	The Permeability-Step Problem	83
	6.3	Propagation of Small Porosity Disturbances	86
	6.4	Magmatic Solitary Waves	88
	6.5	Solitary-Wave Trains	93
	6.6	The Compaction Length in the Asthenosphere	94
	6.7	Literature Notes	96
	6.8	Exercises	97
CHAPTER 7		**Porosity-Band Emergence under Deformation**	100
	7.1	Governing Equations	101
	7.2	Linearized Governing Equations	102
	7.3	Viscosity	105
	7.4	The (In)Stability of Perturbations	105
		7.4.1 Pure Shear	105
		7.4.2 Simple Shear	108
	7.5	Wavelength Selection by Surface Tension	114
	7.6	Literature Notes	116
	7.7	Exercises	119
CHAPTER 8		**Conservation of Energy**	121
	8.1	The Internal-Energy Equation	122
	8.2	The Enthalpy Equation	124
	8.3	The Temperature Equation	126

8.4 The Entropy Equation 126
8.5 Boussinesq and Lithostatic Approximations 127
8.6 Dissipation-Driven Melting and Compaction 130
8.7 Decompression Melting 134
8.8 Literature Notes 137
8.9 Exercises 139

CHAPTER 9 **Conservation of Chemical-Species Mass** 141
9.1 Thermodynamic Components 143
9.1.1 Congruent Melting 145
9.1.2 Incongruent Melting 145
9.2 Trace Elements 146
9.2.1 Equilibrium Transport Model 147
9.2.2 Disequilibrium Transport Model 148
9.3 Radiogenic Trace Elements and Their Decay Chains 151
9.4 Closed-System Evolution of a Decay Chain 153
9.4.1 Evolution with Melting Only 154
9.4.2 Evolution due to Ingrowth Only 154
9.4.3 Evolution by Both Melting and Ingrowth 156
9.5 Literature Notes 158
9.6 Exercises 159

CHAPTER 10 **Petrological Thermodynamics of Liquid and Solid Phases** 161
10.1 The Equilibrium State 162
10.1.1 Partition Coefficients from Ideal Solution Theory 163
10.1.2 Computing the Equilibrium State 167
10.1.3 Application to a Two-Pseudo-Component System 167
10.1.4 Application to a Three-Pseudo-Component System 169
10.1.5 Approaching the Eutectic Phase Diagram 170
10.1.6 Linearizing the Two-Component Phase Diagram 173
10.1.7 Degree of Melting 173
10.2 Thermochemical Disequilibrium and the Rate of Interphase Mass Transfer 174
10.2.1 Affinity as the Thermodynamic Force for Linear Kinetics 175
10.2.2 Linearized Melting Rates 177
10.3 Computing the Melting Rate at Equilibrium 177
10.4 Remarks about Mantle Thermochemistry 179
10.5 Literature Notes 180
10.6 Exercises 181

CHAPTER 11 **Melting Column Models** 182
11.1 Fluid Mechanics 182
11.2 Melting-Rate Closures 186

11.2.1 Prescribed Melting Rate 186
11.2.2 Thermodynamically Consistent Melting Rate 187
11.3 The Visco-Gravitational Boundary Layer 196
11.4 The Decompaction Boundary Layer 199
11.5 Isotopic Decay-Chain Disequilibria in a Melting Column 203
11.5.1 Constant Transport Rates 205
11.5.2 Variable Transport Rates 208
11.6 Literature Notes 210
11.7 Exercises 211

CHAPTER 12 **Reactive Flow and the Emergence of Melt Channels** 214
12.1 Governing Equations 215
12.2 The Melting-Rate Closure 215
12.3 Problem Specification 217
12.4 Scaling and Simplification 218
12.5 Linearized Stability Analysis 221
12.5.1 The Base State 222
12.5.2 The Growth Rate of Perturbations 223
12.5.3 The Large-Damköhler Number Limit 226
12.5.4 A Modified Problem and Its Analytical Solution 227
12.6 Physical Mechanisms 231
12.7 Application to the Mantle 233
12.8 Literature Notes 234
12.9 Exercises 236

CHAPTER 13 **Tectonic-Scale Models and Modeling Tools** 238
13.1 Governing Equations in the Small-Porosity Approximation 239
13.2 Corner-Flow with Magmatic Segregation 241
13.3 Melt Focusing through a Sublithospheric Channel 243
13.3.1 Lateral Transport in Semi-Infinite Half-Space 243
13.3.2 Lateral Transport to a Mid-Ocean Ridge 246
13.4 Coupled Dynamics and Thermochemistry with the Enthalpy Method 252
13.5 Literature Notes 255
13.6 Exercises 257

CHAPTER 14 **Numerical Modeling of Two-Phase Flow** 258
14.1 The One-Dimensional Solitary Wave (Instantaneous) 259
14.1.1 Finite Difference Discretization 260
14.1.2 Finite Element Discretization 263
14.2 The One-Dimensional Solitary Wave (Time-Dependent) 267
14.3 A Two-Dimensional Manufactured Solution (Instantaneous) 269
14.4 Magmatic Solitary Waves as a Benchmark for Numerical Solutions 275
14.5 Porosity Bands as a Benchmark for Numerical Solutions 276
14.6 Literature Notes 277
14.7 Exercises 278

CHAPTER 15 **Solutions to Exercises** 280
15.1 Exercises from Chapter 3: One-Phase Mantle Dynamics 280
15.2 Exercises from Chapter 4: Conservation of Mass and Momentum 284
15.3 Exercises from Chapter 5: Material Properties 287
15.4 Exercises from Chapter 6: Compaction and Its Inherent Length Scale 293
15.5 Exercises from Chapter 7: Porosity-Band Emergence under Deformation 300
15.6 Exercises from Chapter 8: Conservation of Energy 303
15.7 Exercises from Chapter 9: Conservation of Chemical-Species Mass 308
15.8 Exercises from Chapter 10: Petrological Thermodynamics of Liquid and Solid Phases 310
15.9 Exercises from Chapter 11: Melting Column Models 310
15.10 Exercises from Chapter 12: Reactive Flow and the Emergence of Melt Channels 315
15.11 Exercises from Chapter 13: Tectonic-Scale Models 318
15.12 Exercises from Chapter 14: Numerical Modeling of Two-Phase Flow 319

Bibliography 321

Index 337

Preface

The discovery of plate tectonics tectonics in the late 1960s and subsequent advances in seismic tomography, isotope geochemistry, and geodynamic modeling fundamentally altered our understanding of the solid Earth. Together they show that its outer surface, locus of an enviroment crucial for biological evolution, is merely the skin of a writhing beast below. The character and history of the skin can be understood only through a broader knowledge of the beast as a whole: the convecting, heterogeneous mantle, that occupies the outer $\sim$3000 km of Earth. The mantle is understood to produce and interconnect the spatiotemporal signals that we observe from our vantage point outside the skin. From ocean basins to mountain belts, from continental rifts to isolated volcanoes, from mid-ocean ridges to faults that generate the most destructive earthquakes and tsunamis—even to the long-term composition of the soil, atmosphere and occans—modern interpretations of the surface environment recognize the central role of the mantle. The habitability of Earth (and hence our very existence) arises in part because the mantle convects to export radiogenic and primordial heat from Earth's interior.

But convection cycles mass up and down to reduce the internal accumulation of potential energy; it doesn't create the chemical diversity of rocks and environments that comprise the skin of Earth. This diversity arises from rock melting and magmatism,[1] which can separate chemical components from each other and send them off to different destinations. Magma *fractionates* chemical components, creating physical and chemical heterogeneity. Some magma becomes the lava that erupts from volcanos, bringing a sample of the source rocks from the mantle to the surface. This magma fractionates further by releasing its volatile elements into the atmosphere and ocean. Moreover, cycles of partial melting, segregation, and freezing can create rocks of distinct composition, such as the continental crust. Radiogenic, heat-producing elements tend to be concentrated in magmas relative to their solid residues. Hence magmatism is a long-term control on the distribution of heating within Earth, and a feedback on mantle convection.

Primary magmas that rise from the deep mantle are rarely attributable to a single, distinct source, because magma also *mixes* compositions. Because it is a liquid, magma has a lower viscosity and higher chemical diffusivity than rock. So as it segregates from its source, it irreversibly mixes with other magma that may be derived from a chemically distinct source. This mixing obscures the heterogeneity inherited from the mantle source and leads to lava that, although distinct in composition from its source rocks, is generally more homogeneous than the individual melts that were mixed to form it.

[1] As well as weathering, sedimentation and metamorphism, though those will not concern us here.

There is a thin veneer of this homogenized lava that blankets more than half of the solid Earth; it is called the oceanic crust.

But magma also affects the mechanics of plate tectonics because it *weakens* rock, making it more susceptible to deformation. Hence magma production near plate boundaries tends to localize and sharpen those boundaries by concentrating deformation there. Localized deformation is a signature of and, hypothetically, a mechanical requirement for plate tectonics on Earth. For example, intrusion of magma-filled fractures into continental crust makes that crust susceptible to extension under lower forces. Extension promotes further magmatic intrusion, such that a feedback can ensue that aids the rifting of continents. Continental rifting and breakup creates sedimentary basins, releases atmospheric gasses, and shapes the evolution of life.

So to develop an understanding of the character and history of the skin of our planet (and other planetary bodies too), it isn't enough to study the animating force of mantle convection. It is also crucial to study magmatism: its fluid mechanics, thermodynamics and chemistry. And while there are a variety of excellent books that cover the subject of mantle convection for students and researchers,[2] there have been none that are fully dedicated to the physics of magmatism. Yet there are many researchers measuring and documenting the nature and consequences of magmatism. And there is a substantial but dispersed scientific literature that constructs a theoretical framework to quantify and explain magmatism. A key purpose of this book is to collect, condense, organize, splice, elaborate, and present that literature in an accessible and convenient source.

This book is intended to help scientists, including PhD students and advanced undergraduates, to prepare themselves for research on or using the theory of magmatism. It is intended to make the theory more broadly accessible to researchers including geodynamicists and petrologists, but also to planetary scientists, mathematicians, physicists, rock mechanicists, and others who can make physical, mathematical or environmental connections to their research. These investigators will bring new perspectives, ideas, and methods, and make progress by developing (or refuting) parts of the theory. Indeed, and perhaps most importantly, this book is intended to present the physical/mathematical theory of mantle magmatism as a *hypothesis* that, although well established, still requires testing against observations of natural and laboratory systems to establish the limits of its validity.

Supplementary material. An online supplement to the book is available through Princeton University Press at the following URL:

https://press.princeton.edu/books/hardcover/9780691176567
/the-dynamics-of-partially-molten-rock

It provides Jupyter Notebooks containing a Python implementation of the code used to generate the book's quantitative figures. These are offered as a means to deepen understanding, explore parameter space, and template research codes.

Acknowledgments. This book is a product of several decades of my life occupied with learning, researching and, more recently, with writing. The work during those years was supported by teachers, mentors, colleagues, and students. It wouldn't have been possible, let alone fun and satisfying, without that group of people. I can mention only

[2]These include *Mantle Convection in the Earth and Planets* (Schubert et al., 2001), *Dynamic Earth: Plates, Plumes and Mantle Convection* Davies (2000), and *Theoretical Mantle Dynamics* Ribe (2018). The latter also has a chapter on the dynamics of partially molten rock.

some of them here, and this list is weighted to those whose contributions have been more direct and/or more recent.

My first and most important teacher was Piero P. Foà, my grandfather, who inculcated the habit of asking questions about nature, and who revealed how natural complexity can be distilled into simple models. A great physics teacher, Frank Norton, inspired my fascination for Newtonian mechanics and its mathematical expression. I was fortunate to have a PhD supervisor, Marc Spiegelman, who was a pioneer of magma dynamics theory (Marc was a student of Dan McKenzie, arguably the founder of field). During my PhD from 2000 through 2006, he generously shared his time, knowledge, and enthusiasm, and set me on a course that I have followed for 20 years, including as a mentor to my own students. Grae Worster encouraged and guided me to deepen my understanding of fluid mechanics and phase change. He introduced me to methods and topics that have enhanced my research in magma dynamics and have become interests in their own right. Philip England helped me to navigate the waters of academia as a junior professor and to build my research group. He has provided frequent reminders to keep my science close to observations and measurements, where models are most valuable. Yasuko Takei has inspired me with her unwavering integrity, creativity, and intellectual courage, and shown me the richness and interest of rock mechanics. This book wouldn't have been possible without the mentorship of this group; I hope they will take some small pride in it.

I have benefited from the collaboration and generous assistance of a set of colleagues who read drafts and responded to specific questions with research, ideas and mathematical contributions. John Rudge gave invaluable feedback on various chapters and provided the calculations that underpin section 11.4; David Rees Jones led a project (Rees Jones and Katz, 2018, constructively reviewed by Peter Kelemen) that generated most of the content for chapter 12, contributed analysis that lead to the model of section 11.3 and gave helpful suggestions on many other parts of the manuscript; Ian Hewitt suggested the approach taken in section 13.3.2. I am indebted to Peter van Keken and Yanick Ricard for their insightful reviews of a draft of this book. The latter was exceptionally generous with detailed comments and suggestions.

Parts of the book were improved by advice from Paul Asimow (chapters 8 and 10), Tom Breithaupt (section 5.1.1), Nestor Cerpa (section 11.2.2), Tobias Keller (section 4.3.3), Dave May (chapter 14), and Dan McKenzie (throughout, but especially chapter 2). Two undergraduate students, Ben Jackson and Zhen Ning Liu, did extensive proofreading and developed end-of-chapter exercises and their solutions. Kevin Miller and Wen-Lu Zhu provided the graphics used in figures 4.1 and 5.3. The online, supplementary Python notebooks were developed with great care and attention to detail by Frederico Santos Teixeira. I was aided by readers who pointed out mistakes and unclear passages including Tong Bo, Jonathan Burley, Prin Eksinchol, Taras Gerya, Lars Hansen, Marianne Haseloff, Luke Jenkins, Jake Jordan, Teresa Kyrke-Smith, Yuan Li, Parker Liautaud, Adina Pusok, Patrick Sanan, Dan Spencer, Dave Stegman, Meng Tian, Sam Weatherley, and Hanwen Zhang.

I am grateful to Peter Molnar for an invitation to teach at the Abdus Salam International Centre for Theoretical Physics in 2011, which prompted me to write the lecture notes that grew to become this book. The majority of that growth occured during a 2016 sabbatical at the University of Cape Town, on the welcome of Chris Harris and while my Oxford teaching was covered by Philip England. The book was completed with the support of Ingrid Gnerlich and colleagues at Princeton University Press, to whom I extend my thanks. I am deeply grateful to the staff of St. Anne's Nursery, Oxford, who have looked after my two boys with kindness and joy over the past five years, and to Maciej Zdrodowski, who shaped the wonderful home in which my family lives (and, in 2021, works).

Between 2009 and 2021 I have benefited from research grants made by the Natural Environment Research Council (NE/H00081X/1, NE/I026995/1, NE/M000427/1, NE/R000026/1), the European Research Council (279925, 772255), and the Deep Carbon Observatory of the Sloan Foundation. These have enabled me to work with talented students and postdocs, including those mentioned above. I am grateful for the support of the Earthquake Research Institute of the University of Tokyo, which provided support during my 2012 sabbatical in Tokyo to work with Takei-san. My efforts in writing this book, especially during my writing period in Cape Town, were generously supported by the Leverhulme Trust through a 2013 Leverhulme Prize.

Finally, I thank Lucie Cluver, whose love and gentle encouragement has rekindled my motivation to complete this project at key moments.

List of Symbols

Category	Symbol	Comment
Coordinate	$\boldsymbol{x}$	Position vector
	t	Time
	$\hat{\boldsymbol{x}}$	Dimensionless basis vector with unit length
	$\hat{\boldsymbol{n}}$	Dimensionless normal vector with unit length
Mathematical	$\mathrm{d}q$	Infinitesimal increment of any quantity q
	$\boldsymbol{I}$	Identity tensor
	T	Transpose
	ϵ	A dimensionless quantity that is much less than unity
	$[q]$	Characteristic scale of a quantity q, e.g., $[\boldsymbol{v}] = U$
	$\mathcal{O}(q)$	Order of magnitude q, e.g., $\mathcal{O}(\epsilon^2)$
	$q^{(j)}$	jth entry in a series expansion of quantity q
Mechanics	$\boldsymbol{v}$, v	Velocity vector and its magnitude
	$\boldsymbol{\omega}$, ω	Vorticity, vector or scalar (for two-dimensional flows)
	ρ	Density
	$\mathbf{g}$, g	Gravitational acceleration vector and its magnitude
	P	Pressure
	$\boldsymbol{\sigma}$	Stress tensor
	$\boldsymbol{\tau}$	Deviatoric stress tensor
	$\dot{\boldsymbol{e}}$	Strain-rate tensor, $\dot{\boldsymbol{e}} \equiv \frac{1}{2}\left[\boldsymbol{\nabla}\boldsymbol{v} + (\boldsymbol{\nabla}\boldsymbol{v})^{\mathsf{T}}\right]$
	$\dot{\boldsymbol{\varepsilon}}$	Deviatoric strain-rate tensor
		$\dot{\boldsymbol{\varepsilon}} \equiv \frac{1}{2}\left[\boldsymbol{\nabla}\boldsymbol{v} + (\boldsymbol{\nabla}\boldsymbol{v})^{\mathsf{T}} - \frac{2}{3}(\boldsymbol{\nabla}\cdot\boldsymbol{v})\boldsymbol{I}\right]$
	$\dot{\varepsilon}_{II}$	Second invariant of strain rate tensor, $\dot{\varepsilon}_{II} \equiv \sqrt{\dot{\boldsymbol{\varepsilon}} : \dot{\boldsymbol{\varepsilon}}/2}$
	η	Shear viscosity of the solid
	$\boldsymbol{\Psi}$, ψ	Vector potential, vector and scalar forms
	$\mathcal{U}$	Scalar potential
Two-phase	q^{ℓ}	Any quantity q pertaining to the liquid
	q^{s}	Any quantity q pertaining to the solid
	$\overline{q}$	Phase average of any quantity q, $\overline{q} \equiv \phi q^{\ell} + (1-\phi)q^{s}$
	Δq	Phase difference of any quantity q, $\Delta q \equiv q^{s} - q^{\ell}$
	ϕ, ϕ^{ℓ}	Volume fraction of liquid
	$(1-\phi)$, ϕ^{s}	Volume fraction of solid

Category	Symbol	Comment
	Γ	Interphase mass-transfer rate (melting positive)
	$\mathcal{C}$	Compaction rate (decompaction positive), $\mathcal{C} \equiv \nabla \cdot \boldsymbol{v}^s$
	$\boldsymbol{q}$	Segregation flux, $\boldsymbol{q} \equiv -\phi \Delta \boldsymbol{v}$
	$\mathbf{F}$	Interphase force per unit volume
	μ^ℓ	Shear viscosity of the liquid
	k_ϕ	Permeability of the porous solid
	K_ϕ	Mobility of liquid through the pores, $K_\phi \equiv k_\phi / \mu^\ell$
	η_ϕ	Shear viscosity of the liquid–solid aggregate
	ζ_ϕ	Compaction viscosity of the liquid–solid aggregate
	ξ_ϕ	Augmented compaction viscosity, $\xi_\phi \equiv \zeta_\phi + \frac{4}{3}\eta_\phi$
	P^i, P^{lith}	Pressure of phase i and lithostatic
	$\mathcal{P}$	Compaction pressure, $\mathcal{P} \equiv \zeta_\phi \mathcal{C} = -(1-\phi)\Delta P$
	δ	Compaction length, usually $\delta \equiv \sqrt{\xi_\phi k_\phi / \mu^\ell}$
	D	Darcy drag, $D \equiv \phi^2 / K_\phi$
	ϕ_Ξ	Porosity of solid disaggregation
Microscopic	RVE	Representative volume element containing solid grains and melt-filled pores
	Φ^i	Indicator function for phase i
	$\breve{q}$	Any quantity q defined within the RVE
	$q^{\mathcal{I}}$	Any quantity pertaining to the liquid–solid interface
	$\mathrm{d}\boldsymbol{S}$	Infinitesimal surface element of the microscopic space
	δV	Volume of the RVE
	$\mathrm{d}^3\breve{\boldsymbol{x}}$	Infinitesimal volume element of the microscopic space
	δ	Dirac Delta function
Material	d	Grain size
	γ	Surface energy
	p	Grain-growth exponent
	Θ	Dihedral angle
	n	Permeability exponent
	$\mathfrak{n}$	Stress exponent in flow law
	$\mathfrak{m}$	Grain-size exponent in flow law
	λ	Sensitivity of η_ϕ to porosity
Energy	F	Degree of melting
	T_p	Potential temperature
	$\mathbb{Q}$	Mass-specific radiogenic heating rate
	Ψ	Volumetric heat dissipation rate
	c_P	Mass-specific heat capacity
	L	Mass-specific latent heat
	k_T	Thermal conductivity
	κ	Thermal diffusivity
	α_ρ	Thermal expansivity
	ϑ	Clausius Clapeyron slope, $\mathrm{d}P/\mathrm{d}T^m$
	u^i	Mass-specific internal energy of phase i

Category	Symbol	Comment
	h^i	Mass-specific enthalpy of phase i
	H	Volumetric bulk enthalpy
	s^i	Mass-specific entropy of phase i
	$\boldsymbol{J}_u$	Diffusive heat flux
Chemistry	c_j^i	Concentration (mass fraction) of species j in phase i
	Γ_j	Interphase mass-transfer rate of species j
	c_j^Γ	Mass fraction of species j in mass transfer, $c_j^\Gamma \equiv \Gamma_j/\Gamma$
	$\mathcal{D}_j$	Diffusivity of species j in liquid
	$\boldsymbol{J}_j^\ell$	Diffusive flux of species j in liquid
	$\mathcal{X}_j$	Exchange reaction rate of species j
	D_j	Mass partition coefficient of trace-element species j
	λ_j	Decay rate of radiogenic species j
	a_j^i	Activitiy of species j in phase i, $a_j^i \equiv \lambda_j c_j^i$
Petrology	f	Mass fraction of liquid
	μ_j^i	Chemical potential of component j in phase i
	$\mathring{\mu}_j^i$	Standard-state chemical potential of component j in phase i
	M_j	Molar mass of component j
	X_j^i	Mole fraction of component j in phase i
	$\check{q}_j^i$	Equilibrium value of any quantity q pertaining to component j in phase i
	q^m	Any quantity q at the melting point of a pure phase
	$\check{K}_j$	Mass partition coefficient of component j, $\check{K}_j \equiv \check{c}_j^s/\check{c}_j^\ell$
	$\check{K}_j^*$	Molar partition coefficient of component j, $\check{K}_j^* \equiv \check{X}_j^s/\check{X}_j^\ell$
	$T^\mathcal{S}$	Solidus temperature
	$M^\mathcal{S}$	Solidus slope, $M^\mathcal{S} \equiv \mathrm{d}T^\mathcal{S}/\mathrm{d}c^s$
	$T^\mathcal{L}$	Liquidus temperature
	$M^\mathcal{L}$	Liquidus slope, $M^\mathcal{L} \equiv \mathrm{d}T^\mathcal{L}/\mathrm{d}c^s$
Column	W_0	Mantle upwelling speed at bottom of column
	K_0, K_1	Mobility prefactors. $K_0 \equiv K_\phi(\phi = \phi_0)$ and $K_1 \equiv K_\phi(\phi = 1)$
	$\mathcal{Q}$	Column flux parameter, $\mathcal{Q} \equiv K_1 \Delta\rho g/W_0$
	δ_f	Length scale of freezing beneath lithosphere
	q_∞	Vertical liquid flux far from boundary layer, $q_\infty \equiv K_0 \Delta\rho g$
	$G(\zeta)$	Freezing shape function

The Dynamics of Partially Molten Rock

CHAPTER 1

Introduction

1.1 Motivation

If not for *magmatism*, the Earth might still be an undifferentiated ball of early solar-system dust—a humble cosmic dust-bunny—rather than the majestic planet on which (and about which) the author composes these words. Magmatism comprises melting of rocks and minerals and segregation of that melt from the residual solid. Over the age of the Earth, magmatism has produced the compositionally layered, radial structure of the core, mantle, and crust (and the ocean and atmosphere, too).

Magmatism shapes the planet today. This is especially true at plate-tectonic boundaries where crust is produced by partial melting of the mantle. These boundaries are associated with mantle convection: convergent boundaries (subduction zones) where negatively buoyant lithosphere founders and sinks into the mantle; divergent boundaries (mid-ocean ridges) where the lithosphere is rifted apart by far-field tectonic stresses. In each of these settings, mantle rock that underlies the boundary partially melts; buoyancy of that melt causes it to be transported toward the surface; some of the transported melt fuels volcanism and the production of new crust.

Indeed, magmatism is responsible for the long-term stability of the geological substrates of human life: atmosphere, oceans, and continental crust. While many of the details remain obscure, we understand the basic outline of this multistep process. It begins at mid-ocean ridges, where divergent plates drive mantle upwelling and partial melting. The magma rises to the ridge axis and, through a combination of eruption and intrusion, forms the oceanic crust. The oceanic crust moves along the sea floor as part of a tectonic plate. While doing so, it is altered by interaction with sea water, becoming hydrated and carbonated. This altered crust eventually subducts back into the mantle. As it sinks to higher pressures and temperatures at depth, metamorphic reactions release the volatile elements from the minerals. They flow into the mantle and cause melting. The magma rises to yield volcanism, this time in a subduction setting, and returns some of the volatile elements to the atmosphere. Through a complicated and poorly understood process, this magma evolves to be continental crust.

But some of the water and carbon that were transported to depth by the oceanic crust do not return immediately to the surface. Instead they are trapped in the mantle. Over the age of the Earth, this could bury all of the surface water and dry out the oceans. But magmatism is again the means of escape: mantle rock that upwells and melts beneath mid-ocean ridges releases its water and carbon into the magma, which transports them to the surface and exsolves them into the ocean or atmosphere.

Thus the plate tectonic/mantle convection system brings rocks up from depth to the surface and back down to depth, while magmatism couples this physical rock cycle to the surface environment. Magma is a crucial link in the chemical cycles that enable habitability of Earth. Magma transports heat and chemical elements, affecting the composition of the atmosphere, ocean, and soils, thus shaping the surface environment. It leads to volcanic eruption hazards and volcanic resources such as hydrothermal energy and ore formation. At mid-ocean ridges, it "re-paves" more than half of the solid Earth (i.e., the ocean floor) with a thin veneer of mantle-derived melt every ~100 million years.

But since the discovery of plate tectonics, geodynamicists have mostly focused on the plate motions that arise from mantle convection. Despite the importance of magmatism, much less effort has been expended to systematically understand and quantitatively model the genesis, segregation, and emplacement of magma and the chemistry it transports. An aim of this book is to promote and facilitate a correction of that imbalance.

Magmatism is obviously not unique to Earth. An extremely active magmatic system shapes Io, one of Jupiter's moons. And since ice is nothing more than a low-density rock, it is fair to say that Europa (another Jovian moon where some 20 km of ice floats atop a ~100-km-thick ocean), is or was shaped by magmatism. The same probably applies to countless exoplanets. Indeed, the physical and chemical interactions of liquid magma and its solid residue must be common to condensed planets throughout the universe. It is a primary aim of this book to strengthen the foundation on which our understanding of these interactions are built.

The primary means by which we shall pursue this aim is to collect relevant theory, developed in different scientific fields, published in different journals, with diverse notations, styles, and applications. We shall bring this together into a single, coherent, structured framework of knowledge: on the written page and also, we hope, in your mind. Exercises are provided to help you erect this mental framework; codes are provided to help you explore and extend the concepts addressed here. Hopefully, with this framework established, you will more easily read the related scientific literature on magma/mantle dynamics, more rapidly come to understand the cutting edge of research, and more ably contribute to the advance of the cutting edge through your own research. Perhaps you will discover that parts of the framework established here should be modified or entirely rebuilt to better describe the physical reality of magmatism; please email me if so!

1.2 Basic Physical Considerations

Magmatism on Earth (let alone all other planets) is broader than the physical system considered in this book. Magmatism on Earth includes the early magma ocean, the production and migration of the lava erupted in continental flood basalts, the deep source region of komatiites, the emplacement of batholiths into the continental crust, and the melting of the continental crust itself to form granites. Magma may be present between 410- and 660-km depth in the mantle mantle transition zone and, deeper, at ~2800-km depth above the core–mantle boundary. Indeed, separation of metal from mantle silicates to form the core is a sort of magmatism, as is the solidification of the inner core. And while magmatism, as a category, may exclude the formation of sea ice or meltwater generation in temperate glaciers, they are governed by essentially the same physics.

Although this book develops theory that is relevant to a wide range of phenomena, our primary concern is magmatism in the shallow ($\lesssim$300 km) mantle, often known as the *asthenosphere*,[1] especially in environments such as mid-ocean ridges, subduction zones and mantle plumes. In these contexts, it is thought that the volume fraction of melt remains small, except where magma stalls before being erupted or crystallizing. In particular, volume fractions in the melting regions are small enough that the solid residue forms a contiguous skeleton that transmits stress and may thus have a different pressure than the interstitial liquid magma.

The mantle is solid; it transmits elastic shear waves in response to rapid changes in stress associated with earthquakes. Observed over longer time scales, however, it is fluid. The most obvious consequence of this fluidity is postglacial rebound, which arises from flow of mantle rock toward surface depressions left by retreating ice sheets. This fluid behavior of a polycrystalline aggregate occurs through solid-state deformation known as *creep*. There are various microscopic mechanisms of creep that are active at the grain scale (e.g., diffusion of crystal mass, motion of crystal dislocations); we shall not be concerned with their microscopic details. At the continuum scale, the creeping aggregate can be described with a viscosity that relates the deviatoric stress and strain rate of the continuum. The theory for flow of such a viscous, slowly deforming material is based on the Stokes equation.

Magma is liquid; it has a range of viscosity that, at its lower end where it is associated with low-silica basalts, is the same as that of glucose syrup at room temperature. It forms by melting of more fusible minerals that are distributed throughout the polycrystalline mantle. The magma fills the pore space at the junctions between solid grains. Because of the wetting properties of magma with mantle minerals, the pores form an interconnected network that is permeable even at vanishingly small volume-fractions of melt. Within the pores at the microscopic scale, a Stokes balance of forces applies to the magma, controlling the microscopic pressure and flow velocity. At the continuum scale, however, these microscopic variables cannot be resolved; instead, upscaled variables representing volume-averaged melt velocity and pressure satisfy a modified Darcy's law.

To summarize the physical context: the mantle is a high-viscosity, polycrystalline, creeping solid that forms a contiguous but porous skeleton (also called the solid *matrix*); the magma is a low-viscosity liquid that is transported through the interconnected network of pores between solid grains. Both liquid magma and solid mantle are modeled as fluids. As we shall see in subsequent chapters, the *two-phase flow* of partially molten mantle is governed by a Stokes/Darcy system of coupled partial differential equations (PDEs). Associated with this system are the usual material properties of shear viscosity and permeability, which appear as constitutive laws (or closure conditions) in the model formulation.

Both solid mantle and liquid magma are compressible. Indeed, over the full depth of the mantle, density differences due to isentropic compression are almost 50% of the mean density. However, over the pressure range of the asthenosphere, this compression is small. Here, the relevant comparison is to the density difference between solid and liquid phases that drives melt segregation. Hence, for mechanical models of the asthenosphere, we shall make a Boussinesq approximation, neglecting compressibility except in body-force terms (see chapter 4). For some thermal models, we retain the effect of isentropic compressibility on temperature (see chapter 8).

---

[1]Derived from the Greek word *asthenes* meaning weak. The asthenosphere has a lower viscosity that the deeper mantle below it or the lithosphere above it.

It is of fundamental importance that assumptions about (in)compressibility do not preclude *compaction*. Compaction is the consolidation of mantle grains with expulsion of magma; it is represented as a convergent solid flux that is balanced by a divergent liquid flux. Decompaction or dilatation of the mantle is the opposite: solid grains spread apart and magma is locally imbibed, giving a divergent solid flux and convergent liquid flux. To clarify this fundamental concept, consider a mixture of an incompressible solid grains and incompressible liquid. If this mixture is fully enclosed by an impermeable membrane, then it cannot be compacted. However, the same mixture enclosed by a permeable membrane can compact as liquid is expelled through the membrane.

In the context of this book, the resistance to compaction arises from two processes. The first is the expulsion of the liquid phase from the pores between solid grains. Because magma is viscous, its flow is retarded by viscous drag on the surrounding solid; expelling it requires a force that is proportional to its viscosity. However, even if the liquid in the pore network were inviscid (or replaced by a void), a second process would still resist compaction. This process is the grain-scale solid deformation that occurs during closure of pores. The viscous resistance to this deformation is known as the *compaction viscosity*. It is distinct from but related to the better-known *shear viscosity*. Neither the liquid nor the compaction viscosity appear in the theory of single-phase mantle dynamics. In fact, the mechanics of viscous compaction is a special feature of partially molten rock that gives rise to much of its interesting behavior.

Of course melting and solidification, which transfer mass between liquid and solid phases, are also crucial to the physics of the partially molten mantle and they feature prominently in the mathematical description of the system. This brings into play a range of thermal physics and chemical thermodynamics related to phase change. The latter is usually known as petrology; it is a field that is older and more developed than geodynamics. Petrological studies of mantle-derived crystalline rocks and lavas give us detailed knowledge of the chemical and mineralogical composition of the mantle and its melts. The solid mantle is, in fact, a grain aggregate of a variety of mineral phases, each with different properties.

The dominant mineral in the asthenospheric mantle is olivine. Olivine forms a solid solution between a magnesian end member (forsterite, Mg_2SiO_4, about 90% of olivine by mass) and a ferrous end-member (fayalite, Fe_2SiO_4). Forsterite has a very high melting temperature (about 1900 °C at atmospheric pressure) and hence melting of olivine contributes little to magma production. Pyroxenes, although less abundant by volume, are the dominant contributor to mantle partial melting. They are more silicious than olivine and more chemically diverse because they allow more chemical substitutions. Importantly, they can incorporate more water and other impurities than olivine. The magnesium–iron solid-solution series of pyroxene has end members of enstatite ($MgSiO_3$) and ferrosilite ($FeSiO_3$), but pyroxenes typically also include some calcium (e.g., $CaMgSi_2O_6$–$CaFeSi_2O_6$) and up to 10% aluminum. Clinopyroxenes (cpx) and orthopyroxenes (opx) are distinguished by their crystal structure, but also by their fusibility; cpx accommodates calcium and melts more readily than opx. Finally, the asthenospheric mantle contains aluminous minerals including garnet (higher pressure), spinel (intermediate pressure), and plagioclase (lower pressure). Although volumetrically unimportant, these minerals support a huge range of chemical substitutions and hence can exert a disproportionate control on the chemistry of melts.

The relative proportions of olivine, clino- and orthopyroxene, and garnet/spinel/plagioclase vary widely within mantle rocks. Melting tends to strip out the aluminous phase and the clinopyroxene first, then the orthopyroxene, eventually leaving only olivine. Mantle rocks with more cpx are thus considered *fertile* (capable of producing more melt) while those containing mostly olivine are considered *refractory*. A typical fertile peridotite, a spinel lherzolite, might consist of about 66% olivine, 24% opx, 8% cpx, and 2% spinel;[2] in contrast, a refractory peridotite such as dunite would be at least 90% olivine with the remainder composed mostly of opx.

The solid mantle is thus a rather complicated physical entity, comprising multiple mineral phases that vary significantly in their composition and proportion. We shall sidestep much of this complexity by treating the solid mantle as a single, mechanically uniform phase. The mechanical transport properties of that phase are most consistent with olivine, the dominant mineral of the shallow mantle and the mineral commonly used in laboratory deformation experiments. When we consider the thermochemistry of melt production, however, some of the petrological terminology and concepts will reappear in the discussion.

The liquid phase is, in a limited way, simpler than the solid.[3] For most cases of relevance, it is a solution of any chemical components that are rejected by the solid. This includes portions of the oxides of silica, magnesium, iron, calcium, and aluminum, as well as solid-incompatible components such as sodium, potassium, carbon, and water. The magma that is produced by melting of mantle rocks is said to be *basaltic* in composition, meaning that it is the liquid parent of the rock basalt. Basaltic lavas erupt at mid-ocean ridges and ocean islands (e.g., Hawaii), volcanos that directly tap magma produced in the mantle. These lavas carry encoded information about their source and path of transport; they deposit that information where it can be sampled and measured.

The natural system outlined above operates according to a vast and complex set of physical and chemical processes and properties: continuum mechanics, microstructural mechanics, thermal and chemical transport, mineralogy, chemical thermodynamics/petrology, and more. These are embedded in the evolving context of planetary geology and tectonics. Some of the observable consequences of the operation of this system are well understood and explained; many are not. Quantitative theory and models can facilitate quantitative tests of hypothetical explanations of observations. Perhaps more importantly, models can sharpen the questions raised by observations, making hypotheses testable by discovering corollary predictions. However, to be tractable and comprehensible (and hence useful), our theoretical treatment of the magma/mantle system will necessarily simplify aspects, even though more sophisticated theory (i.e., theory that describes the some aspect of the physics in more detail[4]) may be available. It is possible (and, indeed, beneficial) to develop and analyze models more complex than those developed here. But in doing so, it is also possible to create models that cannot be understood in simple terms and hence cannot be validated, even though they may be more "realistic." There is little point in such an exercise. This book aims to provide the theoretical basis for incrementally adding complexity (and realism) to models. An aim of equal importance, however, is to provide a basis of physical understanding of the mathematical models. It is on this basis that we will interpret the model behavior when new physics or chemistry is incorporated.

---

[2] See Wilson [1989] for a introduction to igneous petrology of the mantle.

[3] A caveat: there is no broadly accepted thermochemical model for silicate liquids, whereas for most silicate crystals, a variety of competing thermochemical models exist. Understanding the molecular structure of silicate melts is an area of current research that may help to improve thermochemical theory.

[4] As is the case, for example, of the chemical thermodynamics of mantle melting.

One final, basic physical consideration is important to bear in mind. The theory formulated and analyzed here is a *continuum theory* that is based on volume-averaging of the grain-scale, microscopic physics on a scale much larger than individual mantle grains. It represents fundamental conservation principles (mass, momentum, energy) that are true to an extremely good approximation. But that does not imply that this theory is entirely correct. Representing the microscopic physics at the continuum scale is fraught with the potential for error. For example, approximating the volume average of the product of microscopic variables with the product of their volume averages neglects microscopic correlations between variables. This is a classical problem for mean-field theories, including closures such as the viscosity of an aggregate. Furthermore, in some cases there may be no clear separation of scales between microscopic and continuum. For example, coherent alignment of microscopic pores could lead to anisotropic or discontinuous transport properties at the continuum scale. Moreover, it is plausible that large-scale discontinuities (e.g., magma-filled dikes) could initiate from such features. Such emergent features would have important consequences at the continuum scale but cannot be readily deduced from the continuum model. It is thus important to understand and regularly re-evaluate the assumptions involved in upscaling from the microscopic to the continuum scale.

1.3 Research Questions and Applications

Open questions about terrestrial magmatism motivate this book and the theory that it describes. However, with one exception, it is not our aim to present research that addresses those questions; for that, the reader is directed to the scientific literature and encouraged to take matters into her own hands. But in thinking about the physics and mathematics that is introduced below, it is helpful to have in mind some of the motivating questions. This section provides a brief and incomplete overview.

The exception, alluded to above, is the fundamental question *how does magma move through the asthenosphere and adjust to varying physical and chemical conditions en route?* This book provides a detailed account of a particular hypothesis developed in response to that question. The hypothesis states that magma in the asthenosphere moves by porous flow through a solid matrix that can compact and deform according to viscous rheological laws. This hypothesis is broadly considered to be correct, even if direct evidence to support it is lacking. Of course, direct, in situ evidence of physical processes in the asthenosphere is exceedingly rare. Hence, to test this hypothesis, it is necessary to evaluate its quantitative predictions against the observations that are available.

The porous-flow hypothesis is not the only concept for melt transport in the asthenosphere. Transport through an emergent network of veins and dikes was proposed by Sleep [1984] and developed by Sleep [1988] and, more rigorously, by Rubin [1998]. This hypothesis states that veins form and grow in a partially molten rock under external deviatoric stress. If a vein forms in a plane normal to the least compressive stress (most tensile deviatoric stress), magma flows into the vein due to the difference between the ambient pore pressure and that stress. At the same time, the melt pressure in the vein exceeds the least compressive stress, so the vein dilates and grows. Once the vein has reached a critical vertical extent, the buoyancy of the enclosed melt drives crack opening at the top and crack closing at the bottom; the vein moves upward as a dike (a *magma-fracture*, analogous to a hydrofracture) while it draws additional melt from the surrounding porous medium. These mechanics are certainly relevant for shallow environments with larger deviatoric stress [Rivalta et al., 2015] or capped zones of large magma overpressure [Havlin et al., 2013]. They may also be relevant for discrete heterogeneities [Sleep, 1984] and for mantle depths to about 30 km, depending on the deviatoric stress and background permeability [Nicolas and Jackson, 1982; Nicolas, 1986; Ito and Martel, 2002]. Indeed, seismological evidence in the form of earthquake locations that traverse the subduction-zone mantle wedge support the idea of a brittle mechanism [White et al., 2019]. However, the hypothesis of melt extraction from the asthenosphere by diking has received much less attention than the porous flow theory considered in this book. If diking in the asthenosphere is consistent with observations (more on this below), then its distinct physics must be embedded in the physics considered here.

A second research question, related to the first, is *what is the rate of buoyancy-driven melt transport through the asthenosphere and what mechanics control this rate?* Two lines of observational evidence suggest that this rate is faster than that predicted by diffuse porous flow. The first is disequilibrium in the uranium-series isotopes measured in young lavas.[5] Elemental fractionation of parent and daughter nuclides leads to secular disequilibrium in decay rates, but equilibrium is restored over a time-scale proportional to the half-lives of the elements. Some of these half-lives are of the order of kiloyears to tens of kiloyears, and so observed disequilibria constrain fractionation events to have taken place within that time frame. The hypothesis that fractionation occurs at small melt fractions at great depth in the mantle then suggests that melt transport must be rapid enough to preserve disequilibria. Elliott and Spiegelman [2003] provide an overview and references. U-series disequilibrium in Icelandic lavas indicates melt ascent rates of tens to hundreds of meters per year [Stracke et al., 2006]. Studies for ocean-island basalts and island arcs [Claude-Ivanaj et al., 1998; Turner et al., 2004] come to similar conclusions. There is uncertainty, however, in these interpretations that is associated with the various model assumptions required. For example, in a lithologically heterogeneous mantle, partition coefficients and patterns of melt transport may be substantially more complex than those envisioned by simple models [Weatherley and Katz, 2016].

The second line of observational evidence for fast melt transport is from the reconstruction of eruption rates during and after Icelandic deglaciation. Jull and McKenzie [1996] summarized early observations of enhanced volcanic output and showed that they can be explained by accounting for the mantle decompression melting associated with removal of a 2-km-thick ice sheet. If this ice disappears over $\sim$1000 yr, the melting rate beneath it should increase by a factor of $\sim$30. The lag of the volcanic output time series with respect to deglaciation was used by Maclennan et al. [2002] to constrain the rate of buoyancy-driven melt transport to a minimum of $\sim$50 m/yr; to better fit the observations, their model indicated speeds $>$100 m/yr. Eksinchol et al. [2019] developed a more sophisticated model accounting for spatiotemporal evolution of ice removal and mantle isostatic rebound to predict trace-element concentrations. They obtained a best fit to lanthanum concentration data with an average melt speed of 100 m/yr; larger speeds gave a worse fit. However, both of these models assume quasisteady melt transport—that the time scale for adjustment to steady-state melt transport is negligible relative to the time scale of deglaciation. Using a time-dependent model of melt transport, Rees Jones and Rudge [2020] showed that unsteady effects could be

[5] Models of U-series are developed in section 11.5

mapped onto faster melt speeds in a quasi-steady model. They concluded that the maximum steady-state melt speed beneath Iceland is ∼30 m/yr, and that speeds greater than about 10 m/yr are expected beneath most of the global mid-ocean ridge system. These estimates are consistent with constraints from observed U-series disequilibria.

A null hypothesis for melt transport in the mantle is *diffuse porous flow*. This assumes spatially uniform melting and purely vertical melt transport. Assuming magma is produced by upwelling mantle in which the various minerals are uniformly distributed and in contact with each other, we expect melting to be broadly distributed and to create a pervasive network of interconnected pores. The buoyancy of the melt would drive it to segregate vertically. Simple melting column models (as in Chapter 11) tell us that the ratio of the liquid to solid upwelling speeds scales like the maximum degree of melting (say about 20%) divided by the maximum porosity (say about 1%). Hence for mantle upwelling at 10 cm/yr, diffuse porous flow predicts a melt speed of ∼2 m/yr (see section 11.1). This is an order of magnitude lower than the observational estimates noted above. One possible resolution to this discrepancy is that porous flow is not diffuse, but is instead *channelized*. Such channelized flow has been inferred from geological observations [Kelemen et al., 1995a] and shown to be consistent with uranium-series disequilibria [Jull et al., 2002; Elliott and Spiegelman, 2003]. It was predicted with reactive-flow theory assuming a homogeneous mantle source [Aharonov et al., 1995, and see Chapter 12] and with numerical models of a heterogeneous source [Weatherley and Katz, 2012]. Volatile components such as water and CO_2 in the mantle source may promote deep channelization [Keller and Katz, 2016]. These models all utilize highly simplified mantle thermochemistry, however, leaving open questions of how magmatic channelization works and whether it is quantitatively consistent with observations including that of rapid melt extraction.

A third research question is *by what forces and processes does lateral melt transport occur?* The buoyancy of liquid magma in the asthenosphere (see section 5.4.1) explains its gravity-driven, vertical ascent through the permeable solid. But volumetric and geochemical evidence indicates that the magma erupted from volcanoes is not simply sourced from a vertical column beneath them—it comes from a volume of mantle much broader than the volcano itself [e.g., Behn and Grove, 2015]. The lateral transport associated with this pooling of magma is referred to as *melt focusing* and is often invoked in tectonic-scale models of magmatism and volcanism. However, the mechanics of melt focusing remain a subject of debate, which has played out mostly in the context of mid-ocean ridge models. Early work by Spiegelman and McKenzie [1987] and Morgan [1987] linked magma focusing to the dynamic pressure gradient generated by plate-driven corner flow of the solid phase (see sections 3.3 and 13.2), but this mechanism relies on a large and roughly uniform asthenospheric viscosity that is inconsistent with many estimates. Lateral transport through a high-porosity, high-permeability channel along the sloping base of the lithosphere was proposed by Sparks and Parmentier [1991] and has been a dominant paradigm in recent years; it is described in sections 11.4 and 13.3. In a magnetotelluric study along the Mid-Atlantic Ridge, Wang et al. [2020] obtain inversion results with a striking indication of a sublithospheric channel. In contrast, magnetotelluric tomography on the East Pacific Rise by Key et al. [2013] find no evidence for sublithospheric focusing. The magnetotelluric imaging results from Key et al. [2013] depict a distribution of melt within the asthenosphere that is close to what might be predicted under a third mechanism for focusing, proposed more recently by Turner et al. [2017] and Sim et al. [2020]. In this case, the compaction associated with melt extraction creates pressure gradients that drive lateral flow. Other ideas have been proposed and there is no reason to exclude the possibility that multiple mechanisms contribute.

The research question of *how does heterogeneity of lithology (and hence fusibility) of the mantle affect melting and melt transport?* is something of an "elephant in the room" for all of the questions above. Over hundreds of millions of years (probably a few billions), plate tectonics and magmatism have fractionated mantle components to produce oceanic crust, coated it with sediments, altered it with sea water, and subducted it back into the mantle. This rock, of distinct chemistry and lithology, is stirred by mantle convection but not homogenized. Geochemical evidence shows that it is recycled and remelted beneath plate boundaries and at hot spots [e.g., Stracke, 2012]. The characteristics of this heterogeneity, in terms of its compositions, shapes, and length scales, and volume fraction, are poorly known. Even less clear is what effect it might have on melt transport. Sleep [1984] argued that it could nucleate veins and dikes in the mantle; Richter and Daly [1989] highlighted the potential for reactive porous flow effects. In the petrologically simplified context of a mantle with two chemical components, these effects were explored by Weatherley and Katz [2012] and Jordan and Hesse [2015], who predicted that channelization of flow can occur. The implications of heterogeneity (in two plausible, hypothesized forms drawn from a large possible space) for mid-ocean ridges were investigated by Katz and Weatherley [2012], demonstrating the potential for heterogeneous networks of melt-transport channels to control melting and melt extraction. To what extent these or other results are representative of the natural system of heterogeneous, partially molten asthenosphere remains an open question.

It is broadly clear, however, that melt transport affects how mantle heterogeneity is expressed in basalts that erupt at the surface. Hence petrological and geochemical inferences of mantle heterogeneity rely on assumptions about transport and mixing [e.g., Stracke and Bourdon, 2009]. Thus a fifth research question, which arises in attempting to interpret the chemical signature(s) of erupted basalts, is *what are the distinct contributions of the heterogenous mantle source and the spatially variable extraction process?* This question folds in all of the questions above in that it links the dynamics to the observable chemistry of erupted lava. Channelized melt flow, melt focusing, flow in veins and dikes, and other flow complexities will all have geochemical consequences [e.g., Spiegelman and Kelemen, 2003; Behn and Grove, 2015; Sleep, 1984], and those will play out in the context of the heterogeneity that magma inherits from its mantle source [Weatherley and Katz, 2016]. Addressing this question requires models that couple geochemical transport (see chapter 9) with melting and two-phase flow [Richter, 1986; Navon and Stolper, 1987]. But there is an additional challenge: transport pathways are sensitive to partitioning of elements between the liquid and solid [Spiegelman, 1996]. It is simple to assume that trace-element concentrations in magma are in equilibrium with the solid residue of melting, according to partitioning coefficients, and that these melts are instantaneously extracted with no further interaction with the solid. Although this approach has had success in explaining the systematics of geochemical observations [e.g., White et al., 1992], there are theoretical reasons to doubt its validity. Foremost among these is the extremely slow diffusion of trace elements through the interior of solid grains [van Orman et al., 2001] that explains their observed chemical zonation. Disequilibrium models that account for this diffusion have been formulated [e.g., Kenyon, 1990; Iwamori, 1992; Qin, 1992], but it remains an open question whether their complexity improves the skill of predictions. Moreover, extraction without chemical interaction during transport has obvious conceptual difficulties [Navon and Stolper, 1987]. Hence there remains a great and largely unrealized potential for the systematics residing in abundant geochemical data to constrain the style and pathways of melt transport.

All the melt erupted from volcanoes has traversed the cold thermal boundary layer beneath the surface of the solid Earth. In the lithosphere, temperatures fall below the solidus of rocks of magmatic composition, precluding transport by porous flow. Moreover, lithospheric rocks are cold enough that creeping deformation is negligible on the time scale of melt tranport. Plentiful observations show that under these conditions, magma moves through pressure-driven fractures, dikes, and sills [Rivalta et al., 2015]. What is less clear, however, is *how does transport of magma work at the base of the lithosphere, at the transition between porous flow beneath and brittle fracture above?* This has broad implications that include the geochemistry of lavas, frequency and style of volcanic eruptions, the spatial distribution of volcanism, and the creation of mantle heterogeneity. Two-phase models of viscous deformation predict melt pooling and crystallization at the lithosphere–asthenosphere boundary (LAB) [e.g., Ghods and Arkani-Hamed, 2000; Katz, 2008; Keller et al., 2017]. In contrast, Havlin et al. [2013] hypothesized that accumulated, overpressured melt at the LAB will readily enter the lithosphere through dikes, where it will freeze. One means to addressing this problem is through the use of continuum, two-phase models that capture elastic/brittle deformation in addition to viscous flow. Keller et al. [2013] demonstrated feasibility of such models and showed the emergence of remarkably dike-like features, but further work is needed to validate this approach and explore the behavior of models in tectonically realistic contexts.

Other current research questions are derived from the consideration of particular tectonic environments. Mid-ocean ridges can be modeled with the simplest boundary conditions and material properties; they are the best-studied tectonic environment (see chapters 11and 13). Subduction zones are of great interest because of their role in forming continental crust, element cyling through the deep Earth, and the hazards associated with their volcanoes (and earthquakes). The two-phase dynamics of subduction zones is complicated: a water- and carbon-rich liquid enters the mantle wedge from the relatively cold slab and percolates toward higher temperatures, where it promotes *flux melting* at the volatile-saturated solidus. The aqueous liquid becomes hydrous magma; its viscosity and density increase. It is driven upward by buoyancy and eventually interacts with the lithosphere. Cooler temperatures lead to magma evolution and lateral transport. Some magma is erupted but much is frozen into the lithosphere. This vague but probably accurate overview leaves many open questions. Where in the wedge does flux melting take place? What role does decompression melting play? Is melt transport channelized? How much lateral melt flow occurs and by what mechanisms? What roles do buoyancy forces play in shaping the solid flow? Does the subducting slab itself or its lamination of sediments ever melt? Do the sediments rise into the wedge as diapirs? What is the chemistry of the residue that is dragged back into the mantle? What are the fluxes of volatile elements through subduction zones and, in particular, into the deep mantle? In summary, *what are the two-phase dynamics of subduction zones and how does this relate to chemical cycling through the mantle?* There is a vast literature that considers these and other questions; it will not be reviewed here. Recent numerical models of two-phase flow, however, have developed the framework in which such questions can be studied quantitatively [Wilson et al., 2014; Cerpa Gilvonio et al., 2017]. But these models exclude phase change, considering only transport of the aqueous liquid. Going beyond this requires stable and efficient numerical simulations of the nonlinear interactions between melt flow, chemical reaction, and thermal evolution in the context of large temperature and compositional gradients. Developing these is a major challenge.

Magmatism may also occur at the bottom of the mantle, above the core–mantle boundary [Fiquet et al., 2010]. Stixrude and Karki [2005] showed that melts in the deep mantle are more compressible and probably higher density than the solid residue, so they would be expected to sink toward the core. This has led to the hypothesis that ultralow seismic velocity zones (ULVZ) above the core–mantle boundary are regions of dense, pooled melt [reviewed by McNamara, 2019]. Melt at the core–mantle boundary, if it exists, may be the last remanents of a basal magma ocean that has slowly crystallized over the age of the Earth [Labrosse et al., 2007], or may be the product of mantle dynamics and heat derived from the core. In any case, the physical and chemical conditions just above the core–mantle boundary are highly uncertain, making two-phase models of this region almost entirely unconstrained. In this context where there are many open research questions, anything but the simplest models are potentially misleading.

A colder and more readily observable context where two-phase dynamics may be crucial is in glaciers and ice sheets. Glacial ice is a monomineralic, polycrystalline solid—a rock, broadly construed. When ice melts at grain boundaries, pores form with a small dihedral angle (see chapter 5), creating a permeable network. Indeed, an early derivation of the equations of two-phase flow by Fowler [1984] was motivated by the water/ice problem [see also Schoof and Hewitt, 2016]. Glacial ice at the melting temperature, called *temperate ice*, is now recognized as a significant part of ice sheets. It is found near the bed, where the ice is heated by frictional dissipation and geothermal heating, and insulated from the cold atmosphere [Hewitt and Schoof, 2017]. It is also found at the shear margins of ice streams, where viscous dissipation supplies heat [Jacobson and Raymond, 1998]. Englacial pore water may significantly decrease the viscosity of temperate ice relative to cold ice [Duval, 1977], and this may lead to a positive feedback that sharpens the margins of ice streams [Haseloff et al., 2019]. The behavior of partially molten, temperate ice remains little explored, however, leaving open the broad question *what is the role of temperate ice in the dynamics of glaciers and ice sheets?* This also applies to the multiphase dynamics of firn, the unconsolidated sediment of snow and recrystallized grains that accumulates on the surface of glaciers and ice sheets [Meyer and Hewitt, 2017].

Interaction of liquid and solid phases is ubiquitous in cold, crustal rocks and there is a vast literature on crustal hydrology. In this context, multiphase flow means that there are multiple fluid phases present (the crust is modeled as an elastic solid). Most interest has focused on problems of reactive flow and chemical transport—for example, the motion of contaminants or hydrocarbons through the subsurface. What is less common but more relevant for this book is research that addresses the mechanical interaction between phases. This can occur when reaction causes density changes that promote solid microfracture [Yakobson, 1991], when fluid flow causes solid dissolution and modified permeability [Hoefner and Fogler, 1988; Hinch and Bhatt, 1990], or when liquid overpressure creates macroscopic fractures [Fyfe, 2012]. A research question that relates to the present topic is *under what conditions does crustal flow and reaction enhance permeability and hence promote flow and when does it have the opposite effect?* Evans et al. [2018] developed models extended from the theory of magma/mantle dynamics that address this in the context of carbon sequestration in ultramafic crustal rocks [see also Malthe-Sørenssen et al., 2006; Røyne et al., 2008; Rudge et al., 2010]. In Evans et al. [2020], they followed on to study how the volume change of reaction can cause microfracture that promotes further reaction. The mechanics of the rock in this context is poroelastic [e.g., Biot, 1941; MacMinn et al., 2016], whereas that of the asthenosphere can be considered poroviscous [e.g., McKenzie, 1984; Bercovici et al., 2001a].

In the planetary context, rock is often considered to be simultaneously viscous and elastic. The relative weighting of these mechanisms in the response to stress depends on the time scale of the process that is forcing deformation. On the time scale of seismic waves, elasticity is dominant (but seismic attenuation indicates that viscous mechanisms are present too). On the time scale of mantle convection, viscous flow is dominant. But at the intermediate time scale of tidal deformation, both viscous and elastic mechanisms can play a role (though the details of this combination remain a subject of research [Bierson and Nimmo, 2016; Renaud and Henning, 2018]). Dissipation of heat by tidal deformation attests to the importance of viscous mechanisms. With the increasing number of known exoplanets, an emergent research question is *how do tidal dissipation, mantle convection and magmatism interact to determine the structure and dynamics of tidally heated bodies?* In Jupiter's moon Io, as an extreme example, tidal dissipation causes partial melting throughout the silicate mantle [Peale et al., 1979]. Mantle convection transports heat too slowly to achieve a thermal balance under such rapid heating, but magmatic production and segregation can keep pace [Moore, 2003]. Hence heat export relies on transport of magma across the cold lithosphere and out of volcanoes, a process termed heat-piping [O'Reilly and Davies, 1981]. But heat-piping must be inefficient on Io to prevent lithospheric growth to a thickness much greater than observed [Spencer et al., 2020a]. In this context, where magma transport dominates the heat budget and promotes lithological stratification that is potentially unstable [Spencer et al., 2020b], what is the role of mantle convection?

The questions posed in this section are merely a sample of a much larger set that arises from the interaction of liquids and solids in Earth and planets. Theory developed in this book is directly relevant to some of these questions, and indirectly relevant to others. But the concepts and mathematical framework introduced here can provide a base from which to extend into related areas of theory. And, of course, for any particular question arising from natural observations, the theoretical tools that are readily available may guide—but should never constrain—the hypotheses posed to answer it.

1.4 About This Book

1.4.1 OVERVIEW OF THE ORGANIZATION AND CONTENT

The next chapter provides a brief history of the theory of the two-phase dynamics of partially molten asthenosphere in terms of the key investigators, influential or innovative publications, and ongoing themes.

Chapter 3 is a very brief review of single-phase mantle convection. The governing equations are presented without derivation and it is assumed that the reader is already familiar, having previously studied them elsewhere. They are used to illustrate some physical/mathematical concepts that are relevant for two-phase flow. A solution for isoviscous, incompressible Stokes flow is derived to model mantle flow beneath a mid-ocean ridge. This result is used later in the book as a background on which melt transport is computed.

Chapter 4 considers the equations for conservation of mass and momentum of the partially molten aggregate. Mathematical tools and notation are introduced to describe the physics at the grain scale; volume averages then lead to PDEs for continuum variables. Physical arguments are presented that justify assumptions required for computing these averages. In particular, we discuss the dominant theory for the interphase force, exerted by the liquid on the solid and vice versa. General viscous constitutive laws are derived without specifying how viscosities depend on other parameters. These components are then assembled into the full system of equations representing conservation of mass and momentum for the liquid and solid phases. The chapter concludes with a discussion of special cases in which the full system can be simplified into recognizable forms.

In chapter 5 we consider the material properties of partially molten mantle, providing more detail than was presented in chapter 4 but much less books that focus on, for example, rock mechanics. The chapter begins with a discussion of the physics at the microscopic scale that governs the evolution of grain size and the wetting of olivine grains by basaltic melt. An idealized model of pore geometry is developed. This is followed by short studies of the permeability, the shear and compaction viscosity of the aggregate, and the liquid viscosity. Finally we discuss the melting rate, though a more detailed treatment is deferred to later in the book.

Chapters 6 and 7 concern solutions of the governing equations under idealized conditions. These solutions are useful for developing an understanding of the physical behaviour that is encoded in the governing equations and also for recognizing characteristic solutions that may appear in more complex models. Chapter 6 focuses on elucidating compaction and the compaction length—a length scale that emerges from liquid–solid interaction in the framework of two-phase fluid dynamics. We consider canonical problems including magmatic solitary waves. To focus on the key physics, models presented are one-dimensional.

In chapter 7 we consider the role of shear in establishing pressure gradients that drive liquid segregation. The models are motivated by laboratory experiments on partially molten rocks in which deformation leads to an instability and the emergence of high-porosity sheets oriented at a low angle to the shear plane. Models in this chapter are two-dimensional and hence the sheets appear as bands. Although the compaction length also features prominently, this chapter primarily examines how the viscosity of the two-phase aggregate can feed back into the dynamics.

Chapters 8 and 9 develop theory for conservation of energy and conservation of mass for chemical species. In the former, much of the development is in casting the first law of thermodynamics in terms of different variables: internal energy, enthalpy, temperature and entropy. The latter chapter develops the governing equations for conservation of species mass, looking at the different cases of species: thermodynamic components, trace elements and radiogenic trace elements. In both chapters, relatively simple applications are developed as demonstrations of the physics.

Chapter 10 provides a brief introduction to modeling the thermochemistry of mantle petrology. It sets out a simple but useful approach, based on ideal solution theory, for approximating the equilibrium phase fractions and compositions in a two-phase system with an arbitrary number of chemical components. An extension to disequilibrium thermodynamics is also discussed.

Chapter 11 develops one-dimensional models of melting and buoyancy-driven melt segregation. These *column models* impose upwelling of the solid phase to drive the melting. Conservation of energy and species mass are used to couple the thermochemistry of melting to the mechanics of flow. These models demonstrate that Darcy drag is the key force resisting buoyancy over most of the partially molten region. The chapter also considers the decompaction boundary layer beneath the lithosphere. It concludes with a consideration of uranium-series disequilibrium in the context of column models.

Chapter 12 further analyzes the buoyancy-driven vertical segregation of magma, focusing on the reactive corrosivity of upwelling melts and its dynamical consequences. The analysis demonstrates the tendency for reactive localization of magma into high-flux channels with compacted regions between them.

Chapter 13 concerns application of the foregoing theory to models of tectonic scale processes and, in particular, to mid-ocean ridges. It begins with a rescaling and simplification of the equations, assuming small porosity. This leads to a system of equations with a partial decoupling of compaction from large-scale shear flow, which simplifies calculations and enables re-use of existing codes for single-phase Stokes problems. Armed with this approximate formulation, the chapter revisits the mid-ocean ridge mantle-flow problem from chapter 3 and layers on top of it a calculation of magmatic segregation. It then develops a simple model for melt focusing based on a high-porosity channel along the lithosphere–asthenosphere boundary, where freezing leads to an impermeable barrier to vertical flow. The chapter concludes with a discuss of the enthalpy method, an approach to directly coupling the fluid dynamics with the thermochemistry.

Chapter 14 concludes the book with a brief introduction to the numerical methods that have been used to solve the equations of magma/mantle dynamics. It focuses on the finite difference/volume method but also introduces the finite-element method, demonstrating the their effectiveness on benchmark problems.

1.4.2 REFERENCES TO THE LITERATURE

Most of the chapters of this book are written with minimal reference to the literature. This should not be taken as evidence that the theory and ideas presented originate from the author; most of them do not. Rather, references are omitted from the main text to avoid distraction from the flow of concepts and connections. The hope is that by reading this book, one can rapidly obtain the background knowledge that is required for comprehension of the specialist literature.

A section entitled "literature notes" is provided at the end of each chapter. These sections serve two purposes. First, they cite and describe the publications from which the content of the relevant chapter was derived. Second, they provide some advice about further reading: publications that extend the theory, or that approach it from an experimental or observational perspective.

Decisions about what to include and exclude from the literature notes are subjective and based on incomplete information. There is no clear line dividing what is relevant and what is irrelevant. And while the author has a reasonable knowledge of the literature that is immediately related to the book, there are undoubtedly publications that have escaped his attention. Lastly, of course, the relevant literature evolves as new contributions are added and older ones are reevaluated.

1.4.3 MATHEMATICAL NOTATION

This is, predominantly, a book about fluid dynamics. Thus it is appropriate that we exercise care in our use of the word *fluid*. We shall consider a fluid to be a substance that flows in response to a stress applied over a sufficient duration of time. The mantle, which is a solid, behaves elastically when stress fluctuates over short time scales (e.g., seismic waves).[6] In response to the long-term stresses associated with mantle convection and plate tectonics, the mantle flows: the rate of strain is proportional to the stress, though this relationship is not always linear. Since liquid magma is obviously a fluid, we are dealing with a system of two fluids. It is therefore sensible to avoid using "fluid" as a label for either phase.[7] Instead, we opt for the labels

$$\ell \rightarrow \text{liquid phase, magma;}$$

$$s \rightarrow \text{solid phase, mantle.}$$

This has the potential to become confusing in the context of subduction zones, where the water-rich material released by the subducting slab is referred to as "fluid" to distinguish it from silicate-rich magma. However for present purposes, our priority is to avoid misuse of the term "fluid."

The phase labels will be applied to variables as superscripts. For example, the density of liquid is ρ^ℓ while that of the solid is ρ^s. We can apply this to phase fractions too: ϕ^ℓ and ϕ^s are the phase fractions of liquid and solid, respectively. We shall always assume that the liquid and solid together occupy the full space, meaning that there are no voids or other phases present; hence we take $\phi \equiv \phi^\ell$ such that $\phi^s = (1-\phi)$. The $\equiv$ sign signifies an equality that is true by definition of one of the variables (ϕ in this case). Using the phase fractions, we can define the meaning of the overline notation,

$$\overline{\rho} \equiv \phi\rho^\ell + (1-\phi)\rho^s.$$

The symbol $\overline{\rho}$ thus represents the *phase-averaged* or *bulk* density. An overline on any symbol (or group of symbols) will have the equivalent meaning for that quantity. We also define the phase difference with a Δ,

$$\Delta\rho \equiv \rho^s - \rho^\ell,$$

in terms of the solid-minus-liquid difference. We shall retain this definition in chapter 10 even though it is inconsistent with thermodynamic convention.

Superscripts will also be used to represent exponents. A symbol that would require multiple superscripts will be placed into parentheses, e.g., $\left(q^\ell\right)^2$. Finally, superscripts will be used to index the entries in a series expansion. In this case, the index itself will appear in parentheses, e.g.,

$$q = q^{(0)} + \epsilon q^{(1)} + \epsilon^2 q^{(2)} + \dots$$

is an expansion of q in powers of a small parameter, $\epsilon \ll 1$. We will refer to terms in such power series according to their order, meaning their approximate size, and represent this with the notation $\mathcal{O}()$. Hence the $q^{(0)}$ term is $\mathcal{O}^{(1)}$, the $q^{(1)}$ term is $\mathcal{O}(\epsilon)$, and so on. The same notation will be used more broadly to represent the approximate size of any quantity.

[6]Though there is some component of anelastic deformation in seismic waves, leading to dispersion and attenuation.

[7]In the literature on magma/mantle dynamics, f (fluid) has typically been adopted as the label for the magma and m (matrix) for the mantle.

Vectors and tensors will be represented by bold symbols, with no notational distinction between them. Hence it will be up to the reader to make this distinction according to the context in which the symbol is used. Superscripts on vectors and tensors will typically represent the phase: $\boldsymbol{v}^i$ is the velocity vector of phase i, where i can be s or ℓ. Subscripts are used to denote components of a vector or tensor. Hence v_j^ℓ is the jth component of the liquid velocity. Generally we will avoid index notation for vector and tensor operations, except where it contributes to the clarity of a calculation. Furthermore, it shall be convenient to treat sets of chemical concentrations as pseudo-vectors. For example, c_j^i is the concentration of the jth chemical species in phase i. Concentration pseudo-vectors cannot be considered true vectors because the algebraic operations of vector calculus don't apply (e.g., change of coordinates).

We shall reserve the non-italic symbol d as a prefix to denote infinitesimal quantities. Thus $\mathrm{d}Q$ is an arbitrarily small but nonzero change in Q. Other details of notation will be addressed where they arise in the book.

Occasionally it will be useful to rescale variables, usually with the aim of non-dimensionalizing them. If a variable q has a value that is typically of the same order as $[q]$, then we can define a rescaled, non-dimensional variable $q' \equiv q/[q]$. Hence the notation $[q]$ means a dimensional constant that represents the scale of the variable q. Typically we will drop the $'$ on rescaled variables to avoid clutter.

Table 1 lists important symbols used throughout the book.

1.5 The Way Forward?

Research on the two-phase dynamics of the magma/mantle system is still in its early stages, despite the basic theory having been around for more than 30 years. The field is now advancing rapidly, with progress in the theoretical development, constitutive hypotheses, numerical methods and software, and applications to the Earth and other planets. This book makes no attempt to explain developments at the cutting edge, some of which may soon be superseded by others. Instead it provides a foundation of knowledge from which the reader can access, critically evaluate, and contribute to research at the cutting edge. Thus this book is a starting point and/or reference point for the reader.

There are legitimate scientific reasons to question some of the theory that is developed here. Of course the validity of the basic conservation principles, for present purposes, is entirely assured. However, the physical assumptions that are used to adapt those principles to the magma/mantle system are difficult to robustly validate. This validation must ultimately come from a comparison of model predictions with measurements of the natural system being modelled. Those measurements can be obtained in laboratory experiments or from the natural world. But the solid Earth that we seek to model is a complicated and inaccessible place; our observations of it are made from the surface and are generally only proxies for the dynamics, convolving those dynamics in time or space or both. Interpretation of proxy or laboratory data in terms of the two-phase dynamics in the mantle is far from straightforward, but in fact this should be the goal of geoscientists who work with the theory.

Theory that is inconsistent with the natural system is incorrect or incomplete. Deficiencies of theory, where they can be identified, provide the impetus and direction for developments that bring it closer into line with the truth. However, one should also be cognisant that inconsistency with nature is not always straightforward to prove, especially when the constraints derived from observations are interpretations (i.e., models) themselves. We thus face the challenge of testing theory against observations of complex proxies that are, at best, integral constraints on the dynamics. To overcome this, one must take a multidisciplinary approach; it is insufficient to apply any single observational constraint independently. Instead one should test models of two-phase magma/mantle against a diverse dataset including geophysical, geochemical and laboratory (rock mechanics and petrology) data. But to test the theory, one must generate testable predictions from the theory. Hence it is insufficient for the community of geoscientists working with this theory to solve only the PDEs associated with the fluid dynamics. We must go further to quantitatively predict the implications for, e.g., radiogenic disequilibrium of the uranium-series isotopes, seismic anisotropy of partially molten regions, crustal thickness and trace element composition, porosity distribution in deformation experiments on partially molten rock, and more.

To achieve such predictions we require associated theory, some of which is discussed in the chapters of this book. But we also need multi-pronged analysis of the theory. Numerical solutions are necessary; obtaining them efficiently in the multi-physical, multi-dimensional context of tectonic-scale applications is currently a key challenge. But numerical solutions are not enough. We also need analytical solutions of representative problems of reduced complexity to enable mathematical insights into the problem, and we need basic scaling analysis to explain the robust features of the numerical and analytical models. It is not necessary for each geoscience practitioner to do all of these things simultaneously, but progress will require that all are aware of and knit together the results from each of these perspectives and, crucially, from the observations too—a tall order indeed.

That said, I think the chapters below will also demonstrate the inherent value of and interest in the mathematical analysis and extension of the PDEs. Hopefully this book will make the theory more accessible to analysts who are not particularly interested in partial melting of the Earth (or any other planet). And hopefully their results will provide tools and insights that can be adopted by geoscientists.

For the author (and perhaps for the reader), a fundamental point is as follows. Igneous products (lavas, magmas, residues) that are exposed on the surface of the solid Earth and other planets carry chemical, textural, and thermal information from the planet's interior. Measurements of these lavas are valuable in their own right, but without other constraints, only non-unique inferences about the interior of the planet and its evolution can be drawn from such measurements. The theory of magma/mantle dynamics introduces constraints in the form of conservation of mass, momentum, and energy and it incorporates explicit, quantitative statements of material and transport properties. The inclusion of such information doesn't guarantee a unique fit to the data, but it should reduce the range of possible explanations and rule out those that are physically inconsistent. Moreover, such model calculations can provide corollary predictions that are testable with new observations. This complementarity can sharpen the scientific question and eventually hone in on an answer.

CHAPTER 2

A Condensed History of Magma/Mantle Dynamics

In the study of partially molten rock, there is no sharp line between considerations of physics and chemistry. The two are tightly coupled through the thermochemistry of phase change and through variations in material properties such as density, viscosity, and permeability (by way of grain texture). Consistent with the focus of the book, however, this chapter traces the development of the more physical aspects of the theory.

2.1 Foundation

Even before the theory of plate tectonics was fully established, pioneering work by F. Charles Frank at the University of Bristol [Frank, 1968] noted the potential importance of magma/mantle dynamics and proposed a two-phase flow theory to model it. Frank received a doctorate in engineering from the University of Oxford and is best known for work on dislocation theory of crystals. In 1968 he hypothesized that magma upwells beneath mid-ocean ridges to form the oceanic crust, which then moves laterally toward the continents. He further argued that magma migrates from beneath the continents to the mid-ocean ridges, such that downwelling is a broadly distributed response to removal of magma[1]. While his theory of downwelling was falsified almost immediately after it was published, Frank [1968] correctly supplied the bulk mass-conservation equation of the two-phase aggregate (repeated below in equation (4.12)) and proposed a Darcian segregation flux of magma through a pore network around octahedral grains, driven by the density difference between solid and melt. He assumed zero pressure difference between the liquid and solid phases, setting a precedent that lasted for 16 years.

Frank's theory was adapted to model drainage of intergranular water from glaciers by John F. Nye of the University of Bristol [Nye and Frank, 1973]. In the glaciological community, the interconnection of the pore structure of polycrystalline ice was already recognized at that time. Nye and Frank [1973] applied principles of textural equilibrium of partially molten aggregates that were developed in the material sciences literature [Smith, 1948]. The relevance of this theory to characterizing the microstructure of partially molten rocks remained unrecognized until it was noted by Stocker and Gordon [1975] and Bulau et al. [1979].

[1] In the same paper (and apparently independent of the work of Oliver and Isacks [1967]), he also speculated that the "geo-faults" that parallel ocean trenches are where the oceanic crust returns to the mantle.

In the early 1970s, without a clear understanding of the pore structure of partially molten rocks, authors conceived of a porosity threshold ($\sim$10%) above which the rock becomes permeable.[2] This was at the center of the analysis of Norman H. Sleep at the Massachusetts Institute of Technology [Sleep, 1974]. Sleep adopted a more general theory and simplified it by assuming that the solid phase remains contiguous. He used a scaling analysis to simplify the equations and interpreted the general, interphase drag coefficient to obtain a modified form of Darcy's law. The result is a model of segregation that is similar to that of Frank [1968].

The two-phase flow theory adopted by Sleep was published in Russian in 1966 by Dzharulla F. Faizullaev, director of the Institute of Mechanics and Earthquake-Proof Construction in Tashkent, in the then-Uzbek Soviet Socialist Republic; it was translated into English as Faizullaev [1969]. The theory he develops, especially in chapter III.23 on "two-phase motion of a viscous and an ideal medium," bears a striking resemblance to the modern Stokes/Darcy formulation for magma/mantle dynamics: an interphase drag term depends linearly on the velocity difference between phases; absolute pressure is eliminated from the equations. On the other hand, a constitutive law associating isotropic deformation with the local pressure difference between phases is absent, implying the assumption of zero resistance to compaction. And Faizullaev does not link the volume-averaged continuum model he presents to the physics at the sub-continuum scale. Nonetheless, this was a work ahead of its time—especially in comparison to the development of magma/mantle dynamics.

In the late 1970s our understanding of the microstructure and permeability of partially molten rocks caught up with that of ice. Experimental and theoretical work at Yale University by Harve S. Waff and and J. R. Bulau considered the mechanics and thermodynamics of textural equilibrium under static stress conditions in partially molten rocks [Waff and Bulau, 1979; Bulau et al., 1979]. These authors measured dihedral angles of $<60^\circ$ ($\sim 47^\circ$) and recognized the implication that partially molten mantle could be permeable at all porosities, consistent with the work of Smith [1948]. They were reluctant to make such a claim about the Earth but Walker et al. [1978], citing their results, assumed permeability down to vanishing porosity.

The porosity threshold persisted in the work of Donald L. Turcotte at Cornell University, who clarified the continuum theory proposed by earlier workers. Turcotte and Ahern [1978] invoked the idea of a threshold but formulated a model that excluded it; Ahern and Turcotte [1979] explicitly incorporated a cut-off porosity below which the permeability is zero. They were the first to formally consider a one-dimensional column model of mantle upwelling, melting, and melt segregation in a domain aligned with gravity (as in chapter 11). Their models elegantly coupled Darcian mechanics of melt segregation with an equation for conservation of energy, closed by assuming a Clapeyron slope for single-component melting. Both Turcotte and Ahern [1978] and Ahern and Turcotte [1979] assumed zero pressure difference between the liquid and solid phases, an assumption justified in Ahern and Turcotte [1979] by estimating a length scale of 83 m, over which such differences would be relaxed by solid deformation; this was considered to be negligible.

A simple thought experiment regarding pressure differences between phases was proposed by Waff [1980] and proved to be helpful in clarifying the prevailing issues. It is reproduced here in slightly modified form. Consider a column of partially molten

[2]This threshold was adopted from studies of "filter pressing" that promoted the concept of a critical melt fraction [e.g., Arzi, 1978].

mantle rock with uniform porosity ϕ_0 below a horizon at $z=0$ (the z direction is upward so the partially molten rock is at $z<0$). The magma (ℓ) and mantle (s) have different densities, with $\rho^\ell < \rho^s$. If the liquid phase is static and we assume that the mantle is rigid (infinite resistance to flow), then the liquid pressure is hydrostatic (or *magma-static*),

$$\frac{dP^\ell}{dz} = -\rho^\ell g, \tag{2.1}$$

where g is the acceleration due to gravity; the solid pressure is lithostatic,

$$\frac{dP^s}{dz} = -\rho^s g. \tag{2.2}$$

There is then a pressure difference $\Delta P = -(\rho^s - \rho^\ell)gz$ for $z<0$ in the partially molten region (we assume that the pressures are equal at $z=0$). For depths $-z$, deep below $z=0$, this pressure difference can become extremely large. Moreover, we know that the partially molten mantle is *not* rigid—it is ductile and flows in response to stress. Assuming, then, that the mantle offers zero resistance to flow, the pressures of the co-existing liquid and solid are equal. This pressure must support the full, static weight of the partially molten column above, and hence,

$$\frac{dP^\ell}{dz} = -\left[\phi_0\rho^\ell + (1-\phi_0)\rho^s\right]g. \tag{2.3}$$

But this pressure gradient in the magma is larger than the static pressure gradient from equation (2.1) and hence it can drive the less-dense magma upward. This contradiction tells us something profound about partially molten rock: that if it has finite strength and a permeable pore network, it cannot exist in a static equilibrium—instead magma segregates upward and the solid mantle matrix must compact. Static equilibrium without compaction or segregation can only exist if the permeability is zero at finite porosity.

Invoking petrological inferences of immobile melt, Waff [1980] sought a microstructural argument against permeability. He did not take this pressure difference as a driver of deformation, which he assumed was a negligibly slow response. Instead, Waff [1980] invoked a rapid textural adjustment in which the equilibrium microstructure evolves from an interconnected network of grain-edge tubules (established under zero pressure difference) to a set of disconnected liquid inclusions on grain boundaries. He concluded that the partially molten column breaks up into strata within which the pore network is connected but between which it is disconnected and hence impermeable. According to Waff [1980], the disconnected liquid inclusions between strata (and hence the strata themselves) only become connected when the porosity exceeds a threshold of about 5%–22%. The theory proposed by Waff [1980] gained little acceptance.

The state of thinking in the early 1980s [e.g., Waff, 1980; Stolper et al., 1981; Maaløe and Scheie, 1982; Turcotte, 1982; Nicolas, 1986] was thus characterized by a porosity threshold below which the permeability is zero and above which melt segregates according to Darcy's law (balancing pore-scale viscous drag with buoyancy of the melt). Governing equations were obtained on a somewhat *ad hoc* basis, without rigorous derivation; resistance to compaction was considered to be entirely negligible; the formal, fluid-dynamic connection between mantle convection and magma/mantle dynamics remained obscure.

2.2 Axial Age

This landscape changed drastically in 1984 when two papers independently proposed theory for the two-phase dynamics of partially molten, deformable solids [Fowler, 1984; McKenzie, 1984]. Both theories were rigorously derived using the mathematical framework established by Donald A. Drew of Rensselaer Polytechnic Institute in Drew [1971] and Drew and Segel [1971] (usefully reviewed in Drew [1983]). This approach differs from earlier ones in that it begins with consideration of the microscopic scale of solid grains and liquid-filled pores; inside these microscopic sub-domains, the one-phase continuum models are well known. It then uses carefully constructed averages to derive a model of a two-phase continuum that smooths out the details of the microscopic scale but captures its behavior in an average sense.

In contrast to preceding geoscientific models of two-phase flow, Fowler [1984] and McKenzie [1984] rejected the porosity threshold and invoked a connectivity of pores on the basis of sufficiently small dihedral angle. Both argued that liquid and solid pressures are different, in general, and should be related by a constitutive law. And both incorporated the large-scale deformation of the solid as part of the two-phase formulation. We shall consider each in turn: Fowler [1984] on partially molten glacier ice and McKenzie [1984] on partially molten rock. Although the mathematical formulations developed in these two papers are similar, it is the latter that has become the cornerstone of magma/mantle dynamics and that will be discussed in greater detail here.

Andrew C. Fowler studied mathematics at the University of Oxford and received his doctoral degree there in 1978. Fowler [1984] was written at the Massachusetts Institute of Technology, where he was an instructor and then assistant professor until 1985, when he returned to Oxford. The paper concerns the transport of liquid water in polythermal glaciers—i.e., glaciers that have temperatures that reach the melting point in at least some places. In contrast to the situation for magma and mantle rock, liquid water is more dense than the solid phase. Moreover, glaciers are typically modeled as comprising a single chemical component, pure H_2O, whereas the mantle is fundamentally a multi-component chemical system. Aside from these differences (and a shift in temperature), partially molten ice has much in common with partially molten rock from a mechanical perspective. Both are polycrystalline solids that melt, deform viscously by creep, and host a permeable network of pores at their grain boundaries; both are, to excellent approximation, zero Reynolds-number flows of liquid and solid phases. Fowler [1984] formulated the two-phase flow of partially molten ice using the mixture theory of Drew [1983] in terms of conservation equations for mass, momentum, and energy. Constitutive laws relate the deviatoric stress to the strain rate, the isotropic stress (pressure difference between phases) to the compaction rate, and the pressure gradient to the segregation rate. Fowler [1984] adopts a non-Newtonian viscous flow law for the first of these, referred to as Glen's law by glaciologists. The second constitutive law, for isotropic compaction, had been neglected by previous workers. Fowler wrote,

> Implicit in Ahern and Turcotte's analysis is the assumption that vein closure can occur on a time scale short compared to the long term (solid) viscous flow time scale. However, [...] the strain rate of closure is comparable to the bulk flow strain rate, at least when [the inter-phase pressure difference] is comparable to the bulk shear stress: thus, the time scale of closure is the same as that of the bulk flow: consequently *this effect must be included in the large scale flow dynamics.*

To formulate a constitutive law, Fowler [1984] envisions compaction of the solid as analogous to the creep closure of water tunnels in glaciers, with a closure rate that scales with the pressure difference between the ice and water. This isotropic stress is related to the compaction rate by the compaction viscosity, which Fowler obtains as the solid shear viscosity divided by the volume fraction of water (see also section 5.3.2, below).

Finally, Fowler [1984] considers the inter-phase force density associated with phase segregation. Assuming a microscopic, Poiseuille-like flow through the pores, he relates the force density to the difference in volume-averaged velocity between phases. In this relation, the constant of proportionality is a drag coefficient that is chosen to recover Darcy's law. The permeability factor that appears in the drag coefficient is adopted from Nye and Frank [1973] and accounts for the low-dihedral-angle connectivity of the grain-edge pores. The rest of Fowler [1984] concerns scalings, boundary conditions, and approximate solutions that are appropriate for glaciological problems.

The history of the field of magma/mantle dynamics can reasonably be divided into pre- and post-McKenzie [1984] eras. The former has been reviewed above; the latter will be discussed below. Now we turn our attention to this pivotal paper itself. McKenzie [1984] is largely the basis for this book: a plurality of the ideas discussed here were either developed or noted by McKenzie. Some of the theory in that paper was adopted from preceding work, but it also contains crucial, deliberate deviations and new ideas. The governing equations McKenzie derives are similar to those of the contemporary Fowler [1984]. However, in McKenzie [1984], the discussion of their derivation, closure, and application is carefully framed in the context of partial melting, textural equilibrium, and melt transport in the mantle (as opposed to glacier and ice sheets, as in Fowler [1984]). The mathematical derivations are confined to five appendices, making the paper more accessible. For these reasons, McKenzie [1984] has become the standard reference in magma/mantle dynamics and is rightly considered the root of most modern studies, including this one.

Dan P. McKenzie studied physics at the University of Cambridge. He received a PhD in geophysics from Cambridge in 1966, but his early research was also influenced by a period of eight months spent at Scripps Institute of Oceanography in California. McKenzie is best known for his pioneering work on the theory of plate tectonics, his studies of mantle convection and magmatism, and for developing a theory for the formation of sedimentary basins. He was interviewed for this book and some of his comments are cited below.

McKenzie [1984] begins by asserting that the "need for a simple physical model which can describe the generation of a partially molten rock, and the separation of the melt from the residual solid" is "obvious." And since partial melting is a fundamental physical process that is closely associated with plate tectonics, heat transfer, and chemical segregation, few would disagree with this statement. It was also deemed important that such a model should be expressed in terms of tractable, partial differential equations. But as attested to in the paragraphs above, a number of such models had already been proposed. McKenzie [1984] summarized the state of the field by noting that "[a]ll of this work has been hindered by the difficulty of describing the fluid mechanical behaviour of the crystalline residue." These difficulties included mathematical formulations derived from *ad hoc* assumptions rather than conservation principles [Stolper et al., 1981; Maaløe and Scheie, 1982] and models that neglected the mechanical resistance of the matrix to compaction [Walker et al., 1978; Ahern and Turcotte, 1979]. McKenzie objected to Waff [1980] on physical grounds, arguing that buoyant melt will segregate at any finite melt fraction, consistent with the permeable pore network expected under textural equilibrium, a fact that Waff had helped to establish.

McKenzie [1984] acknowledged the similarity of the equations he derived to those of Sleep [1974], who had adopted them from Faizullaev [1969]. Faizullaev's theory described the flow of interpenetrating (more or less viscous) continua with momentum exchange between them—hence it was the most immediate precedent for the momentum conservation formulated by McKenzie. McKenzie's critique of Sleep [1974] was that his formulation has zero compaction viscosity. Indeed, there is no such viscosity in the formulation of Faizullaev [1969], who mostly considered flows of constant, uniform phase fraction. But for Drew [1983], pressure differences between phases and interfaces (and the resultant flow) was essential to the physical model; these ideas were influential for McKenzie and others.

Following Drew [1983], McKenzie [1984], and Fowler [1984] treat the pressure difference between phases as the cause of isotropic deformation of the solid (compaction or decompaction). This is physically straightforward but raises a subtle question: when the pressure differs between coexisting liquid and solid phases, what is the *thermodynamic* pressure of the system? For example, what pressure should be used to compute the melting point? Fowler [1984] argues that each phase has its own melting temperature and that there exists a porosity-weighted average melting temperature of the mixture. For McKenzie [1984], the thermodynamic pressure is defined *only* when the system is in static equilibrium. This was taken to imply a state of zero compaction rate. In this case, an interphase pressure difference can exist *only* due to forces on the interface between phases (i.e., surface tension). But McKenzie's [1984] "assumption is that the pressure in the melt is the same as that in the matrix, or that the surface energy of the matrix grains has no influence on the dynamical behaviour." For reasons of mathematical simplicity, he neglects interface energy and associated forces.

This assumption is more confusing than it is consequential. There is, of course, a mechanical pressure of the viscous solid inherent in McKenzie [1984], though it is not identified as such. One-third of the trace of the solid stress tensor (in McKenzie's [1984] equations (A14) and (A15)) gives the mean, isotropic stress, which is identified as the solid pressure. This was first noted in the context of magma/mantle dynamics by Scott and Stevenson [1986], though its justification by Fowler [1984] is perhaps more nuanced. Scott and Stevenson [1986] provides an alternative derivation of the governing equations but arrives at the same set as McKenzie [1984]. The solid pressure appears as part of the expected relationship with the compaction rate. Of course McKenzie [1984] recognizes that compaction and melting "are unlikely to be independent, since melt extraction will occur at the same time as melt is being generated." Nonetheless, in McKenzie [1984] the relationship between mechanical pressure associated with deformation and thermodynamic pressure associated with melting is not specified. This sufficed in that paper, where calculation of melting is considered independently of calculation of flow. Some subsequent models, for simplicity, have assumed the thermodynamic pressure to be equal to the lithostatic pressure [e.g., Katz, 2008]. Careful consideration of viscoelastic deviatoric stresses by Jull and McKenzie [1996] showed that the solid pressure is, in fact, an excellent approximation of the thermodynamic pressure.

Another point on which McKenzie [1984] is contradictory is that "the expulsion rate always depends on the [mechanical] properties of the matrix," a critique of earlier work. However in one-dimensional calculations of instantaneous, buoyancy-driven segregation, M^c^Kenzie notes that compaction stresses give rise to boundary layers with a thickness scale that emerges from the two-phase physics: the compaction length.[3] Outside of such layers there are regions "in which negligible compaction is occurring and the gravitational forces are supported by the upward percolation of melt,"—i.e., by Darcy drag. Compaction is negligible here in the sense that the isotropic stress it generates is inconsequential in the balance of forces.

Subsequent work combining melting with compaction by Ribe [1985] showed that the Darcy regime dominates the melting region beneath mid-ocean ridges, and that a steady-state porosity arises from the mass balance of compaction and decompression melting. So is it reasonable to neglect compaction stresses entirely [e.g., Turcotte, 1982]? With prescience, McKenzie [1984] states that

> Compaction boundary layers must occur wherever the permeability changes. The mantle at sufficient depth is generally believed to be solid, so melting beneath ridges must produce compaction boundary layers. Where the lithosphere has formed by heat loss to the earth's surface, the permeability of the lithosphere must be less than that of the underlying asthenosphere. Here another boundary layer must form, which is probably unstable.

Work published since 1984 has reinforced the existence and importance of such boundary layers. Moreover, it has shown that compaction stresses are essential in governing localization processes that occur on the scale of the compaction length. The emergent, localized fluctuations of porosity can affect the behavior at much larger scales.

2.3 Exploration

One such process is the propagation of magmatic porosity waves (chapter 6). These have properties of nonlinear waves known to mathematicians as solitary waves or solitons. Their analytical structure was first obtained by Richter and McKenzie [1984] and Scott and Stevenson [1984]. It was the latter who proposed the name *magmons*, analyzed their dispersion properties, and showed that they can emerge from steady-state boundaries of the melting rate. An even more detailed and thorough analysis of magmatic solitary waves was developed by Barcilon and Richter [1986] who argued that the waves are not, in fact, solitons. This is a semantic point, however, and subsequent authors have continued to consider them so. Scott and Stevenson [1986] observed that one-dimensional solitary wave profiles are unstable in two dimensions, where the solutions are circular. The same is true for two-dimensional solutions in a three-dimensional space [Wiggins and Spiegelman, 1995], where the solutions are spherical.

Marc W. Spiegelman was a doctoral student of McKenzie's in Cambridge and later became a professor at the Lamont Doherty Earth Observatory of Columbia University[4]. He sought to apply magma/mantle theory to problems at the scale of plate tectonics: mid-ocean ridges and subduction zones. In Spiegelman and McKenzie [1987], published concurrently with similar work by Morgan [1987], Spiegelman showed how the dynamic pressure gradient generated by corner flow of partially molten mantle tends

[3] The compaction length is of central importance in understanding the dynamics of the partially molten mantle. It was first noted in Stevenson [1980], a conference abstract. However, the mathematical definition of the compaction length written by Stevenson [1980] gives units of inverse length and contains an incorrect dependence on melt fraction. It is unclear whether these errors are typographical or otherwise. Regardless, there is no evidence that this work was known to McKenzie or Fowler in 1984.

[4] The author of the present book was a doctoral student under Spiegelman at Columbia from 2000-2006.

to focus melt toward the plate boundary. This is where mantle flow turns a sharp corner, giving rise to negative pressures (section 3.3). A negative pressure strong enough to produce lateral melt transport over a significant distance requires that mantle viscosity is large—implausibly large, according to later contributions. An alternative model of magma focusing at mid-ocean ridges was proposed by Sparks and Parmentier [1991]. They invoked (de)compaction boundary layers of higher porosity adjacent to the base of the lithosphere. Magma in these layers is deflected laterally along the sloping permeability barrier toward the ridge axis. Melt transport and focusing beneath mid-ocean ridges is discussed in chapter 13.

Another key insight regarding magmatic localization was made by David J. Stevenson of the California Institute of Technology. In Stevenson [1989] he discussed an instability associated with extensional flow of a partially molten aggregate: if the viscosity of the aggregate is porosity-weakening, tension causes melt to flow into places where it was already concentrated; porosity increases there and decreases in the interleaved, lower-porosity regions (chapter 7). This prediction motivated rock mechanicist David L. Kohlstedt of the University of Minnesota to perform a series of laboratory experiments (spanning some 30 years) in which partially molten aggregates of olivine grains and mid-ocean ridge basalt are deformed to large strains [e.g., Holtzman et al., 2003]. The samples are then quenched, sectioned and examined for evidence of melt localization. Porosity localization into bands is robustly obtained, but the details differ in interesting ways from the predictions of Stevenson [1989]. The interplay between magma/mantle theory and these experiments has become a sub-field of current research with significant, open questions. Regardless of whether the laboratory experiments are representative of processes actually occurring in the mantle, they provide an invaluable opportunity to directly test and validate the theory.

Exhumed sections of mantle rock, *ophiolites*, provide a geological perspective on magma genesis and migration. A key observation from some ophiolites is the presence of alternating regions of *dunite* (almost pure olivine) and *harzburgite* (containing pyroxene). The lavas that overly the ophiolite and were extracted from it (before it was exhumed) are in chemical equilibrium with the dunite, not the harzburgite. A hypothesis for the origin of these features was proposed by Kelemen et al. [1995a]. They argued that the dunite is formed by channelized melt transport and *replacive* melting. Replacive melting occurs when upwelling melts become undersaturated in silica as they depressurize. To re-equilibrate, they react with harzburgite, dissolving pyroxene and replacing it by precipitating olivine. Hence the residuum becomes a dunite. Quantitative support for this came from Aharonov et al. [1995], who coupled reactive flow and compaction theory to show that the channelization could arise by flow instability. Following from this work, again, the interplay between observations and two-phase theory has become a rich and active sub-field of magma/mantle dynamics; we consider it in chapter 12.

The chemical transport associated with melt segregation has been a consistent theme in the literature, going back to McKenzie [1984] and McKenzie [1985], who modeled trace elements alongside the two-phase physics. The possibility of trace-element disequilibrium between melt and its solid residue was proposed and investigated by Kenyon [1990] and Spiegelman and Kenyon [1992]. This disequilibrium arises from the finite size of solid mantle grains and the extremely slow diffusion of trace elements through them. Iwamori [1993a] extended the theory to explicitly include trace-element diffusion through representative, spherical solid grains of constant size. This approach, however, has not persisted in the literature. For a decade after this, the subject was advanced predominantly by Marc Spiegelman. He developed melt-transport models of uranium-series disequilibrium [Spiegelman and Elliott, 1993] and of trace elements beneath mid-ocean ridges [Spiegelman, 1996] and associated with channelized melt extraction [Spiegelman and Kelemen, 2003; Elliott and Spiegelman, 2003]. Although trace elements and their isotopes have no role in melting and two-phase flow, they can be used as a diagnostic proxy for these dynamics [e.g., Spiegelman and Reynolds, 1999], enabling comparisons between models and observations; in this there is significant research potential that remains unrealized.

2.4 Generalization, Extension, and Future History

Since the mid-1990s, the two-phase theory of magma/mantle dynamics has held the interest of a small but persistent community of researchers. An important generalization, developed by David Bercovici of Yale University and Yanick Ricard of the Ecole Normale Supérieure of Lyon, is based on the mathematical symmetry of phases [Bercovici et al., 2001a; Ricard et al., 2001].[5] Here symmetry means that prior to any simplifications on the basis of material properties, each phase obeys the same equations of mass, momentum, and energy conservation as the other. The introduction of interface properties such as surface tension is facilitated by this approach and, indeed, it was this extension that motivated the theoretical development. In Bercovici and Ricard [2003], it is shown that neglecting surface tension and assuming that one of the two phases is much less viscous than the other (as is the case for magma and mantle), the general theory reduces to that proposed by McKenzie [1984]. A summary of this theory is provided by Ricard [2007] in section 7.02.5.2.

Work by Ghods and Arkani-Hamed [2000] and Katz [2008] coupled the canonical two-phase mechanics with energy conservation and multi-component thermodynamics in tectonic-scale models. This was subsequently used to investigate melt transport in the presence of mantle heterogeneity [Weatherley and Katz, 2012; Katz and Weatherley, 2012] and volatile elements [Keller and Katz, 2016]. Other authors have contributed on associated topics—too many to review in this introduction; many more citations can be found in the Literature Notes at the end of each of the chapters in this book.

The composite or complex rheology for partially molten rocks is another area of progress in the recent literature. These include anisotropic viscosity [Takei and Holtzman, 2009a,b,c], damage [Bercovici et al., 2001b], viscoelasticity [Connolly and Podladchikov, 1998, 2007] and viscoelastic plasticity [Keller et al., 2013]. Keller et al. [2013] had the ambitious goal of modeling the propagation of magma from the asthenosphere through the lithosphere and to the surface. They made substantial progress, showing, for example, that dike-like features can emerge from a continuum, two-phase theory. Magma dynamics of the lithosphere is an important area of contemporary research.

There has been a notable evolution of numerical methods and computational strategies applied to solving two-phase flow problems. In the 1980s and 1990s, software codes were based on finite-difference discretizations of the governing equations and included "hand-rolled" solvers that were highly optimized but inextricable from the discretized problem [e.g., Spiegelman, 1993b]. After 2005, investigators increasingly utilized software packages developed by computational scientists for two-phase flow problems. Use of the Portable, Extensible Toolkit for Scientific Computation [(PETSc) Balay et al., 2019, 2020, 1997] was discussed by Katz et al. [2006] and Katz et al. [2007] in the context

[5]See also Bercovici and Ricard [2003], which corrects some equations from Bercovici et al. [2001a].

of a staggered grid, finite-volume discretization. More recently, finite-element methods have been developed to solve the discretized equations of magma/mantle flow [Keller et al., 2013; Wilson et al., 2014; Alisic et al., 2014], some of these facilitated by use of the FEniCS software package [Logg et al., 2012; Alnæs et al., 2015]. Rhebergen et al. [2014] and Rhebergen et al. [2015] were the first papers to analyze stability and convergence of finite-element methods, and to develop high-performance preconditioners.

I interviewed Dan M^c^Kenzie in April 2016 at the second workshop of the Melt in the Mantle program of the Isaac Newton Institute.[6] I asked him to reflect back on the period when he was deriving the theory presented in M^c^Kenzie [1984]. He explained that his motivation, at the time, was the well-documented isotopic diversity of the mantle source,[7] thought to be a consequence of incomplete mantle mixing. The observations suggested that despite the vigorous stirring associated with mantle convection, mantle heterogeneity is preserved. Furthermore, the diversity of isotopic ratios of the source is retained during melt extraction. How can melt transport fail to homogenize the isotopic diversity? How can measurements of lavas be used to infer the characteristics of mantle heterogeneity? M^c^Kenzie thought that a quantitative theory for melt segregation might address this problem, but he was unsatisfied with the existing theory and its *ad hoc* derivation. During a stay at the University of Chicago, M^c^Kenzie found the work of Drew and Segel [1971] on mixture theory. In it, he recognized a way to systematically derive the theory of magma/mantle dynamics.[8] One key consideration in his derivation was the requirement of frame independence for the interphase force, which constrained its formulation to be Galilean-invariant. M^c^Kenzie incorporated these ideas into his 1984 paper. In 2016, at the Isaac Newton Institute programme on Melt in the Mantle, M^c^Kenzie expressed amazement and gratitude for the diversity of research topics spawned from that work. However, it was clear to him that the original, motivating questions about the heterogeneity of the mantle source and how melt extraction preserves isotopic heterogeneity remain unanswered. They are much more difficult than he originally envisioned.

Results from the theory of magma/mantle dynamics have illuminated the source of this difficulty. It arises from the conflicting propensity of magmatism to both fractionate and mix geochemistry. The extent to which geochemical variability is preserved during melt extraction from the mantle is controlled by detailed pathways of melt transport. The melting process can produce increments of melt with radically different chemistry than their solid source. Geochemists have typically assumed that incremental, fractional melts are extracted from the mantle with minimal mixing. Simple physical models of vertical transport, however, suggest that incremental melts should be aggregated and mixed [Ribe, 1985]. This mixing should dampen or eliminate chemical variability inherited from the source heterogeneity [Bo et al., 2018], and yet at least some of this variability appears to be preserved. Variations in long-lived isotope ratios, in particular, may be preserved and recorded in characteristic mixing arrays [Liang, 2020]. More detailed models of magmatic segregation [e.g., Aharonov et al., 1995; Spiegelman et al., 2001], consistent with observations [e.g., Kelemen et al., 1995a], predict channelized transport that isolates deeply sourced melts from shallower magmas. Hence,

[6]Many of the talks presented during the three workshops of this program are available online, in perpetuity. See https://www.newton.ac.uk.

[7]There was extensive discussion at a meeting of the Royal Society on 1 and 2 November 1978. See Bailey and Tarney [1980] for details.

[8]On his return to the UK, Cambridge fluid dynamicists were skeptical; George Batchelor reportedly told M^c^Kenzie "there is no such thing as mixture theory."

Figure 2.1. A photo taken at the Newton Institute, University of Cambridge, during the 2016 Melt in the Mantle workshop, organized by John Rudge. Shown in the photo are, from left, Dan McKenzie, Marc Spiegelman, and the author.

a homogeneous source and channelized melt transport can create chemical variability [Spiegelman and Kelemen, 2003; Elliott and Spiegelman, 2003]. Moreover, if this channelization process is forced by heterogeneity of the mantle source [Weatherley and Katz, 2012; Katz and Weatherley, 2012], observed chemical variability may represent a *convolution* of that derived from source and transport. Their deconvolution, as envisioned by McKenzie ~40 years ago, requires a substantial advance beyond the current cutting edge.

CHAPTER 3

A Review of One-Phase Mantle Dynamics

Mantle convection is the physical context within which melt generation and transport occur. Not surprisingly then, it also establishes the mathematical context: partial differential equations representing conservation of mass, momentum, and energy for viscous, polythermal fluid flows. Many of the ideas that form the basis for the mathematical theory of mantle dynamics translate directly to the study of coupled magma/mantle dynamics. Hence a brief review of mantle dynamics is appropriate. It is assumed that the reader is already acquainted with these concepts. The Literature Notes at the end of this chapter contain references to material to help the reader learn or refresh this material. In the following, we focus on the concepts and models that will be relevant later in the book.

3.1 Governing Equations

The governing equations for mantle dynamics can take various forms, depending on the physical processes that are included or neglected. All formulations are based on the fact that, in the mantle, inertial and Coriolis forces are much weaker than viscous forces. The ratio of inertial to viscous forces is quantified by the Reynolds number,

$$\mathrm{Re} \equiv \frac{\rho U L}{\eta}, \tag{3.1}$$

which is dimensionless. The symbols represent characteristic values of density ρ, flow speed U, length scale over which the speed varies L, and dynamic viscosity η. Taking values relevant for the whole mantle of the Earth gives a Reynolds number of about 10^{-19}—effectively zero. Hence we neglect the inertial terms of the Cauchy momentum equation to give the Stokes equation,

$$\nabla \cdot \boldsymbol{\sigma} + \rho \mathbf{g} = 0. \tag{3.2}$$

Here $\boldsymbol{\sigma}$ represents the viscous stress tensor, ρ is the mantle density, and $\mathbf{g}$ is a vector that represents the acceleration due to gravity. The Stokes equation states that internal stresses balance the gravitational body force.

To relate the stress tensor in (3.2) to mechanical quantities of interest, we first decompose it into its isotropic and deviatoric parts as

$$\boldsymbol{\sigma} = -P\boldsymbol{I} + \boldsymbol{\tau}, \tag{3.3}$$

where $\boldsymbol{I}$ is the identity tensor. The deviatoric stress tensor $\boldsymbol{\tau}$ is defined such that its trace is zero,

$$\boldsymbol{\tau} \equiv \boldsymbol{\sigma} - \tfrac{1}{3}\text{trace}\,(\boldsymbol{\sigma})\boldsymbol{I}. \tag{3.4}$$

Comparing equations (3.4) and (3.3), we identify the fluid-dynamical pressure as the mean isotropic stress, $P \equiv -\text{trace}\,(\boldsymbol{\sigma})/3$. The negative sign in this definition means that pressure has the opposite sign convention as stress: *pressure is positive when it represents compression while stress is positive when it represents tension.* We use this sign convention throughout the book.

Following Newton, we posit a viscous constitutive law that relates the deviatoric stress tensor and the deviatoric strain rate tensor,

$$\boldsymbol{\tau} = 2\eta\dot{\boldsymbol{\varepsilon}}, \tag{3.5}$$

where η represents the mantle viscosity. A non-Newtonian viscosity can be described by this equation as long as η is allowed to be a function of the stress or strain rate. The deviatoric strain rate tensor is defined as the full strain rate tensor minus its isotropic part,

$$\dot{\boldsymbol{\varepsilon}} \equiv \dot{\boldsymbol{e}} - \tfrac{1}{3}\text{trace}\,(\dot{\boldsymbol{e}})\boldsymbol{I}. \tag{3.6}$$

Like the deviatoric stress tensor, this tensor has zero trace and hence no isotropic component. The full strain-rate tensor is

$$\dot{\boldsymbol{e}} \equiv \tfrac{1}{2}\left[\nabla\boldsymbol{v} + (\nabla\boldsymbol{v})^{\mathsf{T}}\right], \tag{3.7}$$

which is the symmetric part of the velocity gradient tensor, $\nabla\boldsymbol{v}$. With the definitions (3.6) and (3.7), the full viscous constitutive law can be expressed as

$$\boldsymbol{\sigma} = -P\boldsymbol{I} + \eta\left[\nabla\boldsymbol{v} + (\nabla\boldsymbol{v})^{\mathsf{T}} - \tfrac{2}{3}\left(\nabla\cdot\boldsymbol{v}\right)\boldsymbol{I}\right]. \tag{3.8}$$

Again, note that a positive value of P indicates compression whereas the sign convention for stress is tension-positive.

Constitutive models more complicated than (3.5) can incorporate elasticity and plasticity. Such complexity is required to produce behavior consistent with plate tectonics, for example. For our present purposes, however, a simple viscous model will suffice.

Conservation of momentum (3.2) must be coupled with conservation of mass to form a complete model of the mechanics. The latter is written

$$\frac{\partial\rho}{\partial t} + \nabla\cdot(\rho\boldsymbol{v}) = 0. \tag{3.9}$$

This equation states that change in the local mass per unit volume is caused by divergence of the flux of mass. If the velocity field has nonzero divergence, the density must vary in space and/or time, according to this equation. Variation in density ρ drives mantle convection; inclusion of such variation is thus crucial in modeling mantle dynamics. However, the density variations are small relative to the density itself. For most purposes, it is advantageous to simplify by making the *Boussinesq approximation*. In this approximation we neglect all variation in density except in the body-force term of (3.2), where it multiplies $\mathbf{g}$. Then the conservation of mass equation takes the form of the incompressible continuity equation,

$$\nabla \cdot \boldsymbol{v} = 0. \tag{3.10}$$

Referring back to equation (3.6) we note that for an incompressible fluid, $\dot{\boldsymbol{\varepsilon}} = \dot{\boldsymbol{e}}$. This simplifies the deviatoric stress tensor for incompressible flow from (3.8) to a form that may be more familiar, $\boldsymbol{\tau} = \eta \left[\nabla \boldsymbol{v} + (\nabla \boldsymbol{v})^{\mathsf{T}} \right]$. It is important to note that an incompressible viscous fluid lacks a bulk viscosity that would be associated with compression; instead, the isotropic stress field (pressure) take values that prevent any isotropic deformation.

Density variations in the body force term of (3.2) drive convection. We have yet to specify, however, what causes those density variations. The simplest approach is to consider variations that are a function of temperature only, as we do below. We hence require a model for temperature variations, which we obtain from the principle of energy conservation as

$$\rho c_P \left[\frac{\partial T}{\partial t} + \nabla \cdot (T\boldsymbol{v}) \right] = \nabla \cdot k_T \nabla T + \rho \mathbb{Q}. \tag{3.11}$$

In this equation we have introduced c_P for mass-specific heat capacity, T for temperature, k_T for thermal conductivity, and $\mathbb{Q}$ for the mass-specific rate of heat production by decay of radioactive elements. Consistent with the Boussinesq approximation, work associated with density change has been neglected. Moreover, for convection, the domain-averaged rates of pressure–volume work and viscous dissipation balance each other. Hence, for consistency of approximation, the viscous dissipation is also neglected in equation (3.11). The Literature Notes at the end of this chapter give a brief discussion of how these non-Boussinesq terms are accounted for in more detailed models.

Equation (3.11) can be rewritten more compactly as

$$\frac{\mathrm{D}T}{\mathrm{D}t} = \kappa \nabla^2 T + \mathbb{Q}/c_P, \tag{3.12}$$

where we have assumed the thermal conductivity is uniform, moved it outside the divergence, and defined $\kappa \equiv k_T/\rho c_P$ as the thermal diffusivity. $D/Dt = \partial/\partial t + \boldsymbol{v} \cdot \nabla$ is a Lagrangian derivative; it represents the rate of change of a quantity (temperature in this case) in a reference frame that is fixed to an infinitesimal parcel of material as it moves with the flow.

3.2 Mantle Convection

Here we consider, in the simplest way possible, how mantle convection is described by the theory above. Gathering equations from section 3.1 and defining a coordinate system in which the z-direction is upward, opposite to the direction of gravity (hence $\mathbf{g} = -g\hat{\mathbf{z}}$), we write the force balance as (see 3.2)

$$\nabla P = \eta \nabla^2 \boldsymbol{v} - \rho g \hat{\mathbf{z}}. \tag{3.13}$$

In this equation, we have assumed a uniform viscosity and taken it outside the derivative. Assuming two-dimensionality of the problem ($\partial \cdot / \partial y = 0$) for simplicity, taking the curl of equation (3.13), and rearranging, gives

$$\nabla^2 \omega = -\frac{g}{\eta} \frac{\partial \rho}{\partial x} \tag{3.14}$$

where $\omega \hat{\mathbf{y}} \equiv \nabla \times \boldsymbol{v}$ is the vorticity of the flow and can be thought of as the overturning circulation. This equation shows that convection is driven by density gradients that are perpendicular to gravity.

Taking this a step further, consider variations of mantle density arising from temperature only. We can assume linear variation of density with temperature,

$$\rho(T) = \rho_0 [1 - \alpha_\rho (T - T_0)], \tag{3.15}$$

where α_ρ is the coefficient of thermal expansion and a subscript 0 represents a reference value. We substitute this *equation of state* into (3.14).

Next we rescale variables and equations into non-dimensional form. This has various motivations, but the most important here is to form a dimensionless group of parameters that, as a single dimensionless number, controls the behavior of the solution. The first step is to identify typical values (usually called *characteristic scales*) for the variables. Here we introduce a useful notation:

$$[\boldsymbol{x}] = L, \qquad [T] = \delta T, \qquad [\boldsymbol{v}] = \kappa / L. \tag{3.16}$$

These equations state, respectively, that a typical length scale is the thickness of the mantle L, a typical temperature scale is the temperature drop across the mantle δT, and a typical velocity is of the same order as a rate of conductive heat transfer κ / L. The vorticity inherits[1] its scaling from these: $[\omega] = [\boldsymbol{v}] / [\boldsymbol{x}] = \kappa / L^2$. We then define non-dimensional versions of all variables by dividing by their charateristic scale: for a variable q we define dimensionless version q' as

$$q' \equiv q / [q]. \tag{3.17}$$

Definitions such as this one are substituted into the governing equations to replace dimensional quantities with their non-dimensional equivalents. The scale constants and problem parameters are collected into non-dimensional groups. Primes on the dimensionless variables are then dropped for concision.

Application of this procedure to the combination of (3.14) with (3.15) gives

$$\nabla^2 \omega = \mathrm{Ra} \frac{\partial T}{\partial x}, \tag{3.18}$$

---

[1] Note that differential operators scale with the quantity in their denominator. Hence, for example, $[\nabla] = 1 / [\boldsymbol{x}]$ and $[\nabla \times \boldsymbol{v}] = [\boldsymbol{v}] / [\boldsymbol{x}]$.

where $\mathrm{Ra} \equiv \rho_0 \alpha_\rho \delta T g L^3/(\eta\kappa)$ is the Rayleigh number, the dimensionless number that controls the vigor of thermal convection. This vorticity equation could be solved with the energy equation (3.11) and appropriate boundary and initial conditions to model mantle convection (although this is probably not a practical approach).

Without solving it, equation (3.18) tells us that lateral temperature gradients are a cause of circulation (i.e., convection) in mantle flow. Moreover, in exercise 3.5 we learn that lateral gradients, and hence mantle convection, can arise from the instability of a stratified state. If our equation of state (3.15) for density had included other dependencies (e.g., composition), we'd have other source terms on the right-hand side of (3.18).

Circulation could also be driven by boundary conditions, but we neglect to specify these for the mantle as a whole. Rather we move on to consider a particular plate-tectonic boundary—a mid-ocean ridge—and consider a model in which buoyancy is neglected and boundary conditions drive the flow.

3.3 Kinematic Solutions for Corner Flow

In Earth, mantle overturning and plate tectonics are both parts of the same thermochemically convecting system. The cold, rigid, oceanic lithosphere contains the negative buoyancy that drives downwelling flow at subduction zones. This lithosphere also elastically transmits stress back to the mid-ocean ridges, where plate divergence drives passive mantle upwelling. Although subduction zones and mid-ocean ridges are part of the dynamically convecting lithosphere–mantle system, the stresses in the plates vary on a much larger scale than the boundaries themselves. Therefore, local to a ridge or subduction zone, we can neglect body forces and consider the mantle flow to be driven by an imposed plate motion that is obtained from observations. This is known as a kinematic model and the solution that is obtained is said to be a kinematic or passive flow field.

Kinematic solutions can be obtained numerically or, with significant simplifications, analytically. Here we present the kinematic corner-flow solution (see the Literature Notes at the end of this chapter). We consider the incompressible, isoviscous, dimensional Stokes equations without body forces,

$$\nabla P - \eta \nabla^2 \boldsymbol{v} = \boldsymbol{0}, \tag{3.19a}$$

$$\nabla \cdot \boldsymbol{v} = 0, \tag{3.19b}$$

again take a coordinate system with z upward, and define a streamfunction $\psi(x,z)$ such that

$$\boldsymbol{v} = \nabla \times (\psi \hat{\boldsymbol{y}}) = -\frac{\partial \psi}{\partial z}\hat{\boldsymbol{x}} + \frac{\partial \psi}{\partial x}\hat{\boldsymbol{z}}. \tag{3.20}$$

We have assumed invariance in the y-direction such that $\partial \cdot /\partial y = 0$, which is appropriate for the idealized geometry considered below.

By substituting (3.20) into (3.19), we see that (3.19b) is automatically satisfied. By then taking the x-derivative of the z-component of (3.19a) and subtracting it from the z-derivative of the x-component, we obtain the biharmonic equation,

$$\nabla^2(\nabla^2 \psi) = 0. \tag{3.21}$$

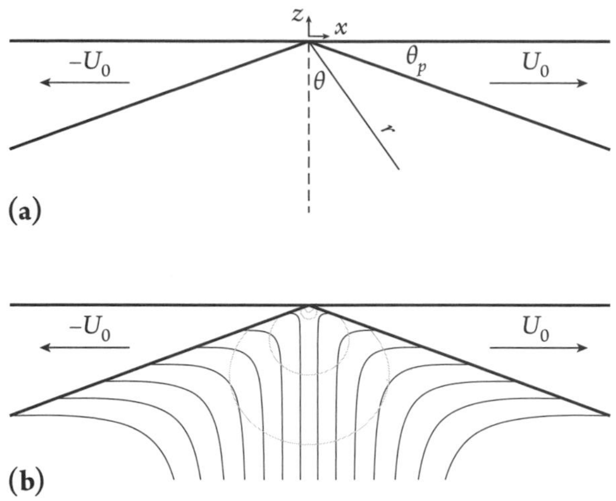

Figure 3.1. Corner-flow kinematic model of mantle flow beneath a mid-ocean ridge. **(a)** Domain geometry and polar coordinate system with $\theta = 0$ pointing downward along the dashed line and θ increasing counterclockwise. The bottom of the lithospheric plates makes an angle θ_p to the horizontal and $\theta = \pi/2 - \theta_p$ to the vertical. **(b)** Streamlines of the solution for $\boldsymbol{v}$ and contours of the solution for P from equations (3.28). For the analytical corner-flow solution, the pressure has a minimum of $-\infty$ at the ridge axis, $r = 0$.

We wish to solve this equation in the wedge-shaped region below the sloping base of the lithospheric plates, as shown in figure 3.1a. We take the origin to be at the ridge axis with $\hat{z}$ pointing upward. Then

$$\boldsymbol{x} = r\left(\hat{\boldsymbol{x}} \sin\theta - \hat{z}\cos\theta\right), \tag{3.22}$$

and $\theta = 0$ is the symmetry axis (dashed line). Unit vectors in the radial and tangential directions are $\hat{\boldsymbol{r}}$ and $\hat{\boldsymbol{\theta}}$ respectively. In this coordinate system, the lithospheric boundary conditions are conveniently imposed at $\theta = \pm(\pi/2 - \theta_p)$, where θ_p is the angle between the horizontal and the base of the lithosphere.

In two-dimensional polar coordinates, the biharmonic equation (not shown) looks messier but has the simple, separable solution

$$\psi = rU_0 f(\theta), \tag{3.23a}$$

$$\boldsymbol{v} = U_0\left[f'(\theta)\hat{\boldsymbol{r}} - f(\theta)\hat{\boldsymbol{\theta}}\right], \tag{3.23b}$$

$$P = -\frac{\eta U_0}{r}\left[f'''(\theta) + f'(\theta)\right], \tag{3.23c}$$

with

$$f(\theta) = C_1 \sin\theta + C_2\theta\sin\theta + C_3\cos\theta + C_4\theta\cos\theta. \tag{3.24}$$

To solve for constants $C_{1\text{-}4}$ we require four boundary conditions. These come from assuming that the lithospheric plates move horizontally away from the ridge axis at speed U_0. In polar coordinates this is described by

$$\boldsymbol{v} = U_0\left(\cos\theta_p\hat{\boldsymbol{r}} \pm \sin\theta_p\hat{\boldsymbol{\theta}}\right) \text{ at } \theta = \pm(\pi/2 - \theta_p). \tag{3.25}$$

Using (3.23b), the boundary conditions imply

$$\left.\begin{aligned} f &= \mp\sin\theta_p \\ f' &= \cos\theta_p \end{aligned}\right\} \text{ at } \pm(\pi/2 - \theta_p), \tag{3.26}$$

which are the four equations needed to determine the constants in (3.24). We obtain

$$C_1 = \frac{2\sin^2\theta_p}{\pi - 2\theta_p - \sin 2\theta_p}, \quad C_2 = C_3 = 0, \quad C_4 = \frac{-2}{\pi - 2\theta_p - \sin 2\theta_p}. \tag{3.27}$$

Combining (3.27) and (3.24) with (3.23b–3.23c) gives the solution

$$\boldsymbol{v} = U_0\left\{\left[(C_1 + C_4)\cos\theta - C_4\theta\sin\theta\right]\hat{\boldsymbol{r}} - (C_1\sin\theta + C_4\theta\cos\theta)\hat{\boldsymbol{\theta}}\right\}, \tag{3.28a}$$

$$P = \frac{2\eta U_0}{r}C_4\cos\theta. \tag{3.28b}$$

Expressing these results in the Cartesian coordinate system of Figure 3.1 we obtain

$$\boldsymbol{v}\cdot\hat{\boldsymbol{x}} = U_0C_4\left(\arctan\frac{x}{z} - \frac{xz}{x^2+z^2}\right), \tag{3.29a}$$

$$\boldsymbol{v}\cdot\hat{\boldsymbol{z}} = U_0C_4\left(\sin^2\theta_p - \frac{z^2}{x^2+z^2}\right), \tag{3.29b}$$

$$P = 2\eta U_0C_4\frac{-z}{x^2+z^2}. \tag{3.29c}$$

Streamlines of the velocity and contours of the pressure from (3.28) (or, equivalently, (3.29)) are shown in Figure 3.1b. Noting that $C_4 < 0$, the solution has negative pressure everywhere. It is most negative as $r \to 0$, where the pressure diverges to $-\infty$ (note that θ is restricted to the interval $[-\pi/2, \pi/2]$).

The corner-flow solution illustrates the important concept of *passive flow*, which is mantle flow in response to kinematic boundary conditions. This is in contrast to *active flow*, flow that is driven by the gravitational body force acting on variations in density. The corner-flow solution is also useful as a basis for two-phase flow models in chapter [13].

3.4 Literature Notes

The most important reference on *Geodynamics* is, without doubt, the book *Geodynamics* by Turcotte and Schubert [2014], now in its third edition. It provides an excellent introduction to the fluid dynamics of mantle flow as well as rheology, heat transfer, and many other topics. It includes a helpful introduction to mantle convection.

A thorough treatment of mantle convection was published in *Mantle Convection in the Earth and Planets* by Schubert et al. [2001]. For a more concise review, the reader is referred to the chapter "The Physics of Mantle Convection" from *Treatise on Geophysics* [Ricard, 2007].

In this chapter we made a strict Boussinesq approximation, neglecting all terms associated with volume change, including the pressure–volume (PV) work in the energy equation. A better model for mantle convection [Turcotte and Oxburgh, 1972] incorporates these terms and hence resolves the mantle isentropic temperature gradient. This more complete model is sometimes known as the *extended Boussinesq approximation*, meaning that it extends the equations to include physics that is neglected in the strict Boussinesq approximation. Jarvis and McKenzie [1980] formulated a non-Boussinesq, compressible model of mantle convection in terms of the anelastic liquid equations. They showed that the extended Boussinesq approximation is valid when the scale height of the isentropic temperature change is large compared to the height of the mantle. They also showed that viscous dissipation generates heat at exactly the rate needed to balance heat absorption by the compressional work cycle. This means that, for convection problems, dissipation and PV work need to be included or neglected in tandem.

A key theme in the recent literature on mantle convection has been the complex rheologies that allow for self-consistent simulation of plate tectonics. Tackley [2000b,a] report key advances on this topic; Bercovici [2003] provides a good review.

The corner-flow solution to the incompressible Stokes equations was first published by Batchelor [1967] and first used for geodynamics by McKenzie [1969]. The treatment in this chapter is based on Spiegelman and McKenzie [1987]. We return to this topic in chapter 13, where we consider the consequences of the corner-flow pressure and flow fields in models of melt transport with constant, uniform porosity.

3.5 Exercises

3.1 Show that any velocity that can be expressed as (3.20) automatically satisfies the incompressible continuity equation.

3.2 Use the incompressible continuity equation (3.10), the assumption of uniform viscosity, and the coordinate system as specified in the text to derive (3.13) from (3.2).

3.3 Investigate the streamfunction for cornerflow. *Hint: The following formulas are for the two-dimensional Laplacian in plane polar coordinates.*

$$\nabla^2 q = \frac{1}{r}\frac{\partial}{\partial r}\left(r\frac{\partial q}{\partial r}\right) + \frac{1}{r^2}\frac{\partial^2 q}{\partial \theta^2},$$

$$\begin{aligned}\nabla^2 \boldsymbol{v} = &\left(\frac{1}{r}\frac{\partial}{\partial r}\left(r\frac{\partial v_r}{\partial r}\right) + \frac{1}{r^2}\frac{\partial^2 v_r}{\partial \theta^2} - \frac{v_r}{r^2} - \frac{2}{r^2}\frac{\partial v_\theta}{\partial \theta}\right)\hat{\boldsymbol{r}} \\ &+ \left(\frac{1}{r}\frac{\partial}{\partial r}\left(r\frac{\partial v_\theta}{\partial r}\right) + \frac{1}{r^2}\frac{\partial^2 v_\theta}{\partial \theta^2} - \frac{v_\theta}{r^2} + \frac{2}{r^2}\frac{\partial v_r}{\partial \theta}\right)\hat{\boldsymbol{\theta}}.\end{aligned}$$

(a) By substituting the streamfunction (3.23a) into the biharmonic equation (3.21), determine the ordinary differential equation (ODE) satisfied by $f(\theta)$.

(b) Using your solution to (a), verify that the pressure field in the kinematic model for corner flow is given by (3.23c). Show that (3.24) is the general solution for $f(\theta)$.

(c) Suppose $\psi = r^n B_0 g(\theta)$, where B_0 is has dimensions to make this equation consistent with (3.20). Find the general solution for $g(\theta)$ where $n \neq 1$ is an integer. Note the particular cases of $n = 0, 2$.

3.4 Corner flow for a subduction zone. Consider a slab of rigid, oceanic lithosphere subducting with speed U_0 at angle θ_0 below the horizontal. Above this lies a second rigid plate at rest along the horizontal. Assume that both plates have zero curvature. Take a plane polar coordinate system such that $\theta = 0$ is the horizontal (aligned with $\hat{\boldsymbol{x}}$), with θ increasing counterclockwise.

(a) By imposing no-flux and no-slip boundary conditions along both plates, determine the velocity field of the kinematic model. You may assume that the streamfunction is of equivalent form to (3.23a).

(b) Suppose that the no-slip boundary condition along the oceanic lithosphere is replaced by imposing a constant tangential shear stress τ. This is achieved by setting $\eta r^{-1} \partial v_r / \partial \theta = \tau$ on the slab top at $\theta = -\theta_0$. In this case, we have $\psi \propto \tau/\eta$ and so dimensional analysis requires $\psi = \tau r^2 g(\theta)/\eta$. Use your answer to exercise 3.3(c) to determine the velocity field.

3.5 The onset of Rayleigh-Bénard convection. Consider a layer of fluid between $z = 0$ and $z = h$ in the (x, z) plane, and assume that the layer has infinite horizontal extent. The system is modeled using mass conservation (3.10), momentum conservation (3.13), and energy conservation (3.12) with $\mathbb{Q} = 0$, uniform thermal conductivity, and the linear equation of state (3.15). The boundary conditions are no normal flux, no shear stress, and fixed temperature on the plates:

$$w = 0, \quad \tau_{xz} = 0, \quad T = T_0 \qquad \text{at } z = h, \tag{3.30a}$$

$$w = 0, \quad \tau_{xz} = 0, \quad T = T_0 + \delta T \qquad \text{at } z = 0. \tag{3.30b}$$

Consider small perturbations about a steady, *base state* of no motion by introducing a small parameter $\epsilon \ll 1$ and expanding the velocity and temperature as follows:

$$\begin{aligned}
\boldsymbol{v} &= \boldsymbol{0} + \epsilon \boldsymbol{v}^{(1)}(\boldsymbol{x}, t) + \mathcal{O}(\epsilon^2), \\
T &= T^{(0)}(z) + \epsilon T^{(1)}(\boldsymbol{x}, t) + \mathcal{O}(\epsilon^2), \\
P &= P^{(0)}(z) + \epsilon P^{(1)}(\boldsymbol{x}, t) + \mathcal{O}(\epsilon^2).
\end{aligned}$$

(a) By substituting the above expansions into the governing equations and discarding terms of $\mathcal{O}(\epsilon^1)$ or higher, determine the $\mathcal{O}(\epsilon^0)$ equations governing the base state. Hence, obtain solutions for $T^{(0)}$ and $P^{(0)}$.

(b) Using the same substitution but balancing terms of $\mathcal{O}(\epsilon^1)$, obtain equations governing perturbations to the base state. In particular, find the two equations relating $T^{(1)}$ to the vertical velocity component $w^{(1)}$.

By non-dimensionalizing the equations using the characteristic scales

$$[\boldsymbol{x}] = h, \quad [\boldsymbol{v}] = \kappa/h, \quad [t] = h^2/\kappa, \quad [T] = \delta T,$$

defining $\hat{w} \equiv w^{(1)}/[\boldsymbol{v}]$ and $\hat{T} \equiv T^{(1)}/[T]$, and rewriting the two governing equations in terms of dimensionless variables we have

$$\nabla^2\nabla^2\hat{w} = -\mathrm{Ra}\frac{\partial^2}{\partial x^2}\hat{T}, \qquad \left(\frac{\partial}{\partial t} - \nabla^2\right)\hat{T} = \hat{w},$$

where Ra is the Rayleigh number, defined in section 3.2. All symbols are now dimensionless, even if only temperature and velocity are marked with a ˆ symbol.

(c) Show that, after accounting for the base-state solution and nondimenzionalizing, the boundary conditions on both the top and bottom of the layer become $\hat{w} = 0$, $\frac{\partial^2\hat{w}}{\partial z^2} = 0$ and $\hat{T} = 0$.

(d) As ansatz, assume solutions[2] of the form

$$\left(\hat{w}, \hat{T}\right) = (w_c, T_c)\sin(n\pi z)\,\mathrm{e}^{ikx+st}$$

with n a positive integer and (w_c, T_c) unknown constants. In the case of *marginal stability*, where the exponential growth rate $s = 0$, determine the Rayleigh number Ra as a function of n and k^2. Plot this Ra versus k for $n = 1, 2, 3$. Determine the minimum (critical) value of Ra over all possible values of n and k^2. Comment on the physical significance of this number.

(e) The critical Rayleigh number increases significantly if either of the stress-free boundary conditions are replaced by no-flow boundary conditions. How can this be understood in physical terms?

[2]These trial solutions comprise separable eigenfunctions of the linear operators that satisfy the boundary conditions.

CHAPTER 4

Conservation of Mass and Momentum

Having reviewed the one-phase flow theory of mantle convection, we now turn our attention to the material central to this book: a two-phase flow theory for the partially molten mantle. In particular, this chapter deals with the fluid dynamics, based on principles of conservation of mass and momentum. Later chapters will add consideration of conservation of energy and composition.

The physical picture to keep in mind as we go through the derivation is of an aggregate of solid grains with liquid melt occupying the interconnected pores between them, as shown in figure 4.1. For a sufficiently large ensemble of randomly oriented grains and pores, the statistical distribution of orientations is uniform and hence the system may be considered isotropic. This will be important when we take volume averages over such ensembles to develop the continuum-scale theory, below.

4.1 The Representative Volume Element and Phase-Averaged Quantities

We pose our derivation in terms of a representative volume element (RVE, figure 4.1) that contains a mixture of the two phases, solid grains and liquid interstitial melt. The RVE is a fundamental concept in continuum theories; it must contain a volume that is large enough to house many microscopic elements (e.g., atoms, molecules, grains) but small enough that variation of any continuum property across it is well approximated as being linear. In the case of two-phase flow, we need the volume to contain many solid grains and liquid-filled pores. This is possible if the RVE has a characteristic dimension that is many times the mean grain size. This means that we are effectively averaging over the volume and so our continuum model cannot resolve any processes at scales smaller than the RVE (e.g., the grain scale).

To derive the continuum equations, we must write down a minimal model of the physics at the grain scale that can then be volume averaged. It is convenient to consider distinct coordinate systems at each scale. At the continuum scale, the coordinate system is associated with a position vector $\boldsymbol{x}$ and a volume element $\mathrm{d}^3\boldsymbol{x}$; at the microscopic scale, the coordinate system has a position vector $\breve{\boldsymbol{x}}$ and a volume element $\mathrm{d}^3\breve{\boldsymbol{x}}$. We can imagine that at every position $\boldsymbol{x}$, there is a finite microscopic space (the RVE) of positions $\breve{\boldsymbol{x}}$.

With this separation of scales in mind, we can write a conservation equation for some volumetric field property Υ (units of stuff per unit volume, e.g., kg m^{-3}) within an RVE

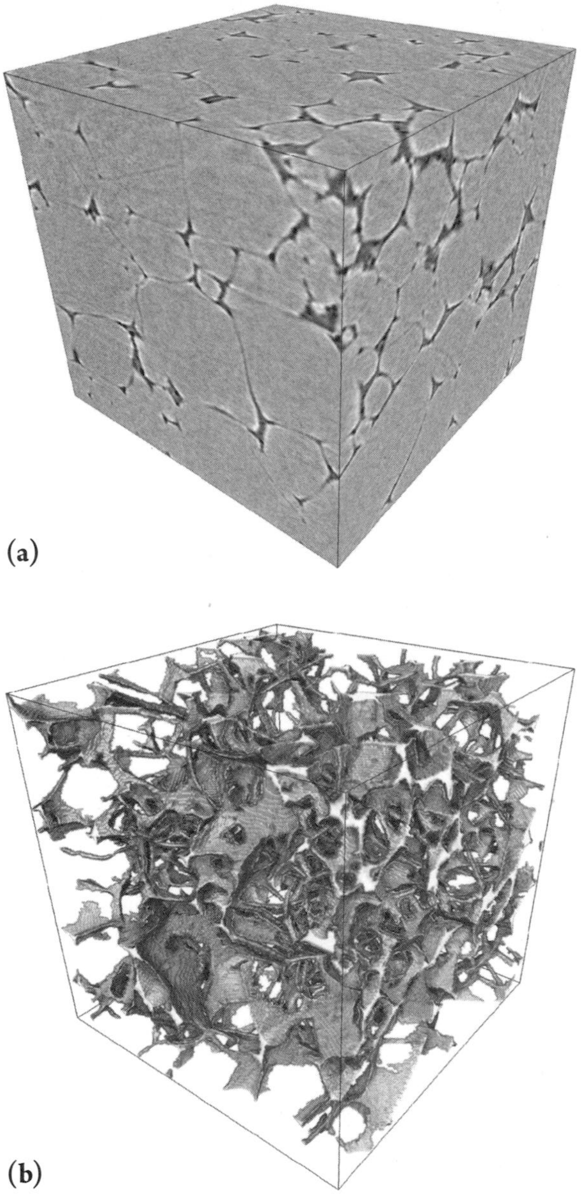

(a)

(b)

Figure 4.1. An x-ray microtomographic image of a cubic volume containing olivine grains and ~5% basaltic melt. The characteristic grain size is ~35 μm and the cube is 300 μm on each side. Hence this volume contains about 1200 grains. This illustrates the concept of an RVE. **(a)** Surface rendering of x-ray intensity. Melt-filled pores appear darker; olivine grains appear lighter. **(b)** Volume rendering of the pore space between olivine grains. Images courtesy of K.J. Miller using data produced for Miller et al. [2015].

that is fixed in space,

$$\frac{\mathrm{d}}{\mathrm{d}t}\int_{\mathrm{RVE}} \Upsilon\, \mathrm{d}^3\breve{\boldsymbol{x}} = -\int_{\partial\mathrm{RVE}} \boldsymbol{J}_\Upsilon \cdot \mathrm{d}\boldsymbol{S} + \int_{\mathrm{RVE}} S_\Upsilon\, \mathrm{d}^3\breve{\boldsymbol{x}}. \tag{4.1}$$

In this equation, $\int_{\mathrm{RVE}} \mathrm{d}^3\breve{\boldsymbol{x}}$ represents an integration over the volume of the RVE, $\int_{\partial\mathrm{RVE}} \mathrm{d}\boldsymbol{S}$ represents an integration over its boundary, where $\mathrm{d}\boldsymbol{S}$ is an infinitesimal surface patch described by its outward-pointing normal vector. The first term in the equation is the rate of change of the amount of Υ in the RVE. It is equal to the rate at which Υ enters the RVE through its boundary (first term on the right-hand side), and the rate at which Υ is produced in the RVE by any possible sources (second term on the right-hand side). $\boldsymbol{J}_\Upsilon$ is the flux of Υ by, for example, advection or diffusion; S_Υ is the volumetric production rate of Υ. S_Υ can be negative, of course, representing a sink of Υ.

To convert (4.1) to a PDE we apply Gauss' theorem,

$$\int_{\partial\mathrm{RVE}} \boldsymbol{J}_\Upsilon \cdot \mathrm{d}\boldsymbol{S} = \int_{\mathrm{RVE}} \nabla\cdot \boldsymbol{J}_\Upsilon\, \mathrm{d}^3\breve{\boldsymbol{x}}, \tag{4.2}$$

to the surface integral. Since the RVE is fixed in space, the time derivative in (4.1) can be moved inside the integral. Then each term is a volume integral over the same volume. These can be evaluated by assuming that the integrand varies no more than linearly or, loosely speaking, dropped by allowing the arbitrary volume to shrink down to a point in the continuum. Hence the general PDE arising from (4.1) is

$$\frac{\partial \Upsilon}{\partial t} + \nabla\cdot \boldsymbol{J}_\Upsilon = S_\Upsilon. \tag{4.3}$$

To be more rigorous about volume-averaging in the context of a multi-phase system, we define an *indicator function* $\Phi^i(\breve{\boldsymbol{x}})$ that identifies the location of phases within the RVE:

$$\Phi^i(\breve{\boldsymbol{x}}) \equiv \begin{cases} 1 & \text{if there is phase } i \text{ at } \breve{\boldsymbol{x}}, \\ 0 & \text{if there is some other phase at } \breve{\boldsymbol{x}}. \end{cases} \tag{4.4}$$

Here and below, the $\breve{}$ symbol denotes a microscopic variable that varies at the scale of grains and pores.

The indicator function allows us to rigorously define the macroscopic phase fraction ϕ^i of phase i as

$$\phi^i \equiv \frac{1}{\delta V}\int_{\mathrm{RVE}} \Phi^i(\breve{\boldsymbol{x}})\, \mathrm{d}^3\breve{\boldsymbol{x}}, \tag{4.5}$$

where $\delta V \equiv \int_{\mathrm{RVE}} \mathrm{d}^3\breve{\boldsymbol{x}}$ is the volume of the RVE.

Suppose the RVE contains N distinct and immiscible phases. If the volume is *saturated* then one of the N indicator functions is non-zero at every point within the RVE such that

$$\sum_{i=1}^{N}\int_{\mathrm{RVE}} \Phi^i(\breve{\boldsymbol{x}})\, \mathrm{d}^3\breve{\boldsymbol{x}} = \delta V \quad \Rightarrow \quad \sum_{i=1}^{N}\phi^i = 1. \tag{4.6}$$

This means that there is no empty space and hence that the sum of the continuum phase fractions is unity. We will always assume a saturated continuum in this book. Since any point in the microscopic space is occupied by exactly one phase, the identity of that phase is determined by evaluating the set of indicator functions at that point. For an N-phase medium, the value of $N-1$ indicator functions are required. In this book we consider only two phases, liquid magma and solid mantle, as shown in figure 4.1. It is sometimes useful to refer to the liquid and solid fractions as ϕ^ℓ and ϕ^s. Typically, however, they will be written as porosity ϕ and solid fraction $1-\phi$, respectively.

The concept of phase-volume fraction is thus defined as a volume average of an indicator function over a representative volume. What about other quantities? The same principle applies, and we can write the integral

$$\phi^\ell \boldsymbol{v}^\ell = \frac{1}{\delta V}\int_{\text{RVE}} \Phi^\ell(\breve{\boldsymbol{x}})\,\breve{\boldsymbol{v}}(\breve{\boldsymbol{x}})\,\mathrm{d}^3\breve{\boldsymbol{x}}, \tag{4.7}$$

where $\breve{\boldsymbol{v}}$ is the microscopic velocity (solid or liquid) that one would measure at a point $\breve{\boldsymbol{x}}$ within the RVE. To evaluate the integral we use the mean-value theorem for definite integrals and assume that the RVE is small enough that the quantity of interest, $\breve{\boldsymbol{v}}$ in this case, varies no more than linearly within it. Multiplication by the indicator function ensures that the integral selects only the microscopic liquid velocity. $\phi^\ell \boldsymbol{v}^\ell$ is then the mean volume flux of liquid; $\boldsymbol{v}^\ell$ is, of course, the mean liquid velocity. A similar though more complicated procedure can be followed for tensor quantities.

At the boundary of any phase in a saturated medium is an *interface* with another phase. In a two-phase medium, we need consider only one interface, with a microscopic position $\breve{\boldsymbol{x}}^{\mathcal{I}}$ and a microscopic velocity $\breve{\boldsymbol{v}}^{\mathcal{I}}$. In this context, the unit vector that is normal to the phase boundary and points outward from phase i is given by $\breve{\boldsymbol{n}}^i \equiv -\boldsymbol{\nabla}\Phi^i/\left|\boldsymbol{\nabla}\Phi^i\right|$.[1]

4.2 Conservation of Mass

We can use the indicator function to write an integral form for the mass per unit volume for either phase $i=\ell, s$,

$$m^i = \frac{1}{\delta V}\int_{\text{RVE}} \Phi^i \breve{\rho}\,\mathrm{d}^3\breve{\boldsymbol{x}}. \tag{4.8}$$

Then equation (4.1) helps us to write an integral form of conservation of mass for phase i,

$$\frac{\mathrm{d}}{\mathrm{d}t}\int_{\text{RVE}} \Phi^i \breve{\rho}\,\mathrm{d}^3\breve{\boldsymbol{x}} = -\int_{\partial\text{RVE}} \Phi^i \breve{\rho}\breve{\boldsymbol{v}}\cdot\mathrm{d}\boldsymbol{S} + \int_{\mathcal{I}\text{RVE}} \breve{\rho}^i(\breve{\boldsymbol{v}}^{\mathcal{I}} - \breve{\boldsymbol{v}}^i)\cdot\breve{\boldsymbol{n}}^i\,\mathrm{d}S_{\mathcal{I}}. \tag{4.9}$$

The first term represents the rate-of-change of the mass of phase i contained by the RVE; the second term represents the rate at which that phase is flowing into the RVE; the last term represents the rate at which mass is added to phase i by motion of the interface relative to the material that it separates. Motion of the interface relative to the phases it separates is the consequence of phase change. For example, melting of a solid grain causes the microscopic liquid–solid interface to move toward the interior of that grain.

[1] As we discuss below, the gradient of the indicator function contains a Dirac delta function exactly on the interface $\delta\left(\breve{\boldsymbol{x}}-\breve{\boldsymbol{x}}^{\mathcal{I}}\right)$. This singular distribution is strictly valid only within an integral; normalizing by the magnitude of the gradient eliminates it. Below we won't be overly concerned with mathematical stricture.

To disallow cavitation and loss of saturation, we require continuity of the normal mass flux across the internal interface at $\breve{\boldsymbol{x}}^{\mathcal{I}}$. Then, noting that $\breve{\boldsymbol{n}}^{\ell}=-\breve{\boldsymbol{n}}^{s}$, it becomes evident that the last term in (4.9) is equal and opposite for the solid and the liquid.

Dividing (4.9) by δV, defining the volumetric melting rates as

$$\Gamma \equiv \frac{1}{\delta V}\int_{\mathcal{I}_{\text{RVE}}} \breve{\rho}^{\ell}(\breve{\boldsymbol{v}}^{\mathcal{I}}-\breve{\boldsymbol{v}}^{\ell})\cdot\breve{\boldsymbol{n}}^{\ell}\,\mathrm{d}S_{\mathcal{I}} = -\frac{1}{\delta V}\int_{\mathcal{I}_{\text{RVE}}} \breve{\rho}^{s}(\breve{\boldsymbol{v}}^{\mathcal{I}}-\breve{\boldsymbol{v}}^{s})\cdot\breve{\boldsymbol{n}}^{s}\,\mathrm{d}S_{\mathcal{I}} \tag{4.10}$$

and using the procedures discussed in section 4.1, we obtain the conservation of mass equations

$$\frac{\partial\phi\rho^{\ell}}{\partial t}+\boldsymbol{\nabla}\cdot\left[\phi\rho^{\ell}\boldsymbol{v}^{\ell}\right]=\Gamma, \tag{4.11a}$$

$$\frac{\partial(1-\phi)\rho^{s}}{\partial t}+\boldsymbol{\nabla}\cdot\left[(1-\phi)\rho^{s}\boldsymbol{v}^{s}\right]=-\Gamma. \tag{4.11b}$$

These equations state that, at a point in the continuum, the change in phase mass per unit volume is due to the divergence of mass flux of that phase and transfer of mass between phases. Note that the effect of melting is equal and opposite for the liquid and the solid. Summing the two phase-specific equations gives an equation for the conservation of bulk mass,

$$\frac{\partial\overline{\rho}}{\partial t}+\boldsymbol{\nabla}\cdot\overline{\rho\boldsymbol{v}}=0, \tag{4.12}$$

where an overline on a quantity indicates that it is phase-averaged: $\overline{q}=\phi q^{\ell}+(1-\phi)q^{s}$. Equation (4.12) should look familiar. It is the two-phase analogue to equation (3.9), mass conservation for a one-phase medium.

On the basis of (4.12), one might expect that there is a two-phase continuity equation similar to $\boldsymbol{\nabla}\cdot\boldsymbol{v}=0$; this is indeed the case. Returning to equations (4.11) and making the Boussinesq approximation (assuming that each phase is independently incompressible with uniform density) means that we can move density out of the derivatives and over to the right-hand side. With this approximation, summing the phase-specific equations gives

$$\boldsymbol{\nabla}\cdot\overline{\boldsymbol{v}}=-\Gamma\Delta(1/\rho). \tag{4.13}$$

In this equation and below, we take $\Delta q=q^{s}-q^{\ell}$ for any quantity q; hence $\Delta(1/\rho)=1/\rho^{s}-1/\rho^{\ell}$. The left-hand side of (4.13) is the divergence of the phase-averaged velocity. The right-hand side has a volume-source term proportional to the melting rate. Assuming that the melt is less dense than its solid residue, melting causes a volume expansion. In this case, a positive melting rate will cause a divergence of the total volume flux.

How large is the flux divergence caused by melting? Typical mantle values for solid and liquid density are 3300 and 2800 kg/m^3 respectively, so $\Delta(1/\rho)\approx-0.08$ m^3/kg, which turns out to be rather small in most contexts. (This doesn't mean that the volume change on melting never plays a role in the dynamics; see literature notes below.) It is therefore common to extend the Boussinesq approximation to apply to density differences outside of body-force terms, and hence to neglect the right-hand side of (4.13) giving

$$\nabla \cdot \overline{v} = 0. \tag{4.14}$$

This is a two-phase continuity equation analogous to equation (3.10) which states that the phases are independently incompressible. The velocity of either phase can be divergent only if the velocity of the other phase is convergent (or, in other words, if there is a non-zero segregation flux $\boldsymbol{q} \equiv \phi(\boldsymbol{v}^\ell - \boldsymbol{v}^s)$ — see exercise 4.3).

Having assumed constant density of the solid phase under the Boussinesq approximation, the mass-conservation equation for the solid (4.11b) can be rewritten in a form that is physically informative and mathematically useful. It reads

$$\frac{\mathrm{D}_s\phi}{\mathrm{D}t} = \Gamma/\rho^s + (1-\phi)\mathcal{C}, \tag{4.15}$$

where the Lagrangian derivative is associated with the solid velocity and we have introduced the compaction rate,

$$\mathcal{C} \equiv \nabla \cdot \boldsymbol{v}^s, \tag{4.16}$$

a shorthand notation that is used in the literature on magma/mantle dynamics[2]. Equation (4.15) states that as a parcel of solid moves with the mantle flow, its porosity changes due to melting and to (de-)compaction.

4.3 Conservation of Momentum

The phase velocities that appear repeatedly above must be determined by a model of the fluid dynamics. This arises by consideration of the principle of conservation of momentum. We leap over many of the details of that argument, which can be found in standard fluid dynamics texts, and focus on the two-phase aspect of the problem. In particular, we must account for the force imposed by each phase on the other phase. This is the interphase force and, consistent with Newton's third law of motion, its effect is equal and opposite on the phases, as we shall see below.

Returning to our concept of the RVE, we construct the integral representation of momentum conservation as

$$\begin{aligned}\frac{\mathrm{d}}{\mathrm{d}t}\int_{\mathrm{RVE}} \Phi^i \breve{\rho}\breve{\boldsymbol{v}}\,\mathrm{d}^3\breve{\boldsymbol{x}} + \int_{\partial\mathrm{RVE}} \Phi^i\left(\breve{\rho}\breve{\boldsymbol{v}}\right)\breve{\boldsymbol{v}}\cdot\mathrm{d}\boldsymbol{S} + \int_{\mathcal{I}_{\mathrm{RVE}}}\left(\breve{\rho}^i\breve{\boldsymbol{v}}^i\right)\left(\breve{\boldsymbol{v}}^i - \breve{\boldsymbol{v}}^{\mathcal{I}}\right)\cdot\breve{\boldsymbol{n}}^i\mathrm{d}S_{\mathcal{I}} \\ = \int_{\mathrm{RVE}} \Phi^i\breve{\rho}\boldsymbol{g}\,\mathrm{d}^3\breve{\boldsymbol{x}} + \int_{\partial\mathrm{RVE}} \Phi^i\breve{\boldsymbol{\sigma}}\cdot\mathrm{d}\boldsymbol{S} + \int_{\mathcal{I}_{\mathrm{RVE}}} \breve{\boldsymbol{\sigma}}^i\cdot\breve{\boldsymbol{n}}^i\,\mathrm{d}S_{\mathcal{I}},\end{aligned} \tag{4.17}$$

where, for completeness, we have included inertial terms. Note that the total momentum of either phase within the RVE is $\int_{\mathrm{RVE}} \Phi^i\breve{\rho}\breve{\boldsymbol{v}}\,\mathrm{d}^3\breve{\boldsymbol{x}}$. The terms of the left-hand side of the equation represent the rate of change of momentum, advective transport of momentum, and interphase transfer of momentum. It is already known to the reader that the Reynolds number (equation (3.1)) for the solid mantle is zero. For the magma, we consider flow through pores of size 10^{-4} m at a speed of 10 m/yr with density 3000 kg/m^3

[2] The term *compaction* rate is actually a misnomer: a positive divergence of the solid corresponds to local de-compaction of the two-phase system and, in the absence of freezing, to an increase in porosity. We should actually refer to $\nabla \cdot \boldsymbol{v}^s$ as the *decompaction rate*. However, it is customary in the literature to call the solid divergence the compaction rate $\mathcal{C}$ and we will follow this pattern.

and viscosity 1 Pa-s. Based on these upper-bound parameter values, the pore-Reynolds number is then $\sim 10^{-9}$ and so inertia of the magma is also negligible. Hence we discard terms of the left-hand side of equation (4.17).

The terms on the right-hand side of equation (4.17) represent the gravitational body force, viscous stresses on the exterior surface of the RVE, and stresses acting on the internal interface $\mathcal{I}$ between the phases. Missing from this equation is a term representing a line force acting on the intersection between the interface $\mathcal{I}$ and the faces of the RVE. Surface tension would give rise to such a term, but it is neglected here.

Dividing equation (4.17) by δV, defining the volumetric interphase force as

$$\mathbf{F} \equiv -\frac{1}{\delta V}\int_{\mathcal{I}_{\text{RVE}}} \breve{\boldsymbol{\sigma}}^{\ell}\cdot\breve{\boldsymbol{n}}^{\ell}\,\mathrm{d}S_{\mathcal{I}} = \frac{1}{\delta V}\int_{\mathcal{I}_{\text{RVE}}} \breve{\boldsymbol{\sigma}}^{s}\cdot\breve{\boldsymbol{n}}^{s}\,\mathrm{d}S_{\mathcal{I}}, \tag{4.18}$$

and applying methods discussed above (section 4.1), we can rewrite the integral equation in differential form

$$\nabla\cdot\left(\phi^{\ell}\boldsymbol{\sigma}^{\ell}\right) + \phi^{\ell}\rho^{\ell}\mathbf{g} - \mathbf{F} = \mathbf{0}, \tag{4.19a}$$

$$\nabla\cdot\left(\phi^{s}\boldsymbol{\sigma}^{s}\right) + \phi^{s}\rho^{s}\mathbf{g} + \mathbf{F} = \mathbf{0}. \tag{4.19b}$$

Note that in the absence of accelerations or interfacial tension arising from surface energy, the interphase force **F** must be equal in magnitude and opposite in direction between the two phases (as was assumed in equation (4.18)). With the choice of sign in equations (4.19), **F** represents the force per unit volume that the liquid phase exerts on the solid phase.

Equations (4.19) can be summed to give the bulk conservation of momentum equation

$$\nabla\cdot\overline{\boldsymbol{\sigma}} + \overline{\rho}\mathbf{g} = \mathbf{0}. \tag{4.20}$$

Note the similarity between this equation and equation (3.2) for one-phase mantle convection. To make progress in our derivation, we now need to unpack the stress tensors and the interphase force.

4.3.1 STRESS AND PRESSURE

Each of the two phases has its own mechanical pressure, defined in the usual way as the mean normal stress,

$$P^{\ell} = -\tfrac{1}{3}\operatorname{trace}\boldsymbol{\sigma}^{\ell}, \tag{4.21a}$$

$$P^{s} = -\tfrac{1}{3}\operatorname{trace}\boldsymbol{\sigma}^{s}. \tag{4.21b}$$

These macroscopic pressures can be straightforwardly related to their microscopic counterparts according to volume averages, as parallel to equation (4.7).

The pressure can be separated from the total stress $\boldsymbol{\sigma}$ to leave the deviatoric stress $\boldsymbol{\tau}$,

$$\boldsymbol{\sigma}^{\ell} = -P^{\ell}\boldsymbol{I} + \boldsymbol{\tau}^{\ell}, \tag{4.22a}$$

$$\boldsymbol{\sigma}^{s} = -P^{s}\boldsymbol{I} + \boldsymbol{\tau}^{s}. \tag{4.22b}$$

Again, definitions of these stresses can be written in terms of volume averages of stress at the microscopic scale.

With the foregoing, we can write the bulk stress in terms of pressures and deviatoric stresses as

$$\overline{\boldsymbol{\sigma}} = \phi\left[-P^{\ell}\boldsymbol{I} + \boldsymbol{\tau}^{\ell}\right] + (1-\phi)\left[-P^{s}\boldsymbol{I} + \boldsymbol{\tau}^{s}\right], \tag{4.23}$$

$$\overline{P} = \phi P^{\ell} + (1-\phi)P^{s}. \tag{4.24}$$

Pressure and deviatoric stress must be related to compaction and flow. We return to this in section 4.3.3, below.

We again note the choice of sign convention here, which is the same as was used in chapter 3. Positive values of P^i indicate compression whereas positive values in the tensor $\boldsymbol{\sigma}^i$ indicate tension. This is why a negative sign appears in front of pressure in the definitions above.

4.3.2 THE INTERPHASE FORCE

Underlying and unresolved by the continuum formulation are two immiscible phases separated by a spatially complex interface. Forces are transmitted across this interface in two ways: mechanical stress and surface tension. In the present formulation we neglect surface tension (though see the Literature Notes at the end of this chapter). Although it is possible to understand the interphase force through detailed analysis, the simpler, canonical approach is to look for a form that is consistent with Darcy's law for flow of a liquid through the interconnected pore space of a granular solid medium.

The location and orientation of the liquid–solid interface within the RVE is selected by the gradient of the indicator function $\nabla\Phi^i = -\delta\left(\breve{\boldsymbol{x}} - \breve{\boldsymbol{x}}^{\mathcal{I}}\right)\breve{\boldsymbol{n}}^i$, where $\delta(\boldsymbol{x})$ is the Dirac delta function. Because of its argument in this case, the delta function is non-zero only on the microscopic position of the interface. $\breve{\boldsymbol{n}}^i$ is a unit-normal vector rooted on the interface and pointing outward from phase i. Hence the origin of the interphase force term of equation (4.17) can be understood by considering

$$\begin{aligned} -\frac{1}{\delta V}\int_{\text{RVE}} \breve{\boldsymbol{\sigma}}^i \cdot \nabla\Phi^i \, \mathrm{d}^3\breve{\boldsymbol{x}} &= \frac{1}{\delta V}\int_{\text{RVE}} \breve{\boldsymbol{\sigma}}^i \cdot \breve{\boldsymbol{n}}^i\,\delta\left(\breve{\boldsymbol{x}} - \breve{\boldsymbol{x}}^{\mathcal{I}}\right)\mathrm{d}^3\breve{\boldsymbol{x}}, \\ &= \frac{1}{\delta V}\int_{\mathcal{I}\text{RVE}} \breve{\boldsymbol{\sigma}}^i \cdot \breve{\boldsymbol{n}}^i \, \mathrm{d}S_{\mathcal{I}}, \end{aligned}$$

where the delta function converts the integral over the volume to an integral over the interface surface. In this integral, the stress tensor that is selected is that which is effective exactly on the interface. By virtue of mechanical equilibrium in the absence of surface tension, stress must be continuous across the interface. Therefore, noting that $\nabla\Phi^s = -\nabla\Phi^{\ell}$,

$$\breve{\boldsymbol{\sigma}}^s \cdot \nabla\Phi^s + \breve{\boldsymbol{\sigma}}^{\ell} \cdot \nabla\Phi^{\ell} = \boldsymbol{0}, \tag{4.25}$$

and so we could choose the stress tensor of either phase. Making the canonical choice of $\breve{\boldsymbol{\sigma}}^{\ell}$ as the interface stress and focusing on the interphase term for the liquid, we can write

$$-\frac{1}{\delta V}\int_{\mathcal{I}_{\mathrm{RVE}}} \breve{\boldsymbol{\sigma}}^{\ell}\cdot\breve{\boldsymbol{n}}^{\ell}\,\mathrm{d}S_{\mathcal{I}} = \frac{1}{\delta V}\left[\int_{\mathcal{I}_{\mathrm{RVE}}} \breve{P}^{\ell}\breve{\boldsymbol{n}}^{\ell}\,\mathrm{d}S_{\mathcal{I}} - \int_{\mathcal{I}_{\mathrm{RVE}}} \breve{\boldsymbol{\tau}}^{\ell}\cdot\breve{\boldsymbol{n}}^{\ell}\,\mathrm{d}S_{\mathcal{I}}\right]. \tag{4.26}$$

Both terms on the right-hand side are written in terms of microscopic variables and need to be re-expressed in terms of macroscopic variables. To do this we make assumptions based on physical reasoning. We take each term in turn.

First we consider the liquid pressure acting normal to the interface. Inspired by the characteristics of partially molten mantle, we assume that the liquid viscosity is relatively small and that neighboring pores are well connected. From this we infer that neighboring pores will be in mechanical equilibrium, with approximately equal microscopic pressure. Hence microscopic pressure must vary on scales much larger than the grain size. We therefore replace the microscopic liquid pressure with the macroscopic, which is independent of $\breve{\boldsymbol{x}}$ and therefore can be moved outside the integral of equation (4.26). Then, using a result from mixture theory (and see exercise 4.4), we write $\int_{\mathcal{I}_{\mathrm{RVE}}} \boldsymbol{\nabla}\Phi^{\ell}\,\mathrm{d}S_{\mathcal{I}} = \delta V\,\boldsymbol{\nabla}\phi$ such that we have

$$\frac{1}{\delta V}\int_{\mathcal{I}_{\mathrm{RVE}}} \breve{P}^{\ell}\breve{\boldsymbol{n}}^{\ell}\,\mathrm{d}S_{\mathcal{I}} \approx -P^{\ell}\boldsymbol{\nabla}\phi. \tag{4.27}$$

Physical intuition can be gained by imagining the simple scenario of a planar interface between a pure liquid at $x<0$ and a pure solid at $x>0$. Then $\boldsymbol{\nabla}\phi$ evaluates as a delta function (units of inverse length) times a unit vector $-\hat{\boldsymbol{x}}$. The force per unit volume associated with the liquid pressure is then $P^{\ell}\delta(x)\hat{\boldsymbol{x}}$, with the delta function requiring that this force apply exactly on the interface. In a more general situation, the sharp transition from liquid to solid is replaced by a porosity gradient.

Second we consider the liquid traction on the interface. Here we make a key assumption that the interface has a convoluted, *tortuous*[3] shape that has no preferred orientation. In this case, the stress that the liquid exerts on the interface will average to zero over the RVE—except in the direction of the mean segregation flow. This is analogous to the viscous drag on a sphere, which is directed opposite the relative motion of the sphere[4]. Hence for a Newtonian fluid,

$$-\frac{1}{\delta V}\int_{\mathcal{I}_{\mathrm{RVE}}} \breve{\boldsymbol{\tau}}^{\ell}\cdot\breve{\boldsymbol{n}}^{\ell}\,\mathrm{d}S_{\mathcal{I}} \approx -\mu^{\ell}\frac{\Delta\boldsymbol{v}}{L_{\mathrm{pore}}}\frac{\delta A_{\mathcal{I}}}{\delta V}, \tag{4.28}$$

where μ^{ℓ} is the viscosity of the liquid phase, L_{pore} is a characteristic pore diameter (hence $\mu^{\ell}\Delta\boldsymbol{v}/L_{\mathrm{pore}}$ is a characteristic shear stress) and $\delta A_{\mathcal{I}}$ is the interfacial area within the RVE of volume δV.

To develop this argument, we assume that pores are tubes of uniform cross section along grain edges, which is approximately valid for porosity below a few percent and small dihedral angle (see chapter 5). If we define d as the characteristic grain size, the cross-sectional area of the pores scales as $d^2\phi$. This suggests that a characteristic pore diameter scales as $L_{\mathrm{pore}} \propto d\phi^{1/2}$. Moreover, for this pore geometry, the area of

[3]Tortuosity measures the twistyness of a curve. It is sometimes defined as the arc length divided by the distance spanned between two points on the curve.

[4]Indeed when considering the segregation of gas bubbles or isolated solid crystals from a viscous fluid, it is appropriate to use a Stokeslet model for the interface stress. This fits into the same continuum framework but gives a different result for **F**.

solid–liquid contact per grain can be shown to scale as $A_{\mathcal{I}} \propto d^2\phi^{1/2}$. The volume of the solid–liquid system, using the assumption of small porosity, scales as d^3 per grain. The ratio of these quantities gives us the volumetric interfacial area as

$$\frac{\delta A_{\mathcal{I}}}{\delta V} \propto d^{-1}\phi^{1/2}, \tag{4.29}$$

and hence we can write

$$\frac{1}{L_{\text{pore}}}\frac{\delta A_{\mathcal{I}}}{\delta V} = Cd^{-2}, \tag{4.30}$$

where C is a proportionality constant associated with details of the pore geometry. Using this equation in the right-hand side of equation (4.28) gives $\mu^\ell \Delta \boldsymbol{v} C/d^2$.

The process that generates the interphase force is flow of magma through the permeable pore network of the solid matrix. Hence we expect that the matrix permeability k_ϕ should appear as part of the formulation of that force. Details of the permeability are discussed in the next chapter (section 5.2); here we simply introduce the permeability law that is derived from the model of pores as tubes of uniform cross section. This permeability can be written as

$$k_\phi = \frac{d^2\phi^2}{C}. \tag{4.31}$$

Using this formulation allows us to re-write C/d^2 as ϕ^2/k_ϕ. Combining this with equations (4.28) and (4.27), the interphase force per unit volume becomes

$$\mathbf{F} = -\left(\frac{\mu^\ell\phi^2}{k_\phi}\Delta\boldsymbol{v} + P^\ell\nabla\phi\right). \tag{4.32}$$

We can identify $D \equiv \mu^\ell\phi^2/k_\phi$ as the canonical Darcy drag coefficient. This formulation of the interphase force has the important property of *Galilean invariance*: $\mathbf{F}$ is the same in any inertial reference frame. While certain details of the present formulation of $\mathbf{F}$ may be debated, any formulation must have this basic property.

4.3.3 CONSTITUTIVE EQUATIONS IN THE MAGMA–MANTLE LIMIT

It seems appropriate that the viscous relationship between macroscopic stress and strain rate be derived from a volume averaging of the same relationship at the microscopic scale. Indeed it is straightforward to write down this averaging as

$$\phi^i\boldsymbol{\tau}^i = \mu^i\frac{1}{\delta V}\int_{\text{RVE}} \Phi^i\left[\nabla\breve{\boldsymbol{v}}^i + (\nabla\breve{\boldsymbol{v}}^i)^{\mathsf{T}}\right]\mathrm{d}^3\breve{\boldsymbol{x}}, \tag{4.33}$$

where μ^i for $i = s, \ell$ is the pure-phase viscosity for deformation at the microscopic scale, where both the liquid and solid are incompressible fluids. There is no unique or proven approach to evaluating this integral; physical and mathematical arguments only lead to loose constraints on the form of the result, requiring arbitrary assumptions to close the model (see Literature Notes for further discussion and references). Hence we follow previous authors (and related subjects) in making basic assumptions about the macroscopic constitutive equations directly. The reader should recognize here the hypothetical nature of the theory. Indeed, the adequacy of a purely viscous rheology as a description of the asthenospheric solid matrix is an assumption that should be critically evaluated through research.

In the asthenosphere, the porosity is generally small. The pores have the form of slender tubes connecting grain triple junctions or thin sheets between grain faces. We cannot expect to resolve, within our continuum model framework, the deviatoric stress $\breve{\boldsymbol{\tau}}^\ell$ that fluctuates on the microscopic length scale within pores (these spaces are much smaller than the RVE). Furthermore, we expect that the Darcy drag force from section 4.3.2 will be in approximate balance with the liquid pressure gradient and body force. Hence it is appropriate to assume that the volume average of the microscopic deviatoric stress is negligible at the continuum scale, i.e.,

$$\boldsymbol{\tau}^\ell \approx \mathbf{0}. \tag{4.34}$$

As we shall see below, this canonical assumption leads to a conservation of momentum equation for the liquid that takes the form of Darcy's law. We now have the simple relation

$$\boldsymbol{\sigma}^\ell = -P^\ell \boldsymbol{I} \tag{4.35}$$

and, since the liquid is independently incompressible, we do not require further constitutive information about it. We do, however, require constitutive equations relating stresses in the bulk, two-phase medium to the strain rate of that medium.

It is intuitive that deformation of the bulk, two-phase aggregate should occur in response to the bulk, two-phase stress of equation (4.23). A simple thought experiment helps us to refine this concept: a handful of sand grains responds identically to deviatoric stress whether it is at the bottom of a bathtub full of water or at the bottom of the ocean. This is because the liquid pressure, irrespective of its magnitude, pushes on the grains isotropically and causes no deformation.[5] Therefore, to formulate a rheology for a liquid/solid aggregate, we should consider only the stress that is effective in causing deformation. This effective stress is the bulk stress after removal of the liquid pressure. We can therefore define the *effective stress tensor* as

$$\overline{\boldsymbol{\sigma}}^{\text{eff}} \equiv \overline{\boldsymbol{\sigma}} + P^\ell \boldsymbol{I} \tag{4.36}$$

noting that, because of the opposite sign conventions for pressure and stress, removal of the pressure from the stress tensor is achieved by addition. The effective stress tensor is frequently used in soil mechanics and poroelasticity, where it is (equivalently) defined as $\overline{\boldsymbol{\sigma}}^{\text{eff}} \equiv (1-\phi)\left(\boldsymbol{\sigma}^s + P^\ell \boldsymbol{I}\right)$. Using equations (4.22), the effective stress can be written in terms of the deviatoric stress of the solid as

$$\overline{\boldsymbol{\sigma}}^{\text{eff}} = (1-\phi)\left[-\Delta P \boldsymbol{I} + \boldsymbol{\tau}^s\right], \tag{4.37}$$

where $\Delta P \equiv P^s - P^\ell$. This tensor has two parts, the isotropic part associated with the pressure difference between phases and the deviatoric part associated with the deviatoric stress tensor of the solid. The factor of $(1-\phi)$ signifies that these are stresses per unit volume of the solid phase. We have assumed that the solid is the phase that resists deformation, which is true when two essential conditions are met:

[5]This is known in the soil mechanics literature as *Terzaghi's principle*.

- first, that the solid viscosity is much greater than that of the liquid;
- second, that the solid forms a contiguous skeleton that transmits stress at the continuum scale.

The partial differential equations derived below are formally invalid if either of these conditions are not met. For further discussion of this issue see section 4.4 and the Literature Notes at the end of this chapter.

An *effective pressure* can be defined in terms of the isotropic part of the effective stress tensor in the usual way,

$$\mathcal{P}^{\text{eff}} \equiv -\tfrac{1}{3}\text{trace}\left(\overline{\boldsymbol{\sigma}}^{\text{eff}}\right) = (1-\phi)\Delta P. \tag{4.38}$$

This can be used to define the deviatoric part of the effective stress.

As we concluded from our thought experiment above, the effective stress is the quantity that is associated with deformation of the aggregate. Hence a viscous constitutive law should be of the form $\overline{\boldsymbol{\sigma}}^{\text{eff}} = f(\dot{\boldsymbol{e}}^s)$, where f is a (possibly nonlinear) function of the full strain-rate tensor of the solid. In particular, we expect the isotropic part of $\overline{\boldsymbol{\sigma}}^{\text{eff}}$ (the effective pressure) to be a function of the isotropic part of $\dot{\boldsymbol{e}}^s$; likewise for the deviatoric parts. Assuming a simple proportionality, we can write

$$\text{trace}\left(\overline{\boldsymbol{\sigma}}^{\text{eff}}\right) \propto \text{trace}\left(\dot{\boldsymbol{e}}^s\right) \tag{4.39a}$$

$$\overline{\boldsymbol{\sigma}}^{\text{eff}} - \tfrac{1}{3}\text{trace}\left(\overline{\boldsymbol{\sigma}}^{\text{eff}}\right)\boldsymbol{I} \propto \dot{\boldsymbol{e}}^s - \tfrac{1}{3}\text{trace}\left(\dot{\boldsymbol{e}}^s\right)\boldsymbol{I}, \tag{4.39b}$$

with $\dot{\boldsymbol{e}}^s \equiv \frac{1}{2}\left[\boldsymbol{\nabla}\boldsymbol{v}^s + (\boldsymbol{\nabla}\boldsymbol{v}^s)^{\mathsf{T}}\right]$.

Taking the isotropic part first, we posit a constant of proportionality of $3\zeta_\phi$ such that the constitutive law (4.39a) becomes

$$(1-\phi)\Delta P = -\zeta_\phi \mathcal{C}, \tag{4.40}$$

where ζ_ϕ is the *compaction viscosity*. This viscosity has commonly been referred to in the literature as the *bulk viscosity* [e.g., McKenzie, 1984]. We avoid that terminology here for two reasons. First, we will use the term *bulk* to refer to properties of the two-phase mixture (e.g., bulk density $\overline{\rho} = \phi\rho^\ell + (1-\phi)\rho^s$). And second, in elasticity theory, the bulk modulus quantifies the elastic compressibility of a material. In this book we consider both phases to be individually incompressible. However, the subscript on ζ_ϕ reminds us that it is a property of the two-phase aggregate that is expected to vary with porosity.

In the isotropic constitutive relation (4.40), $\mathcal{C}$ is the compaction rate that was defined in equation (4.16). Equation (4.40) states that when the solid phase is overpressured with respect to the liquid, $\Delta P > 0$ acts as an isotropic stress that causes the solid to compact and expel the liquid. Decompaction results when the liquid is overpressured with respect to the solid ($\Delta P < 0$). This relationship has led previous authors to define a variable called the *compaction pressure* $\mathcal{P} \equiv \zeta_\phi \mathcal{C} = -(1-\phi)\Delta P$, which we shall use throughout the book. The attentive reader will notice that $\mathcal{P} = -\mathcal{P}^{\text{eff}}$.

Now we consider the deviatoric part of the effective stress. The left-hand side of (4.39b) is expanded using equation (4.37); the right-hand side is the deviatoric strain rate tensor

$$\dot{\boldsymbol{\varepsilon}}^s \equiv \tfrac{1}{2}\left[\nabla \boldsymbol{v}^s + \left(\nabla \boldsymbol{v}^s\right)^\mathsf{T} - \tfrac{2}{3}\mathcal{C}\boldsymbol{I}\right]. \tag{4.41}$$

Combining these with a constant of proportionality of $2\eta_\phi$ gives the viscous constitutive model

$$(1-\phi)\boldsymbol{\tau}^s = \eta_\phi\left[\nabla \boldsymbol{v}^s + \left(\nabla \boldsymbol{v}^s\right)^\mathsf{T} - \tfrac{2}{3}\mathcal{C}\boldsymbol{I}\right]. \tag{4.42}$$

The shear viscosity of the two-phase aggregate is η_ϕ, where the subscript reminds us that the formulation of this viscosity must account for the presence of a low-viscosity melt on the grain boundaries. It can be shown (4.5) that the deviatoric stress and strain rate have zero trace and therefore they do not describe compaction.

In (4.39) we have assumed linear, scalar proportionality between effective stress and strain rate. The resulting constitutive law can be generalized to non-Newtonian viscosity by simply taking ζ_ϕ and η_ϕ to be functions of the stress or strain rate. Furthermore, to model an anisotropic viscosity, the scalar proportionality constant $2\eta_\phi$ could be replaced with a fourth-rank viscosity tensor. Discussion of anisotropic viscosity is beyond the current scope, but see the Literature Notes at the end of this chapter for further information.

The two constitutive equations for isotropic (4.40) and deviatoric (4.42) stress fully express our physical assumptions about the material response to stress. We can use them to rewrite the effective stress from equation (4.37) in terms of strain rates:

$$\overline{\boldsymbol{\sigma}}^{\text{eff}} = \zeta_\phi \mathcal{C}\boldsymbol{I} + 2\eta_\phi \dot{\boldsymbol{\varepsilon}}^s. \tag{4.43}$$

Furthermore, we can consistently express the solid stress tensor as

$$\boldsymbol{\sigma}^s = -P^\ell \boldsymbol{I} + \frac{\zeta_\phi}{1-\phi}\mathcal{C}\boldsymbol{I} + \frac{\eta_\phi}{1-\phi}\left[\nabla \boldsymbol{v}^s + \left(\nabla \boldsymbol{v}^s\right)^\mathsf{T} - \tfrac{2}{3}\mathcal{C}\boldsymbol{I}\right]. \tag{4.44}$$

Note that in this form, the solid stress tensor is written without reference to a solid pressure. To make it explicit, recall that $P^s = -\text{trace}\,(\boldsymbol{\sigma}^s)/3$. Applying this to (4.44) gives

$$P^s = P^\ell - \frac{\zeta_\phi}{1-\phi}\mathcal{C}, \tag{4.45}$$

which is consistent with the isotropic constitutive equation (4.40), as it should be.

4.4 A Note about Disaggregation

In the foregoing derivations, it has been implicitly assumed that the solid phase forms a contiguous skeleton of grains. The theory is based on the presence of a solid phase that can sustain deviatoric stresses much greater than those of the liquid—hence we are able to take $\boldsymbol{\tau}^\ell = 0$ and model the liquid using Darcy's law. This is a good approximation for liquid fractions below what is called the *disaggregation threshold* ϕ_Ξ. For $\phi > \phi_\Xi$, the solid skeleton breaks down and the two-phase medium becomes a slurry of noncontiguous solid grains suspended in liquid. The assumptions made in deriving the equations, and hence the equations themselves, become formally invalid.[6] It is thought that for the minerals that comprise the asthenospheric mantle, $0.21 \le \phi_\Xi \le 0.3$.

4.5 The Full Mechanical System, Assembled

Combining equations (4.19a), (4.32), and (4.35), gives a modified form of Darcy's law representing conservation of momentum for the liquid (see exercise 4.6). Writing $\overline{\boldsymbol{\sigma}} = \overline{\boldsymbol{\sigma}}^{\text{eff}} - P^\ell \boldsymbol{I}$ and combining this with (4.43) and (4.20) gives an equation representing conservation of momentum for the bulk, two-phase aggregate. These two equations plus conservation of mass with the Boussinesq approximation give us the full system of partial differential equations that govern the fluid mechanics:

$$\phi\left(\boldsymbol{v}^\ell - \boldsymbol{v}^s\right) + \frac{k_\phi}{\mu^\ell}\left(\nabla P^\ell - \rho^\ell \mathbf{g}\right) = \mathbf{0}, \tag{4.46a}$$

$$-\nabla P^\ell + \nabla\cdot 2\eta_\phi \dot{\boldsymbol{\varepsilon}}^s + \nabla \zeta_\phi \mathcal{C} + \overline{\rho}\mathbf{g} = \mathbf{0}, \tag{4.46b}$$

$$\frac{\mathrm{D}_s\phi}{\mathrm{D}t} - \Gamma/\rho^s - (1-\phi)\mathcal{C} = 0, \tag{4.46c}$$

$$\nabla\cdot\left[\phi\boldsymbol{v}^\ell + (1-\phi)\boldsymbol{v}^s\right] = 0. \tag{4.46d}$$

This system of four equations contains four unknowns: $\boldsymbol{v}^\ell$, $\boldsymbol{v}^s$, P^ℓ, ϕ. To solve it we require closure conditions for the shear η_ϕ and compaction viscosity ζ_ϕ of the aggregate, the liquid viscosity μ^ℓ, the permeability k_ϕ, and the melting rate Γ, all of which will be discussed in forthcoming pages.

A substantial simplification is possible, however, by eliminating the liquid velocity. Taking the divergence of equation (4.46a) then using equation (4.46d) to eliminate the divergence of the segregation flux (see exercise 4.7) allows us to rewrite equations (4.46) as

$$-\mathcal{C} + \nabla\cdot\frac{k_\phi}{\mu^\ell}\left(\nabla P^\ell - \rho^\ell\mathbf{g}\right) = 0, \tag{4.47a}$$

$$-\nabla P^\ell + \nabla\cdot 2\eta_\phi \dot{\boldsymbol{\varepsilon}}^s + \nabla \zeta_\phi \mathcal{C} + \overline{\rho}\mathbf{g} = \mathbf{0}, \tag{4.47b}$$

$$\frac{\mathrm{D}_s\phi}{\mathrm{D}t} - \Gamma/\rho^s - (1-\phi)\mathcal{C} = 0. \tag{4.47c}$$

We now have a much simpler system of three equations in three unknowns. Equation (4.47a) states that the compaction rate is driven by the divergence of the Darcy flux. Equation (4.47b) is a two-phase Stokes equation and states that gradients of liquid pressure, shear, and compaction stresses balance the gravitational body force. Equation (4.47c) is a porosity-evolution equation that describes how melting and compaction modify the porosity of a Lagrangian parcel of the solid.

There are, of course, other ways to formulate this system of equations, making fewer (or more, or different) assumptions, including physics that was neglected here,

[6] A system of equations that is valid across the disaggregation threshold would, for starters, need to retain the deviatoric liquid stress, which becomes important for a suspension of solid grains. The interphase force would require modification as well. See the Literature Notes at the end of this chapter.

or by simply recombining the equations by algebraic manipulation. For example, some authors have written the equations in terms of the mean velocity $\overline{\boldsymbol{v}}$ and the segregation flux $\boldsymbol{q}$. While this rearrangement may be physically appealing, it does not enable any decoupling or simplification of the system of PDEs.

4.6 Special, Limiting Cases

It is instructive to consider the governing equations in a few special cases.

4.6.1 NO POROSITY, NO MELTING

When the porosity and the melting rate are everywhere zero, we expect the system of equations to reduce to the incompressible, single-phase Stokes equations. Taking $\phi = \Gamma = \mathcal{C} = 0$ we see that equations (4.46a) and (4.46c) are satisfied trivially, while (4.46b) and (4.46d) become

$$-\nabla P^\ell + \nabla \cdot \eta_\phi \left[\nabla \boldsymbol{v}^s + (\nabla \boldsymbol{v}^s)^\mathsf{T}\right] + \rho^s \mathbf{g} = \mathbf{0}, \tag{4.48a}$$

$$\nabla \cdot \boldsymbol{v}^s = 0. \tag{4.48b}$$

Furthermore, equation (4.40) tells us that $P^\ell = P^s$ and hence equations (4.48) are reduced to the one-phase system from chapter 3.

4.6.2 PARTIALLY MOLTEN, RIGID MEDIUM

When the porosity is nonzero and constant (though not necessarily uniform), the equations reduce to another simple limit. Here we set $\boldsymbol{v}^s = \mathbf{0}$ and $\mathcal{C} = \Gamma = 0$. Equation (4.46c) is trivially satisfied. Equations (4.46a) and (4.46d) then become

$$\phi \boldsymbol{v}^\ell = -\frac{k_\phi}{\mu^\ell}\left(\nabla P^\ell - \rho^\ell \mathbf{g}\right), \tag{4.49a}$$

$$\nabla \cdot (\phi \boldsymbol{v}^\ell) = 0. \tag{4.49b}$$

These equations represent Darcian porous flow in a rigid, saturated permeable medium. In the case of no liquid flow with a nonzero permeability, $\nabla P^\ell = \rho^\ell \mathbf{g}$.

4.6.3 CONSTANT, UNIFORM SOLID VISCOSITY

If the viscosities η_ϕ and ζ_ϕ are taken to be constant and uniform (η_0 and ζ_0), equation (4.47b) can be written (see exercise 4.9)

$$\nabla P^\ell = -\eta_0 \nabla \times \nabla \times \boldsymbol{v}^s + \left(\zeta_0 + \tfrac{4}{3}\eta_0\right) \nabla \mathcal{C} + \overline{\rho}\mathbf{g}. \tag{4.50}$$

In this equation we see that the stresses associated with deformation are cleanly decomposed into two parts: stresses arising from the solenoidal (divergence-free) solid shear and stresses arising from irrotational solid compaction. This motivates a Helmholtz decomposition of the velocity

$$\boldsymbol{v}^s = \nabla \times \boldsymbol{\Psi} + \nabla \mathcal{U}, \tag{4.51}$$

where $\boldsymbol{\Psi}$ is a vector potential and $\mathcal{U}$ is a scalar potential. The first term represents the solenoidal part of the flow while the second term is the irrotational part. Using this definition, equations for the potentials and the compaction rate are

$$\nabla^2\mathcal{U} = \mathcal{C}, \tag{4.52a}$$

$$-\nabla\cdot\frac{k_\phi}{\mu^\ell}\left(\zeta_0 + \tfrac{4}{3}\eta_0\right)\nabla\mathcal{C} + \mathcal{C} = \nabla\cdot\frac{k_\phi}{\mu^\ell}\left[\eta_0\nabla\times\nabla^2\boldsymbol{\Psi} - (1-\phi)\Delta\rho\mathbf{g}\right], \tag{4.52b}$$

$$\eta_0\nabla^4\boldsymbol{\Psi} = -\nabla\times\left(\overline{\rho}\mathbf{g}\right). \tag{4.52c}$$

where $\nabla^4 \equiv \nabla^2\nabla^2$ is the biharmonic operator. The second of these three equations is the compaction equation, a scalar Helmholtz equation for the compaction rate that is of the form $-\nabla^2\mathcal{C} + \mathcal{C} =$ (driving terms). It is obtained by combining the decomposition in (4.51) with equation (4.50) and using this to eliminate ∇P^ℓ from equation (4.47a). The third equation is obtained by taking the curl of equation (4.50). The system (4.52) provides a basis of linear PDEs that can be solved for variables $\mathcal{C}$, $\boldsymbol{\Psi}$, and $\mathcal{U}$ given the porosity field ϕ. Using this solution, the porosity can be updated with equation (4.47c).

We can gain insight into the effect of the body force on the solid flow by further consideration of equation (4.52c). Assuming a two-dimensional problem in which $\mathbf{g} = -g\hat{\mathbf{z}}$, $\partial\cdot/\partial y = 0$, and $\boldsymbol{\Psi} = \psi\,\hat{\mathbf{y}}$, we can write the vorticity as $\omega\hat{\mathbf{y}} \equiv \nabla\times\boldsymbol{v}^s = \nabla\times\nabla\times(\psi\,\hat{\mathbf{y}})$. Then for $\phi = \phi(x, z)$, equation (4.52c) becomes

$$\nabla^2\omega = \frac{\Delta\rho g}{\eta_0}\frac{\partial\phi}{\partial x}. \tag{4.53}$$

This equation states that the lateral gradients in bulk density create circulation. For a constant and uniform $\Delta\rho$, this is equivalent to a porosity-driven convection. The physics becomes clearer by noting the similarity between equations (4.52c) and (3.18) for thermal convection.

The full system in potential form (4.52) is rarely used because variations in shear viscosity are considered to be important for most problems of interest. However, in some applications, the solenoidal flow can be discarded by setting $\boldsymbol{\Psi} = \mathbf{0}$, leaving only the scalar equations (4.52a) and (4.52b). These are easier to solve than the coupled system for native variables (4.47).

4.7 Literature Notes

The foundation of the theory for magma/mantle dynamics is methods of volume averaging for two-phase media. These were developed by Drew and Segel [1971] and Drew [1971] among others. Drew [1983] reviews this field. The approach taken here was influenced by Le Bars and Worster [2006].

The derivation of the equations representing conservation of mass and momentum given here draws on the approaches of Bercovici et al. [2001a], Rudge et al. [2011] and Fowler [2011]. The results are consistent with M^c^Kenzie [1984], Fowler [1985], and Scott and Stevenson [1986]. Reviews of early work are given by Ribe [1987] and Stevenson and Scott [1991]. Volume averaging is replaced by ensemble averaging by Oliveira et al. [2018]. In Simpson et al. [2010a], formal homogenization of the microscale problem is used to derive the macroscopic equations.

An outstanding challenge in the derivation of the equations governing the mechanics is how to volume-average the microscopic stresses (equation (4.33)). The issues are well described in section 4.2 of Bercovici et al. [2001a], where the authors ultimately resort to assumptions justified by the resulting form of the continuum equations. Their equations differ by a factor of $1 - \phi$ in the deviatoric stress of the solid phase from those obtained here.

The permeability model used in the derivation of the interphase force was proposed by Frank [1968] and reviewed by Rudge [2018a].

Some authors have incorporated additional physics into the system of equations. The series Bercovici et al. [2001a], Ricard et al. [2001], and Bercovici and Ricard [2003] consider the dynamics of a more general, two-fluid system where the viscosities of the phases need not be distinct. They consider the physics of surface tension and introduce a formulation for damage. Their formulation is founded on a strict symmetry between the two phases. This symmetry is obviously broken in the case of the partially molten asthenosphere, where one phase is a liquid and the other a polymineralic solid with grain–grain phase boundaries.

Other authors have sought to extend the rheological formulation. Connolly and Podladchikov [1998] considered viscoelastic compaction, and Connolly and Podladchikov [2007] extended this to viscoelastic–plastic compaction, obtaining a compaction viscosity that depends on the sign of the compaction rate. These studies, however, excluded shear deformation. A more general treatment was developed by Keller et al. [2013], which broadened the rheological model of the solid to include a viscoelastic–plastic rheology for compaction and shear. This has subsequently been considered by Yarushina et al. [2020].

Motivated by problems of volcanology, other authors have extended the theory to model the segregation of compressible gas bubbles. Bercovici and Michaut [2010] develop theory and Michaut et al. [2013] applies it to understanding cylical patterns in volcanic eruptions. Three-phase theory for segregation of iron from a partially molten mantle was developed by Ricard et al. [2009]. A recent attempt to develop a fully general theory for an arbitrary system of N phases was made by Keller and Suckale [2019].

Most authors have made the Boussinesq approximation before obtaining solutions. The role of the density change during melting was considered by Šrámek et al. [2007]. They show that at the initiation of melting, when porosity is very low, a volume increase can lead to magma overpressure and, if the region is connected to one with higher porosity, to a "squirting boundary layer" from which melt is expelled. This layer is inconsequential and can be neglected. Vestrum and Butler [2020] considered the role of volume expansion in a stability analysis and found that with variable compaction viscosity, a relatively weak instability growth is predicted.

Our discussion of the potential formulation of the governing equations under constant viscosity follows Spiegelman [1993a], who introduced the Helmholtz decomposition for velocity. Another velocity decomposition was proposed by Scott [1988] that used the center-of-mass and segregation velocities. However, the native velocities are simply linear combinations of these and, on this basis, the governing equations cannot be decomposed.

Some workers have obtained numerical solutions of the governing equations in which isolated patches of the domain reach porosities above the disaggregation limit [Katz, 2008; Keller et al., 2013]; Their argument has been that although the computed balance of forces in these regions is unphysical, mass is conserved and the solution is consistent with the rest of the domain, where the equations remain valid. Hence the isolated, disaggregated patches don't pollute the solution where porosity is small. Keller and Suckale [2019] developed a framework for consistent modeling across the full range of porosity, but computational tractability requires simplifications or parameter restrictions. Mixing length theory has been proposed by Bower et al. [2018] for large-scale transport in a dissaggregated two-phase suspension of solid crystals in liquid magma.

4.8 Exercises

4.1 Write a definition of the solid velocity that is similar to equation (4.7).

4.2 A small parcel of mantle with density of 3300 kg/m^3 is melting at a rate of 1 kg/m^3/yr. If it has a porosity of 0.02 in the steady state, what is the associated compaction rate? Note the units of compaction rate.

4.3 Show that by extending the Boussinesq approximation to neglect interphase density differences outside of body-force terms, we can write

$$\mathcal{C} = -\nabla \cdot \left[\phi\left(\boldsymbol{v}^\ell - \boldsymbol{v}^s\right)\right] = -\nabla \cdot \boldsymbol{q}.$$

The term in square brackets is called the *segregation flux*, $\boldsymbol{q}$.

4.4 In section 4.1 we introduced the concept of the representative volume element (RVE).

(a) Given a position $\boldsymbol{x}$, are volume-averaged quantities such as the phase fraction $\phi^i(\boldsymbol{x})$ dependent on the choice of RVE? Justify your answer briefly.

(b) By writing the phase fraction evaluated at $\boldsymbol{x}$ as

$$\phi^i(\boldsymbol{x}) = \frac{1}{\delta V}\int_{\text{RVE}} \Phi^i(\boldsymbol{x}+\check{\boldsymbol{x}})\,\mathrm{d}^3\check{\boldsymbol{x}},$$

use your answer to part (a) to show that

$$\nabla\phi(\boldsymbol{x}) = \frac{1}{\delta V}\int_{\mathcal{I}_{\text{RVE}}} \nabla\Phi^\ell(\boldsymbol{x}+\check{\boldsymbol{x}})\,\mathrm{d}S_{\mathcal{I}}.$$

Hint: by showing that the integral limits may be considered as $\boldsymbol{x}$-independent, you may take the gradient operator inside the integral.

4.5 From equation (4.42), identify the deviatoric strain-rate tensor and show that it has zero trace.

4.6 Verify that equations (4.19a), (4.32), and (4.35) can be combined to give equation (4.46a).

4.7 Show how (4.47a) can be derived from (4.46d) and (4.46a).

4.8 Obtain a version of equations (4.47) without extending the Boussinesq approximation, i.e., by retaining terms in $\Delta(1/\rho)$.

4.9 Show that (4.47b) can be written as (4.50) when η_ϕ and ζ_ϕ are constant. You may wish to use the vector identity $\nabla^2\boldsymbol{v} = \nabla(\nabla\cdot\boldsymbol{v}) - \nabla\times\nabla\times\boldsymbol{v}$.

4.10 Pure-shear flow on the plane has the form $\boldsymbol{v}^s = \dot{\gamma}(x\hat{\boldsymbol{x}} - y\hat{\boldsymbol{y}})$, where $\dot{\gamma}$ is constant.

(a) Determine the two-dimensional potentials $\boldsymbol{\Psi}$ and $\mathcal{U}$ associated with $\boldsymbol{v}^s$ using the Helmholtz decomposition given by (4.51).
(b) For some cases we can decompose the velocity as $\boldsymbol{v}^s = \frac{\partial H}{\partial y}\hat{\boldsymbol{x}} - \frac{\partial H}{\partial x}\hat{\boldsymbol{y}}$. Determine $H(x, y)$ (up to an arbitrary constant) for the case of pure shear.
(c) By considering $\boldsymbol{v}^s \cdot \nabla H$, comment on and sketch the pure shear flow trajectories.

CHAPTER 5

Material Properties

The system of governing equations (4.47) can be solved for the primary variables if well-behaved models are specified for the material properties: permeability k_ϕ, viscosities η_ϕ, ζ_ϕ, and μ^ℓ, and melting rate Γ. These are typically called *closures*. Much of the interesting research into two-phase dynamics of the mantle involves understanding the consequences of different closures that represent hypotheses about the material properties. In this book, our goal is not to review all the extant research on closures for k_ϕ, η_ϕ, ζ_ϕ, and μ^ℓ; it is instead to provide a basis on which to understand the relevant literature. To this end, there are some fundamental theories and accepted ideas about how these material properties can be quantified. Before presenting them, we consider the question of the geometry and evolution of an aggregate of mantle grains plus magma-filled pores at the microscopic scale.

5.1 Microstructure

Microstructure is the grain-scale geometrical arrangement of solid grains and liquid melt. It is defined by the location of the interface between two solid grains (*ss*) or between a liquid magma and a solid grain (ℓs). These interfaces have an energy per unit area γ_{ss} and $\gamma_{\ell s}$, which arise from the bonds between atoms in the mineral lattice that are disrupted at the interface. This stored energy has the potential to do work; it is therefore considered as potential energy. Its derivative with respect to surface length is a force. Therefore, surface energy per unit area is equivalent to surface force per unit length, a quantity we refer to as *surface tension*. This tension is embedded in and parallel to the interface. We neglected it from the force balance that was considered in chapter 4.3 because it is only a significant driver of melt segregation at mantle scales of order 1 cm or smaller.[1]

Surface energy (or tension) is an important driver of change in microstructure, which evolves to minimize potential energy. In the context of the grain–melt system

[1]This assertion can be quantified by considering a balance between gravity-driven melt segregation and surface tension. We assume spherical, millimetric grains at small porosity and consider spatial variations in porosity of $\sim$0.05 that cause variations in surface energy per unit volume. A balance can exist if the length scale of porosity variations is $\Delta z \sim 2\Delta\phi\gamma/(\Delta\rho g\langle d\rangle) \approx 3.3$ cm, where $\gamma = 1$ J-m^2 is the surface energy per unit area and $\langle d\rangle$ is the mean grain size. Hence we expect surface tension to play a role in the dynamics of melt segregation only at very small length scales. For more information, see references cited in the Literature Notes at the end of this chapter.

considered here, this evolution occurs in two ways: grain growth and textural equilibration. Although both are driven by the same potential, they are treated independently. We discuss each in turn, breifly. For more information, we refer the reader to the Literature Notes and onward from there.

5.1.1 GRAIN-SIZE CHANGE

Changes in grain size (at constant melt fraction) occur by transfer of mass between grains and by subdivision of grains to create new grain boundaries. In the former case, most research has gone into *normal grain growth*, which is growth of the mean grain size driven by reduction in potential energy. Subdivision of grains is the outcome of *dynamic recrystallization*. This occurs when deformation at high stress introduces a large and heterogeneous density of dislocations into grains; those dislocations self-organize to form new grain boundaries. Most theories for grain-size change have their origins in materials science, where the materials of interest are chemically pure and relatively simple in their crystallographic structure. We have only a primitive understanding of grain-size change in geological systems. While the theory discussed below will apply to some extent, it is inadequate for most natural cases of interest.

We first consider normal grain growth and, to illustrate, we consider a toy system of two grains and no melt. For convenience, we assume that the grains are spherical with surface energy per unit area γ. They have radii d_a and d_b and hence their total surface energy E and total volume V are

$$E = 4\pi\gamma\left(d_a^2 + d_b^2\right), \tag{5.1}$$

$$V = \frac{4}{3}\pi\left(d_a^3 + d_b^3\right). \tag{5.2}$$

We assume that density is constant and hence exchange of mass is equivalent to exchange of volume. We are interested in changes $\mathrm{d}d_a$ and $\mathrm{d}d_b$ such that the evolution is spontaneous ($\mathrm{d}E < 0$) and volume-conserving ($\mathrm{d}V = 0$). Applying the latter constraint to equation (5.2) gives

$$\frac{\mathrm{d}d_a}{\mathrm{d}d_b} = -\frac{d_b^2}{d_a^2}. \tag{5.3}$$

Taking the total derivative of the energy from equation (5.1), dividing through by $\mathrm{d}d_b$ and using (5.3) we obtain

$$\frac{\mathrm{d}E}{\mathrm{d}d_b} = 8\pi\gamma d_b\left(1 - d_b/d_a\right),$$

$$\Rightarrow \frac{\mathrm{d}E}{\mathrm{d}V_b} = -2\gamma\left(\frac{1}{d_a} - \frac{1}{d_b}\right). \tag{5.4}$$

where $V_b = 4\pi d_b^3/3$ is the volume of grain b. This equation shows that when $d_b > d_a$, the total energy decreases when grain b grows ($\mathrm{d}V_b > 0$). This toy model allows us to infer that larger grains grow spontaneously by taking mass away from the smaller grains. It also shows that some grains vanish while others grow, leading to an increase in the mean grain size.

A classical model for normal grain growth postulates that a grain within a system of many grains (of the same composition) grows if its surface energy per volume is smaller than the mean of the population.[2] Since surface area is $\propto d^2$ and volume is $\propto d^3$, their ratio is proportional to $1/d$, which is the curvature of a sphere. Hence for grain i in a population containing many grains, the growth model reads

$$\frac{\mathrm{d}d_i}{\mathrm{d}t} = \alpha\gamma \left(\frac{1}{\langle d \rangle} - \frac{1}{d_i} \right), \tag{5.5}$$

where $1/\langle d \rangle$ is the mean grain curvature and α is a kinetic coefficient subsuming information about grain shape and diffusivity (mass moves between grains by diffusion). Equation (5.5) states that larger grains grow at the expense of smaller grains, which shrink. Although this equation appears simple, it masks various difficulties. Most importantly, the mean grain size $\langle d \rangle$ is changing with time in an unknown manner. Indeed, the rate of change of $\langle d \rangle (t)$ is the quantity of principal interest in geodynamics.[3]

The appearance in equations (5.4) and (5.5) of a curvature difference can be understood in terms of the consequent difference in pressure between grains. Just as with a soap bubble, the tension within a curved grain surface increases the pressure inside the grain (as described by Kelvin's equation, not shown here). This tension scales with the curvature of the surface, which is $1/r$ for a sphere of radius r. Hence small grains have higher pressure; this higher pressure confers a higher chemical potential. Differences in chemical potential then drive matter diffusion away from small grains and toward large grains.

A simple approach to understanding the growth of a *population* of grains (in terms of their mean grain size) is to assume that it is driven by the reduction of an energy potential, the surface energy per unit volume. This gives rise to

$$\frac{\mathrm{d}\langle d \rangle}{\mathrm{d}t} = \frac{k_g \gamma}{p} \langle d \rangle^{1-p}, \tag{5.6}$$

$$\Rightarrow \langle d \rangle^p - \langle d \rangle_0^p = k_g \gamma t, \tag{5.7}$$

where k is a kinetic factor that is considered to have an Arrhenius dependence on temperature, and $\langle d \rangle_0$ is the mean grain size at $t = 0$. Taking the grain-growth exponent $p = 2$ obtains a growth rate proportional to the surface energy per unit volume $\gamma / \langle d \rangle$ in equation (5.6). Laboratory data on monomineralic systems tend to be fit by $p > 2$; however, such fits are poorly constrained, in general, because very long experiments are required to tightly constrain the exponent (due to slow growth and the need to exclude initial transients).

Grain growth in the presence of other solid phases or impurities is a topic of current research. These additives give rise to *Zener pinning* whereby grain growth is limited at physical obstacles that separate grains of the same composition. Matter diffusion

---

[2]More accurately, the theory states that it is the difference between the grain curvature and a "critical curvature," closely related to the mean, that drives grain-size change.

[3]Geodynamic models also calculate the spatial variation of mean grain size $\langle d \rangle (\boldsymbol{x}, t)$. Hence the population of grains that is represented by the mean at a position $\boldsymbol{x}$ becomes localized to some finite region around $\boldsymbol{x}$; grains grow in response to their "local" environment. This raises the question of the true length scale over which a grain interacts with other grains. See the Literature Notes at the end of this chapter for further discussion.

must then circumnavigate the obstacle on a much longer path. Hence, in the context of polymineralic solids, grain growth is rate-limited by diffusion over a length scale proportional to the mean grain size; therefore, $p \approx 3$ is expected. Grain growth under these circumstances is also known as *Ostwald ripening*.

Grain-size reduction by dynamic recrystallization can balance grain growth and lead to a quasi-steady grain size. Dynamic recrystallization is also a topic of active research, especially in rock physics. Dislocations in the crystal lattice of solid grains are introduced as the grains deform at relatively high stress. These dislocations interact and align, forming subgrain boundaries. With deformation at sufficient stress, these sub-grain boundaries become new grain boundaries and hence the number of grains increases and the average grain size decreases. When considering these processes in detail, there are variations and complexities that require further analysis; it is beyond the current scope to review this field.

A simplifying approach that avoids these detailed considerations is to assume that a fraction of the rate of work done in viscous deformation pays the energetic cost of creating new grain boundaries and increasing the surface energy density. The energetic cost is the rate at which surface-energy density increases due to the formation of new grain-surface area

$$\dot{E} \propto \frac{\gamma}{\langle d \rangle^3} \frac{\mathrm{d}\langle d \rangle^2}{\mathrm{d}t}, \tag{5.8}$$

where the volume and area of the mean grain are expressed as relevant powers of the mean grain size. Introducing a constant of proportionality k_r associated with the grain geometry ($k_r = 6$ for spheres, 12 for cubes) and differentiating gives $\dot{E} = k_r \gamma \langle \dot{d} \rangle / \langle d \rangle^2$. The supply of energy (in the form of work that creates and organizes dislocations) available to pay this energetic cost is some fraction of the total rate of work by viscous mantle flow. The rate of work is given by the stress tensor contracted with the strain-rate tensor, which is the continuum analogue of force dotted with velocity. Assuming zero melt fraction for present purposes, the work rate available to drive dynamic recrystallization is $\chi \boldsymbol{\tau} : \dot{\boldsymbol{\varepsilon}}$, where χ is a positive fraction (the *Taylor-Quinney fraction*) less than unity. It should be nonzero only when stresses are large enough to activate some form of dislocation-accomodated creep.[4] The determination of an appropriate value for χ has no broadly accepted solution; see the Literature Notes at the end of this chapter for further discussion.

Equating the (strictly positive) energy supply rate with the energetic cost from equation (5.8) and rearranging for the rate of grain-size change gives

$$\frac{\mathrm{d}\langle d \rangle}{\mathrm{d}t} = -\chi \frac{\boldsymbol{\tau} : \dot{\boldsymbol{\varepsilon}}}{k_r \gamma} \langle d \rangle^2, \tag{5.9}$$

where the negative sign ensures that positive energy leads to a decrease in the mean grain size. It is important to bear in mind that equation (5.9) is one of several competing hypotheses currently debated, although it has generally been favored for use in geodynamic models.

In the geodynamic context, it is typical to assume that dynamic recrystallization acts simultaneously with and independently of normal grain growth, and hence the total

[4]Even in cases where irreversible work is done by dislocation-accomodated creep, it is empirically known that most of this is dissipated as an increase in temperature.

rate of change is the sum of the two:

$$\frac{\mathrm{D}\langle d\rangle}{\mathrm{D}t} = \frac{k_g\gamma}{p}\langle d\rangle^{1-p} - \chi\frac{\boldsymbol{\tau}:\dot{\boldsymbol{\varepsilon}}}{k_r\gamma}\langle d\rangle^2. \tag{5.10}$$

The Lagrangian derivative indicates that the mean grain size is advected with the continuum velocity of the solid phase. Evidently, for a Lagrangian parcel of rock, larger surface tension leads to larger grains, all else being equal.

Coupling grain-size evolution with geodynamic models is a recent development in geoscientific research. The control of grain size on permeability and viscosity is well established (see below), but less is known about how grain-size feedbacks might affect the behavior of geodynamic systems. Studies have typically used equation (5.10) with the assumption of a steady state and neglecting advection of grain size:

$$\frac{k_g\gamma}{p}\langle d\rangle^{1-p} = \chi\frac{\boldsymbol{\tau}:\dot{\boldsymbol{\varepsilon}}}{k_r\gamma}\langle d\rangle^2, \tag{5.11a}$$

$$\Rightarrow \langle d\rangle^{1+p} = \frac{1}{\chi}\frac{k_g k_r\gamma^2}{2p\eta}\frac{1}{\dot{\boldsymbol{\varepsilon}}:\dot{\boldsymbol{\varepsilon}}}, \tag{5.11b}$$

where, in solving for the mean grain size, we have used equation (3.5), the constitutive law for an incompressible, Newtonian fluid. This type of equation relates mean grain size to the deviatoric stress under which it equilibrated. It is called a *paleopiezometer* because it hypothetically allows for an interpretation of grain size recorded in the geological record in terms of the stress at which the rock last deformed (subject to obvious assumptions).

But complexities of a polymineralic mantle that undergoes solid–solid phase transformations and deforms by both dislocation- and diffusion-accommodated creep mechanisms are unresolved. Hence it remains unclear whether equation (5.10) represents a valid approach to predicting the mean grain size. Indeed there may be situations where grains larger and smaller than the mean behave differently from each other and thus it is necessary to model the evolution of the full grain-size distribution.

At present, there is little work (or consensus) on models coupling variable grain size with dynamics of the partially molten mantle. We neglect grain-size variations in subsequent chapters.

5.1.2 TEXTURAL EQUILIBRATION

For a system comprising a set of grains and mass of melt, spontaneous evolution to minimize the energy density has another effect besides changing the grain size. The arrangement of grains and melt also evolves. Suppose that grain size could be held constant while this rearrangment takes place to minimize energy. This pseudo-minimum energy state is called *textural equilibrium*.[5] It is impossible, in practice, for an undeforming system containing a huge number of grains and their interstitial melt to rearrange to attain a global energy minimum. If it did, however, the result would have the form of

[5] In the case that there is nonnegligible elastic-strain energy stored in the solid grains, the state of textural equilibrium is an energy minimization with respect to both strain and interfacial energy. We will ignore this possibility here.

one huge grain per mineral type, plus one pool of melt. However, on the scale of a cluster of grains plus melt, the microstructure can evolve toward a local energy minimum. In this section we make some basic calculations that expose the characteristics of the local textural equilibrium of partially molten mantle.

Olivine is the most abundant mineral in the asthenosphere and generally the most readily deformed. Therefore it is the dominant control on the properties of partially molten mantle. We consider a simplified system composed of olivine grains and interstitial basaltic melt. The olivine grains are assumed to have isotropic surface energy and a space-filling, highly symmetric shape. In three dimensions, the olivine is idealized as a tetradecahedron, as shown in Figure 5.1(a). In two dimensions, the appropriate space-filling shape is a hexagon. Tessellation of a plane with hexagons leads to two-grain interfaces along lines and three-grain junctions at points, as in panel (b). In three dimensions, the two-grain interfaces extend as flat surfaces while the three-grain junctions are lines that terminate at four-grain junction points.[6]

Melting of grains occurs on the grain boundaries and, more specifically, in regions with the highest concentrations of dangling bonds: at the grain corners that form triple and quadruple junctions. These crystal corners are the first to dissolve, creating rounded edges and the pore structure shown in Fig. 5.1(a). A key characteristic of this pore structure is the *dihedral angle*, which is defined in a plane normal to the triple-junction line (i.e., the plane of the page in fig. 5.1(b)). It is the angle formed by a liquid-filled pore, at the point where it terminates along a grain–grain interface. The size of the angle is determined by a balance of forces. The forces, shown in Figure 5.2(a), are the surface tensions along each interface that terminates at the grain–grain–magma junction. Each of these (one solid–solid and two liquid–solid) exerts a tension that is equivalent to its surface energy. Balancing forces perpendicular to the direction of γ_1 gives $\gamma_2 \sin(\pi - \Theta_3) = \gamma_3 \sin(\pi - \Theta_2)$; similarly, in the direction perpendicular to γ_2 the balance is $\gamma_1 \sin(\pi - \Theta_3) = \gamma_3 \sin(\pi - \Theta_1)$. Therefore we find that

$$\frac{\gamma_1}{\sin(\pi - \Theta_1)} = \frac{\gamma_2}{\sin(\pi - \Theta_2)} = \frac{\gamma_3}{\sin(\pi - \Theta_3)}. \tag{5.12}$$

Now we take $\gamma_3 = \gamma_{ss}$ and $\gamma_1 = \gamma_2 = \gamma_{\ell s}$. By symmetry, this implies that $\Theta_1 = \Theta_2 = \pi - \Theta_3/2$ and hence we can eliminate angles Θ_1 and Θ_2 and drop the subscript on Θ_3. Making these substitutions and rearranging equation (5.12) gives an equation for the dihedral angle Θ as

$$\cos\frac{\Theta}{2} = \frac{1}{2}\frac{\gamma_{ss}}{\gamma_{\ell s}}. \tag{5.13}$$

Inspection of this result shows that if the tension on solid–solid and liquid–solid interfaces are equal, the dihedral angle is 120°; for $\gamma_{\ell s} \to \infty$, the dihedral angle goes to 180°. Dihedral angles of less than 60° produce a textural equilibrium in which the pores span the length of triple junctions and are interconnected to other pores at the quadruple junctions. Hence, the pores form an interconnected, permeable network, even at vanishingly small porosity (fig. 5.1(a)). If the dihedral angle is larger than 60°, textural equilibrium has isolated pores at the quadruple junctions and the rock is impermeable for porosities below a threshold value. This threshold is zero at $\Theta = 60°$ and increases monotonically with Θ.

[6]It can be shown that in three dimensions, four-phase junctions are stable only at a point. Grain-boundary junctions that form a line must therefore be triple junctions. See Bulau et al. [1979] and references therein.

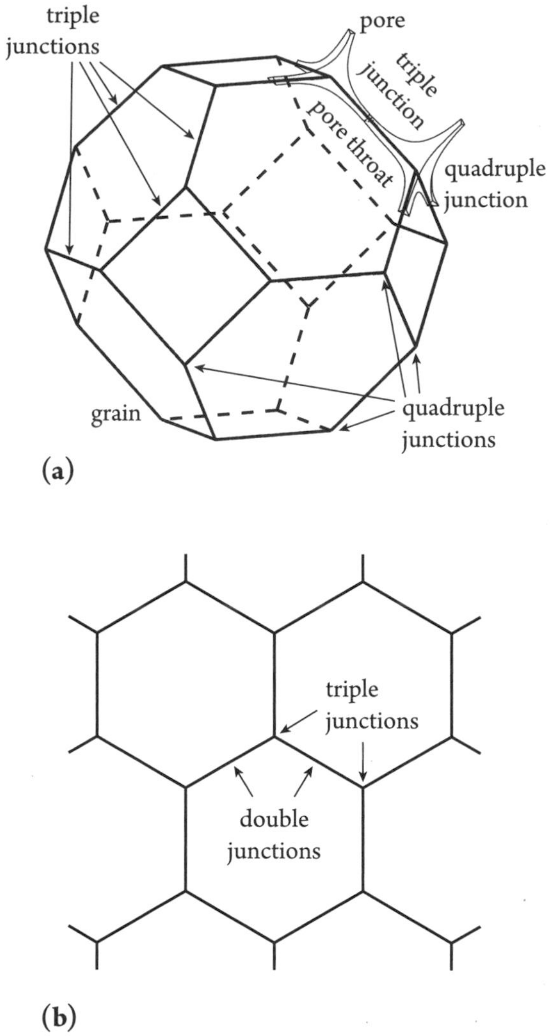

Figure 5.1. Grain geometry. **(a)** An idealized, three-dimensional grain in the shape of a tetradecahedron. This shape is chosen because it is space filling in three dimensions and satisfies the rules of grain-boundary connectivity: double junctions on planes; triple junctions along lines, quadruple junctions at points. An idealized pore is shown, extending along a triple junction between quadruple junctions. Figure adapted from Zhu and Hirth [2003] with permission of Elsevier. **(b)** Tessellation of a plane with idealized, two-dimensional grains in the shape of hexagons. The hexagon is space filling and satisfies the two-dimensional version of the grain-boundary connectivity rules. No pores are shown.

Measurements of the dihedral angle for basaltic magma between olivine grains gives $35^\circ < \Theta < 50^\circ$, which implies that $\gamma_{\ell s} < \gamma_{ss}$ (note that for $\Theta \to 0$ we require $\gamma_{\ell s} \to \gamma_{ss}/2$ from above).

Given a dihedral angle, the three-dimensional shape of a pore is determined by the thermodynamics of the interface. In particular, the chemical potential at a point on the boundary of a crystal is a function of the local curvature of the boundary. A detailed model of this is beyond the present scope, but the important result (also discussed above in the context of grain-size change) is that variations in curvature along a grain boundary give rise to gradients in chemical potential that drive mass transport by diffusion. This causes the surface to evolve toward uniform curvature, which defines the equilibrium state. For a two-dimensional surface embedded in a three-dimensional space, the curvature at a point on the surface can be expressed as the radii of two circles, r_1 and r_2. At that point, the circles are perpendicular to each other and tangent to the surface.

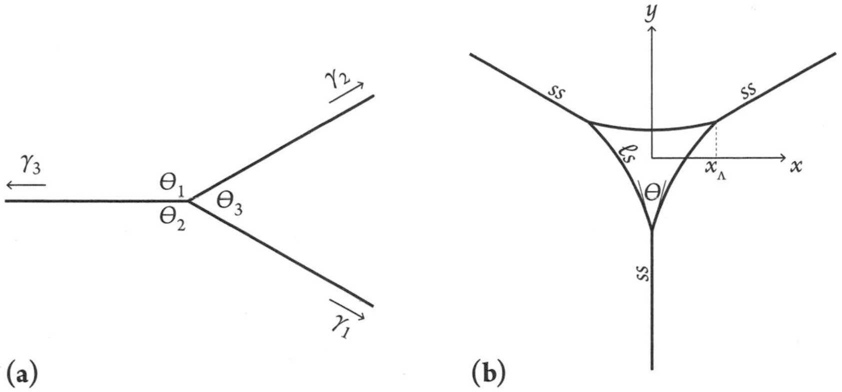

Figure 5.2. The dihedral angle. **(a)** Interfaces oriented according to angles Θ_k and interfacial tensions σ_k. **(b)** The outline of a two-dimensional pore at a triple junction. This can be considered to be a slice through the three-dimensional pore shown in Fig. 5.1(a), taken at the pore throat, halfway along the triple junction. The dihedral angle is $\Theta = 30°$. Interfaces are labeled as liquid–solid (ℓs) or solid–solid (ss).

Uniform curvature then requires that

$$\frac{1}{r_1} + \frac{1}{r_2} = \text{const.} \tag{5.14}$$

Solving this equation for the form of the liquid–solid interface is challenging. For simplicity, we limit our attention to a pore throat, halfway along the triple junction where $r_1 \ll r_2$ (Fig. 5.1(a)). Hence we can approximate the surface as having $r_2 \to \infty$ and seek the shape, in two dimensions, that has $1/r_1 = \text{const.}$ This shape is an arc-segment of a circle that spans two grain–grain boundaries, intersecting each of them at the dihedral angle.

By orienting our coordinate system as shown in figure 5.2(b), we can represent the upper liquid–solid interface as a function $y(x)$. In particular, $y(x)$ is the circular arc segment with its center at $(0, y_0)$ and a radius r_1,

$$y(x) = y_0 \pm \sqrt{r_1^2 - x^2}, \tag{5.15a}$$

$$y'(x) = \mp \frac{x}{\left(r_1^2 - x^2\right)^{1/2}}, \tag{5.15b}$$

where the sign ($\pm$) is determined by whether the liquid–solid interfaces are concave ($\Theta < 60°$) or convex ($\Theta > 60°$).

Values for r_1 and y_0 are obtained by matching the boundary conditions at the pore corners where the interfaces meet at the dihedral angle,

$$y(x_\Lambda) = x_\Lambda \tan(\pi/6) \equiv y_\Lambda, \tag{5.16a}$$

$$y'(x_\Lambda) = \tan(\pi/6 - \Theta/2) \equiv y'_\Lambda. \tag{5.16b}$$

Here, x_Λ is the positive x-position at which the liquid–solid boundary terminates at the solid–solid boundary. Assuming a dihedral angle $\Theta \neq 60°$ and using equations (5.15b) and (5.16b) we find that

$$r_1 = \frac{x_\Lambda}{y'_\Lambda}\sqrt{1 + {y'_\Lambda}^2}. \tag{5.17}$$

Then, by geometry, $y_0 = y_\Lambda + \mathrm{sgn}(y'_\Lambda)\sqrt{r_1^2 - x_\Lambda^2}$. These values and equation (5.15a) are used to plot the upper interface in figure 5.2b. The other two interfaces are obtained by rotating the upper interface by $\pm 120°$ about the origin.

Real olivine–basalt pore networks, such as the one shown in figure 4.1, are much more complicated than this simple theory would suggest. However they are consistent with theory in that there is no apparent percolation threshold: pores are interconnected and liquid can segregate from solid down to vanishingly small porosity. Moreover, real pore networks have a property called *tortuosity*, which is a measure of their curviness. There are a variety of proposed measures of tortuosity and a lack of consensus as to the best choice. The simplest measure is the ratio of arc length along a streamline of the microscopic flow to the start-to-end distance of the streamline. This is difficult to measure in natural rocks, for obvious reasons. The tortuosity has implications for the dispersion of chemistry by a flow.

5.2 Permeability

The permeability of porous media is a subject of broad interest and an extensive literature. It is not the purpose of this book to review that literature, but rather to discuss one theoretically justified (and empirically calibrated) mathematical model. The model can be understood in terms of geometrically simple, periodic arrays of pores and grains, or it can be shown to be consistent with average properties derived from simulations of flow through measured, three-dimensional pore spaces between olivine grains, as illustrated in figure 5.3(a).

Our simple model of permeability is written as

$$k_\phi = \frac{d^2\phi^n}{C}, \tag{5.18}$$

where d is the grain size. C and n are dimensionless constants that depend on the wetting properties of the melt—in particular, on the dihedral angle. They may also depend on the rate of deformation, if it is large enough to move the aggregate away from textural equilibrium. It is straightforward to show that that n must be greater than zero; the definition of the Darcy drag $D \equiv \mu\phi^2/k_\phi$ tells us that we must have $n \geq 2$ such that the drag coefficient does not increase with increasing porosity.

Equation (5.18) is closely related to the Kozeny-Carman equation. Kozeny-Carman contains an additional factor of $(1-\phi)^{-m}$, which allows it to be valid to large porosity. However, since our concern here is with small porosities below the disaggregation limit, we drop the dependence on solid fraction. Figure 5.3(b) shows that this is an acceptable approximation.

The question of mantle grain size and its distribution in space and time is largely unresolved (though see the Literature Notes at the end of this chapter). Hence we lump it into a generic prefactor k_0, which represents the permeability at some reference value

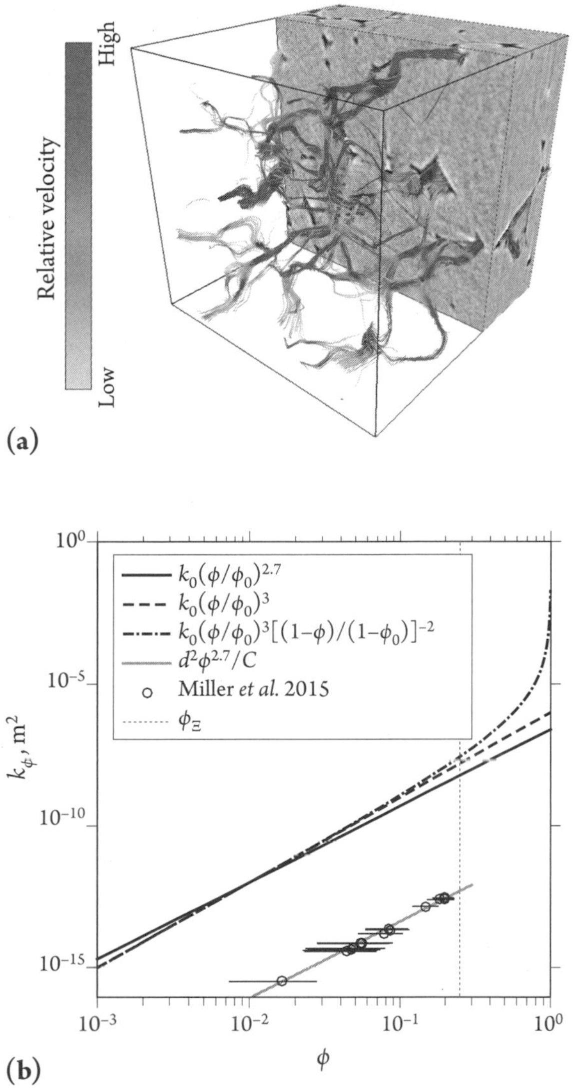

Figure 5.3. Permeability arising from the grain-scale connectivity of pores in a partially molten aggregate. **(a)** Simulated, microscopic liquid streamlines due to a pressure gradient imposed in the simulation. The calculations are based on the pore microstructure in an olivine–basalt aggregate, measured *in situ* with X-ray microtomography; the gray-scale image represents a portion of that data. Image reprinted from Miller et al. [2015] with permission of Elsevier. **(b)** Comparison of porosity–permeability relations. Circles represent data from Miller et al. [2015] with uncertainties in porosity. The fit to data (gray line) is based on a grain size of $d = 35$ μm and $C = 58$ [Miller et al., 2014]. Black lines plot equation (5.19) with $\phi_0 = 0.01$ and $k_0 = 10^{-12}$ m^2. For $C = 58$, these reference values are equivalent to a grain size of 7.6 mm when $n = 3$ and 3.8 mm when $n = 2.7$.

of porosity ϕ_0. Making this simplification we have

$$k_\phi = k_0(\phi/\phi_0)^n. \tag{5.19}$$

This equation is plotted in figure 5.3 for two values of n. Also shown are permeabilities computed on the basis of simulation of flow through pores of a measured pore geometry, displayed in figure 5.3(a) (see also fig. 4.1). The vertical offset between the best fit to this data and the plots of equation (5.19) comes from the difference in grain size: approximately 35 μm as measured in the experiments versus 5 mm for the model curves.

The line of best fit in figure 5.3 corresponds to a permeability exponent of $n = 2.7$. The data on which this fit is based extend only to a minimum porosity of $\sim 2\%$. Theoretical work on textural equilibrium of olivine with basalt suggests that at porosity of less than about 1%, the exponent decreases to $n \approx 2$. The consequences of this difference are small in models analyzed in this book.

5.3 Viscosity

Recall that the constitutive model used throughout this book, which quantifies the relationship between stress and deformation, is a viscous one. Hence the stress is a function of the *rate* of strain. This was most succinctly expressed in equation (4.43) where we introduced the compaction viscosity ζ_ϕ and the shear viscosity η_ϕ of the two-phase aggregate. Now we consider these two properties in more detail, as well as noting the liquid viscosity μ.

5.3.1 SHEAR VISCOSITY OF THE AGGREGATE

The shear viscosity of the mantle is, like permeability, the subject of a large literature. Much is known about creeping mantle flow; this knowlege is generally derived from information about physical mechanisms at the microscopic scale, empirical quantification of flow laws obtained by laboratory deformation of mantle rock, and model-based inversions of geophysical observations such as postglacial rebound. The subject is too broad to review in detail here.

Instead we discuss a generic flow law that, with appropriate parameter choices, can apply to several deformation mechanisms, including diffusion and dislocation creep, in the presence of partial melt. This is written as

$$\eta_\phi^k = A_k\, d^{\mathfrak{m}_k/\mathfrak{n}_k} \exp\left[\frac{1}{\mathfrak{n}_k}\left(\frac{E_k^* + PV_k^*}{RT} - \lambda_k\phi\right)\right] \dot{\varepsilon}_{II}^{(1-\mathfrak{n}_k)/\mathfrak{n}_k}, \tag{5.20}$$

where d is grain size; E_a and V_a are the activation energy and volume, respectively; R is the universal gas constant; λ is a constant; $\dot{\varepsilon}_{II} \equiv \sqrt{\dot{\boldsymbol{\varepsilon}} : \dot{\boldsymbol{\varepsilon}}/2}$ is a tensor invariant that measures the strain-rate magnitude; and A, λ, $\mathfrak{m}$, and $\mathfrak{n}$ are empirical constants. The superscript or subscript k indexes a set of deformation mechanisms (e.g., diffusion creep, dislocation creep, diffusion-accomodated grain-boundary sliding); parameter values may differ for each mechanism. For N simultaneously active deformation mechanisms responding to the same stress, the composite viscosity is given as the harmonic mean,

$$\eta_\phi = \left(\sum_{k=1}^{N} \frac{1}{\eta_\phi^k} \right)^{-1}. \quad (5.21)$$

This formulation of the composite viscosity arises from the assumption that the deformation mechanisms are independent of each other.

As we did above for permeability, it is convenient to lump the grain-size variation with the prefactor and provide reference values (subscript 0) for all of the variables,

$$\eta_\phi^k = \eta_0^k \exp\left\{ \frac{1}{\mathfrak{n}_k} \left[\frac{E_k^* + PV_k^*}{R} \left(\frac{1}{T} - \frac{1}{T_0} \right) - \lambda_k(\phi - \phi_0) \right] \right\} \left(\frac{\dot{\varepsilon}_{II}}{\dot{\varepsilon}_{II0}} \right)^{(1-\mathfrak{n}_k)/\mathfrak{n}_k}. \quad (5.22)$$

The reference value for pressure is zero. When all variables take values equal to their reference values, $\eta_\phi^k = \eta_0^k$.

The porosity and strain rate are determined by the solution to the fluid dynamical problem outlined in chapter 4. A model of the temperature that affects the viscosity will only be developed where we discuss conservation of energy in chapter 8, below. Until that point, we will assume isothermal systems. The pressure that appears in equations (5.20) and (5.22) is usually taken to be the static pressure due to the weight of the overlying rock (the *lithostatic pressure*), rather than the total pressure. Moreover, V^* is poorly constrained but is generally small and often assumed to be negligible over the shallow mantle. For purposes of exploring the fluid dynamics, we'll consider the simplified form

$$\eta_\phi = \eta_0 \exp\left[-\lambda(\phi - \phi_0)/\mathfrak{n} \right] \left(\frac{\dot{\varepsilon}_{II}}{\dot{\varepsilon}_{II0}} \right)^{(1-\mathfrak{n})/\mathfrak{n}}. \quad (5.23)$$

In most of this book, we simplify further by taking $\mathfrak{n} = 1$. We explore the consequences of $\mathfrak{n} \neq 1$ in section 7.4.2.

5.3.2 COMPACTION VISCOSITY OF THE AGGREGATE

A mechanical model for viscosity associated with compaction[7] should, in general, depend on the viscosities of the two phases present. However in the context of a system in which one phase is at least 17 orders of magnitude more viscous than the other, we can safely consider it to be a property of the more viscous phase. Moreover, it should be independent of the permeability that controls the transport of liquid. In soil mechanics, the bulk viscosity of the solid, associated with compaction, is formulated for the *drained* state of the solid, meaning that the pores remain at a state of zero pressure during deformation. Likewise, we'll derive an expression based on the assumption that the pores are filled with an inviscid liquid (melt) that contributes no resistance to deformation. Therefore, we interpret the compaction viscosity as arising solely from the grains' resistance to the deformation required to accomodate compaction.

The solid phase is independently incompressible. The compaction viscosity measures the resistance to compaction of the aggregate of solid and liquid. This resistance

[7]This viscosity, which we call the compaction viscosity, is referred to in the literature as the *bulk viscosity*. See the paragraph containing equation (4.40) for a discussion of terminology.

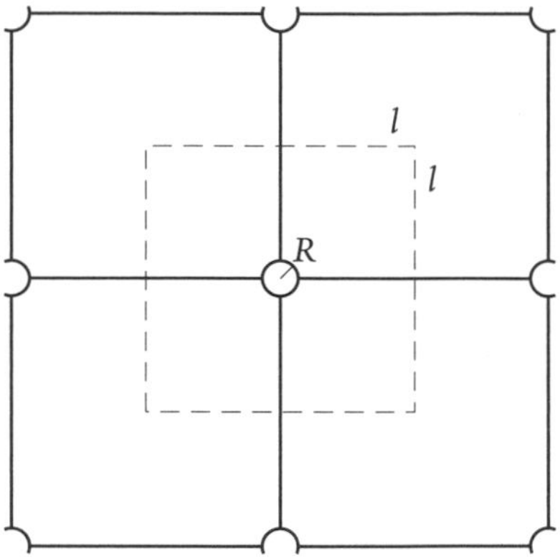

Figure 5.4. Schematic diagram of the micromechanical model used to study the compaction viscosity. The lattice extends periodically in all directions, including forward and backward out of the plane of the drawing.

should therefore diverge as the porosity goes to zero—to be consistent with incompressibility of the solid phase.

We consider a micromechanical model in which a cylindrical melt tubule (pore) appears on a four-grain edge of a regular lattice. A cross section of the lattice is shown in figure 5.4. The tubule is centered in a microscopic cube with length l in each direction. This volume represents a single lattice cell and hence the porosity is given by $\phi = l\pi R^2/l^3$, where R is the tubule radius. The porosity then evolves according to the rate of change of the tubule radius,

$$\frac{\mathrm{d}\phi}{\mathrm{d}t} = \frac{2l\pi R}{l^3}\frac{\mathrm{d}R}{\mathrm{d}t}. \tag{5.24}$$

The problem is then to solve for $\mathrm{d}R/\mathrm{d}t$. We do this in the context of a cylindrical coordinate system that is aligned with the tubule. We consider two models for deformation, by solid flow and by solid diffusion.

Mechanical model. First we consider the case that (de)compaction is accomodated by Newtonian deformation of solid grains. Assuming azimuthal symmetry around the grain tubule, this incompressible flow is purely radial with the form $\breve{\boldsymbol{v}}^s = u_r\hat{\boldsymbol{r}}$. The incompressible continuity equation $\boldsymbol{\nabla}\cdot\breve{\boldsymbol{v}}^s = 0$ then requires that $\mathrm{d}(ru_r)/\mathrm{d}r = 0$. This is satisfied by

$$u_r = -\frac{CR^2}{2}\frac{1}{r}, \tag{5.25}$$

where the constant $-CR^2/2$ is chosen with foresight to simplify the derivation. Kinematics requires that $\mathrm{d}R/\mathrm{d}t = u_r|_{r=R}$ and then, by substituting equation (5.25) into (5.24) and rearranging, we find that

$$C = -\frac{1}{\phi}\frac{\mathrm{d}\phi}{\mathrm{d}t}. \tag{5.26}$$

To determine C in terms of physical forces and material properties we invoke mechanical equilibrium.

We assume that the liquid is inviscid and flows down the tubule without resistance. As in section 4.3.2, force balance on the interface requires that $\breve{\boldsymbol{\sigma}}^\ell\cdot\breve{\boldsymbol{n}}^\ell + \breve{\boldsymbol{\sigma}}^s\cdot\breve{\boldsymbol{n}}^s = 0$.

Therefore we have

$$\left(\breve{P}^s - \breve{P}^\ell\right)\breve{\boldsymbol{n}}^s = \breve{\boldsymbol{\tau}}^s \cdot \breve{\boldsymbol{n}}^s\Big|_{r=R},$$
$$= 2\eta\frac{\mathrm{d}u_r}{\mathrm{d}r}\Big|_{r=R} = \eta C. \qquad (5.27)$$

Combining this result with the kinematics from equation (5.26) gives

$$-\Delta P = \frac{\eta}{\phi}\frac{\mathrm{d}\phi}{\mathrm{d}t}, \qquad (5.28)$$

where we have applied a straightforward upscaling of microscopic pressures to the continuum scale (see justification in section 4.3.2).

Our development has implicitly taken the frame of reference of the solid. We can therefore replace the derivative in equation (5.28) with a Lagrangian derivative in the frame of the solid,

$$\frac{\mathrm{D}_s\phi}{\mathrm{D}t} = -\frac{\phi}{\eta}\Delta P. \qquad (5.29)$$

Using the mass conservation equation (4.46c) with $\Gamma = 0$ to eliminate the Lagrangian derivative of porosity, and using the constitutive law (4.40) to eliminate ΔP, we arrive at

$$\zeta_\phi = \frac{\eta}{\phi}(1-\phi)^2 \approx \frac{\eta}{\phi}, \qquad (5.30)$$

where the approximate result holds if the porosity is below the disaggregation threshold $\phi < \phi_\Xi$ (see section 4.4). Recall that η is the intrinsic viscosity of the solid phase, which is distinct from the viscosity of the aggregate η_ϕ when $\phi > 0$. Furthermore, note that the compaction viscosity diverges as $\phi \to 0$, as required to recover incompressibility of the solid phase.

Diffusion model. The model developed above assumes that compaction occurs by viscous deformation of the solid grains. Grains can also change their shape in response to deviatoric stress by diffusion of matter from more compressive regions to less compressive regions (or, equivalently, by diffusion of lattice vacancies in the opposite direction). This diffusion can pass through the interior of the grain, which gives rise to Nabarro-Herring creep, or it can pass around the grain boundary and lead to Coble creep.

Here we develop a simplified model of (de)compaction accomodated by Nabarro–Herring creep. Microstructural mechanics of crystals tells us that the concentration of lattice vacancies in a grain, linearized around a reference state, is

$$c = c_0\left(1 - \frac{\Omega P}{kT}\right), \qquad (5.31)$$

where c_0 is the reference concentration; Ω is the atomic volume; P and T are pressure and temperature, respectively; and k is the Boltzman constant. Vacancies move by random walk through the lattice and hence their flux is

$$\boldsymbol{J} = -D\nabla c, \tag{5.32}$$

with D a diffusivity. At steady state, $\nabla \cdot \boldsymbol{J} = 0$ and hence $\nabla^2 c = 0$. The Laplace equation in cylindrical coordinates with azimuthal symmetry has the general solution

$$c = A \ln r + B. \tag{5.33}$$

We determine the constants A, B by imposing boundary conditions according to the geometry shown in figure 5.4: the edge of the tubule is at $r = R$ and the interior of the grain is at $r \approx l$. For the former, we again assume that $P = P^\ell$ and for the latter that $P = P^s$. Imposing and combining these two conditions gives

$$A \ln(R/l) = c_0 \frac{\Omega \Delta P}{kT}. \tag{5.34}$$

To obtain A we note that the grain grows into the pore ($\dot{R} < 0$) as matter diffuses from the interior of the grain and plates onto the boundary. This requires that vacancies diffuse away from the boundary $\boldsymbol{J} \cdot \breve{\boldsymbol{n}}^s < 0$. The plating rate and the flux are related by the volume of vacancies and hence we have $\dot{R} = -\Omega \boldsymbol{J} \cdot \hat{\boldsymbol{r}}|_R$ (note that in fig. 5.4, $\breve{\boldsymbol{n}}^s = -\hat{\boldsymbol{r}}$). Then, using equations (5.32) and (5.33), this becomes $\dot{R} = \Omega DA/R$. We write this result in terms of the porosity by rearranging the kinematic equation (5.24) to give

$$A = \frac{l^2 \dot{\phi}}{2\pi \Omega D}. \tag{5.35}$$

Moreover, we note that since $\phi \propto (R/l)^2$, we can write $\ln \phi \propto 2 \ln(R/l)$. Using this and (5.35) in (5.34) gives

$$\frac{kTl^2}{4\pi c_0 \Omega^2 D} \ln \phi \frac{\mathrm{d}\phi}{\mathrm{d}t} = \Delta P. \tag{5.36}$$

Then, by comparison with equation (5.28) and the result obtained for mechanical compaction, we can infer that

$$\zeta_\phi \approx -\frac{kTl^2}{4\pi c_0 \Omega^2 D} \ln \phi, \tag{5.37}$$

which is always positive because $0 \le \phi \le 1$.

Note that the prefactor in (5.37) corresponds to the shear viscosity of a melt-free crystal lattice deforming by Nabarro–Herring creep. Therefore we can write

$$\zeta_\phi \approx -\eta \ln \phi. \tag{5.38}$$

The logarithmic singularity in this relation diverges to infinity with $\phi \to 0$ much slower than the inverse-porosity singularity in equation (5.30). This could have important consequences for melt segregation at small porosity.

Models of Coble creep are more complicated in that grain materials diffuse through the interconnected pore system from high-stress crystal facets to low-stress facets. A singularity for $\phi \to 0$ exists in this case, too. Although the analytical form of the singularity is unknown, it is localized to a range of porosities that are small—much less than 1%. At porosities above this range, the compaction viscosity is predicted to be a constant multiple $\sim 5/3$ of the shear viscosity of the aggregate. See the Literature Notes for a discussion.

5.3.3 SHEAR VISCOSITY OF THE LIQUID

The viscosity of magma μ^ℓ will receive little attention in this book—we will treat it as a constant. This is primarily because the consequences of variation in melt viscosity have not been carefully studied in the context of magma/mantle dynamics. Analysis of the behavior associated with variable melt viscosity is a current research challenge.

We do know, however, that composition is an important control on magma viscosity insomuch as it affects the polymerization. Basaltic melts typically have a moderate viscosity of 1–10 Pa-sec. Increasing silica content causes greater polymerization of the melt and hence increases its viscosity. With increasing volatile content, polymerization and viscosity are reduced. Granitic melts in the crust, which have high silica content, are so viscous that their segregation rate is neglibible—unless they are rich in water.

Better models of the molecular structure of silicate melts as a function of their composition are required to predict the viscosity as well as the thermodynamic stability of silicate melts.

5.4 Thermodynamic Properties

5.4.1 DENSITY

The buoyancy of partial melts is proportional to their difference in density with the solid mantle: magma produced in the shallow mantle rises because it is less dense than its residue ($\Delta\rho > 0$). In our derivation of the governing equations in chapter 4, we made the simplifying Boussinesq approximation of constant and uniform phase densities, and hence of constant and uniform density difference and buoyancy. In reality, both magma and solid mantle are compressible; this leads to changes in density with depth. Solid minerals also undergo solid–solid phase transitions at increasing pressure to higher-density states. We briefly examine the consequences of compressibility; for references and further discussion see the Literature Notes at the end of this chapter.

The relative compressibilities of mantle solid and liquid phases are most relevant here. The most comprehensively studied system is $MgSiO_3$, a major component of the mantle that forms pyroxene at low pressure (and transitions to majorite, perovskite, and postperovskite at higher pressures deep in the mantle). At atmospheric pressure, $MgSiO_3$ liquid is $\sim$20% less dense than solid of the same composition—this is the *structural* density difference. However, experiments and first-principles molecular dynamics simulations demonstrate that $MgSiO_3$ liquid is substantially more compressible than $MgSiO_3$ solid. The density difference between them decreases with increasing pressure, approaching $\sim$ 4% at 140 GPa, a pressure roughly corresponding to the bottom of the mantle.

Partial melts of the mantle differ in composition from their solid residue. The density difference is most affected by the partitioning of iron. Iron is concentrated in the melt by a factor of about 2.5$\times$ greater than magnesium. Because iron is a heavier element than magnesium, iron-rich melts have a *chemical* density difference with solids that is opposite in sign to that of their structural density difference. This chemical density difference is roughly independent of pressure in the asthenosphere. Therefore, because the structural density difference is diminished with increasing pressure, the total density difference between solid and partial melt can become negative. At some depth, mantle partial melts are denser than their residual solids and hence would segregate downwards.

Flotation experiments on olivine grains immersed in peridotitic melt show that a density crossover occurs at approximately 10 GPa or about 300 km depth. At this pressure, structural compression of the melt has increased its density to the point that it is equal to that of the solid. At higher pressures, melt becomes the denser phase and would segregate downwards. This could occur down to depths where the solid mineralogy changes to make iron more compatible in the solid residue. However, the greater structural compressibility of the melt means that partial melts of perovskite enriched with iron would become denser than solid perovskite near the bottom of the mantle.

Our focus in this book is on melt transport in the asthenosphere and hence we will assume positive $\Delta\rho$ throughout. Models of magmatic segregation in the transition zone or the bottom of the mantle would need to modify this.

5.4.2 SOLID–LIQUID PHASE CHANGE

The melting rate Γ is another key parameter in the equations; underlying it is the entire field of mantle petrology, which includes thermodynamic theories of great complexity (and that is just to describe the *equilibrium* thermodynamics). Indeed, there are two general approaches to modeling the melting rate: equilibrium laws and kinetic laws. The former assumes that at any point in space and time, the petrological system is in thermodynamic equilibrium. Hence the rate of melting is associated with the rate of change of the equilibrium melt fraction, which in turn depends on the rate of change of the thermodynamic state (e.g., pressure P, temperature T, and bulk composition $\overline{c}_i$). In contrast, kinetic laws assume that phase change is rate-limited by physics at the grain scale, and hence that thermodynamic equilibrium is not achieved in general. The melting rate in kinetic laws is typically forced by the amount of chemical disequilibrium and can depend on thermodynamic variables. We discuss thermodynamic and kinetic laws for phase change in more detail in chapter 10.

In either case, for the applications considered in this book we seek simple parameterizations of the petrology that capture the essential features,[8] but neglect most of the details. For example, as a thermodynamic law for what is loosely referred to as adiabatic decompression melting, we could prescribe a closure of

$$\Gamma \approx \left.\frac{\mathrm{d}F}{\mathrm{d}P}\right|_{\mathrm{S}} \frac{\mathrm{D}_s P}{\mathrm{D}t} = -\rho^s \boldsymbol{v}^s \cdot \hat{\mathbf{g}} \left.\frac{\mathrm{d}F}{\mathrm{d}z}\right|_{\mathrm{S}}, \tag{5.39}$$

where F is the degree of melting, $\hat{\mathbf{g}}$ is a dimensionless unit vector pointing in the direction of gravity, and $\mathrm{d}F/\mathrm{d}z|_{\mathrm{S}}$ is the isentropic productivity of decompression melting. The latter can be approximated as being constant or can take a more complicated,

[8] We will define *essential* on a case-by-case basis, depending on the physics and phenomena that we seek to model. A note of caution, however: simplifying the model behaviour a priori, on the basis of expectations of model sensitivity (or lack thereof), can exclude important or interesting results. Models that incorporate more realistic petrology and thermochemistry should be developed.

parameterized form to capture, for example, variations with pressure or extent of melting. Equation (5.39) is not a bad approximation for decompression melting, but it excludes reactive melting and any phase change due to nonisentropic changes in temperature. To address those aspects, we require theory for the thermochemistry of magma/mantle dynamics, which is developed in chapters 8–10. In the preceding chapters we will take $\Gamma = 0$ and examine the behavior of solutions with a fixed total porosity.

5.5 Literature Notes

The literature on grain-size evolution has mostly been written by materials scientists, but Karato [2008] provides a helpful overview and starting point. Its discussion begins with theories for normal grain growth. The development of average grain-growth kinetics, as in equations (5.6) and (5.7) is credited to Burke and Turnbull [1952]. Equation (5.5) for the growth of an individual grain is generally referred to as the *Hillert model* [Hillert, 1965]. Other, less-accepted models exist; see Atkinson [1988] for a review. Although normal grain growth is a classical subject and the focus of many studies, there remain notable discrepancies between observed grain-size distributions and model predictions [Rios and Zöllner, 2018]. Breithaupt [2021] showed that these can be substantially reduced if individual grains respond to their local environment (the surrounding grains), rather than to the global mean grain size. That work provides a brief review of the relevant literature from materials science.

The study of dynamic recrystallization is substantially more complicated than the brief overview given in this chapter. Again, Karato [2008] provides more details and references. Adequately large strains at a constant stress may lead to a steady-state grain size under dynamic recrystallization. This steady grain size can be expressed as a function of stress [e.g., Twiss, 1977], which has been used as a *paleopiezometer* (a measure of the stress recorded in the texture of deformed rocks). The theory discussed above [Austin and Evans, 2007] builds on these ideas to proposed a paleo-wattmeter in which recrystallized grain size is dynamically related to the energy dissipation rate. Behn et al. [2009] compared two models of dynamic recrystallization against experimental data.

Another strand of theoretical development has cast grain-size variation in terms of a damage model [e.g., Bercovici et al., 2001a; Bercovici and Ricard, 2005; Ricard and Bercovici, 2009; Rozel et al., 2011]. This thermodynamics-based approach consistently incorporates grain-size variations into the overall energy budget. The simplified result for mean-grain-size evolution is consistent with that obtained by Austin and Evans [2007]. In both cases, a major uncertainty is the partitioning of work between dissipative heating and the production of dislocations of new grain boundaries, a problem with a long history [Farren and Taylor, 1925; Taylor and Quinney, 1934] that was reconsidered by Bercovici and Ricard [2003]. However, note that the thermodynamics of microstructural evolution remains a subject with fundamental open questions, such as how to quantify the entropy [Pawan et al., 2020].

Geodynamic studies have incorporated mean-grain-size dynamics into flow models. Hall and Parmentier [2003] considered the consequences for convective instability; Turner et al. [2015] modeled the mean-grain-size field beneath a mid-ocean ridge; Turner et al. [2017] extended this to consider magmatic segregation with dynamic grain size. Wada et al. [2011] considered grain-size dynamics in subduction zones; Cerpa Gilvonio et al. [2017] extended that model to include magma. There is a considerable literature on grain-size reduction in ductile shear zones [e.g., Montési and Hirth, 2003]. Grain-size reduction was investigated as a feedback involved in the formation and persistence of tectonic plate boundaries by Landuyt et al. [2008]. Other studies that model the consequences of grain-size variation for mantle convection include Bercovici et al. [2001b], Dannberg et al. [2017] and Rozel [2012].

The textural equilibrium of partially molten mantle rock has been studied both theoretically and experimentally. Bulau et al. [1979] adapted theory from materials science to model the equilibrium shape of pores for a given dihedral angle and grain shape. A companion experimental study by Waff and Bulau [1979] considered olivine–basalt in conditions close to textural equilibrium, finding a high degree of interconnectedness of pores, as predicted by theory. Experiments by Vaughan et al. [1982] on grains of $\sim$10 μm established an upper limit of 200 hours for the time over which textural equilibrium is reached. This time should be $<$1000 h for grains of order 1 mm [McKenzie, 1984]—still fast relative to other time scales associated with melt segregation. However, work by Faul et al. [1994] sought to challenge the idea that the computed textural equilibrium is an accurate description of the actual microstructure.

Various complexities have been considered. Toramaru and Fujii [1986] measured the dihedral angle of basalt with orthopyroxene. Measurements of the anisotropic surface energy of olivine crystals were made by Cooper and Kohlstedt [1982] and the role of anisotropic surface energy on wetting and pore alignment was investigated experimentally by Jung and Waff [1998]. Experiments by Yoshino et al. [2007] provide evidence that the dihedral angle between forsteritic olivine and its hydrous melt can approach zero at high pressure.

The theory and calculations of pore geometry in textural equilibrium were further developed by von Bargen and Waff [1986] and Cheadle et al. [2004]. Detailed calculations of minimum-energy pores by Rudge [2018a] provide a complete map of pore geometries around tetrakaidecahedral grains at all porosities.

In systems evolving by static grain growth, Garapic et al. [2013] found that grain faces remain wetted by films of magma that are $\lesssim$100 nm thick. Jin et al. [1994] and Daines and Kohlstedt [1997] considered nonequilibrium pore structures under deformation.

Microstructural models were used by, for example, von Bargen and Waff [1986] and Cheadle et al. [2004] to calculate estimates of permeability. Zhu and Hirth [2003] used network models of pores surrounding olivine grains. Simpson et al. [2010a,b] applied homogenization methods on simple unit cells. Better constraints come from experimental measurements such as by Riley and Kohlstedt [1991], Wark and Watson [1998], Liang et al. [2001], Renner et al. [2003] and Connolly et al. [2009]. More recently, Zhu et al. [2011] and Miller et al. [2014] measured and discretized the three-dimensional pore space of a texturally equilibrated olivine–basalt aggregate. They computed the flow through this pore system for a fixed pressure gradient and determined its permeability and electrical conductivity [Miller et al., 2015].

There has been some debate in the literature as to whether there is a minimum-porosity threshold for nonzero permeability. This argument goes back at least to the 1960s and was given its most detailed theoretical justification by Waff [1980]. The idea of a threshold fell from favor during the later 1980s and 1990s, but was revived by Faul [1997] and then disputed by Wark et al. [2003] and others. It is now generally thought that melts can segregate down to vanishingly small porosities, though Miller et al. [2016] showed that this may depend on the lithology of the residue.

Permeability may be sensitive to deviatoric stress and deformation. This was demonstrated by Xiao et al. [2006] in experiments on calcite and by Ghanbarzadeh et al. [2015] for porous flow through a salt formation. Hier-Majumder [2011] developed theory to model permeability of a deforming, two-phase aggregate and found that anisotropic permeability can result from the deviatoric strain rate.

The shear viscosity of the mantle is the subject of an enormous literature that combines solid-state physics, experimental rock mechanics, materials science, and other fields. A useful introduction is provided in Turcotte and Schubert [2014], and an advanced treatment is given by Karato [2008]. Highly cited reviews include Karato and Wu [1993] and Hirth and Kohlstedt [2003]. Reviews that focus on partially molten mantle include Hirth and Kohlstedt [1995a,b]. Experiments by Mei et al. [2002] and Zimmerman and Kohlstedt [2004] constrained the sensitivity of shear viscosity to melt fraction [see also Kelemen et al., 1997]. Scott and Kohlstedt [2006] extended the range of porosity in experiments beyond the disaggregation threshold.

There is increasing interest in the viscous anisotropy of the mantle, arising from lattice-preferred orientation as well as melt-preferred orientation. For the former, Hansen et al. [2012] reported the result of experiments on olivine aggregates. Viscous anisotropy due to the coherent alignment of melt pockets (pores) between grains has been studied experimentally by Takei [2010] and a theory was derived for diffusion creep by Takei and Holtzman [2009a,b,c].

The viscosity of basaltic liquids is perhaps less complex than that of the crystalline mantle and is more completely understood. Dingwell et al. [2004] measured the viscosity of peridotite-derived melts. Giordano et al. [2008] proposed a general model for magma that accounts for temperature and composition dependence. It was calibrated on an experimental database of some 1700 laboratory measurements. Bottinga and Weill [1972] is an older, highly cited reference that proposes a viscosity model for magma.

The porosity dependence of compaction viscosity, also known as the *bulk viscosity*, was recognized by M^c^Kenzie [1984] (he referred to it as the bulk viscosity). In appendix C of that work, M^c^Kenzie adopted a model with a logarithmic singularity as $\phi \to 0$. Elsewhere in the paper, solutions to the governing equations were obtained that assumed ζ_ϕ to be constant. Subsequent workers including Fowler [1985] and Scott and Stevenson [1986] adopted theory from Batchelor [1967] that employs a dependence on ϕ^{-1}. The micromechanical models of Sleep [1988], Bercovici et al. [2001a], Hewitt and Fowler [2008], Simpson et al. [2010a,b] and others arrive at the same conclusion. These models are summarized and elaborated by Schmeling et al. [2012].

Theory for diffusion-accomodated compaction was developed by Takei and Holtzman [2009a,b] and Rudge [2018b] building on, for example, Cooper and Kohlstedt [1984]. They considered Nabarro–Herring and Coble creep. The model of Nabarro–Herring creep introduced above is simplified from Rudge [2018b]. For Coble creep, diffusion is dominantly through the interconnected melt phase that fills the pores between grain. This diffusion transports components of the crystal that are dissolved at high-stress faces and precipitated at low-stress faces of solid grains. Chemical diffusivity in the melt is much larger than in the grain, so in Takei and Holtzman [2009a] and Rudge [2018b] it is assumed to be infinite. They derive expressions for the compaction viscosity that have a weak dependence on porosity and remain finite at $\phi = 0$. The latter result stems from the assumption of infinite diffusivity. Takei and Holtzman [2009b] developed a refined model of compaction under Coble creep that includes finite diffusivity in the melt; numerical solutions display a singularity for vanishing porosity.

Measurement of the compaction viscosity in experiments is difficult and there are few published studies on the topic. Renner et al. [2003] performed experiments with a low-viscosity, lithium silicate melt to isolate the effect of compaction viscosity. They fit a rather complicated model to their data and cite various previous studies from the materials sciences.

There is a substantial literature on the density of mantle solids and melts as a function of pressure and temperature (i.e., their equations of state) that will not be summarized here. Poirier [2000] gives an introduction to the physical basis of equations of state. The sign-change of the density difference between mantle melts and their residues, as a function of pressure, was predicted by Stolper et al. [1981] and verified experimentally by Agee and Walker [1993] and Ohtani et al. [1995]. First-principles molecular dynamics simulations by Stixrude and Karki [2005] showed the greater compressibility of perovskite liquid than solid in the lower mantle and enabled a prediction of the basal-mantle solidus.

Melt production and the thermodynamics of mantle melting is the subject of a vast literature. We return to the subject in chapters 8, 9, and 10 and cite the relevant literature in those chapters.

5.6 Exercises

5.1 (a) Without calculating, draw the shape of a two-dimensional pore with a dihedral angle of 60°.
(b) Write down the dihedral angle that corresponds to circular pores in two dimensions.

5.2 Explain and justify the use of the boundary conditions (5.16).

5.3 Derive an expression for the two-dimensional pore half-width x_Λ in terms of the volume-fraction of liquid ϕ and dihedral angle Θ. Assume the grains are regular hexagons with facets of length d_f.

5.4 Consider the micromechanical model illustrated in figure (5.4), where cylindrical melt tubules sit along the four-grain junctions of a cubic lattice. Assume l and R are constant throughout this question, and assume that the flow is driven by the gravitational body force (i.e., no pressure gradient is applied).

(a) Use the isoviscous, incompressible form of the Stokes equation (3.13) and appropriate boundary conditions to determine the fluid velocity parallel to the tubules. You may assume that the cubic lattice is aligned with gravity.
(b) By integrating the velocity over the cross-sectional area of a tubule, determine the mean velocity through the tubule. Hence determine the mean (Darcy) velocity per unit cross-sectional area in the lattice in terms of ϕ and l.
(c) By relating your solution from part (b) to equation (4.46a), determine the permeability k_ϕ.
(d) Suppose we tilt the lattice so tubules are no longer aligned with the direction of gravity. Do you expect the permeability to vary as the lattice is tilted? Explain.

5.5 Consider two-dimensional simple shear flow given (to leading order) by $\boldsymbol{v}^{(0)} = \dot{\gamma} y \hat{\boldsymbol{x}}$.

(a) Determine the leading-order deviatoric strain rate $\dot{\varepsilon}^{(0)}$. Hence obtain the tensor invariant $\dot{\varepsilon}_{II}^{(0)}$.

The shear flow slowly begins to overturn, yielding a first-order correction to the velocity, $\epsilon \boldsymbol{v}^{(1)} = -\epsilon \dot{\gamma} x \hat{\boldsymbol{y}}$.

(b) Compute the first-order correction to the strain rate $\dot{\boldsymbol{\varepsilon}}^{(1)}$.

(c) Use your solution to (b) to determine $\dot{\varepsilon}_{II}$ to $\mathcal{O}(\epsilon^2)$. Hence, show that the shear viscosity given by (5.23) remains unchanged at first order.

5.6 If a set of N mechanisms of mantle deformation all operate simultaneously in response to the same stress, show that the composite viscosity is given by equation (5.21).

5.7 A change in sea-level of $\sim$100 m occured over $\sim 10^4$ years during the last deglaciation; similar changes are recorded in the geological record. Using a mantle upwelling rate of 3 cm/y beneath a mid-ocean ridge, develop a scaling argument to approximate the percentage deviation of Γ that would be induced by such sea-level variation. Assume that mantle melting is due to adiabatic decompression.

CHAPTER 6

Compaction and Its Inherent Length Scale

Compaction and compaction stress are the most basic and important features that are novel to the theory of magma/mantle dynamics. For this reason, the system of PDEs representing the fluid dynamics (4.47) are sometimes referred to as the *compaction equations*. Compaction is also the physical process that is least familiar to people on their first approach to the subject at hand, and hence we introduce it in some detail in this chapter.

To understand compaction, it is helpful to separate it from the other physical processes that are described by the full governing equations. Compaction can occur in one dimension and so, for simplicity, this chapter will consider only one-dimensional domains. We will call this dimension z and choose $\hat{z}$ to be opposite to gravity (upward) such that $\mathbf{g} = -g\hat{z}$. Furthermore we will define symbols $W \equiv \boldsymbol{v}^s \cdot \hat{z}$ and $w \equiv \boldsymbol{v}^\ell \cdot \hat{z}$ to be the vertical component of the solid and liquid velocities.

Compaction and melting are the two processes that change the porosity within a parcel of partially molten rock (in the Lagrangian frame moving with the solid). Our focus in this chapter is on compaction and hence we take $\Gamma = 0$ to remove from consideration the effects of melting.

Finally, we will consider the viscosities (η_ϕ, ζ_ϕ, μ^ℓ) to be constant and uniform. To remind us that this is the case, we refer to the shear and compaction viscosity of the aggregate as η_0 and ζ_0.

With these simplifications, we can rewrite the more general system (4.47) (which already includes the Boussinesq approximation) as

$$-\frac{\mathrm{d}W}{\mathrm{d}z} + \frac{k_0}{\mu^\ell}\frac{\mathrm{d}}{\mathrm{d}z}\left[\left(\frac{\phi}{\phi_0}\right)^n\left(\frac{\mathrm{d}P^\ell}{\mathrm{d}z} + \rho^\ell g\right)\right] = 0, \tag{6.1a}$$

$$-\frac{\mathrm{d}P^\ell}{\mathrm{d}z} + \xi_0\frac{\mathrm{d}^2 W}{\mathrm{d}z^2} - \overline{\rho}g = 0, \tag{6.1b}$$

$$\frac{\mathrm{D}_s\phi}{\mathrm{D}t} - (1-\phi)\frac{\mathrm{d}W}{\mathrm{d}z} = 0, \tag{6.1c}$$

where $\xi_0 \equiv \zeta_0 + 4\eta_0/3$.

The system (6.1) can be simplified further if we neglect body forces and apply boundary conditions that force the dynamics. This is our approach in the next section, where we solve what might be the simplest problem involving compaction.

6.1 The Compaction-Press Problem

In this problem we consider a semi-infinite region of $z \leq 0$ that contains partially molten rock with porosity ϕ_0. This region is bounded by a barrier at $z=0$. The liquid phase can cross the barrier with no resistance but the solid phase is impeded. Across the barrier at $z>0$, there is a liquid-filled region with pressure $P^\ell=0$. Motion of the barrier into the two-phase region at velocity $v=-W_0$ is imposed. This motion causes compaction of the partially molten rock: liquid crosses the barrier and solid is forced to move toward $-\infty$.

To what distance into the partially molten rock does compaction occur? At what distance $z<0$ is the aggregate effectively unaware that the barrier is pushing inward? To answer these questions we seek a solution for the compaction rate *at the instant* that the barrier begins pushing. We take the reference frame that is fixed to the barrier. In this frame, the far-field solid and liquid are moving at a speed W_0 toward the barrier.

Neglecting the body force and noting that $\phi/\phi_0=1$, equations (6.1) become

$$-\frac{\mathrm{d}W}{\mathrm{d}z}+\frac{k_0}{\mu^\ell}\frac{\mathrm{d}^2P^\ell}{\mathrm{d}z^2}=0, \tag{6.2a}$$

$$-\frac{\mathrm{d}P^\ell}{\mathrm{d}z}+\xi_0\frac{\mathrm{d}^2W}{\mathrm{d}z^2}=0. \tag{6.2b}$$

Recall that the first of this set represents compaction caused by the divergence of the segregation flux, whereas the second equation represents force balance of the two-phase aggregate.

Eliminating the pressure gradient from the system (6.2) gives a third-order ordinary differential equation for W,

$$-\frac{\mathrm{d}W}{\mathrm{d}z}+\frac{k_0\xi_0}{\mu^\ell}\frac{\mathrm{d}^3W}{\mathrm{d}z^3}=0. \tag{6.3}$$

This third-order equation requires three boundary conditions. Far from the barrier that forces the problem we can assume that

$$W=W_0 \text{ and } \frac{\mathrm{d}^2W}{\mathrm{d}z^2}=0 \text{ as } z\to-\infty. \tag{6.4}$$

At the permeable barrier,

$$W=0 \text{ on } z=0. \tag{6.5}$$

Note that these conditions are formulated in the reference frame that is moving with the barrier.

Integrating (6.3) once and using boundary conditions (6.4) we obtain

$$-W+\frac{k_0\xi_0}{\mu^\ell}\frac{\mathrm{d}^2W}{\mathrm{d}z^2}=W_0. \tag{6.6}$$

A solution to this equation that satisfies the boundary conditions at $z=0$ and $z\to-\infty$ is

$$W(z)=W_0\left(1-\mathrm{e}^{z/\delta_0}\right), \tag{6.7}$$

where

$$\delta_0 \equiv \sqrt{\frac{\xi_0 k_0}{\mu}} \tag{6.8}$$

and has units of length. The solution (6.7) can be checked by direct substitution. Using the solution, the compaction rate can be obtained as

$$\mathcal{C}(z) = -\frac{W_0}{\delta_0}\, \mathrm{e}^{z/\delta_0} \quad \text{for } z \leq 0. \tag{6.9}$$

This function is plotted in figure 6.1. It is important to note the decay length δ_0 of the compaction rate. This is the length scale over which the disturbance, introduced by the moving boundary, is "felt" by the two-phase aggregate. At distances more than a few times δ_0 from the barrier, the disturbance is small; a few more δ_0 beyond that, it is almost entirely gone. The quantity δ_0 is a fundamental length scale that *emerges* from the physics (and its expression in the PDEs); it is not introduced by any aspect of the domain or the boundary conditions. We call it the *compaction length* and refer to it repeatedly throughout the book.

It is interesting to consider the pressures that are associated with the compacting region. Using equation (6.2b) and the boundary condition $P^\ell = 0$ at $z = 0$ we can obtain

$$P^\ell(z) = \frac{\xi_0 W_0}{\delta_0}\left(1 - \mathrm{e}^{z/\delta_0}\right) \tag{6.10}$$

for the liquid pressure. The constitutive law (4.40) that relates isotropic stress and compaction requires $\Delta P = -\zeta_\phi \mathcal{C}/(1-\phi)$ and hence we obtain the solid pressure as

$$P^s(z) = P^\ell(z) - \frac{\zeta_0}{1-\phi_0}\mathcal{C}(z). \tag{6.11}$$

Far from the barrier, there is no compaction and the solid pressure and the liquid pressure are equal. In contrast, at $z = 0$, $\mathcal{C} = -W_0/\delta_0$ and the solid pressure is

$$P^s(0) = \frac{\zeta_0 W_0}{\delta_0(1-\phi_0)}. \tag{6.12}$$

The role of $1 - \phi_0$ is to increase the pressure as the area of solid in contact with the barrier decreases.

The instantaneous speed of the liquid phase can be calculated using the pressure gradient, obtained by differentiating the liquid pressure (eqn. (6.10)), and the force-balance equation (4.46a) for the liquid phase. It is

$$w(z) = W(z) + \frac{W_0}{\phi_0}\mathrm{e}^{z/\delta_0}. \tag{6.13}$$

This is consistent with the absence of segregation of liquid and solid far from the permeable barrier ($z \to -\infty$).

These results tell us something of fundamental importance: the compaction length represents a characteristic distance over which perturbations to the compaction pressure $\mathcal{P} \equiv \zeta_\phi \mathcal{C}$ are transmitted by the two-phase aggregate. As an analogy, imagine a

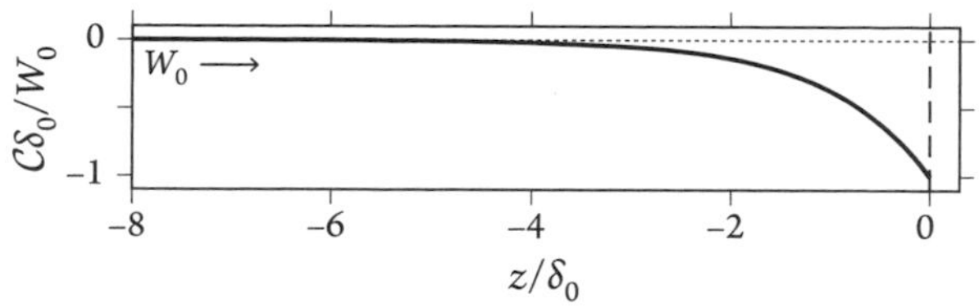

Figure 6.1. The compaction-rate solution of the compaction-press problem, plotted from equation (6.9). Coordinate z is dimensional and has the same units (e.g., meters) as δ_0.

boat moving forward through still water. Ahead of the boat the water surface is displaced vertically by the bow wave; the gradient in surface height causes a forward flow. Ahead of the bow wave, the water surface remains flat, unaffected by the presence of the boat. The physics of compaction is different but the concept is similar: there is a finite decay length of mechanical perturbations associated with the isotropic (compaction) stress. Where they are isolated features in a larger domain, these stresses can lead to narrow boundary layers. Indeed we can think of the compacting region in figure 6.1 as the boundary layer of a semi-infinite domain.

The solution that we computed was instantaneous—it did not resolve the evolution of porosity over a finite time interval. If we tried to do this and continued to assume a constant compaction viscosity, we'd find that the solution would blow up when the pores adjacent to the boundary compacted to zero. This is unphysical, however, because the resistance to compaction increases as the porosity decreases. For a compaction viscosity that increases with porosity, a compacted boundary layer with nonzero porosity and permeability can form and grow. See the Literature Notes at the end of this chapter for references.

6.2 The Permeability-Step Problem

The gravitational body force that causes buoyancy-driven segregation of liquid magma from the solid matrix is restored to the equations in this problem. We consider a column of partially molten rock occupying $z \in (-\infty, \infty)$. Melt flows upward (in the direction of $\hat{z}$) due to buoyancy. At $z = 0$ there is a step-change in porosity (and hence permeability). We will not be concerned with why the step exists or how it evolves—the solution we seek is the instantaneous compaction rate $\mathcal{C}(z)$ at $t = 0$.

The porosity is given by the piecewise constant function (shown in figure 6.2(a) and (c), below),

$$\phi(z) = \begin{cases} f_p\phi_0 & \text{for } z > 0, \\ f_m\phi_0 & \text{for } z \leq 0, \end{cases} \tag{6.14}$$

where f_i $(i = p, m)$ are constants that multiply the reference porosity, chosen such that $|f_i - 1| < 1$. Substituting this distribution into (6.1a) and using (6.1b) to eliminate the pressure gradient, we have the equation governing compaction (sometimes called the compaction equation),

$$\delta_0^2 \frac{\mathrm{d}}{\mathrm{d}z}\left(f_i^n \frac{\mathrm{d}\mathcal{C}}{\mathrm{d}z}\right) - \mathcal{C} = \frac{k_0 \Delta\rho g}{\mu^\ell} \frac{\mathrm{d}}{\mathrm{d}z}\left[f_i^n(1 - f_i\phi_0)\right]. \tag{6.15}$$

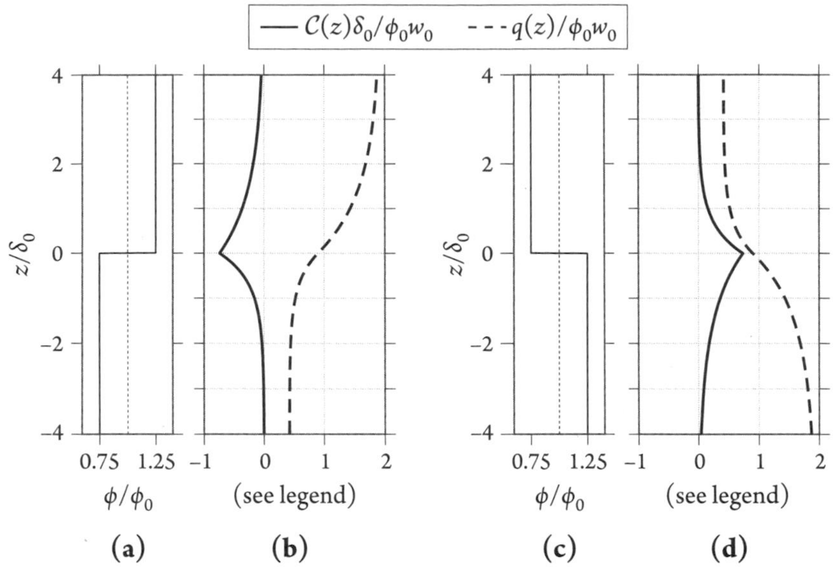

Figure 6.2. Porosity, compaction rate, and segregation flux for the permeability-step problem. **(a)** A porosity increase with z: $f_m = 0.75$, $f_p = 1.25$. **(b)** Scaled compaction rate and segregation flux for the porosity increase. **(c)** A porosity decrease with z: $f_m = 1.25$, $f_p = 0.75$. **(d)** Scaled compaction rate and segregation flux for the porosity decrease.

Instead of solving this equation on the full domain, we will solve on both half-domains and match them with boundary conditions at $z = 0$. Then the permeability is constant within each semi-infinite subdomain and (6.15) becomes

$$(\delta_i)^2 \frac{\mathrm{d}^2\mathcal{C}}{\mathrm{d}z^2} - \mathcal{C} = 0, \tag{6.16}$$

where

$$\delta_i = \sqrt{\frac{k_0 f_i^n \xi_0}{\mu^\ell}} \tag{6.17}$$

is the compaction length for each of the two subdomains. Equation (6.16) admits the general solution

$$\mathcal{C}(z) = A \exp\left(z/\delta_i\right) + B \exp\left(-z/\delta_i\right). \tag{6.18}$$

Boundary conditions are required to solve for the unknown constants A and B.

We expect that the largest disturbance will be found at the permeability step, $z = 0$, and that any compaction-rate disturbance will decay to zero for $z \to \pm\infty$. Furthermore, we expect the compaction rate to be continuous across $z = 0$. Hence we have

$$\mathcal{C} = \mathcal{C}^* \text{ at } z = 0, \tag{6.19a}$$

$$\mathcal{C} = \frac{\mathrm{d}\mathcal{C}}{\mathrm{d}z} = 0 \text{ for } z \to \pm\infty, \tag{6.19b}$$

where $\mathcal{C}^*$ is the extremal compaction rate with a value to be determined. The conditions at $z \to \pm\infty$ require that $A = 0$ for the solution on $z > 0$ and $B = 0$ for the solution on $z \leq 0$. Applying the condition at $z = 0$, the solution becomes

$$\mathcal{C}(z) = \mathcal{C}^* \times \begin{cases} \exp\left(-z/\delta_p\right) & \text{for } z > 0, \\ \exp\left(z/\delta_m\right) & \text{for } z \leq 0. \end{cases} \tag{6.20}$$

To determine the extremal compaction rate $\mathcal{C}^*$, we integrate equation (6.15) over an interval from $-\epsilon$ to ϵ and consider the limit as $\epsilon \to 0$. The integral is

$$\left[\delta_i^2 \frac{\mathrm{d}\mathcal{C}}{\mathrm{d}z}\right]_{-\epsilon}^{\epsilon} - \int_{-\epsilon}^{\epsilon} \mathcal{C}\mathrm{d}z = \frac{k_0 \Delta\rho g}{\mu^\ell} \left[f_i^n (1 - f_i\phi_0)\right]_{-\epsilon}^{\epsilon}. \tag{6.21}$$

The second term goes to zero in the limit because $\mathcal{C}$ is continuous across zero. The derivative of compaction rate can be evaluated using (6.20) to give

$$\frac{\mathrm{d}\mathcal{C}}{\mathrm{d}z} = \mathcal{C}^* \begin{cases} -1/\delta_p & \text{at } z = 0^+, \\ +1/\delta_m & \text{at } z = 0^- \end{cases} \tag{6.22}$$

on either side of zero. Using (6.22) in (6.21) gives us

$$\mathcal{C}^* = \frac{k_0 \Delta\rho g}{\delta_0 \mu^\ell} \tilde{f}, \tag{6.23}$$

where

$$\tilde{f} = \frac{f_m^n (1 - f_m\phi_0) - f_p^n (1 - f_p\phi_0)}{f_m^{n/2} + f_p^{n/2}}. \tag{6.24}$$

The combination of equations (6.23) and (6.20) is a complete solution to the permeability step problem.

The quantity $\tilde{f}$ in equation (6.23) is dimensionless and therefore the fraction that precedes it must have dimensions of compaction rate (i.e., inverse time). We can clarify the meaning of this group by looking back at the modified Darcy's law (4.46a), assuming a lithostatic pressure gradient $\nabla P^\ell \sim \rho^s \mathbf{g}$ and a negligible solid velocity to obtain the relationship $\phi w \sim k_0 \Delta\rho g/\mu^\ell$. Hence we identify an inherent velocity scale for buoyancy-driven segregation as

$$w_0 \equiv \frac{k_0 \Delta\rho g}{\phi_0 \mu^\ell}. \tag{6.25}$$

Using this characteristic scale in equation (6.23) and combining with equation (6.20) we have

$$\mathcal{C}(z) = \frac{\phi_0 w_0}{\delta_0} \tilde{f} \begin{cases} \exp\left(-z/\delta_p\right) & \text{for } z > 0, \\ \exp\left(z/\delta_m\right) & \text{for } z \leq 0. \end{cases} \tag{6.26}$$

This solution is plotted in Figure 6.2. Figures 6.2(a) and (b) show an upward step in porosity with z ($f_p > 1$, $f_m < 1$), whereas figures 6.2(c) and (d) show a downward step ($f_p < 1$, $f_m > 1$).

Note that the compaction rate takes an extremal value at $z = 0$ and decays back to zero with distance from there. This decay is roughly on the scale of the reference compaction length, as it was in section 6.1. Closer inspection, however, shows that decay is more rapid where the porosity is smaller than ϕ_0 and more gradual where the porosity is larger than ϕ_0. This is because the actual compaction length is $\delta_i = \sqrt{f_i^n}\delta_0$. It is important to note that the reference compaction length is only a good approximation of the actual compaction length when and where permeability and viscosity are close to their reference values.

Moreover, we note from the solution (6.26) that the compaction rate has a characteristic scale of $\phi_0 w_0/\delta_0$ for this buoyancy-driven problem. Compare this to the characteristic scale of W_0/δ_0 that we obtained for the compaction-press problem. In that case, the dynamics were forced by the boundary conditions, which introduced a speed W_0.

The one-dimensional segregation flux $q \equiv \phi(w - W)$ is given by combining equations (6.1a) and (6.1b) to eliminate P^ℓ. Then, using the solution for compaction rate (6.26) to obtain $\mathrm{d}\mathcal{C}/\mathrm{d}z$ (see exercise 6.1), we find that

$$q = \phi_0 w_0 \begin{cases} \left[f_p^n(1 - f_p\phi_0) + \tilde{f} f_p^{n/2} \exp(-z/\delta_p)\right] & z > 0, \\ \left[f_m^n(1 - f_m\phi_0) - \tilde{f} f_m^{n/2} \exp(z/\delta_m)\right] & z \leq 0. \end{cases} \tag{6.27}$$

The segregation flux $q(z)$ is plotted as dashed lines in figure 6.2 along with the compaction rate $\mathcal{C}(z)$. Figure 6.2(a) shows a porosity distribution with a step to higher porosity at $z = 0$. This change is associated with a step to higher permeability and hence reduced Darcy drag on the liquid. Far from the permeability step ($|z| \gg \delta_i$), Darcy drag alone balances buoyancy; this determines the background segregation flux $\phi_0 w_0 f_i^n(1 - f_i\phi_0)$ shown in panel (b). In the vicinity of the step ($|z| \lesssim \mathcal{O}(\delta_i)$), the segregation flux must transition continuously between the far-field values. The increase of segregation flux with z is associated with a liquid underpressure ($P^\ell < P^s$), and hence a convergence of the solid phase ($\mathcal{C} < 0$). According to equation (6.1c), this causes the porosity to decrease.

The situation is reversed in figures 6.2(c) and (d), which show a step to lower porosity and its consequences for compaction rate and segregation flux. The upward segregation flux decreases with z within a few compaction lengths of the step. This causes a *liquid overpressure* ($P^\ell > P^s$), which drives a divergence of the solid phase. With time, this would cause an increase in the local porosity.

If we examine the flux q in the limit of zero step in porosity or permeability $f_m = f_p = 1$, we find that equation (6.27) uniformly reduces to $q = \phi_0(1 - \phi_0)w_0$. For small porosity, this is approximated as $q \sim \phi_0 w_0$, which is a reference buoyancy-driven vertical liquid flux.

The permeability-step problem is constructed to illustrate features of buoyancy-driven compaction and the compaction length; we do not compute the evolution of porosity. In the next section, however, we consider a time-dependent problem where the porosity evolution equation plays a central role.

6.3 Propagation of Small Porosity Disturbances

In permeability-step problem we saw that, when driven by magma buoyancy, an upward-increasing segregation flux causes compaction $\partial\phi/\partial t < 0$ and an upward-decreasing flux causes decompaction $\partial\phi/\partial t > 0$. This suggests that if we place a porosity (and flux) increase immediately below a flux decrease, we might obtain a stable porosity wave that moves upward, decompacting ahead and compacting behind. In the current section we consider small-amplitude porosity perturbations and linearize the governing equations. In section 6.4 below, we extend our analysis to finite amplitude, nonlinear waves.

For this analysis we need both the compaction equation (from the combination of (6.1a) and (6.1b)) and the porosity evolution equation (6.1c),

$$-\delta_0^2 \frac{\mathrm{d}}{\mathrm{d}z}\left[\left(\frac{\phi}{\phi_0}\right)^n \frac{\mathrm{d}\mathcal{C}}{\mathrm{d}z}\right] + \mathcal{C} = -\phi_0 w_0 \frac{\mathrm{d}}{\mathrm{d}z}\left[\left(\frac{\phi}{\phi_0}\right)^n (1-\phi)\right], \tag{6.28a}$$

$$\frac{\mathrm{D}_s\phi}{\mathrm{D}t} = (1-\phi)\mathcal{C}, \tag{6.28b}$$

It is helpful to rescale the variables in this system to simplify the equations and identify the controlling, nondimensional groups of parameters. We use the foregoing sections to identify the appropriate characteristic scales.

Clearly the scale for distance should be the reference compaction length δ_0 and the scale for liquid velocity should be the characteristic buoyancy speed w_0. These suggest a time scale of δ_0/w_0. Our solution for the compaction rate in the permeability step problem (6.26) motivates a characteristic scale for the compaction rate of $\phi_0 w_0/\delta_0$, which in turn suggests that the solid velocity (that is associated with compaction) should scale with $\phi_0 w_0$. These scalings can be summarized using a notation we shall employ throughout the book:

$$[z] = \delta_0, \qquad [t] = \delta_0/w_0, \qquad [\mathcal{C}] = \phi_0 w_0/\delta_0, \qquad [W] = \phi_0 w_0. \tag{6.29}$$

Finally, since the permeability law already includes a reference porosity ϕ_0, we define a normalized porosity $\varphi \equiv \phi/\phi_0$. Substitution of this definition and rescaling of dimensional variables with scales, (6.29) leads to the system

$$-\left(\varphi^n \mathcal{C}_z\right)_z + \mathcal{C} = -\left[\varphi^n(1-\phi_0\varphi)\right]_z, \tag{6.30a}$$

$$\varphi_t + \phi_0 W \varphi_z = (1-\phi_0\varphi)\mathcal{C}. \tag{6.30b}$$

where all variables (i.e., $z, t, \mathcal{C}, W$, and φ) are now dimensionless, rescaled versions of those that appeared previously. Derivatives have been rewritten using subscripts that denote the independent variable.

To simplify the equations (6.30), we recognize that three terms are of order ϕ_0 whereas the others are $\mathcal{O}(1)$. For cases in which $\phi_0 \ll 1$, it is appropriate to drop the $\mathcal{O}(\phi_0)$ terms. Doing so gives us the simpler, nondimensional system

$$-\left(\varphi^n \mathcal{C}_z\right)_z + \mathcal{C} = -\left(\varphi^n\right)_z, \tag{6.31a}$$

$$\varphi_t = \mathcal{C}. \tag{6.31b}$$

These equations can be combined to eliminate the compaction rate, giving

$$-\left(\varphi^n \varphi_{tz}\right)_z + \varphi_t = -\left(\varphi^n\right)_z. \tag{6.32}$$

This equation is nonlinear because of the permeability relation φ^n.

We now assume that the background porosity is perturbed by an arbitrarily small traveling wave,

$$\varphi(z,t) = 1 + \epsilon e^{i(kz-\omega t)}, \tag{6.33}$$

where k is the wavenumber, ω is the frequency, and $\epsilon \ll 1$. Because the disturbance is small, it is appropriate to linearize the equations using a Taylor series expansion,

$$\varphi^n \approx 1 + \epsilon n e^{i(kz-\omega t)} + \mathcal{O}(\epsilon^2). \tag{6.34}$$

Substituting equations (6.33) and (6.34) into equation (6.32) and dropping terms of $\mathcal{O}(\epsilon^2)$ we obtain the dispersion relation

$$\omega k^2 + \omega = nk, \tag{6.35}$$

which leads to the phase and group velocities

$$c_p = \omega/k = \frac{n}{k^2+1}, \tag{6.36a}$$

$$c_g = \mathrm{d}\omega/\mathrm{d}k = c_p - \frac{2nk^2}{(k^2+1)^2}. \tag{6.36b}$$

The phase and group velocities are plotted in Figure 6.3 as a function of the wavelength $\lambda = 2\pi/k$. Considering first the phase velocity (thick lines), we see that $c_p \sim n$ at large wavelength. At small wavelength ($\lambda \to 0$), equation (6.36a) tells us that $c_p \sim n/k^2$ (i.e., the phase speed goes to zero as k^2). The transition between the large-wavelength and small-wavelength regimes occurs around a dimensionless $\lambda = 2\pi$. Recalling that lengths are scaled by the compaction length, this means that waves much longer than the compaction length move fast, while those much smaller than the compaction length move slowly. Compaction stresses, relevant on length scales smaller than the compaction length, retard the propagation of waves.

The group velocity c_g (solid lines) has similar behavior at large and small wavelengths, but becomes negative for $\lambda < 2\pi$ (nondimensionalized according to (6.29)). Hence we see that over some intermediate range, a small-amplitude wave packet could potentially move downward.

By linearizing the permeability with equation (6.34), this development has neglected a nonlinearity between porosity and permeability that is of fundamental importance to magma/mantle dynamics. Taking this into account affects the form of waves that is predicted, as we will see in the next section. The harmonic waveforms proposed above would be unstable under nonlinear evolution according to the full governing equations (6.31).

6.4 Magmatic Solitary Waves

The small-porosity governing equations (6.31) admit a nonlinear *solitary wave* solution. To find it, we assume that it exists and moves at a constant velocity. Then we solve for its shape in terms of porosity and compaction rate.[1]

[1]These magmatic solitary waves are sometimes referred to as *magmons* or, more prosaically, as *compaction waves*.

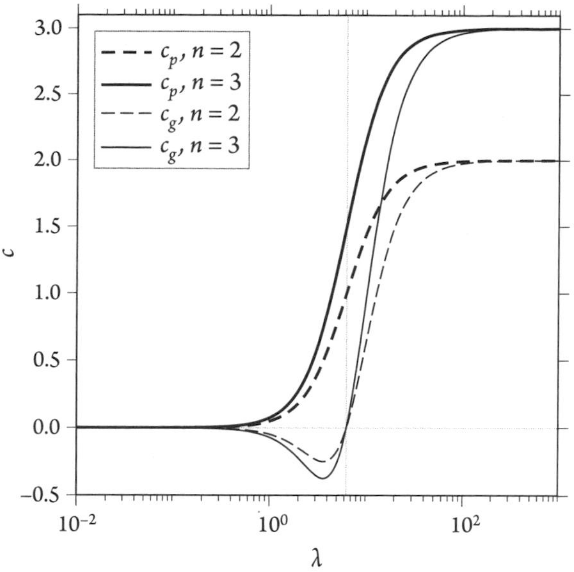

Figure 6.3. Phase (c_p) and group (c_g) velocities of small-amplitude porosity waves for $n=2,3$, plotted as a function of wavelength $\lambda=2\pi/k$ from equations (6.36). The vertical gray line is at $\lambda=2\pi$.

Based on our expectation, developed in section 6.3 above, of traveling-wave behavior, we look for solutions of the form

$$\varphi=\Lambda(Z),\quad \text{where } Z=z-vt \tag{6.37}$$

is a coordinate system that travels with the wave and v is the speed of the wave, a constant to be determined. Our goals are to determine $\Lambda(Z)$ and v as a function of problem parameters.

Equations (6.31) can be transformed into the moving coordinate system by noting that

$$\frac{\partial}{\partial t}=\frac{\partial Z}{\partial t}\frac{\partial}{\partial Z}=-v\frac{\partial}{\partial Z}, \tag{6.38a}$$

$$\frac{\partial}{\partial z}=\frac{\partial Z}{\partial z}\frac{\partial}{\partial Z}=\frac{\partial}{\partial Z}. \tag{6.38b}$$

Combining equations (6.38), (6.37), and (6.31) we have

$$-\left(\Lambda^n \mathcal{C}_Z\right)_Z+\mathcal{C}=-\left(\Lambda^n\right)_Z, \tag{6.39a}$$

$$-v\Lambda_Z=\mathcal{C}. \tag{6.39b}$$

Combining these to eliminate $\mathcal{C}$ we obtain a third-order, nonlinear differential equation for Λ,

$$v\left(\Lambda^n\Lambda_{ZZ}-\Lambda\right)_Z=-\left(\Lambda^n\right)_Z. \tag{6.40}$$

To integrate (6.40) we need boundary conditions. We expect the solitary wave to be localized around $Z=0$ and hence we assume that as $Z \to \pm\infty$, the porosity is flat at the background level, i.e.,

$$\Lambda = 1 \text{ and } \Lambda_{ZZ} = 0 \text{ as } Z \to \pm\infty. \tag{6.41}$$

Integrating (6.40) once and applying these boundary conditions we obtain

$$\upsilon\left(\Lambda^n \Lambda_{ZZ} - \Lambda\right) = -\Lambda^n + (1-\upsilon). \tag{6.42}$$

Instead of trying to integrate (6.42), we convert it to a first-order equation in $\mathcal{C}$ by using the first derivative of the mass conservation equation (6.39b) to give

$$\mathcal{C}_Z = 1 + (\upsilon - 1)\Lambda^{-n} - \upsilon\Lambda^{1-n}. \tag{6.43}$$

To make this equation integrable, we must recast it in terms of $\mathcal{C}_\Lambda (= \mathrm{d}\mathcal{C}/\mathrm{d}\Lambda)$. This is achieved by noting that

$$\frac{\mathrm{d}\mathcal{C}}{\mathrm{d}Z} = \frac{\mathrm{d}\mathcal{C}}{\mathrm{d}\Lambda}\frac{\mathrm{d}\Lambda}{\mathrm{d}Z} = -\frac{\mathcal{C}}{\upsilon}\frac{\mathrm{d}\mathcal{C}}{\mathrm{d}\Lambda}, \tag{6.44}$$

where we have made use of (6.39b). Substituting this into (6.43) gives

$$\mathcal{C}\,\mathrm{d}\mathcal{C} = -\upsilon\left[1 + (\upsilon-1)\Lambda^{-n} - \upsilon\Lambda^{1-n}\right]\mathrm{d}\Lambda. \tag{6.45}$$

Equation (6.45) is integrable. The constant of integration is determined by noting that far from the solitary wave, the compaction rate goes to zero as the porosity goes to its background level, $\mathcal{C}(\Lambda = 1) = 0$. Integrating equation (6.45), we obtain

$$\mathcal{C}^2 = -2\upsilon \begin{cases} \left[\Lambda - \upsilon \ln \Lambda - (\upsilon - 1)/\Lambda + (\upsilon - 2)\right] & \text{for } n = 2, \\ \Lambda^{1-n}\left[\Lambda^n - \frac{n^2-2n+\upsilon}{(n-1)(n-2)}\Lambda^{n-1} + \frac{\upsilon}{n-2}\Lambda + \frac{1-\upsilon}{n-1}\right] & \text{for } n > 2, \\ (\Lambda - 1)^2\Lambda^{-2}\left[\Lambda - \frac{1}{2}(\upsilon - 1)\right] & \text{for } n = 3. \end{cases} \tag{6.46}$$

A reminder of our goals: to determine the shape $\Lambda(Z)$ and the wave speed υ of the magmatic solitary wave. The amount of buoyancy in a solitary wave will depend on the amplitude of the wave, Λ^*, and hence the speed the solitary wave travels should also depend on the amplitude. We thus seek $\upsilon(\Lambda^*)$, the dispersion relation. The maximum porosity disturbance Λ^* occurs at the point of transition between decompaction and compaction, where $\mathcal{C} = 0$. Applying this criterion to (6.46) and solving for υ gives

$$\upsilon(\Lambda^*) = \begin{cases} (\Lambda^* - 1)^2 / \left(\Lambda^* \ln \Lambda^* - \Lambda^* + 1\right) & \text{for } n = 2, \\ 2\Lambda^* + 1 & \text{for } n = 3. \end{cases} \tag{6.47}$$

A general solution for integer $n > 3$ exists but is not of use here. Curves for $n = 2, 3$ are plotted in figure 6.4. Note that a solitary wave with infinitesimal amplitude ($\Lambda^* \to 1$) above the background porosity travels with a finite speed, equal to the long-wavelength speed for linearized porosity waves (section 6.3, fig. 6.3).

We next seek the form of the solitary wave. Ideally we would solve for an explicit solution, $\Lambda(Z)$. Unfortunately, the best we can do is an implicit solution, $Z(\Lambda)$. We

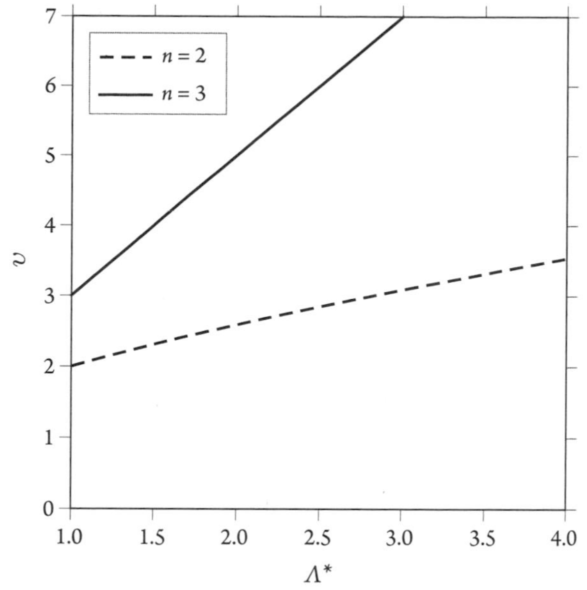

Figure 6.4. Dimensionless solitary wave speed υ computed with equation (6.47) as a function of wave amplitude Λ^* relative to the background porosity.

obtain this by rearranging equation (6.39b) and integrating forward from the peak of the wave $\Lambda(0)=\Lambda^*$,

$$Z=-\upsilon\int_{\Lambda^*}^{\Lambda}\frac{\mathrm{d}\Lambda}{\mathcal{C}}. \tag{6.48}$$

Here we restrict our focus to $n=3$. Combining (6.46), (6.47), and (6.48) to eliminate $\mathcal{C}$ and υ, integration gives

$$Z(\Lambda)=\left(\Lambda^*+\tfrac{1}{2}\right)^{\frac{1}{2}}\left[\frac{1}{(\Lambda^*-1)^{\frac{1}{2}}}\ln\left(\frac{(\Lambda^*-1)^{\frac{1}{2}}-(\Lambda^*-\Lambda)^{\frac{1}{2}}}{(\Lambda^*-1)^{\frac{1}{2}}+(\Lambda^*-\Lambda)^{\frac{1}{2}}}\right)-2\left(\Lambda^*-\Lambda\right)^{\frac{1}{2}}\right]. \tag{6.49}$$

This can be inverted for $\Lambda(Z)$ numerically by using, for example, Newton's method.

Profiles of Λ and $\mathcal{C}$ for $n=3$ and various Λ^* are plotted in figure 6.5; (a) shows the normalized porosity profile $\Lambda=\phi/\phi_0$ of the solitary wave as a function of the nondimensional, traveling coordinate Z. Panel (b) shows that there is decompaction on the leading side of the wave and compaction on the trailing side. The antisymmetry of the pattern of $\mathcal{C}(Z)$ indicates that a porosity profile given by (6.49) will move upward (in the direction opposite to $\mathbf{g}$) without change of shape.

The speed with which the solitary waveform moves upward, υ, is distinct from the speed $w(Z)$ at which magma ascends. Is υ larger or smaller than $\max w(Z)$? We determine this by first solving for the magma speed w. To this end, we use (6.39b) and the dimensionless continuity equation (4.14) to write $[\Lambda(w-W)]_Z=\upsilon\Lambda_Z$. We then integrate, using the nondimensional condition $\Lambda(w-W)=1$ at $Z\to\infty$, to obtain

$$\Lambda(w-W)=\upsilon\Lambda+(1-\upsilon). \tag{6.50}$$

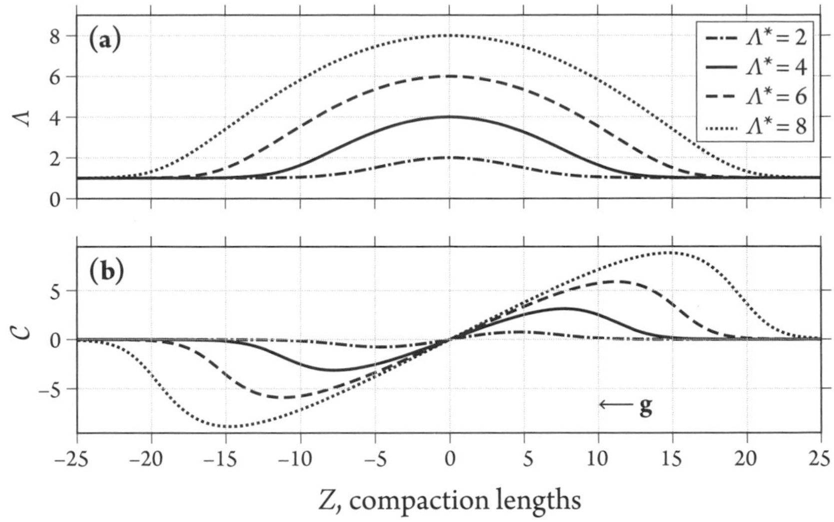

Figure 6.5. Profiles of porosity perturbation and compaction rate for solitary waves of various amplitude. $n=3$ in all cases. **(a)** Profiles of normalized porosity $\Lambda=\phi/\phi_0$. **(b)** Profiles of compaction rate. The gravity vector g points to the left, as indicated in panel (b).

It is interesting to note that despite the nonlinearity of the equations, the segregation flux scales linearly with the normalized porosity Λ. To solve (6.50) for w we need to know W, which we obtain (for $n=3$) using

$$W=\phi_0\int_{-\infty}^{Z} \mathcal{C}\,\mathrm{d}Z=\phi_0\int_{1}^{\Lambda}\mathcal{C}\frac{\mathrm{d}Z}{\mathrm{d}\Lambda}\,\mathrm{d}\Lambda, \tag{6.51a}$$

$$=-\upsilon\phi_0\int_{1}^{\Lambda}\mathrm{d}\Lambda, \tag{6.51b}$$

$$=-\upsilon\phi_0\,(\Lambda-1)\le 0. \tag{6.51c}$$

To go from (6.51a) to (6.51b), we rearrange the second equality of equation (6.44), obtain $Z_\Lambda=-\upsilon/\mathcal{C}$ and substitute.

Substituting the result (6.51c) into (6.50), again using (6.47) and rearranging, we find that the wave speed relative to the maximum speed of melt segregation is

$$\upsilon-w(0)=2\left[1+\phi_0\left(\Lambda^*-1\right)\left(\Lambda^*+\tfrac{1}{2}\right)\right] \tag{6.52}$$

for $n=3$. Since the right-hand side is always positive, we learn that magmatic solitary waves in one dimension always move faster than the magma within them.

Magmatic solitary waves exist in two and three dimensions too, but they have a different form in each case. This means that if the one-dimensional solitary wave pattern obtained above is used to initialize a solitary wave layer in higher dimensions (i.e., $\varphi(\boldsymbol{x})=\Lambda(z)$), this layer will break up into solitary waves that are appropriate to the spatial dimension (circular in two dimensions; spherical in three dimensions). Solutions for these forms and their associated wave speeds exist and are useful for benchmarking numerical solutions.

In two and three dimensions, solitary waves contain a core of liquid that travels at the wave speed. Hence, the wave has the potential to trap and advect melt in isolation from surrounding melts (but not in isolation from the solid). The Literature Notes at the end of this chapter provide further discussion and references.

6.5 Solitary-Wave Trains

In section 6.2 we saw that a step in porosity/permeability can cause a local, instantaneous perturbation in compaction rate. How would such a step evolve over time? In section 6.4, we learned that there are specific porosity profiles that move as waves without changing shape. Do such forms emerge spontaneously or do they only appear when imposed as an initial condition on models? To address both of these questions we introduce the following problem: consider two regions of constant porosity with a smooth transition between them; compute the evolution of porosity according to equations (6.39).

This problem cannot be solved analytically. The results of a numerical solution are shown in figure 6.6. We consider a downward transition in normalized porosity $\varphi = \phi/\phi_0$ of amplitude 0.5. The transition is imposed with a hyperbolic tangent function at $t = 0$, where the width of the transition is set to be $\sim$10 compaction lengths. There are two phases of evolution evident in figure 6.6.

The initial phase, which occurs during $t \lesssim 8$, involves steepening of the transition. This occurs because the melt flux in the lower region is greater than that in the upper region. Melt moves toward the step faster than it moves away, increasing the porosity within the transition. However, in this phase the transition remains long compared to the compaction length. Therefore, the first term in equation (6.39a), representing the gradient in compaction stress, remains small. The segregation rate is established by the balance of buoyancy and Darcy drag. The compaction rate is then a kinematic consequence of the flux divergence,

$$\mathcal{C} \sim -\left(\Lambda^n\right)_Z. \tag{6.53}$$

This is known as the *zero-compaction-length* (ZCL) *approximation*. It is strictly valid only when all features of interest in the calculation are much larger than the compaction length. However, it is a singular perturbation to the system of PDEs and hence its application leads to the loss of any compacting boundary layers (such as that of fig. 6.1). Without those boundary layers, a porosity profile that decreases upward can evolve into a propagating discontinuity in porosity: a shock wave. Despite this and because the ZCL approximation reduces an elliptic equation to a much simpler algebraic one, it has been widely used and discussed.

The second phase of the evolution shown in figure 6.6 begins at $t \approx 8$, when the width of the transition shortens to $\sim$3 compaction lengths. Then the transition begins to "feel" like the sharp step that we imposed in the permeability-step problem (section 6.2). This change occurs because the first term in equation (6.39a), representing the viscous resistance to (de)compaction, begins to contribute to balancing buoyancy. As in figure 6.2(b) for a downward step, the compaction rate is positive; porosity accumulates at the step, forming a porosity pulse. The pulse takes a shape similar to a solitary wave and grows as it moves upward. On its trailing edge is an upward step to larger porosity, as in figure 6.2(a). This causes a divergent melt flux and negative compaction rate, so porosity decreases locally. Behind this negative pulse is a downward step in porosity, where melt

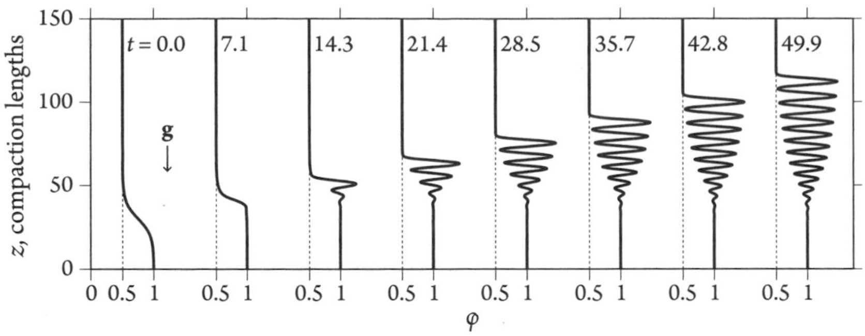

Figure 6.6. Development of a train of magmatic solitary waves from a smoothed step in porosity. This calculation uses $n=3$. The gravity vector g points to the bottom of the page, as indicated in the figure.

accumulates into a positive pulse ... and so on. The result is a train of solitary waves that propagate upward.

There is much more that is known and could be discussed here regarding the detailed mechanics of solitary waves and wave trains. The present treatment is meant as an introduction to the subject, however. The interested reader is referred to the Literature Notes at the end of this chapter, and onward from there to the literature.

6.6 The Compaction Length in the Asthenosphere

Our analysis in this chapter exposed two important, characteristic scales that emerge from the equations governing the mechanics. The first is the compaction length, introduced above, in equation 6.8, as

$$\delta_0 \equiv \sqrt{\frac{(\zeta_0 + 4\eta_0/3)k_0}{\mu^\ell}}, \tag{6.54}$$

the length scale over which a pressure difference between phases causes compaction. The second is the characteristic Darcy speed of the melt, introduced above, in equation 6.25, as

$$w_0 \equiv \frac{k_0 \Delta\rho g}{\phi_0 \mu^\ell}, \tag{6.55}$$

the buoyancy-driven segregation rate of magma. The actual compaction length δ and Darcy speed $|\boldsymbol{v}^\ell|$ vary in space and time according to local conditions. However, for a particular region or model of interest, it is usually possible to conceive of characteristic values.

Putting tight numerical constraints on those values is more challenging. We require estimates of the characteristic matrix viscosity η_0, magma viscosity μ^ℓ, permeability k_0, mantle porosity ϕ_0, gravitational acceleration g, and density difference between phases $\Delta\rho$. These will vary between locations within Earth's mantle, and even more so across the universe. To fix ideas, we focus on the asthenosphere beneath terrestrial mid-ocean ridges. Parameter values or ranges are reported in table 6.1. We first consider w_0.

At moderate depths ($\sim$30 km) beneath mid-ocean ridges, the magma is basaltic, with a composition that is well constrained by experiments and observations. Therefore

Table 6.1. Parameters for rescaling compaction problems to dimensional quantities, based on basaltic melt in an olivine-rich mantle rock at a depth of ~30 km beneath an intermediate-spreading-rate mid-ocean ridge. The permeability, shear, and compaction viscosity are evaluated at $\phi = \phi_0$. Reference values for w_0 and δ_0 are computed using reference values in the lines above. The reference value for w_0 is about 1 m/yr. Ranges are computed with extremal values from the ranges above. The reference values are, in an informal sense, the author's guess of the most probable value within the range, given current understanding. The range should thus be understood as including all plausible values, even if some choices are very unlikely to be correct.

Quantity	Reference value	Range	Units
μ^ℓ	1	1–10	Pa-sec
$\Delta\rho$	500		kg/m^3
g	10		m/sec^2
ϕ_0	0.01	0.003–0.03	
k_0	10^{-12}	10^{-13}–10^{-11}	m^2
η_0	10^{19}	10^{18}–10^{20}	Pa-sec
ζ_0	10^{21}	10^{18}–10^{22}	Pa-sec
w_0	5×10^{-7}	10^{-9}–10^{-5}	m/sec
δ_0	10^4	10^2–10^5	m

the liquid viscosity and density difference are also well constrained. To calculate the reference Darcy speed w_0 we also need a characteristic porosity ϕ_0 and permeability k_0. These are not as well constrained. Seismic tomography indicates that porosities are of order 1% whereas interpretations of uranium-series disequilibria and Iceland deglacial magmatism suggest values about an order of magnitude smaller. The characteristic permeability must be large to satisfy these same constraints by diffuse Darcy flow (as illustrated in chapter 11). To constrain the permeability on the basis of laboratory experiments requires a knowledge of the mantle grain size. This isn't well known and, moreover, the square of its value appears in the permeability. Grain sizes in the range of 0.5–5 mm are associated with two orders of magnitude variation in mantle permeability. In table 6.1 we choose a moderate value of k_0 with the assumption that melt speeds are boosted from $w_0 \approx 1$ m/yr by reactive channelization of melt transport (chapter 12) and thereby become consistent with observational constraints.

The compaction length in the partially molten asthenosphere also suffers from a large uncertainty. In part, this is due to our poor constraints on the shear viscosity of the asthenosphere. Postglacial rebound estimates typically are of order 10^{21} Pa-sec, but these average the viscosity over a depth range of the entire upper mantle, a range that may include sharp viscosity variations (e.g., in the lithosphere). Inferences from post-seismic stress relaxation suggest much lower values of order 10^{17} Pa-sec in the months immediately following the earthquake, but these are likely associated with transient deformation rather than sustained flow. An even larger uncertainly however, is associated with the compaction viscosity. This property is notoriously difficult to measure in the laboratory, especially at small porosity. Theoretical models can make predictions, but these are sensitive to assumptions about the style of grain-scale deformation that accomodates compaction.

If compaction is accommodated by mechanical deformation of grains (section 5.3.2), this suggests that $\zeta_\phi \propto \eta/\phi$, where η is the shear viscosity of the aggregate at zero porosity. Assuming a constant of proportionality of order unity, we determine that $\zeta_\phi \gg \eta_\phi$ and that the compaction length is controlled by the compaction viscosity $\delta_0 \sim \sqrt{\zeta_0 k_0/\mu^\ell}$.

If, instead, compaction is accomodated by diffusion creep (matter diffusion from high-stress to low-stress patches on grains[2]), then the scaling with porosity is $\zeta_\phi \propto -\eta \ln \phi$. As with ϕ^{-1}, $\ln \phi$ is also singular for $\phi \to 0$, but the singularity is weaker. For a constant of proportionality of $\mathcal{O}(1)$ and a porosity of 1%, the compaction viscosity is only a factor of about five times greater than the shear viscosity. For Coble creep at larger porosity, the ratio of compaction to aggregate shear viscosity may be independent of porosity and approximately 5/3. Hence, assuming $\zeta_\phi \sim \eta_\phi$ as a lower bound, we have $\delta_0 \sim \sqrt{2\eta_0 k_0/\mu^\ell}$.

The range of possible values of the compaction length in table 6.1 accounts for this unfortunate lack of empirical constraints. Assuming that the inverse-porosity dependence is correct and $\phi_0 = 0.01$, we expect a reference compaction length that is of order 10 km. This should be considered to be an upper bound.

It is important to note that the parameters and scales discussed in this section pertain to the partially molten asthenosphere sufficiently far from sharp permeability changes (e.g., beneath the cold lithosphere) that a reference state can be defined. However, the compaction length is an important control on the size of such boundary layers, and reference values for an outer region may not be applicable in a region that is near to the boundary. They are thus indicative but not definitive and care must be taken to assess the validity of these estimates in specific applications.

6.7 Literature Notes

The first paper to recognize compaction as a physical process in the magma/mantle system was M^c^Kenzie [1984], who introduced the concept and definition of the compaction length. The equations he derived were immediately recognized as permitting solitary wave solutions and a surge of interest in this area ensued.

Early treatments of magmatic solitary waves by Richter and M^c^Kenzie [1984] and Scott and Stevenson [1984] contained some basic results, but the landmark study is Barcilon and Richter [1986]. The treatment given here is largely drawn from Spiegelman [1993b]. Scott et al. [1986], Olson and Christensen [1986], and Helfrich and Whitehead [2006] showed that solitary waves appear in laboratory experiments on conduits of buoyant fluid in a more viscous fluid.

The instability of lower-dimensional solitary waves in higher-dimensional spaces was first considered by Barcilon and Lovera [1989], who showed the existence of distinct forms at each dimension. Three-dimensional numerical models illustrating this instability and evolution were presented by Wiggins and Spiegelman [1995].

The mathematical properties of the solitary-wave problem with porosity-dependent compaction viscosity were systematically explored in a broad parameter space by Simpson et al. [2007] (see also Richard et al. [2012]). Later, Simpson and Spiegelman [2011] formulated high-accuracy solutions suitable for numerical benchmarking of simulation

[2] This diffusion can occur through the interior of the grains, which is called Nabarro–Herring creep and is discussed in section 5.3.2. Or it can occur along the grain boundaries and through the melt, which is called Coble creep. The latter is the more efficient pathway, especially when melt is present.

codes. A mathematical treatment of the wave-train envelope is given by Marchant and Smyth [2005].

Spiegelman [1993b] provides a thorough investigation of solitary waves and wave trains. That work touches on the ZCL approximation and shock waves, the permeability-step problem, the role of melting and freezing, and other topics. It includes a two-dimensional analysis of the linearized governing equations that simplifies to our one-dimensional treatment in section 6.3. A similarly comprehensive overview of the dynamics is given by Ricard et al. [2001].

Viscoelastic-plastic models of porosity waves were developed and analyzed by Connolly and Podladchikov [1998], Connolly and Podladchikov [2007], and Yarushina et al. [2015]. Porosity waves in the context of damage rheology were studied by Cai and Bercovici [2013] and Cai and Bercovici [2016] in one and two dimensions, respectively. In Hier-Majumder et al. [2006], the authors considered the effect of surface tension on solitary waves.

Compaction may be an important process in the formation of adcumulate rocks, formed at the base of a magma chamber where crystals settle onto a pile. McKenzie [2011] modeled the adcumulates of the Skaergaard intrusion with compaction theory, but this interpretation has been disputed on the basis of textural evidence by Holness et al. [2017].

Compaction has also been studied in the context of gas transport through a volcanic conduit [Michaut et al., 2009; Bercovici and Michaut, 2010]. Solitary wave solutions are obtained and may explain cyclic features of eruptions [Michaut et al., 2013]. A distinct and interesting feature of this two-phase system is the large compressibility of the gas phase.

Compaction of partially molten rocks has physical similarities to consolidation of sediments [Buscall and White, 1987] and pressure filtration of wood pulp [Hewitt et al., 2016] that result in mathematical similarities among the models. Hewitt et al. [2016] provides a comprehensive study of the finite-time compaction press problem and explores the roles of permeability and compaction viscosity in determining the form of the solution.

6.8 Exercises

6.1 Using (4.46a), (6.1b), and (6.26), derive the expression for the segregation flux (6.27).

6.2 Consider a layer of partially molten rock of thickness $\mathcal{H}$ and initial porosity ϕ_0. The layer is contained between two plates, one of which is permeable (at $z=-\mathcal{H}$) and the other is impermeable (at $z=0$). Solid is moving through the permeable barrier at nondimensional speed W_0/w_0. Neglect body forces.

(a) Solve for the compaction rate at $t=0$.
(b) Show that if $\delta \gg \mathcal{H}$, the compaction rate is uniform across the layer.
(c) Show that if $\delta \ll \mathcal{H} \to \infty$, your solution from the first part of the problem becomes equal to that for the infinite half-space in (6.9).

6.3 Consider a three-layered column of partially molten rock occupying $z \in (-\infty, \infty)$. The column consists of a finite layer of porosity $\phi = f_2\phi_0$ occupying $|z| \leq \mathcal{H}/2$, bounded by two semi-infinite layers of porosity $\phi = f_1\phi_0$. The problem can be treated as one-dimensional; multipliers f_i are constant and uniform.

(a) Noting that $\phi(-z)=\phi(z)$, by making the transformation $z \to -z$, use (6.16) and (6.21) to show that the compaction rate is antisymmetric, i.e., that, $\mathcal{C}(-z)=-\mathcal{C}(z)$.
(b) Determine the initial compaction rate.
(c) By making the substitution $\xi = z - \mathcal{H}/2$ and taking the limit $\mathcal{H} \to \infty$, verify that the solution given by (6.20) and (6.23) is recovered.

6.4 Consider the initial compaction of a uniform-porosity layer in the half-space $z \geq 0$. The layer sits on an impermeable surface at $z=0$ on which the velocities $\boldsymbol{v}^s = W\hat{\boldsymbol{z}} = \boldsymbol{0}$ and $\boldsymbol{v}^\ell = w\hat{\boldsymbol{z}} = \boldsymbol{0}$. The layer compacts under its own weight.[3]

(a) Using the extended Boussinesq approximation and assuming uniform porosity, show that

$$\phi \frac{\partial w}{\partial z} = -(1-\phi)\frac{\partial W}{\partial z}.$$

Hence, using (6.1), deduce that

$$\frac{\partial^3 W}{\partial z^3} = \frac{1}{\delta_0^2}\frac{\partial W}{\partial z}, \tag{6.56a}$$

$$\left(\zeta_\phi + \tfrac{4}{3}\eta_\phi\right)\frac{\partial^2 W}{\partial z^2} = -\frac{\mu\phi}{k_\phi}(w-W) + (1-\phi)\Delta\rho g. \tag{6.56b}$$

Note: The factor $(1-\phi)$ *appears in* (6.56b) *because, for this question, we choose to not approximate* $(1-\phi)\approx 1$ *(though we still require the porosity to be below disaggregation).*
(b) Determine the initial solid and liquid velocity profiles. Hence, find the initial rate of change of porosity.

6.5 (a) By neglecting all gravitational forces and assuming that both k_ϕ and μ^ℓ are constant, use (4.47) to derive the following equations coupling $\boldsymbol{v}^s$ and P:

$$\nabla\cdot\left(\boldsymbol{v}^s - \frac{k_\phi}{\mu^\ell}\nabla P\right) = 0, \tag{6.57a}$$

$$-\nabla P + (\zeta_\phi + \tfrac{4}{3}\eta_\phi)\nabla\left(\nabla\cdot\boldsymbol{v}^s\right) - \eta_\phi\nabla\times\left(\nabla\times\boldsymbol{v}^s\right) = 0. \tag{6.57b}$$

(b) Use the length and velocity scalings given by (6.29) together with an appropriate scaling for the pressure to obtain the nondimensional form of equations (6.57a) and (6.57b). Your answer should include the dimensionless ratio of viscosities $\nu_\phi = \zeta_\phi/\eta_\phi + 4/3$.
(c) Hence, show that the solid vorticity $\boldsymbol{\omega}^s = \nabla\times\boldsymbol{v}^s$ satisfies the harmonic equation, and that $\mathcal{C}$ satisfies the Helmholtz equation

$$-\nabla^2\mathcal{C} + \mathcal{C} = 0. \tag{6.58}$$

6.6 In this problem we determine the velocity and pressure fields induced by a point injection of melt of strength Q.

[3] This calculation was proposed by McKenzie [1984, Appendix B.].

(a) Given that $G(r)=\frac{1}{4\pi r}$ satisfies $-\nabla^2 G(r)=\delta^{(3)}(\boldsymbol{x})$ (where $r=|\boldsymbol{x}|$ and $\delta^{(3)}(\boldsymbol{x})$ is the three-dimensional Dirac delta function), show that $\mathcal{C}(r)=G(r)e^{-r}$ satisfies (6.58) for $r>0$.

For irrotational flows it can be shown that the solid velocity has the form

$$\boldsymbol{v}^s=\boldsymbol{\nabla}\,(\chi+\mathcal{C}),$$

where χ is a harmonic function. For the remainder of the question, take $\chi=\frac{A}{r}$ and $\mathcal{C}=\frac{B}{r}e^{-r}$, where A and B are constants.

(b) Determine the velocity field $\boldsymbol{v}^s$ and verify that $\boldsymbol{\omega}^s=\boldsymbol{0}$. Hence, use the dimensionless form of equation (6.57b) to determine the pressure field P.

(c) It can be shown that the mean stress tensor has the form

$$\overline{\boldsymbol{\sigma}}=\frac{2}{\nu_\phi}\left(-\mathcal{C}\boldsymbol{I}+\frac{1}{2}\left(\boldsymbol{\nabla}\boldsymbol{v}^s+(\boldsymbol{\nabla}\boldsymbol{v}^s)^{\intercal}\right)\right).$$

Use your solution to part (b) to show that

$$\overline{\boldsymbol{\sigma}}\cdot\frac{\boldsymbol{x}}{r}=\frac{4}{\nu_\phi}\left(\frac{A}{r^2}+\frac{Be^{-r}}{r^2}(1+r)\right)\frac{\boldsymbol{x}}{r^2}.$$

A sphere of melt with radius R forms due to a point source of strength Q. The boundary conditions associated with this system at $r=R$ are given by

$$\boldsymbol{v}^s-\boldsymbol{\nabla}P=\frac{Q\boldsymbol{x}}{4\pi R^3},$$
$$\overline{\boldsymbol{\sigma}}\cdot\frac{\boldsymbol{x}}{R}=-P\frac{\boldsymbol{x}}{R}.$$

(d) Use the above boundary conditions in part (c) to determine the constants A and B.

(e) By taking the limit $R\to 0$, obtain the velocity and pressure fields induced by a point injection of melt of strength Q.

6.7 Use equations (6.46), (6.47), and (6.48) to derive the implicit solution for the shape described by equation (6.49).

Hint: Use partial fractions, keeping in mind the form of $Z(\Lambda)$. Note that

$$(\Lambda-1)=\left((\Lambda^*-1)^{1/2}-(\Lambda^*-\Lambda)^{1/2}\right)\left((\Lambda^*-1)^{1/2}+(\Lambda^*-\Lambda)^{1/2}\right).$$

6.8 In the limit $(\Lambda^*-\Lambda)/(\Lambda^*-1)\ll 1$, we can invert (6.49) to obtain a first-order solution of the form $\Lambda=\Lambda^*-AZ^m$. Determine the constants A and m.

CHAPTER 7

Porosity-Band Emergence under Deformation

In the preceding chapter we considered a one-dimensional system in which the solid phase underwent compaction but not shear deformation. This allowed us to explore the relationships among magma buoyancy, permeability, interphase pressure differences, and compaction rate. In the current chapter, we consider how compaction can also be driven by shear of the solid phase. Two-dimensional, pattern-forming behavior emerges and we elucidate the role of the shear viscosity of the two-phase aggregate in shaping the dynamics.

Furthermore, the problems considered in this chapter allow us to make a comparison with laboratory experiments. These experiments, shown schematically in figure 7.1, are performed on small samples of mantle material at asthenospheric pressure and temperature (see caption of fig. 7.1 for details). Samples are prepared with a period of static annealing to ensure that they are close to textural equilibrium, with a roughly uniform initial porosity and no air bubbles.[1] Temperature is raised such that the samples partially melt ($\phi \approx 0.05$), with the melt residing in the pore space between grains. The samples are then subjected to shear deformation. The imposed strain rates ($\approx 10^{-4}$/s) are very rapid compared to those typical of mantle convection ($\approx 10^{-14}$/s). Upon reaching the targeted total strain, the samples are quenched, sectioned, and polished. Those that have undergone sufficiently large deformation have developed an anastomosing pattern of high-porosity sheets that are normal to the plane containing the principle axes of stress (the plane of the cuts seen in fig. 7.1). On these sections, the high-porosity sheets appear as high-porosity bands surrounded by compacted, low-porosity lenses. A robust feature of the pattern, observed in the experimental section in figure 7.2, is that the bands are oriented at an angle of $15° \pm 5°$ to the shear plane. These angles are observed in all experiments that produce bands and represent a challenge for models.

There are other measurable quantities associated with the emergent patterns in experiments. For example, the maximum porosity reached in bands saturates after a strain of approximately 2; band width and spacing varies with the rate of deformation; there is a radially inward flow of melt in torsion experiments (see the Literature Notes at the end of this chapter). The robust features of the experimental patterns provide a context for testing the validity of the governing equations of chapter 4 and, importantly, the material properties and constitutive relations in chapter 5. Hence, the laboratory results

[1] In the rock mechanics literature, the word *porosity* refers to the volume fraction of air bubbles, whereas the volume fraction of silicate melt is the *melt fraction*.

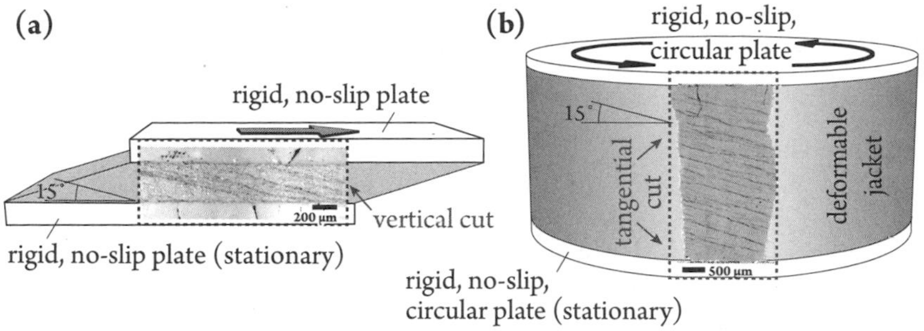

Figure 7.1. Schematic diagrams of laboratory experiments in which partially molten rock is subjected to imposed shear deformation. The shear plane is parallel to the top and bottom plates of the experiments. The solid phase is composed of olivine grains with sizes between 1 and 10 μm and, in some cases, a small fraction of added chromite grains; the liquid phase is typically basalt. **(a)** Simple-shear deformation geometry. The height of the experimental sample is about 0.5 mm. **(b)** Torsional deformation geometry. The height of the experimental sample is about 1 cm. Figure from Takei and Katz [2013] (© Cambridge University Press 2013; reprinted with permission). References to papers describing experimental materials, methods and results are provided in the Literature Notes at the end of this chapter.

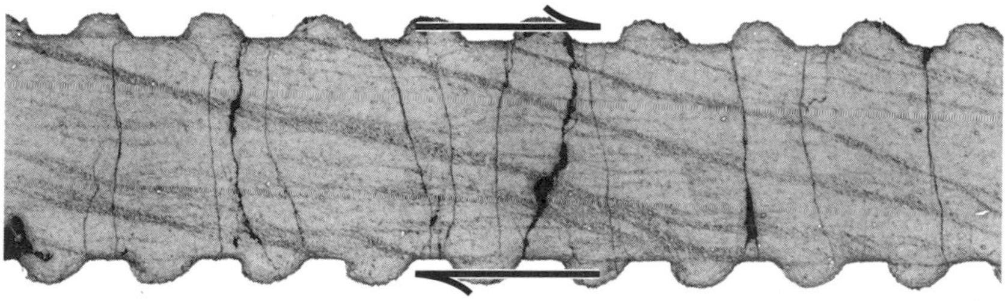

Figure 7.2. Cross section of a quenched experiment. Vertical extent is 0.5 mm. Subvertical black marks are cracks caused by quench of the experiment and should be ignored. Arrows show the imposed simple-shear deformation. A vector normal to the shear plane would point toward the top of the page. Dark bands have a larger fraction of quenched melt; lighter regions are compacted and contain little melt. The initial composition is 4 vol.% basaltic melt, evenly distributed in the pores of a matrix composed of 80% olivine and 20% chromite. Total shear strain is 3.4. Adapted from Katz et al. [2006].

are driving theoretical validation, refinement, and extension. But it is also worth noting that the instability giving rise to bands was predicted by Stevenson [1989] before it was recognized experimentally by Holtzman et al. [2003]. Although much theoretical progress has been made since the instability was discovered, aspects of these patterns are an ongoing challenge. This chapter provides an introduction to the approaches taken.

7.1 Governing Equations

Here we adapt the governing equations of magma dynamics to model the experimental system. In the experiments, the temperature is such that the basalt is molten and the olivine is solid. No melting takes place during the experiments, and freezing only occurs when the experiment is rapidly quenched. Hence, we can simplify the equations by imposing $\Gamma = 0$. Furthermore, because deformation in the experiments is rapid, we expect that segregation of melt is driven by compaction stresses and dynamic pressure gradients, rather than by buoyancy.

The importance of buoyancy can be quantified by the following scaling analysis. Shear stress scales as viscosity times the rate of shear strain $\eta_\phi \dot{\gamma}$ whereas buoyancy scales as $g\Delta\rho h_e$, where h_e is the height of the experiment. A dimensionless number that approximates the importance of buoyancy is then $\mathrm{B} \equiv g\Delta\rho h_e/\eta_\phi \dot{\gamma}$. Taking typical values for a torsional experiment of $\Delta\rho = 300$ kg/m^3, $h_e = 1$ cm, $\eta_\phi = 10^{12}$ Pa-s, and $\dot{\gamma} = 10^{-4}$/s, we obtain $\mathrm{B} = 3 \times 10^{-7} \ll 1$. Hence, it is a good approximation to neglect buoyancy of melt. A similar argument can be applied to the body force in the momentum equation for the two-phase aggregate. Note, however, that in the mantle where $\dot{\gamma} \approx 10^{-14}$/s, we expect $\mathrm{B} \sim \mathcal{O}(10^3)$ and hence a very different regime. We should therefore not expect the experimental results or the solutions obtained here to apply directly to the mantle.

With these considerations, equations (4.47) become

$$-\mathcal{C} + \boldsymbol{\nabla} \cdot K_\phi \boldsymbol{\nabla} P = 0, \tag{7.1a}$$

$$-\boldsymbol{\nabla} P + \boldsymbol{\nabla} \cdot 2\eta_\phi \dot{\boldsymbol{\varepsilon}} + \boldsymbol{\nabla} \zeta_\phi \mathcal{C} = \mathbf{0}, \tag{7.1b}$$

$$\frac{\partial \phi}{\partial t} + \boldsymbol{v} \cdot \boldsymbol{\nabla} \phi - (1-\phi)\mathcal{C} = 0, \tag{7.1c}$$

where, for concision, we have defined the *mobility* $K_\phi \equiv k_\phi/\mu^\ell$ and dropped superscripts indicating solid and liquid phases. These will be restored where needed to avoid ambiguity. We can eliminate pressure from equation (7.1a) by substituting equation (7.1b). And we can then eliminate pressure from equation (7.1b) by taking the curl (recall that $\boldsymbol{\nabla} \times \boldsymbol{\nabla} q = 0$ for any scalar quantity q). We then have

$$-\mathcal{C} + \boldsymbol{\nabla} \cdot K_\phi \left[\boldsymbol{\nabla} \cdot 2\eta_\phi \dot{\boldsymbol{\varepsilon}} + \boldsymbol{\nabla} \zeta_\phi \mathcal{C}\right] = 0, \tag{7.2a}$$

$$\boldsymbol{\nabla} \times \boldsymbol{\nabla} \cdot 2\eta_\phi \dot{\boldsymbol{\varepsilon}} = \mathbf{0}, \tag{7.2b}$$

$$\dot{\phi} + \boldsymbol{v} \cdot \boldsymbol{\nabla} \phi - (1-\phi)\mathcal{C} = 0, \tag{7.2c}$$

where we have written the partial derivative of porosity with respect to time as $\dot{\phi}$.

7.2 Linearized Governing Equations

The system of equations (7.2) above is nonlinear and coupled, making analytical solutions difficult or impossible to obtain. However, we infer that, because temperature and pressure are constant throughout the sample, the aggregate viscosity can depend only on porosity and strain rate. Since these are constrained by the solution to the system of PDEs, the system is closed by prescribing the variation of viscosity with porosity and strain rate. Such closures, described in chapter 5, are nonlinear, adding to the nonlinearity already present through the porosity–permeability relationship. Nonetheless, the system (7.2) could be solved numerically to expose its behavior.

To gain mechanical insight, we develop an approximate analysis here. Our strategy is to investigate the emergence of the band pattern from a nominally uniform porosity field. When the porosity field is truly uniform, our model should predict a uniform, steady flow with no compaction or porosity change; this becomes a reference state. When the porosity field is very close to the reference state with only a small perturbation, the governing equations can be linearized with respect to the reference state to derive equations that govern the perturbation. This system of equations, because it is linear, will admit normal-mode solutions. Our analysis will quantify the growth or decay of the normal modes as a function of their characteristics (exercise 3.5 applies this method to convection). This is called *linearized stability analysis.*

We begin by expanding all variables as the power series

$$\phi = \phi^{(0)} + \epsilon\phi^{(1)} + \mathcal{O}(\epsilon^2), \tag{7.3a}$$

$$\mathcal{C} = \mathcal{C}^{(0)} + \epsilon\mathcal{C}^{(1)} + \mathcal{O}(\epsilon^2), \tag{7.3b}$$

$$\boldsymbol{v} = \boldsymbol{v}^{(0)} + \epsilon\boldsymbol{v}^{(1)} + \mathcal{O}(\epsilon^2), \tag{7.3c}$$

$$\dot{\boldsymbol{\varepsilon}} = \dot{\boldsymbol{\varepsilon}}^{(0)} + \epsilon\dot{\boldsymbol{\varepsilon}}^{(1)} + \mathcal{O}(\epsilon^2), \tag{7.3d}$$

$$K_\phi = K_\phi^{(0)} + \epsilon K_\phi^{(1)} + \mathcal{O}(\epsilon^2), \tag{7.3e}$$

$$\eta_\phi = \eta_\phi^{(0)} + \epsilon\eta_\phi^{(1)} + \mathcal{O}(\epsilon^2) \tag{7.3f}$$

$$\zeta_\phi = \zeta_\phi^{(0)} + \epsilon\zeta_\phi^{(1)} + \mathcal{O}(\epsilon^2). \tag{7.3g}$$

In these expansions, because $\epsilon \ll 1$, each term is much smaller than the previous one and therefore we have $1 \gg \epsilon \gg \epsilon^2 \gg \ldots$. The *leading-order* or *zeroth-order terms*, which are the first entries in the series, represent the reference state. We also refer to the reference state as the *base state*. We make one critical assumption about the base state: that it has a spatially uniform porosity, compaction rate, and deviatoric strain rate. Hence by construction, $\phi^{(0)}, \dot{\boldsymbol{\varepsilon}}^{(0)}$, and $\mathcal{C}^{(0)}$ are all independent of position (and hence so are $\eta_\phi^{(0)}, \zeta_\phi^{(0)}$, and $K_\phi^{(0)}$).

The next step in analyzing the equations is to substitute the expansions (7.3) into the governing equations (7.2) and balance terms at the zeroth and then the first order of ϵ. The first-order equations will tell us how the system responds to the first-order perturbation to the base state.[2] We won't be concerned with higher orders of ϵ (e.g., $\mathcal{O}(\epsilon^2)$).

Substituting (7.3) into (7.2), requiring uniformity of $\phi^{(0)}, \dot{\boldsymbol{\varepsilon}}^{(0)}$, and $\mathcal{C}^{(0)}$, and balancing terms at leading order gives

$$\mathcal{C}^{(0)} = 0, \tag{7.4a}$$

$$\nabla \times \nabla \cdot \dot{\boldsymbol{\varepsilon}}^{(0)} = \mathbf{0}, \tag{7.4b}$$

$$\dot{\phi}^{(0)} - \left(1 - \phi^{(0)}\right)\mathcal{C}^{(0)} = 0. \tag{7.4c}$$

[2]If this feels like magic, consider that in the case of zero perturbation, the leading-order terms in the series expansion must represent a solution to the governing equations. But linear(ized) equations allow for superposition of solutions and hence, the perturbation terms, when they are nonzero, must independently balance.

The first of these equations states that there is no compaction at leading order. Using this result, the third equation then tells us that the base-state porosity is constant. The second equation places a constraint on the form of the base-state solid flow field. However, we had already assumed a uniform base-state strain rate (velocity that varies linearly with position such that $\nabla \boldsymbol{v}^{(0)}$ is uniform) and hence equation (7.4b) is satisfied by construction.

Returning to our substitution of the expansion in equations (7.3) into the governing equations (7.2) and balancing terms at first order in ϵ, we have

$$-\delta_0^2 \nabla^2 \mathcal{C}^{(1)} + \mathcal{C}^{(1)} = 2K_\phi^{(0)} \dot{\boldsymbol{\varepsilon}}^{(0)} : \nabla\nabla \eta_\phi^{(1)}, \tag{7.5a}$$

$$-\eta_\phi^{(0)} \nabla \times \nabla \cdot \dot{\boldsymbol{\varepsilon}}^{(1)} = \nabla \times \left(\dot{\boldsymbol{\varepsilon}}^{(0)} \cdot \nabla \eta_\phi^{(1)} \right), \tag{7.5b}$$

$$\dot{\phi}^{(1)} + \boldsymbol{v}^{(0)} \cdot \nabla \phi^{(1)} = \left(1 - \phi^{(0)}\right) \mathcal{C}^{(1)}, \tag{7.5c}$$

where

$$\delta_0 = \left[K_\phi^{(0)} \left(\zeta_\phi^{(0)} + \tfrac{4}{3} \eta_\phi^{(0)} \right) \right]^{1/2} \tag{7.6}$$

is the base-state, reference compaction length. Here we have used the assumption of a linear velocity field to eliminate terms in $\nabla \cdot \dot{\boldsymbol{\varepsilon}}^{(0)}$. Also note the use of dyadic notation,

$$\dot{\boldsymbol{\varepsilon}}^{(0)} : \nabla\nabla \eta_\phi^{(1)} \equiv \dot{\varepsilon}_{ij}^{(0)} \frac{\partial}{\partial x_i} \frac{\partial}{\partial x_j} \eta_\phi^{(1)},$$

where the Einstein summation convention applies to the right-hand side. Since both i and j are repeated indices with implied sums, the full expression reduces to a scalar value (which it must, for consistency with eqn. (7.5a)).

Motivated by the experimental configuration shown in figure 7.1a, we assume two-dimensionality of the flow and choose a Cartesian coordinate system aligned such that all variations are in the x–y plane and all z-derivatives are zero (the x–y plane is assumed to be parallel to the section shown in figs. 7.1a and 7.2). We then expand the first-order velocity with a Helmholtz decomposition (as we did in section 4.6.3, where the associated viscosity was also constant),

$$\boldsymbol{v}^{(1)} = \nabla \mathcal{U} + \nabla \times (\psi \hat{\boldsymbol{z}}). \tag{7.7}$$

Here ψ can be identified as the streamfunction of the incompressible part of the perturbation velocity. After substitution of (7.7) into (7.5b) and laborious simplification (exercise 7.2) we obtain

$$\hat{\boldsymbol{z}} \eta_\phi^{(0)} \nabla^4 \psi = 2\nabla \times \left(\dot{\boldsymbol{\varepsilon}}^{(0)} \cdot \nabla \eta_\phi^{(1)} \right). \tag{7.8}$$

This equation is opaque but governs the perturbations to the solid flow associated with the perturbed viscosity field.

7.3 Viscosity

To go further, we examine the details of the shear viscosity and its expansion. Following from the general viscous flow law of equation (5.23) with constant temperature and pressure, we adopt a model in which the viscosity depends on porosity and strain rate,

$$\eta_\phi = \eta_\phi^{(0)} \exp\left[-\tfrac{\lambda}{\mathrm{n}}\left(\phi - \phi^{(0)}\right)\right]\left(\dot{\varepsilon}_{II}/\dot{\varepsilon}_{II}^{(0)}\right)^{-\mathcal{N}}, \tag{7.9}$$

where $\mathcal{N} \equiv (\mathrm{n}-1)/\mathrm{n}$ and we have used the base-state solution to normalize variations. This is a nonlinear relationship that cannot be used directly; it must be linearized for our purposes.

To that end, the second invariant of the strain-rate tensor is expanded as

$$\dot{\varepsilon}_{II} = \dot{\varepsilon}_{II}^{(0)} + \epsilon\dot{\varepsilon}_{II}^{(1)} + \mathcal{O}(\epsilon^2). \tag{7.10}$$

Then, substituting (7.3a) and (7.10) into (7.9), expanding nonlinear terms in Taylor series, and collecting terms at each order we have (see exercise 7.3)

$$\eta_\phi = \eta_\phi^{(0)}\left(1 - \epsilon\frac{\lambda\phi^{(1)}}{\mathrm{n}} - \epsilon\frac{\mathcal{N}\dot{\varepsilon}_{II}^{(1)}}{\dot{\varepsilon}_{II}^{(0)}}\right) + \mathcal{O}(\epsilon^2). \tag{7.11}$$

This equation linearly relates viscosity variations to perturbations in porosity and deviatoric strain rate. Below, it is combined with the first-order system (7.5) to eliminate $\eta_\phi^{(1)}$.

7.4 The (In)Stability of Perturbations

In this section we analyse the linearized model of perturbation growth/decay under two linear flows: pure shear and simple shear. In both cases, motivated by the observation of porosity bands in experiments, our perturbation will take the form of harmonic plane waves,

$$\phi^{(1)}(\boldsymbol{x}, t) \propto \mathrm{e}^{i\boldsymbol{k}\cdot\boldsymbol{x}+s(t)}. \tag{7.12}$$

The orientation of these waves is described by an angle θ to the x-axis, as shown in figure 7.3. By describing the perturbations with complex exponential functions (eigenfunctions of the linear equations), we ensure that we can find solutions. Departing from a strict replication of experiments, we choose an infinite domain to avoid specification of boundary conditions. In this context, the question of stability is rigorously expressed as: do the harmonic porosity perturbations exponentially grow or decay? To answer this, we solve for the exponential growth rate $\dot{s}$ under each of the two base-state flows.

7.4.1 PURE SHEAR

Pure shear is the simplest linear flow and was used by Stevenson [1989] to demonstrate the instability to growth of high-porosity bands (sheets in three dimensions).

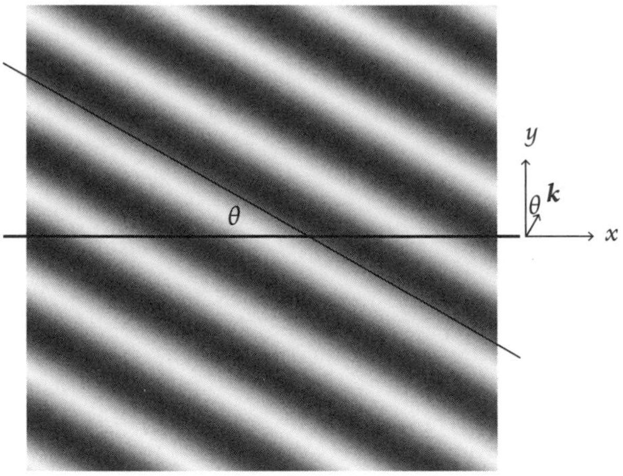

Figure 7.3. Schematic diagram showing harmonic perturbation with wavevector $\boldsymbol{k}$. The wavevector is oriented at an angle θ to the y-axis such that the lines of constant phase-angle make an angle θ to the x-axis. The pattern is assumed to extend to infinity in all directions. The shear plane is defined by its normal, $\hat{\boldsymbol{y}}$.

It is described by the velocity field and strain-rate tensor

$$\left.\begin{aligned} \boldsymbol{v}^{(0)} &= \dot{\gamma}\,(x\hat{\boldsymbol{x}} - y\hat{\boldsymbol{y}}) \\ \dot{\boldsymbol{\varepsilon}}^{(0)} &= \dot{\gamma}\begin{pmatrix} 1 & 0 \\ 0 & -1 \end{pmatrix} \end{aligned}\right\} \quad \text{pure shear.} \tag{7.13}$$

Note that the flow is extending in the x-direction, contracting in the y-direction, and has zero rotation.

For this analysis we'll take $\mathrm{n} = 1$, corresponding to Newtonian viscosity. Hence, from equation (7.11), we have $\eta_\phi^{(1)} = -\eta_\phi^{(0)}\lambda\phi^{(1)}$. Using this, we can compute two required quantities

$$\dot{\boldsymbol{\varepsilon}}^{(0)} : \boldsymbol{\nabla}\boldsymbol{\nabla}\eta_\phi^{(1)} = -\lambda\eta_\phi^{(0)}\dot{\gamma}\left(\frac{\partial^2\phi^{(1)}}{\partial x^2} - \frac{\partial^2\phi^{(1)}}{\partial y^2}\right), \tag{7.14a}$$

$$\boldsymbol{\nabla}\times\left(\dot{\boldsymbol{\varepsilon}}^{(0)}\cdot\boldsymbol{\nabla}\eta_\phi^{(1)}\right) = 2\dot{\gamma}\eta_\phi^{(0)}\lambda\frac{\partial^2\phi^{(1)}}{\partial x\partial y}\hat{\boldsymbol{z}}. \tag{7.14b}$$

Substituting these into (7.5) and (7.8), we obtain

$$-\delta_0^2\nabla^2\mathcal{C}^{(1)} + \mathcal{C}^{(1)} = -2\lambda K_\phi^{(0)}\eta_\phi^{(0)}\dot{\gamma}\left(\frac{\partial^2\phi^{(1)}}{\partial x^2} - \frac{\partial^2\phi^{(1)}}{\partial y^2}\right), \tag{7.15a}$$

$$\eta_\phi^{(0)}\nabla^4\psi = 4\lambda\eta_\phi^{(0)}\dot{\gamma}\frac{\partial^2\phi^{(1)}}{\partial x\partial y}, \tag{7.15b}$$

$$\dot{\phi}^{(1)} = \left(1 - \phi^{(0)}\right)\mathcal{C}^{(1)}. \tag{7.15c}$$

The first of these, (7.15a), is the compaction equation and governs isotropic deformation (compaction); the second equation is derived from the equation for two-phase Stokes and governs shear deformation; the third is an evolution equation for porosity. In the latter, we have neglected the advection of porosity for simplicity. We will reintroduce this term when we consider simple shear, below.

To solve this system we require the form of the porosity perturbation. We propose that

$$\phi^{(1)} = Ae^{\dot{s}t}\begin{cases}\sin(2\pi x/l), & \text{for } \theta = \pi/2\\ \sin(2\pi y/l), & \text{for } \theta = 0,\end{cases} \tag{7.16}$$

where A is a constant and uniform prefactor of $\mathcal{O}(1)$, $\dot{s}$ is the growth rate, and l is the wavelength of oscillation. These represent wavefronts that are perpendicular and parallel, respectively, to the extensional direction $\hat{\boldsymbol{x}}$. Substituting equation (7.16) into equation (7.15c) we find that the compaction rate is related to the porosity perturbation as

$$\mathcal{C}^{(1)} = \dot{s}\phi^{(1)}/\left(1-\phi^{(0)}\right). \tag{7.17}$$

Substituting (7.17) into the compaction equation (7.15a), defining and substituting $\nu_\phi^{(0)} \equiv \zeta_\phi^{(0)}/\eta_\phi^{(0)} + 4/3$, and solving for $\dot{s}$ gives

$$\dot{s} = \pm 2\left(1-\phi^{(0)}\right)\frac{\lambda\dot{\gamma}}{\nu_\phi^{(0)}}\frac{(2\pi\delta_0/l)^2}{1+(2\pi\delta_0/l)^2}, \tag{7.18}$$

where the + sign applies for $\theta = \pi/2$ (perpendicular to the extension) and the − sign applies for $\theta = 0$ (parallel to the extension). This tells us that if the partially molten solid has extension across the weak, higher-porosity bands, the bands will grow. By "grow" we mean that liquid moves from regions of lower porosity to regions of higher porosity. This causes the higher-porosity regions to increase their porosity, making them weaker—a self-reinforcing feedback or *instability*. Mathematically, this appears as exponential growth of $\phi^{(1)}$ with time.

Given this explanation, we might expect the reference permeability to appear in equation (7.18), as is it controls the Darcy drag experienced during liquid segregation. Instead we find the compaction viscosity (within ν_ϕ), indicating that the resistance to compaction/decompaction is a control on the growth rate. We can make sense of this by returning to equation (7.1b) and hypothesizing a stress balance between the incompressible shear flow acting on gradients of the viscosity and gradients in the compaction rate. In particular, we postulate that $\nabla P \sim 0$ and that

$$2\dot{\boldsymbol{e}}^s \cdot \nabla\eta_\phi \sim -[\zeta_\phi + 4\eta_\phi/3]\nabla\mathcal{C}, \tag{7.19}$$

where $\dot{\boldsymbol{e}}^s$ is the symmetric part of the solid velocity gradient (the full strain-rate tensor). Assuming $\theta = \pi/2$, we have $\nabla\eta_\phi = \hat{\boldsymbol{x}}\partial\eta_\phi/\partial x = \hat{\boldsymbol{x}}\mathrm{d}\eta_\phi/\mathrm{d}\phi \times \partial\phi/\partial x$. And using mass conservation we have $\mathcal{C} \sim \dot{\phi} \sim \dot{s}\phi$, where we have assumed $(1-\phi) \sim 1$ and neglected advection. Then, using these in equation (7.19), with the Newtonian shear viscosity of equation (7.9), we obtain

$$2\dot{\gamma}\eta_0\lambda\frac{\partial\phi}{\partial x} \sim \dot{s}[\zeta_0 + 4\eta_0/3]\frac{\partial\phi}{\partial x}. \tag{7.20}$$

This asymptotic approximation can be simplified to give $\dot{s} \sim 2\dot{\gamma}\lambda/\nu_\phi$. Comparing this result to (7.18), we see that our hypothesis can explain the growth-rate prefactor. The permeability is also a control, however, and it appears via the reference compaction length in (7.18). We discuss this next, in terms of the perturbation wavelength, which is measured in comparison to the compaction length.

The wavelength dependence of the growth rate, plotted in figure 7.4, remains to be explained. If the wavelength is much greater than the compaction length, i.e., $l \gg \delta_0$, then $(2\pi\delta_0/l)^2 \ll 1$. In this case, equation (7.18) states that $|\dot{s}| \sim 0$. This is consistent with what we learned in chapter 6: that variations of porosity on a length scale much larger than the compaction length do not generate compaction stresses. The other extreme case is $l \ll \delta_0$, where the perturbation wavelength is smaller than the compaction length. In this case, $|\dot{s}| \sim 2\left(1-\phi^{(0)}\right)\lambda\dot{\gamma}/\nu_\phi^{(0)}$, consistent with the asymptotic result in equation (7.20). So our theory predicts that interphase pressure differences become important when porosity variations have a length scale that is similar to or smaller than the compaction length. A more subtle point is that the liquid pressure is (asymptotically) uniform whereas the solid pressure varies spatially.

Figure 7.4 shows that the growth rate is highest and constant for wavelength $l \to 0$. This is a model prediction that all wavelengths that are sufficiently smaller than the compaction length grow at the same rate, including vanishingly small wavelengths. But as the wavelength approaches the grain size, we would expect some physical mechanism to reduce the growth rate of the instability. Our result suggests that some important physics is missing from our model. We discuss a wavelength-limiting mechanism below in section 7.5.

7.4.2 SIMPLE SHEAR

A pure-shear flow is illustrative but it isn't a close match to the laboratory experiments. In those, the overall pattern of flow is close to simple shear,[3] another linear flow that is given by

$$\left.\begin{aligned} \boldsymbol{v}^{(0)} &= \dot{\gamma} y \hat{\boldsymbol{x}} \\ \dot{\boldsymbol{\varepsilon}}^{(0)} &= \frac{\dot{\gamma}}{2}\begin{pmatrix} 0 & 1 \\ 1 & 0 \end{pmatrix} \end{aligned}\right\} \quad \text{simple shear.} \tag{7.21}$$

A rotation of the coordinate system by 45° counterclockwise would diagonalize the strain-rate tensor, aligning it with the directions of extension and contraction. Unlike pure shear, however, the rotation matrix for simple shear is not zero. We shall see the effects of this rotation below.

Assuming the full non-Newtonian, porosity-dependent viscosity from equation (7.9), we have $\eta_\phi^{(1)} = -\eta_\phi^{(0)}\left(\lambda\phi^{(1)}/\mathrm{n} + \mathcal{N}\dot{\varepsilon}_{II}^{(1)}/\dot{\varepsilon}_{II}^{(0)}\right)$. Substituting this into equations (7.5) and (7.8) we obtain

$$-\delta_0^2\nabla^2\mathcal{C}^{(1)} + \mathcal{C}^{(1)} = -2K_\phi^{(0)}\eta_\phi^{(0)}\dot{\gamma}\left[\frac{\partial^2}{\partial x\partial y}\left(\frac{\lambda}{\mathrm{n}}\phi^{(1)} + \frac{\mathcal{N}}{\dot{\varepsilon}_{II}^{(0)}}\dot{\varepsilon}_{II}^{(1)}\right)\right], \tag{7.22a}$$

[3] Torsional flow is like simple shear wrapped around a cylinder.

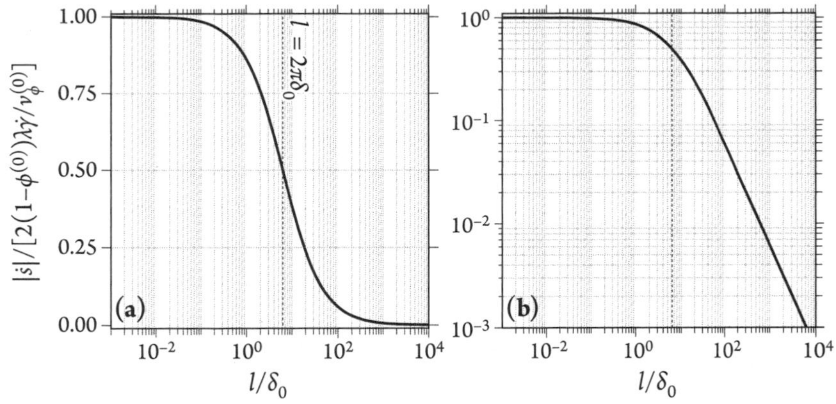

Figure 7.4. Dispersion curves showing normalized growth rate of perturbations versus perturbation wavelength. Note that the growth rate is positive when porosity perturbations are oriented perpendicularly to the direction of extension and negative when they are parallel. **(a)** On log–linear axes. **(b)** On log–log axes.

$$\nabla^4\psi = -\dot{\gamma}\left[\left(\frac{\partial^2}{\partial x^2} - \frac{\partial^2}{\partial y^2}\right)\left(\frac{\lambda}{n}\phi^{(1)} + \frac{\mathcal{N}}{\dot{\varepsilon}_{II}^{(0)}}\dot{\varepsilon}_{II}^{(1)}\right)\right], \tag{7.22b}$$

$$\dot{\phi}^{(1)} + \dot{\gamma}y\frac{\partial\phi^{(1)}}{\partial x} = \left(1-\phi^{(0)}\right)\mathcal{C}^{(1)}. \tag{7.22c}$$

Note that in (7.22c) we have included the advection term on the left-hand side, which describes transport of the porosity perturbations by the base-state (simple shear) flow.

Porosity advection by the base-state. To understand the effect of the advection term in equation (7.22c), it is helpful to consider the homogeneous problem

$$\dot{\phi}^{(1)} + \dot{\gamma}y\frac{\partial\phi^{(1)}}{\partial x} = 0. \tag{7.23}$$

This is a transport equation and has the general solution

$$\phi^{(1)}(\boldsymbol{x},t) = f[x - (\dot{\gamma}y)t,\ y], \tag{7.24}$$

where $f(x,y)$ is the initial porosity distribution. For the analysis ahead, it is convenient to define the initial porosity perturbation as

$$\phi^{(1)}(\boldsymbol{x},0) = A\mathrm{e}^{i\boldsymbol{k}_0\cdot\boldsymbol{x}}, \tag{7.25}$$

where A is again a constant and uniform prefactor of $\mathcal{O}(1)$ and $\boldsymbol{k}_0 = \boldsymbol{k}(t=0) = k_{0x}\hat{\boldsymbol{x}} + k_{0y}\hat{\boldsymbol{y}}$ is the initial wavevector. According to equation (7.24), and in the absence of compaction (for now), this condition evolves as

$$\phi^{(1)}(\boldsymbol{x},t) = A\exp\left[i(k_{0x}(x-\dot{\gamma}yt) + k_{0y}y)\right], \tag{7.26a}$$

$$= A\exp\left[i\boldsymbol{k}(t)\cdot\boldsymbol{x}\right]. \tag{7.26b}$$

Comparison of (7.26) with (7.24) indicates that

$$\boldsymbol{k}(t) = k_{0x}\hat{\boldsymbol{x}} + (k_{0y} - \dot{\gamma} k_{0x} t)\,\hat{\boldsymbol{y}}. \tag{7.27}$$

This is a time-dependent wavevector that rotates with the flow and changes length.

Figure 7.5(a) shows a wave front with wavevector $\boldsymbol{k}$ at time t. It makes an angle to the shear plane of

$$\theta(t) = \tan^{-1} \frac{k_{0x}}{k_{0y} - \dot{\gamma} k_{0x} t}. \tag{7.28}$$

A plot of band angle from equation (7.28) as a function of progressive shear strain (on the y-axis) is shown in figure 7.5(b). We shall refer to the curves on this plot as *passive advection trajectories* because they represent the evolution of the angle of a hypothetical band that is being advected by simple-shear flow, but that is not modifying the flow. It is important to note the difference between rotation due to simple shear and that due to solid-body rotation. For solid-body rotation, all initial wavefront angles change at a uniform rate. In contrast, figure 7.5(b) shows that wave fronts that start at a low angle rotate slowly whereas those at high angles (approaching 90°) rotate most rapidly.

Growth of porosity bands when $\mathfrak{n} = 1$. Just as for pure shear, we expect porosity perturbations to grow if they undergo extension across their wave fronts. We also expect perturbations to be rotated along passive advection trajectories according to the wavevector defined in (7.27). Hence, we define the porosity perturbation as

$$\phi^{(1)}(\boldsymbol{x}, t) = A \exp\left[i\boldsymbol{k}(t) \cdot \boldsymbol{x} + s(t)\right], \tag{7.29}$$

where $s(t)$ is the exponential amplitude and $\dot{s}$ is the growth rate. Note that we consider all possible angles between wave fronts and the shear plane. This is in contrast to our analysis of pure shear, where we considered only two angles.

Substituting equation (7.29) into equation (7.22c) we again find that $\mathcal{C}^{(1)} = \dot{s}\phi^{(1)}/\left(1 - \phi^{(0)}\right)$, for pure shear (note that this is because of our careful choice of $\boldsymbol{k}(t)$). Substituting this result into (7.22a) with $\mathfrak{n} = 1$ and $\mathcal{N} = 0$, and rearranging for $\dot{s}$, gives

$$\dot{s} = 2\left(1 - \phi^{(0)}\right) \frac{\lambda\dot{\gamma}}{\nu_\phi^{(0)}} \cdot \frac{\delta_0^2 k_x k_y}{1 + (\delta_0 k)^2}, \tag{7.30}$$

where $k^2 = \boldsymbol{k} \cdot \boldsymbol{k} = |\boldsymbol{k}|^2$. Using $k_x = |\boldsymbol{k}| \sin\theta$, $k_y = |\boldsymbol{k}| \cos\theta$, and a double-angle formula, (7.30) can be written as

$$\dot{s} = \left(1 - \phi^{(0)}\right) \frac{\lambda\dot{\gamma}}{\nu_\phi^{(0)}} \cdot \frac{(\delta_0 k)^2}{1 + (\delta_0 k)^2} \sin 2\theta. \tag{7.31}$$

Note the similarity to equation (7.18), which tells us that we have the same dependence on wavelength as in figure 7.4. Now, however, we take the angle of perturbations into account and find an angular dependence of $\sin 2\theta$.

In figure 7.6, we plot equation (7.31) for perturbation wavelengths that give a saturated growth rate (i.e., those with $\delta_0 k \gg 1$). The peak growth rate is at 45°, the angle at which perturbation wave fronts are oriented normal to the direction of most rapid

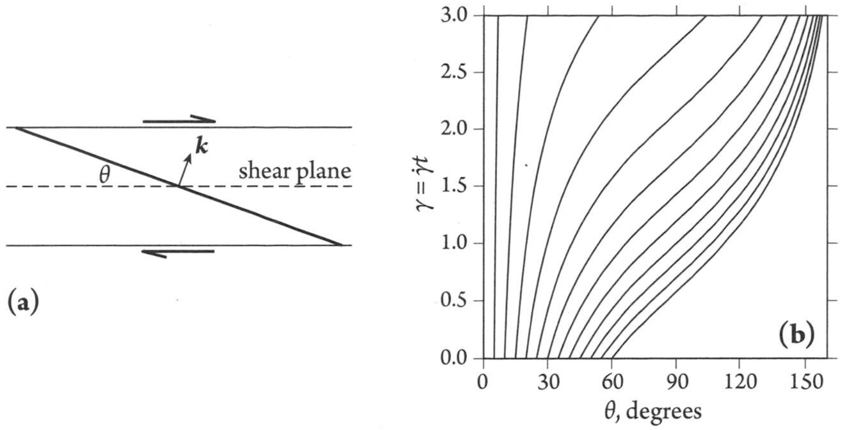

Figure 7.5. Porosity advection by the base-state simple-shear flow. **(a)** Schematic diagram normal to the shear plane showing a representative high-porosity band oriented at an angle θ with normal $\boldsymbol{k}$. The band is being rotated to higher θ by the simple-shear flow. **(b)** Passive advection trajectories that track the evolution of θ with progressive shear strain γ.

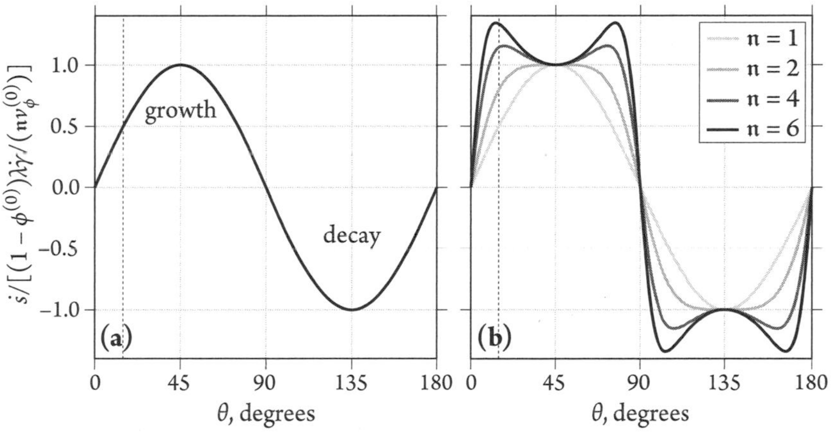

Figure 7.6. Normalized growth rate of small-wavelength ($l \ll \delta_0$) porosity bands under simple shear. **(a)** Newtonian viscosity. **(b)** Non-Newtonian viscosity with various values of $\mathfrak{n}$. The vertical dotted lines mark $\theta = 15°$.

extension for simple shear. This relationship is consistent with what we found in the case of pure shear.

Growth of porosity bands when $\mathfrak{n} \geq 1$. Having analyzed the equations for simple shear with Newtonian viscosity, we now consider the case where $\mathfrak{n} \geq 1$ and hence viscosity may be non-Newtonian. This couples the strain rate, via the viscosity, to the compaction rate. So whereas for the previous examples we have not solved the biharmonic equation governing ψ, in this case we'll need ψ to compute $\dot{\varepsilon}_{II}^{(1)}$. We'll simplify this calculation slightly by assuming that the compaction rate itself doesn't affect the viscosity—hence, we don't need the compaction potential $\mathcal{U}$ in the strain-rate invariant.

Then, using $\dot{\gamma} y \hat{\mathbf{x}} + \epsilon \nabla \times (\psi \hat{\mathbf{z}})$ as the effective solid velocity, we find that (exercise 7.4),

$$\dot{\varepsilon}_{II} = \frac{\dot{\gamma}}{2} + \frac{\epsilon}{2}\left(\frac{\partial^2 \psi}{\partial y^2} - \frac{\partial^2 \psi}{\partial x^2}\right) \tag{7.32}$$

to an accuracy that is first order in ϵ. From this result we deduce that

$$\frac{\dot{\varepsilon}_{II}^{(1)}}{\dot{\varepsilon}_{II}^{(0)}} = \frac{1}{\dot{\gamma}}\left(\frac{\partial^2 \psi}{\partial y^2} - \frac{\partial^2 \psi}{\partial x^2}\right). \tag{7.33}$$

Because our system of equations is linear, we can assume that the eigenfunction solutions for each of the fields are linearly related to each other by (possibly complex) coefficients. Hence, considering ψ as an example, we can take $\psi = \hat{\psi}\phi^{(1)}$, where $\hat{\psi}$ is an unknown, constant, complex coefficient. The perturbation to viscosity is then written as

$$\eta_\phi^{(1)} = -\eta_\phi^{(0)}\left[\frac{\lambda}{\mathrm{n}} + \frac{\mathcal{N}\hat{\psi}}{\dot{\gamma}}\left(k_y^2 - k_x^2\right)\right]\phi^{(1)}. \tag{7.34}$$

Note the difference between the porosity-weakening term, with no dependence on the wavenumber, and the shear-weakening term, which includes k_x and k_y. The form in which the wavenumber appears, $k_y^2 - k_x^2$, indicates that, for a fixed $k_y \neq 0$, weakening is greatest when $k_x = 0$—i.e., for porosity perturbations that are parallel to the shear plane.

The equations governing the perturbation quantities (7.22) become

$$\dot{s}\left(\delta_0^2 k^2 + 1\right) = 2\left(1 - \phi^{(0)}\right) K_\phi^{(0)} \eta_\phi^{(0)} k_x k_y \left[\frac{\lambda\dot{\gamma}}{\mathrm{n}} + \hat{\psi}\mathcal{N}\left(k_y^2 - k_x^2\right)\right], \tag{7.35a}$$

$$\hat{\psi} k^4 = \left(k_y^2 - k_x^2\right)\left[\frac{\lambda\dot{\gamma}}{\mathrm{n}} + \hat{\psi}\mathcal{N}\left(k_y^2 - k_x^2\right)\right]. \tag{7.35b}$$

Eliminating $\hat{\psi}$ from these equations and writing wavenumber components in terms of the angle of porosity bands gives the growth rate as

$$\dot{s} = \left(1 - \phi^{(0)}\right)\frac{\lambda\dot{\gamma}}{\mathrm{n}\upsilon_\phi^{(0)}} \cdot \frac{(\delta_0 k)^2}{1 + (\delta_0 k)^2}\left(\frac{\sin 2\theta}{1 - \mathcal{N}\cos^2 2\theta}\right). \tag{7.36}$$

The systematics with wavenumber are exactly as we saw for Newtonian viscosity: magnitude of growth rate goes to zero when the wavelength of perturbations is much larger than the compaction length. The systematics with band angle have changed, though. Recall that $\mathcal{N} \equiv (\mathrm{n} - 1)/\mathrm{n}$; since $\mathrm{n} \geq 1$, we have $0 \leq \mathcal{N} < 1$. Therefore, when the viscosity is non-Newtonian, the denominator is reduced near the maxima of $\cos^2 2\theta$: at $0°$ and $90°$. Note also that the denominator never becomes negative. Figure 7.6 illustrates that the consequence of non-Newtonian viscosity is a split in the growth-rate peak toward lower and higher angles. For sufficiently non-Newtonian viscosity ($\mathrm{n} \sim 5$), the low-angle peak in growth rate is reduced to approximately $15°$, close to the mean angle observed in experiments. However, a typical value of the stress exponent for dislocation creep of olivine is 3.5. Moreover, in situ measurements from laboratory experiments indicate a stress exponent closer to 1.5 (see the Literature Notes).

The behavior of the non-Newtonian model can be readily understood in physical terms. As previously noted, the $\sin 2\theta$-dependence arises from the orientation of porosity perturbations with respect to the principal axes of the base-state strain-rate tensor (which is aligned to the stress tensor). The deviatoric tensile stress across bands is maximized when the bands are oriented perpendicularly to the direction of maximum extension rate. The $\cos^2 2\theta$ term has an origin in the strain-rate weakening of the medium. The porosity bands are inherently weaker by virtue of their higher porosity and can therefore localize deformation. Under non-Newtonian viscosity, localized deformation weakens the two-phase aggregate and reduces its resistance to extension. Deformation is most effectively localized onto bands when they are aligned parallel to the shear plane (or to the conjugate shear at 90° to the shear plane). However, at 0° and 90°, $\sin 2\theta$ is zero and there is no extension across bands. Equation (7.36) thus represents a balance, controlled by $\mathcal{N}$, between orientation for maximal extension and for maximal shear.

Growth-rate curves for large $\mathfrak{n}$ in figure 7.6(b) show an obvious symmetry between low (~15°) and high (~75°) angle bands. In experiments such as that shown in figure 7.2, the absence of a high-angle set of bands is striking. A possible means of reconciling this disagreement emerges when we consider finite strain.

Extending the analysis to finite strain. The method of linearized stability analysis is strictly valid only when the perturbations to the base state are small, i.e., when $\epsilon \exp s(t) \ll 1$. This is true by construction at $t = 0$. It should persist to finite strain, depending on the initial amplitude of perturbations and the growth rate, but the limits of validity are unclear. In this case, it is informative to put aside such concerns and proceed boldly by integrating $\dot{s}$ over time to obtain $s(t)$. This integration is complicated by the fact that the band angle θ is a function of time.

Restricting our focus to the case of wavelengths much smaller than the compaction length, we can obtain s by integrating

$$s(\theta, t) = \left(1 - \phi^{(0)}\right) \frac{\lambda \dot{\gamma}}{\mathfrak{n} \nu_\phi^{(0)}} \int_0^t \left(\frac{\sin 2\theta(t')}{1 - \mathcal{N} \cos^2 2\theta(t')} \right) \mathrm{d}t', \tag{7.37}$$

where t' is a dummy variable of integration, to distinguish it from the (variable) upper limit of integration, t. Evaluation of this integral requires specification of $\theta(t = 0)$ and the use of equation (7.28) to compute $\theta(t)$. The integral in (7.37) is evaluated numerically. Results are shown in figure 7.7 as a function of shear strain γ and angle θ for $\mathfrak{n} = 1$ in panel (a) and $\mathfrak{n} = 6$ in panel (b).

A comparison of panels in figure 7.7 emphasizes the difference between Newtonian and non-Newtonian viscosity. At a strain of 3, the linearized analysis predicts bands at about 70° for Newtonian viscosity and 18° for $\mathfrak{n} = 6$. The difference arises both from the difference in the spectrum of growth rate (fig. 7.6) as well as the differential rotation rate of bands, shown by the dotted passive advection trajectories. These trajectories also explain the asymmetry between low and high-angle bands in figure 7.7(b). Low-angle bands rotate slowly and hence remain in an optimal orientation for growth; in contrast, high-angle bands are rapidly rotated out of a favorable orientation toward angles where the growth rate is negative. This geometric result is specific to simple shear, but applies for any choice of rheological law.

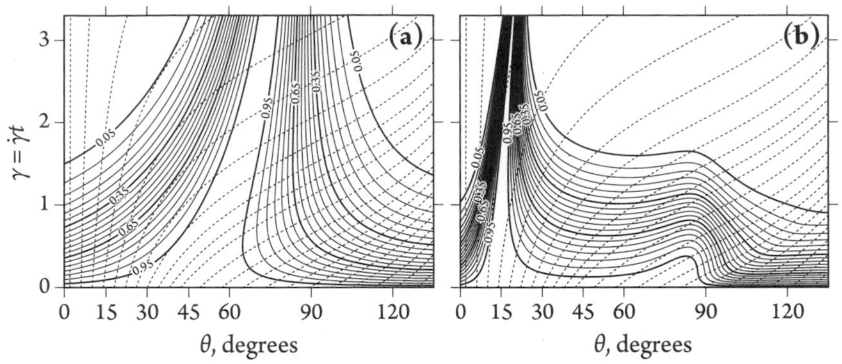

Figure 7.7. Amplitude of porosity perturbations $e^{s(t)}$ as a function of angle and strain $\gamma = \dot{\gamma}t$. The amplitude has been normalized at each level of strain by the maximum amplitude at that strain. Solid lines are equally spaced growth-rate isopleths going from zero (black) to one (white). Dotted curves are passive advection trajectories from equation (7.28). **(a)** Newtonian viscosity ($n = 1$, $\mathcal{N} = 0$). **(b)** Non-Newtonian viscosity ($n = 6$, $\mathcal{N} = 5/6$).

7.5 Wavelength Selection by Surface Tension

Previous sections show that for a viscosity of the two-phase aggregate that weakens with porosity, the system is linearly unstable to harmonic perturbations of porosity. This may explain the existence of high-porosity bands in experiments (e.g., fig. 7.2). However, as we showed in figure 7.4, the stability analysis presented so far does not help to understand the characteristic spacing or width of bands. Instead it predicts the growth rate at all wavelengths smaller than the compaction length. The bands in experiments are clearly broader than the scale of solid grains, so we can infer that physics is missing from the governing equations.

One hypothesis for the force that provides mode selection, *regularizing* the growth-rate spectrum, is *capillarity*. Pore-scale capillarity arises from surface tension acting within a medium that has both liquid–solid (pore walls) and solid–solid interfaces (grain boundaries). When the surface energy of solid–solid interfaces is larger than that of liquid–solid interfaces, the dihedral angle is small and melt is pulled onto grain faces (see section 5.1.2). This process is spontaneous and reduces the surface energy per unit volume. At the continuum level, capillarity effectively manifests as a pressure gradient driving melt from higher to lower porosity. This can be shown to reduce the maximum growth rate of porosity bands, but does so independently of wavenumber (exercise 7.8). A more detailed consideration of capillarity shows that it drives flow up gradients in curvature of porosity, $\nabla(\nabla^2\phi)$. Bercovici and Rudge [2016] discuss details of the physics and mathematics of this process, which we review here in simplified form.

An effective model that captures mode selection by surface tension is

$$-\mathcal{C} + \nabla\cdot K_\phi\left[\nabla\cdot 2\eta_\phi\dot{\boldsymbol{\varepsilon}} + \nabla\zeta_\phi\mathcal{C} - \mathcal{S}\nabla\left(\nabla^2\phi\right)\right] = 0, \tag{7.38}$$

where $\mathcal{S}(\phi) = (1-\phi)\gamma/\mathcal{A}$ is proportional to the surface tension γ on the interface between phases divided by the interfacial area density $\mathcal{A}$. Hence, the last term in (7.38) is a pressure gradient arising from surface tension. In particular, it represents a gradient in surface tension across a *diffuse interface* that separates a region of higher porosity from a region of lower porosity. We have neglected other surface tension-related terms that are not responsible for mode selection.[4]

Linearizing equation (7.38) using the expansions in equations (7.3) gives

$$-\delta_0^2\nabla^2\mathcal{C}^{(1)}+\mathcal{C}^{(1)}=K_\phi^{(0)}\left(2\dot{\boldsymbol{\varepsilon}}^{(0)}:\nabla\nabla\eta_\phi^{(1)}-\mathcal{S}^{(0)}\nabla^4\phi^{(1)}\right). \tag{7.39}$$

Taking simple shear as the base-state flow with a viscosity model given by equation (7.34) we obtain

$$\dot{s}\frac{(\delta_0 k)^2+1}{\left(1-\phi^{(0)}\right)}=K_\phi^{(0)}\left[2\eta_\phi^{(0)}k_xk_y\left(\frac{\lambda\dot{\gamma}}{\mathrm{n}}+\hat{\psi}\mathcal{N}\left(k_y^2-k_x^2\right)\right)-\mathcal{S}^{(0)}k^4\right], \tag{7.40}$$

where $\hat{\psi}$ is determined, as before, by (7.35b). Eliminating $\hat{\psi}$ and solving for $\dot{s}$ gives

$$\dot{s}=\left(1-\phi^{(0)}\right)\frac{\lambda\dot{\gamma}}{\mathrm{n}\nu_\phi^{(0)}}\cdot\frac{(\delta_0 k)^2}{1+(\delta_0 k)^2}\left[\left(\frac{\sin 2\theta}{1-\mathcal{N}\cos^2 2\theta}\right)-\mathrm{D}_\mathcal{I}\left(\delta_0 k\right)^2\right], \tag{7.41}$$

where

$$\mathrm{D}_\mathcal{I}=\frac{\gamma\mathrm{n}\left(1-\phi^{(0)}\right)}{\dot{\gamma}\lambda\,\delta_0^2\mathcal{A}^{(0)}\eta_\phi^{(0)}} \tag{7.42}$$

is a dimensionless number expressing the effectiveness of the diffuse interface force at homogenizing porosity against the porosity-localizing force of the shear stress.

Equation (7.41) has an important difference from its predecessor, equation (7.36): the second term, which is negative, causes the growth rate to decrease at sufficiently large wavenumber. This decrease is independent of band angle θ and viscosity exponent. To isolate and understand the behavior of the growth rate including this term, we can take $\mathrm{n}=1$, $\mathcal{N}=0$, and $\theta=45°$ and define dimensionless symbols $k_*=\delta_0 k$ and $\dot{s}_*=\dot{s}/\left[\left(1-\phi^{(0)}\right)\lambda\dot{\gamma}/(\nu_\phi^{(0)}\mathrm{n})\right]$ to give (see exercise 7.5)

$$\dot{s}_*=\frac{k_*^2\left(1-\mathrm{D}_\mathcal{I}\,k_*^2\right)}{k_*^2+1}. \tag{7.43}$$

The asymptotic behavior is clear: for $k_*\to\infty$, we find that $\dot{s}_*\sim 1-\mathrm{D}_\mathcal{I}\,k_*^2$ whereas for $k_*\to 0$, we obtain $\dot{s}_*\sim k_*^2$. Since both of these are diminishing, this dispersion relationship has a maximum at finite wavenumber. A plot of growth rate versus wavelength for various values of $\mathrm{D}_\mathcal{I}$ is given in figure 7.8.

The location of the peak growth rate is attained when $\mathrm{d}\dot{s}_*/\mathrm{d}k_*=0$. Using this stationary point to solve for the fastest-growing mode as a function of $\mathrm{D}_\mathcal{I}$ leads to

$$k_*^{\max}=\sqrt{(1+1/\mathrm{D}_\mathcal{I})^{1/2}-1}, \tag{7.44a}$$

$$\dot{s}_*^{\max}=1-2\mathrm{D}_\mathcal{I}\left(\sqrt{1+1/\mathrm{D}_\mathcal{I}}-1\right). \tag{7.44b}$$

[4]One of these two terms, representing simple capillarity, would appear in equation (7.38) as $\nabla\cdot\nabla\phi$ and hence modifies only the absolute rate of the instability; the other is nonlinear and enters the analysis at $\mathcal{O}(\epsilon^2)$. See exercise 7.8 and Bercovici and Rudge [2016] for full details.

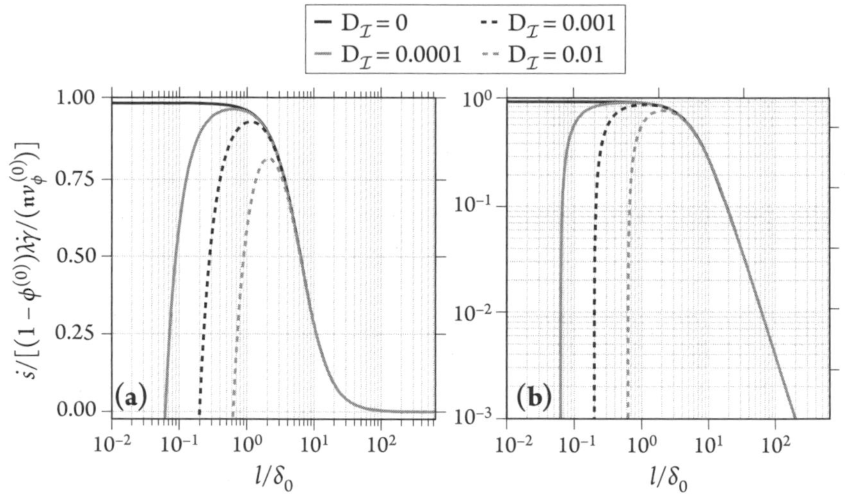

Figure 7.8. Growth rate of porosity bands versus wavelength, calculated with equation (7.43). Increasing values of D$_\mathcal{I}$ represent an increasing strength of surface-tension driven segregation. **(a)** Log-linear plot. **(b)** Log-log plot.

The second of these equations, for $\dot{s}_*^{\max}$, is the result of substituting equation (7.44a) into equation (7.43).

The wavelength and rate of growth of the fastest-growing (and hence dominant) perturbation are plotted as a function of D$_\mathcal{I}$ in figure 7.9. For D$_\mathcal{I}$ of about 10^{-3}, the dominant mode has a wavelength that is approximately one compaction length. This mode grows almost as fast as it would in the absence of surface-energy driven flow. The wavelength of the dominant mode decreases with diminishing importance of surface tension, but only very slowly: $l^{\max}/\delta_0$ remains of $\mathcal{O}(1)$ even when D$_\mathcal{I} \approx 10^{-4}$. Figure 7.9(b) shows that the growth rate of the dominant perturbation drops off sharply when surface tension dampens wavelengths up to and larger than the compaction length. The remaining, longer-wavelength modes are inherently more stable (fig. 7.4) and so $\dot{s}_*^{\max}$ drops off sharply with D$_\mathcal{I}$.

7.6 Literature Notes

After the pioneering work of Stevenson [1989] first recognizing the instability, there were two other early models that preceded an experimental demonstration: Richardson [1998] included the role of buoyancy and Hall and Parmentier [2000] considered the influence of water content.

The Kohlstedt laboratory at the University of Minnesota has been responsible for effectively all of the rock deformation work that has followed. The first experimental study to demonstrate the existence of the instability was Holtzman et al. [2003], which deformed samples in simple-shear geometry and added chromite grains to the starting material to reduce the compaction length. In a follow-up, Holtzman and Kohlstedt [2007] varied the applied stress and strain rate in a series of experiments, showing effects on band thickness and spacing. The systematics of these results have not been explained quantitatively.

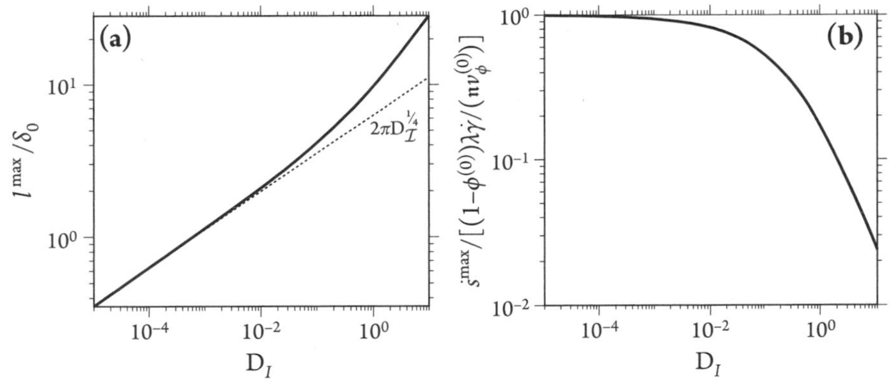

Figure 7.9. Characteristics of the dominant perturbation as a function of $D_{\mathcal{I}}$. **(a)** Wavelength with the largest growth rate $l^{\max}/\delta_0$. In the limit of $D_{\mathcal{I}} \to 0$, the asymptote is $k_*^{\max} \sim D_{\mathcal{I}}^{-1/4}$. **(b)** Growth rate of the dominant perturbation $s_*^{\max}$.

King et al. [2010] was the first to use torsion to deform the sample, allowing for a larger sample size and greater total strain. The results of these experiments in terms of band angle and spacing are consistent with previous experimental results. Qi et al. [2013] used torsion experiments with spherical inclusions to investigate the development of pressure shadows. King et al. [2011b] extended the torsion experiments to use bi-lithologic samples, showing that deformation can enhance reactive melt infiltration.

Theory has advanced in parallel with experiments. The study by Spiegelman [2003] extended the work of Stevenson [1989] to simple shear by proposing the time-dependent wave vector discussed above. Low-angle bands were not obtained by Spiegelman and Kelemen [2003], but were predicted in linearized analysis and numerical solutions by Katz et al. [2006], who included non-Newtonian viscosity but required $n \approx 6$—higher than expected for dislocation creep. Butler [2009] applied a similar analysis to pure shear and proposed a hypothesis for the evolution of band wavelength. Alisic et al. [2016] incorporated non-Newtonian viscosity in simulations of torsion experiments with spherical inclusions. That paper also developed a linearized stability analysis for a torsional base-state flow (absent inclusions).

An in situ experimental measurement of the strain exponent by King et al. [2010] ($n \approx 1.5$) cast doubt on the hypothesis of a large strain-rate exponent due to Katz et al. [2006]. A subsequent study by Rudge and Bercovici [2015] included a model for grain-size variation and interface roughness. They found that under diffusion creep (with Newtonian viscosity), the dynamic recrystallization mechanics create a high *effective* strain-rate dependence and attendant low-angle bands. But again, such a non-Newtonian viscosity is not measured experimentally.

Meanwhile, an alternative explanation was proposed to reconcile low-angle bands with Newtonian viscosity. Takei and Holtzman [2009a,b,c] derived and tested a model for anisotropic viscosity. The theory is based on a microscopic model of diffusion around an idealized, spherical grain that contacts neighboring grains on circular patches. Diffusion is fastest through the melt, so increased wetting of grain boundaries (and hence decreased contiguity of grains) speeds the diffusive response to stress. Anisotropy of grain wetting by melt (usually known as *melt-preferred orientation* (MPO)), which is quantified by the *contiguity tensor*, leads to anisotropy of the fast diffusion pathways, and hence anisotropy of the viscous response to stress. Takei and Holtzman [2009a] proposed and Takei [2010] confirmed experimentally that anisotropy of contiguity can arise when deviatoric stress is applied to the two-phase aggregate. The first numerical simulations that included anisotropic viscosity were presented by Butler [2012]. The theory was consolidated and thoroughly studied by Takei and Katz [2013] and Katz and Takei [2013], which considered simple shear, torsional and Poiseuille flow of the aggregate (for Poiseuille flow see also Allwright and Katz [2014]). Two novel predictions arose: (*i*) the existence of base-state melt segregation toward the central axis in torsional deformation, which is not predicted to occur for any isotropic viscosity, and (*ii*) lower band angle when the anisotropy is allowed to evolve dynamically with the local stress tensor, even when the magnitude of anisotropy is diminished.

The existence of base-state melt segregation in torsion was documented experimentally by Qi et al. [2015], lending support to the theory that anisotropic viscosity is an explanation for the low-angle bands observed in experiments. Theoretical work by Takei and Katz [2015] provided a detailed explanation for the difference between static, imposed anisotropy and dynamic anisotropy. Progress in refining the theory of anisotropic viscosity will be based on a better understanding of the emergence and evolution of MPO, and in considering the role of anisotropic contiguity under dislocation creep.

That melt-preferred orientation has consequences for permeability is known from theoretical [e.g Hier-Majumder, 2011] and experimental [e.g Daines and Kohlstedt, 1997] studies. The consequences of anisotropic permeability for porosity-band formation under simple shear were considered by Taylor-West and Katz [2015].

At least two experimental papers have considered the effect of surface-tension driven flow erasing the band pattern [Parsons et al., 2008; King et al., 2011a]. Bercovici and Rudge [2016] were the first to show how this might stabilize the growth of porosity bands at small wavelength. The only other study to address this problem theoretically was Takei and Hier-Majumder [2009] (though Stevenson [1989] recognized the issue and suggested the grain scale to be a cut-off wavelength).

An open and important question is whether this mechanical instability is active in the mantle beneath mid-ocean ridges (or in other partially molten, deforming regions of the asthenosphere). Butler [2009] performed a stability analysis and numerical simulations and suggested that bands might form, but would align differently than in the laboratory case of negigible buoyancy. Various issues were left unresolved by Butler [2009]. These include the relatively low total strain achieved beneath mid-ocean ridges ($\lesssim 1$) and the strong effect of buoyancy, which would cause incipient bands to move vertically as porosity waves (sections 6.3 and 6.4). The issue of low total strain was considered by Gebhardt and Butler [2016], who embedded a stability analysis into a corner-flow model by assuming a separation of scales between the perturbation wavelength and the tectonic scale of corner flow. They showed that amplitude growth can be significant along corner-flow streamlines. The role of buoyancy was studied by Vestrum and Butler [2020] and shown to cause propagation of porosity bands as waves, without affecting their growth rate. But buoyancy-driven flow can lead to a reaction–infiltration instability (see chapter 12), which could interact with (or preempt) the emergence of shear-driven bands. Rees Jones et al. [2020] used the separation of scales proposed by Gebhardt and Butler [2016] to develop a combined analysis of both instability mechanisms. They predicted that the reactive instability is dominant over most of the melting regime, establishing the orientation of bands/channels. Their amplitude can be modified by shear-driven growth near the lithosphere–asthenosphere boundary.

7.7 Exercises

7.1 Find the strain rate tensor $\dot{\boldsymbol{\varepsilon}}^{(0)}$ and its divergence $\nabla \cdot \dot{\boldsymbol{\varepsilon}}^{(0)}$ for each of the following flows.

(a) Pure shear

$$\boldsymbol{v}^{(0)} = \dot{\gamma}\,(x\hat{\boldsymbol{x}} - y\hat{\boldsymbol{y}}).$$

(b) Simple shear

$$\boldsymbol{v}^{(0)} = \dot{\gamma} y \hat{\boldsymbol{x}}.$$

(c) Poiseuille flow

$$\boldsymbol{v}^{(0)} = \frac{G}{2\mu^{\ell}} y(h-y)\hat{\boldsymbol{x}}.$$

Note that G is a constant.

7.2 Verify equation (7.8), noting that ψ is independent of z.

7.3 By substituting (7.3a) and (7.10) into (7.9) and then performing a Taylor expansion, derive the linear approximation in (7.11).

7.4 For a simple shear with a small perturbation, the effective solid velocity is

$$\boldsymbol{v} = \dot{\gamma} y \hat{\boldsymbol{x}} + \epsilon \nabla \times (\psi \hat{\boldsymbol{z}}).$$

Verify the first-order approximation

$$\dot{\varepsilon}_{II} = \frac{\dot{\gamma}}{2} + \frac{\epsilon}{2}\left(\frac{\partial^2 \psi}{\partial y^2} - \frac{\partial^2 \psi}{\partial x^2}\right).$$

Recall that $\dot{\varepsilon}_{II} \equiv \sqrt{\dot{\boldsymbol{\varepsilon}} : \dot{\boldsymbol{\varepsilon}}/2}$.

7.5 Consider the scalings $k_* = \delta_0 k$ and $\dot{s}_* = \dot{s}/\left[\left(1-\phi^{(0)}\right)\lambda\dot{\gamma}/(v_{\phi}^{(0)}\mathfrak{n})\right]$.

(a) By rescaling (7.41) and setting $\mathfrak{n} = 1$, and $\mathcal{N} = 0$, and $\theta = 45°$, obtain (7.43).
(b) Find k_*^{max}, at which we achieve the maximum growth rate $\dot{s}_*^{max}$.

7.6 Extend the analysis of pure shear to perturbation of arbitrary angle θ.

(a) Determine the growth rate $\dot{s}$ for the first-order porosity perturbation $\phi^{(1)} = Ae^{i\boldsymbol{k}\cdot\boldsymbol{x}+\dot{s}t}$ in the case of pure shear flow, where A is constant and $\boldsymbol{k}\cdot\boldsymbol{x} = 2\pi\left(x\sin\theta + y\cos\theta\right)/l$.
(b) Verify that your solution is consistent with (7.18).
(c) Give a physical interpretation as to why marginal stability ($\dot{s} = 0$) occurs at $\theta = \pm\pi/4$.

7.7 Consider pure shear, but this time including the advection term in (7.1c),

$$\dot{\phi}^{(1)} + \dot{\gamma}\left(x\frac{\partial \phi^{(1)}}{\partial x} - y\frac{\partial \phi^{(1)}}{\partial y}\right) = \left(1-\phi^{(0)}\right)\mathcal{C}^{(1)}. \tag{7.45}$$

The homogeneous problem is

$$\dot{\phi}^{(1)} + \dot{\gamma}\left(x\frac{\partial \phi^{(1)}}{\partial x} - y\frac{\partial \phi^{(1)}}{\partial y}\right) = 0, \tag{7.46}$$

which admits a solution of the form

$$\phi^{(1)}(\boldsymbol{x}, t) = f(r, s), \tag{7.47}$$

where $r \equiv x\exp(-\dot{\gamma}t)$ and $s \equiv y\exp(\dot{\gamma}t)$.

(a) Verify that (7.47) satisfies (7.46).
(b) Assuming that the initial porosity perturbation is of the form (7.25), find the wavevector $\boldsymbol{k}$ as a function of time t. Also find the angle that the wave front makes with the shear plane.
(c) For the porosity perturbation of the form (7.29), use (7.45) to verify that the compaction rate satisfies (7.17).

7.8 Bercovici and Rudge [2016] propose a more complicated form for the pressure gradient arising from surface tension than that considered in section 7.5. Their formulation can be written as

$$\nabla P^{\ell}\Big|_{\text{surface tension}} = \mathcal{S}\left[\mathcal{A}\mathcal{A}''\nabla\phi - \nabla\left(\nabla^2\phi\right) + \mathcal{A}^{-1}\mathcal{A}'\left(\nabla^2\phi\right)\nabla\phi\right], \tag{7.48}$$

where a $\mathcal{A}'$ and $\mathcal{A}''$ are the first and second derivatives of $\mathcal{A}$ with respect to porosity. Repeat the stability analysis of section 7.5, showing how this modifies the growth rate.

CHAPTER 8

Conservation of Energy

We begin this chapter by deriving an equation for conservation of internal energy identical to that of McKenzie [1984]. The equation can be rewritten using other *thermodynamic potentials* that, in some contexts, are more appropriate or convenient. These transformations are achieved using the definitions of the potentials and the *Maxwell relations* of thermodynamics. The present chapter will be much more comprehensible to a reader who is familiar with these; they are covered in most thermodynamics textbooks.

In the first recasting of the conservation equation, we express the internal energy as enthalpy and obtain the enthalpy equation. Then, from the enthalpy formulation, we write an equation for the evolution of temperature. We lastly consider an equation for the evolution of entropy. In each of these cases, we neglect the contribution of the partial specific energies of chemical components in each phase (the chemical potentials).[1] Because the equations for internal energy, enthalpy, temperature, and entropy all arise from the principle of conservation of energy, there are important similarities between them that help make recognizable the physical meaning of their terms.

A discussion of the energetics of two-phase flow will, by necessity, consider the question of thermodynamic equilibrium between the solid and liquid phases. This question can be broken down into three key parts: thermal equilibrium, pressure equilibrium, and chemical equilibrium. Thermal equilibrium requires that that the temperatures of liquid and solid phases occupying the same representative volume element (RVE) are identical. Since temperature diffuses rapidly at the millimetric length scale of solid grains, thermal equilibrium between phases is an excellent assumption where melt transport is by porous flow.

The pressure, however, is a much more difficult question. It has been a fundmental assumption that, in general, liquid and solid phases have distinct pressures within the RVE; this pressure difference is what drives (de)compaction. Since the phases are (potentially) in equilibrium at their interface, the thermodynamic pressure should hold precisely there. What, then, is the *thermodynamic pressure* of equilibrium between the phases that would be measured at their interface? Various workers have taken different

[1]This assumption is considered problematic near the onset of melting, where the melt composition varies sharply with the degree of melting. Large changes in chemical potential in this context lead to large changes in the entropy of fusion Δs [e.g., Hirschmann et al., 1999]. We proceed with it nonetheless for simplicity of exposition, consistency with thermochemical approximations made elsewhere in this book, and our general focus on the fluid dynamics.

views on this question; these are reviewed briefly in the Literature Notes at the end of this chapter. One perspective is that there is, in general, disequilibrium between coexisting phases that is associated with differences in pressure and chemical potential. While this may be the most general assumption, it leads to the most complicated model of the energetics (and melting).

Differences in pressure between phases, however, are likely to be very small relative to the static overburden of rock. Consider, for example, the pressure difference associated with solitary waves. Assuming a small background porosity, we can take $1-\phi \sim 1$ and hence

$$\Delta P \sim \zeta_\phi \mathcal{C} \sim \phi_0 w_0 \zeta_0/\delta_0,$$

where we have taken $\mathcal{C} \sim \phi_0 w_0/\delta_0$ as a typical compaction rate associated with the passage of a solitary wave and ζ_0 as the compaction viscosity at porosity ϕ_0 in the asthenosphere. Supposing $\phi_0 = 0.01$, $w_0 = 1$ m/yr, $\zeta_0 = 10^{20}$ Pa-s, and $\delta_0 = 1$ km, we have $\Delta P \approx 3 \times 10^7$ Pa. At 30-km depth in the mantle, the lithostatic pressure is about 10^9 Pa (1 GPa). Hence, pressure differences between the solid and liquid are relatively small.

Nonetheless, because chemical diffusion at the grain scale is much slower than thermal diffusion, differences in chemical potential between phases may be significant. We do not consider this possibility in the current chapter (but see section 10.2).

To make the development of theory in this chapter more accessible, we assume thermodynamic equilibrium between coexisting phases holds. Moreover, we assume that it holds at an interface pressure that is approximately the liquid pressure P^ℓ. This result stems from the very low viscosity of the liquid relative to the solid. Then, having derived the equations governing conservation of energy (in terms of internal energy, enthapy, temperature, and entropy) on this basis, we further simplify by making two more key assumptions. The first is the Boussinesq approximation, which holds that densities are constant except in terms associated with the gravitational body force. The second is that the thermodynamic pressure, which controls the thermodynamic equilibrium, is equal to the lithostatic pressure. This means that dynamic pressures (including ΔP) are assumed to be negligible in the thermodynamics. We make a third, minor assumption by taking material properties of thermal expansivity, specific heat capacity, and thermal conductivity to be equal between phases.

To illustrate the physics of energy conservation we consider two applications at the end of the chapter. In section 8.6 we analyze a model in which a single-component solid (e.g., water ice) is at its melting temperature and undergoes deformation in a controlled experiment (as could be performed in a laboratory). The heat from viscous dissipation drives melting. In section 8.7 we consider mantle rock undergoing isentropic decompression, as might occur in an upwelling flow of a plume or beneath a mid-ocean ridge. The melting temperature decreases with material ascent; this drives melting.

8.1 The Internal-Energy Equation

The first law of thermodynamics states that the change in internal energy of a system E is equal to the sum of the net addition of heat Q, the net amount of work done on the system W, and the net addition of chemical potential energy. Neglecting the latter contribution and taking the derivative with respect to time, the first law is written

$$\frac{\mathrm{d}E}{\mathrm{d}t} = \frac{\mathrm{d}Q}{\mathrm{d}t} + \frac{\mathrm{d}W}{\mathrm{d}t}. \tag{8.1}$$

Applying the first law to the RVE (fig. 4.1), heat can be added to the system by advection, diffusion, and internal heat generation. Work can be done on the system by body forces (gravity) and by surface forces (stress). Instead of considering the total internal energy E, we model the volume-averaged internal energy per unit mass for each phase u^i. As we did in chapter 4, we start with a balance of microscopic properties integrated over the RVE. On the time scales of interest, we can assume that the solid and the liquid are in local thermal equilibrium: at any location in the two-phase continuum, they have the same temperature and internal energy. Hence we consider both phases simultaneously, $\Phi^s + \Phi^\ell = 1$, and write down an equation for the bulk conservation of internal energy. This is

$$\frac{\mathrm{d}}{\mathrm{d}t}\int_{\mathrm{RVE}} \breve{\rho}\breve{u}\,\mathrm{d}^3\breve{\boldsymbol{x}} = \int_{\mathrm{RVE}} \breve{\rho}\breve{\mathbb{Q}}\,\mathrm{d}^3\breve{\boldsymbol{x}} - \int_{\partial\mathrm{RVE}} \breve{\boldsymbol{J}}_u\cdot\mathrm{d}\boldsymbol{S} - \int_{\partial\mathrm{RVE}} \breve{\rho}\breve{u}\breve{\boldsymbol{v}}\cdot\mathrm{d}\boldsymbol{S} + \int_{\partial\mathrm{RVE}} (\breve{\boldsymbol{\sigma}}\cdot\breve{\boldsymbol{v}})\cdot\mathrm{d}\boldsymbol{S} + \int_{\mathrm{RVE}} \breve{\rho}\mathbf{g}\cdot\breve{\boldsymbol{v}}\,\mathrm{d}^3\breve{\boldsymbol{x}}, \quad (8.2)$$

where $\mathbb{Q}$ is the mass-specific radiogenic heating rate and $\boldsymbol{J}_u$ is the diffusive heat flux. The left-hand side represents the rate of change of internal energy. This change is caused by net additions of heat (first line on the right-hand side): radiogenic heat production, convergence of the diffusive heat flux, and convergence of advective heat transport. The second line of the right-hand side contains terms representing net work done on the system: stresses at the RVE surface and the gravitational body force in the RVE interior. Transfer across the interface between phases within the RVE does not enter this statement of bulk energy conservation for the two-phase aggregate.

Applying Gauss' theorem, dividing by δV, and allowing the arbitrary volume of the RVE to shrink to a point in the two-phase continuum, we obtain

$$\frac{\partial\overline{\rho u}}{\partial t} + \nabla\cdot\overline{\rho u\boldsymbol{v}} = \overline{\rho\mathbb{Q}} - \nabla\cdot\overline{\boldsymbol{J}_u} + \nabla\cdot\overline{\boldsymbol{\sigma}\cdot\boldsymbol{v}} + \overline{\rho\boldsymbol{v}}\cdot\mathbf{g} \quad (8.3)$$

(recall that $\overline{q} = \phi q^\ell + (1-\phi)q^s$). Then, using the conservation of mass equations (4.11), we can rewrite the left-hand side of (8.3),

$$\phi\rho^\ell\frac{\mathrm{D}_\ell u^\ell}{\mathrm{D}t} + (1-\phi)\rho^s\frac{\mathrm{D}_s u^s}{\mathrm{D}t} - \Gamma\Delta u = \overline{\rho\mathbb{Q}} + \nabla\cdot\overline{k_T}\nabla T + \nabla\cdot\overline{\boldsymbol{\sigma}\cdot\boldsymbol{v}} + \overline{\rho\boldsymbol{v}}\cdot\mathbf{g}. \quad (8.4)$$

On the right-hand side of (8.4) we have used Fourier's law of heat transport,

$$\boldsymbol{J}_u^i = -\phi^i k_T^i \nabla T, \quad (8.5)$$

for each phase (having already assumed that the phases are in local thermal equilibrium, $T^\ell = T^s \equiv T$). Consistent with previous usage, $\Delta u \equiv u^s - u^\ell$.

We now focus on simplifying the work terms in equation (8.4). These terms can be re-expressed by substitution of the dot product of the liquid and solid velocities with their respective equations from the system (4.19) to give

$$\begin{aligned}\nabla\cdot\overline{\boldsymbol{\sigma}\cdot\boldsymbol{v}} + \overline{\rho\boldsymbol{v}}\cdot\mathbf{g} &= \mathbf{F}\cdot\left(\boldsymbol{v}^\ell - \boldsymbol{v}^s\right) + \overline{\boldsymbol{\sigma}:\nabla\boldsymbol{v}},\\ &= -P^\ell\nabla\cdot\overline{\boldsymbol{v}} + \frac{\mu^\ell\phi^2}{k_\phi}\left(\boldsymbol{v}^\ell - \boldsymbol{v}^s\right)^2 + \zeta_\phi\mathcal{C}^2 + 2\eta_\phi\left(\dot{\boldsymbol{\varepsilon}}^s\right)^2,\end{aligned} \quad (8.6)$$

where the notation showing as squares of a vector and tensor are shorthand for $\boldsymbol{v}^2 \equiv \boldsymbol{v} \cdot \boldsymbol{v}$ and $\dot{\boldsymbol{\varepsilon}}^2 \equiv \dot{\boldsymbol{\varepsilon}} : \dot{\boldsymbol{\varepsilon}}$, respectively. The second line of (8.6) is obtained by using equations (4.32), (4.35), (4.40), and (4.44) (exercise 8.1). The squared terms in equation (8.6) are always positive and can hence be identified as comprising the mechanical energy dissipation,

$$\Psi \equiv K_\phi^{-1} \boldsymbol{q}^2 + \zeta_\phi \mathcal{C}^2 + 2\eta_\phi \left(\dot{\boldsymbol{\varepsilon}}^s\right)^2, \tag{8.7}$$

where $K_\phi \equiv k_\phi / \mu^\ell$ is the mobility and $\boldsymbol{q} = \phi \left(\boldsymbol{v}^\ell - \boldsymbol{v}^s\right)$ is the segregation flux. The terms in the two-phase dissipation are associated with Darcy flow of the liquid, compaction of the solid and shear of the solid, respectively.[2]

To eliminate $\nabla \cdot \overline{\boldsymbol{v}}$ from (8.6) we manipulate the (non-Boussinesq) mass conservation equations (4.11) to obtain

$$\nabla \cdot \overline{\boldsymbol{v}} = \phi \rho^\ell \frac{\mathrm{D}_\ell (1/\rho^\ell)}{\mathrm{D}t} + (1-\phi)\rho^s \frac{\mathrm{D}_s (1/\rho^s)}{\mathrm{D}t} - \Gamma \Delta \left(1/\rho\right). \tag{8.8}$$

Using this equation, equations (8.6) and (8.7), the conservation of energy equation (8.4) can be rewritten as

$$\phi \rho^\ell \frac{\mathrm{D}_\ell u^\ell}{\mathrm{D}t} + (1-\phi)\rho^s \frac{\mathrm{D}_s u^s}{\mathrm{D}t} - \Gamma \Delta u = \overline{\rho \mathbb{Q}} + \nabla \cdot \overline{k_T} \nabla T + \Psi$$
$$- P^\ell \left[\phi \rho^\ell \frac{\mathrm{D}_\ell (1/\rho^\ell)}{\mathrm{D}t} + (1-\phi)\rho^s \frac{\mathrm{D}_s (1/\rho^s)}{\mathrm{D}t} - \Gamma \Delta \left(1/\rho\right) \right]. \tag{8.9}$$

Note that P^ℓ in this equation is the full pressure, including the lithostatic part. Deep in the asthenosphere, the full magma pressure is very large, which is an issue for numerical solutions of this equation. Indeed, this equation is never used in simulations, but is the basis for deriving more useful forms, as follows below.

8.2 The Enthalpy Equation

In chapter 4, where we formulate the equations governing the fluid dynamics, each phase has its own pressure; the interphase pressure difference is what drives compaction. In formulating the energetics, however, we must recognize that thermodynamic equilibrium between phases occurs when there is a single, common pressure. This pressure would be measured at the interface between phases. In a more general theory this pressure would be $P^{\mathcal{I}}$ and distinct from the liquid and solid pressures. However, in the particular geodynamic limit of $\mu^\ell \ll \eta_\phi$, it is appropriate to assume that the interface pressure (i.e., the *thermodynamic pressure*) is equal to the liquid pressure. Hence, in the thermodynamic relations between enthalpy and internal energy, we use the liquid pressure for both the liquid and solid phases. A discussion of the issue of the thermodynamic pressure in deforming, partially molten rocks is included in the Literature Notes at the end of this chapter, along with references to more general treatments.

[2]Note that although we have approximated the thermodynamic pressure at the interface as P^ℓ, we retain the compaction work done by the pressure difference ΔP in the mechanical dissipation.

In the geodynamic limit, specific enthalpy h^i of a phase is related to its specific internal energy as

$$h^i = u^i + P^\ell/\rho^i, \tag{8.10}$$

which can be written in terms of Lagrangian derivatives as

$$\frac{\mathrm{D}_i h^i}{\mathrm{D}t} = \frac{\mathrm{D}_i u^i}{\mathrm{D}t} + \frac{1}{\rho^i}\frac{\mathrm{D}_i P^\ell}{\mathrm{D}t} + P^\ell \frac{\mathrm{D}_i(1/\rho^i)}{\mathrm{D}t}. \tag{8.11}$$

We also require the enthalpy difference Δh between coexisting solid and melt,

$$\Delta h = \Delta u + P^\ell \Delta(1/\rho). \tag{8.12}$$

Consistent with the assumption that both phases share the interface (liquid) pressure at the interface between phases, we have taken $\Delta P = 0$ in equation (8.12).

Substitution of equations (8.11) and (8.12) into (8.9) leads to

$$\phi\rho^\ell \frac{\mathrm{D}_\ell h^\ell}{\mathrm{D}t} + (1-\phi)\rho^s \frac{\mathrm{D}_s h^s}{\mathrm{D}t} - \Gamma\Delta h = \frac{\partial P^\ell}{\partial t} + \overline{\boldsymbol{v}} \cdot \nabla P^\ell + \overline{\rho \mathbb{Q}} + \nabla \cdot \overline{k_T} \nabla T + \Psi. \tag{8.13}$$

This equation could be used to calculate Γ if all other terms are known. However, if the phase-specific enthalpies are also unknown, it makes this form particularly impractical.

A further manipulation of the enthalpy equation uses the conservation of mass equations (4.11) to return to an Eulerian form for the bulk enthalpy $H \equiv \overline{\rho h}$. The result is

$$\frac{\partial H}{\partial t} + \nabla \cdot \overline{\rho h \boldsymbol{v}} = \frac{\partial P^\ell}{\partial t} + \overline{\boldsymbol{v}} \cdot \nabla P^\ell + \overline{\rho \mathbb{Q}} + \nabla \cdot \overline{k_T} \nabla T + \Psi, \tag{8.14}$$

where $H \equiv \overline{\rho h}$ is the bulk, volumetric enthalpy. It is convenient to rewrite the enthalpy flux divergence in terms of the transport of latent heat. This is achieved by using a relationship from thermodynamics (and again assuming that the thermodynamic pressure in the two-phase system is that of the liquid),

$$\mathrm{d}h^i = c_P^i\,\mathrm{d}T + (1-\alpha_\rho^i T)\,\mathrm{d}P^\ell/\rho^i. \tag{8.15}$$

In the context of phases comprising chemical components of variable concentration, this equation should also contain a term relating compositional changes to changes in enthalpy. We drop this term for simplicity and because two-phase-flow models of silicate systems do not typically include detailed compositional models. Equation (8.15) can be used to formulate the gradient of the phase-specific enthalpy,

$$\nabla h^i = c_P^i \nabla T + \frac{(1-\alpha_\rho^i T)}{\rho^i}\nabla P^\ell. \tag{8.16}$$

This expression can be substituted into (8.14) and, after expanding some derivatives, we obtain

$$\nabla \cdot \overline{\rho h \boldsymbol{v}} = -\Delta h \nabla \cdot \phi\rho^\ell \boldsymbol{v}^\ell + h^s \nabla \cdot \overline{\rho \boldsymbol{v}} + \overline{\rho c_P \boldsymbol{v}} \cdot \nabla T + \overline{(1-\alpha_\rho T)\boldsymbol{v}} \cdot \nabla P^\ell, \tag{8.17}$$

where we have added and subtracted $h^s \nabla \cdot \phi\rho^\ell \boldsymbol{v}^\ell$ to achieve a particular form. We introduce the latent heat $L \equiv h^\ell - h^s = -\Delta h$, which is the enthalpy change associated with melting and freezing. Now assuming that thermal properties c_P^i, α_ρ^i, and k_T^i are equal between phases, we use the bulk mass conservation equation (4.12) and substitute equation (8.17) into (8.14) to give

$$\frac{\partial H}{\partial t} + L\nabla\cdot\phi\rho^\ell\boldsymbol{v}^\ell + c_P\overline{\rho\boldsymbol{v}}\cdot\nabla T = \nabla\cdot k_T\nabla T + \overline{\rho\mathbb{Q}} + \Psi$$
$$+\alpha_\rho T\overline{\boldsymbol{v}}\cdot\nabla P^\ell + \frac{\partial P^\ell}{\partial t} - h^s\frac{\partial\overline{\rho}}{\partial t}. \tag{8.18}$$

This evolution equation states that changes in bulk enthalpy are driven by the divergence of the latent heat flux, advection and diffusion of sensible heat, radiogenic and viscous-dissipative heating, and terms associated with variations in pressure and density (i.e., work terms). This latter set, on the second line of equation (8.18), can be simplified by making two reasonable approximations. We apply these in section 8.5.

Note that the equation for bulk enthalpy evolution (8.18), does not contain the melting rate. If the melting rate is unknown but of secondary interest, this equation usefully avoids having to compute it. However, if energy conservation is to be used as an explicit expression of the melting rate, it is more convenient to cast it in terms of temperature.

8.3 The Temperature Equation

The thermodynamic relation for enthalpy found in (8.15) can also be written in terms of Lagrangian derivatives as

$$\frac{\mathrm{D}_i h^i}{\mathrm{D}t} = c_P\frac{\mathrm{D}_i T}{\mathrm{D}t} + \frac{1-\alpha_\rho T}{\rho^i}\cdot\frac{\mathrm{D}_i P^\ell}{\mathrm{D}t}. \tag{8.19}$$

Here we continue to assume that the heat capacity and thermal expansivity are equal between phases and, further, that the enthalpy is independent of any compositional changes to the phase. Returning to equation (8.13) and substituting (8.19) to eliminate h^i gives an equation for the evolution of temperature,

$$c_P\left[\phi\rho^\ell\frac{\mathrm{D}_\ell T}{\mathrm{D}t} + (1-\phi)\rho^s\frac{\mathrm{D}_s T}{\mathrm{D}t}\right] - \alpha_\rho T\left[\frac{\partial P^\ell}{\partial t} + \overline{\boldsymbol{v}}\cdot\nabla P^\ell\right] =$$
$$-L\Gamma + \nabla\cdot k_T\nabla T + \overline{\rho\mathbb{Q}} + \Psi, \tag{8.20}$$

where we again use $L \equiv -\Delta h$ to represent the latent heat or enthalpy of fusion. This equation states that temperature changes are driven by pressure–volume (PV) work, conversion of sensible heat to and from latent heat, diffusion of sensible heat, and radiogenic and viscous-dissipative heating. As with the enthalpy equation (8.13) (and (8.18)), this equation can be simplified by making a couple of reasonable approximations. We leave this for section 8.5.

8.4 The Entropy Equation

Although entropy is a less-familiar physical quantity, its governing equation has the simplest form of any of the statements of energy conservation. It can be derived by expanding the enthalpy with another relation from thermodynamics,

$$\frac{\mathrm{D}_i h^i}{\mathrm{D}t} = T\frac{\mathrm{D}_i s^i}{\mathrm{D}t} + \frac{1}{\rho^i}\frac{\mathrm{D}_i P^\ell}{\mathrm{D}t}, \tag{8.21}$$

where s^i is the mass-specific entropy of phase i and we have again assumed that the thermodynamic pressure is that of the liquid phase. Here, again, we neglect the contributions of chemical potentials that associate changes in phase composition with internal energy. Substituting this into (8.13) and simplifying gives

$$\phi\rho^\ell \frac{\mathrm{D}_\ell s^\ell}{\mathrm{D}t} + (1-\phi)\rho^s \frac{\mathrm{D}_s s^s}{\mathrm{D}t} - \Gamma\Delta s = \frac{1}{T}\left(\nabla\cdot\overline{k_T}\nabla T + \overline{\rho\mathbb{Q}} + \Psi\right), \tag{8.22}$$

where we have used $\Delta h = T\Delta s$. Using the conservation of mass equations (4.11), this can also be written in terms of phase-averaged properties as

$$\frac{\partial\overline{\rho s}}{\partial t} + \nabla\cdot\overline{\rho s \boldsymbol{v}} = -\nabla\cdot\boldsymbol{J}_s + \Sigma, \tag{8.23}$$

where $\boldsymbol{J}_s \equiv \boldsymbol{J}_u/T$ is the diffusive entropy flux, $\boldsymbol{J}_u$ is given by Fourier's law, and

$$\Sigma \equiv \frac{1}{T}\left(\overline{k_T}\frac{|\nabla T|^2}{T} + \overline{\rho\mathbb{Q}} + \Psi\right) \tag{8.24}$$

is the entropy production rate. This expression of entropy production assumes that the liquid and solid are in thermodynamic equilibrium, and hence any phase change occurs so slowly as to be reversible. Real phase change involves some degree of disequilibrium, and in this case Σ contains an additional term (although it may be vanishingly small). See section 10.2.1 for further details.

The second law of thermodynamics requires that $\Sigma \geq 0$, i.e., that the total entropy can only increase. It is evident by inspection of terms in (8.24) (and by referring back to (8.7) for Ψ) that the second law is satisfied.

There is an important difference between equation (8.23) for entropy and all the other equations representing energy conservation: the entropy equation does not have a term that depends on pressure or pressure gradients. In fact, (8.23) shows that, in the limit of reversible phase change, no melt segregation ($\overline{\boldsymbol{v}} = \boldsymbol{v}^s$), no radiogenic and viscous-dissipative heating, and no heat flow by conduction, an upwelling parcel of partially molten mantle experiences zero change in entropy. This highly idealized case is known as *isentropic* because entropy is conserved. A more general case that allows for irreversible (dissipative) phase change is known as *adiabatic*.[3]

8.5 Boussinesq and Lithostatic Approximations

Two important simplifications can be made to the energy conservation equation (8.9). They also apply to the conservation statement written in terms of enthalpy, (8.13) and (8.18), as well as temperature, (8.20).

[3]Throughout this book we neglect the dissipation associated with phase change and density change. In this context, the terms "adiabatic" and "isentropic" may be considered equivalent.

The first is the Boussinesq approximation, which was discussed in the context of the mechanics in chapters 3 and 4. This states that density variations (and density differences) can be neglected except in body-force terms. Hence, except within those terms, we can assume that $\rho^s \approx \rho^\ell \equiv \rho$ and therefore $\Delta\rho \approx 0$. In this case, equation (8.17) becomes $\boldsymbol{\nabla}\cdot\overline{\boldsymbol{v}} = 0$.

The second approximation is to the liquid pressure P^ℓ that appears in the enthalpy and temperature equations. This is the full pressure, including both the static and dynamic components. Deformation in the mantle is everywhere very slow and the pressure due to the rock overburden is extremely large; hence, it is a good approximation to take $P^\ell \approx P^{\text{lith}}$, the *lithostatic pressure*. The latter is given by the state of no flow ($\boldsymbol{v}^s = \boldsymbol{v}^\ell = \boldsymbol{0}$) and hence equation (4.47b) becomes

$$\boldsymbol{\nabla} P^{\text{lith}} \equiv \overline{\rho}\mathbf{g} \approx \rho\mathbf{g}, \tag{8.25}$$

where the approximation arises from the Boussinesq assumption above.

In applying these approximations, we also assume that the heat capacity, thermal expansivity, and thermal conductivity are equal between phases. The resulting equations, while still capturing all of the physics at leading order for most problems, are in a form that is more amenable to numerical solution (further simplifications are usually needed for analytical solution).

It should be noted that these approximations, while convenient for obtaining model solutions, are formally inconsistent with some of the physics. For example, we assume that pressure in both phases is lithostatic but retain the dissipation term associated with compaction, which is caused by pressure differences between the phases. In models of flow driven by gravity acting on density variations or differences, there is a correspondence between viscous dissipation and adiabatic PV work (see the Literature Notes at the end of this chapter); care must be taken to retain or neglect these terms as a pair.

Internal energy. Terms on the right-hand side of (8.9) involving density variations are zero under the Boussinesq approximation. Hence we have

$$\phi\rho\frac{\mathrm{D}_\ell u^\ell}{\mathrm{D}t} + (1-\phi)\rho\frac{\mathrm{D}_s u^s}{\mathrm{D}t} - \Gamma\Delta u = \rho\overline{\mathbb{Q}} + \boldsymbol{\nabla}\cdot k_T\boldsymbol{\nabla}T + \Psi. \tag{8.26}$$

This equation is independent of the liquid pressure.

Enthalpy. Substituting the lithostatic pressure into equation (8.13) and applying the Boussinesq approximation gives

$$\phi\rho\frac{\mathrm{D}_\ell h^\ell}{\mathrm{D}t} + (1-\phi)\rho\frac{\mathrm{D}_s h^s}{\mathrm{D}t} - \Gamma\Delta h = \overline{\boldsymbol{v}}\cdot\rho\mathbf{g} + \rho\overline{\mathbb{Q}} + \boldsymbol{\nabla}\cdot k_T\boldsymbol{\nabla}T + \Psi. \tag{8.27}$$

Written in terms of the bulk enthalpy H, this equation becomes

$$\frac{\partial H}{\partial t} - \rho L\boldsymbol{\nabla}\cdot(1-\phi)\boldsymbol{v}^s + \rho c_P\boldsymbol{\nabla}\cdot\overline{\boldsymbol{v}}T = \boldsymbol{\nabla}\cdot k_T\boldsymbol{\nabla}T + \rho\overline{\mathbb{Q}} + \Psi + \rho\alpha_\rho T\overline{\boldsymbol{v}}\cdot\mathbf{g}, \tag{8.28}$$

where we dropped Eulerian time derivatives of the lithostatic pressure and the Boussinesq density. The divergence of the latent heat flux is written in terms of the solid velocity by expanding $\boldsymbol{\nabla}\cdot\overline{\boldsymbol{v}} = 0$ and substituting. Equation (8.28) is written in a form that is amenable to finite-volume discretization. In section 13.4, we discuss how the enthalpy equation can be coupled with fluid dynamics, chemistry, and petrology to create tectonic-scale simulations.

In equations (8.27) and (8.28), we retain the term containing the gravity as this is arguably a body-force term: it relates changes in gravitational potential energy to PV work and hence to sensible heat. This level of approximation is referred to as the *extended Boussinesq approximation* in the literature on mantle convection (see the Literature Notes at the end of this chapter Notes). In our discussion of the temperature equation, below, we demonstrate that it gives rise to the isentropic temperature gradient in the mantle. For problems associated with the asthenosphere, this term contributes relatively little to the overall heat budget and is sometimes neglected.

Temperature. As above, we assume that the thermodynamic pressure is that of the liquid, and that it is well approximated by the lithostatic pressure under the Boussinesq approximation. We also use the Boussinesq approximation to arrive at $\nabla \cdot \overline{\boldsymbol{v}} = 0$. Then, assuming material properties are uniform and equal between phases, equation (8.20) becomes

$$\frac{\partial T}{\partial t} + \nabla \cdot \overline{\boldsymbol{v}} T = \kappa \nabla^2 T + \frac{1}{c_P}\left(-L\Gamma/\rho + \alpha_\rho T \overline{\boldsymbol{v}} \cdot \mathbf{g} + \overline{\mathbb{Q}} + \Psi/\rho\right). \tag{8.29}$$

This evolution equation states that changes in temperature are due to the divergence of the advective and diffusive flux of sensible heat, the consumption of latent heat by melting, work done by gravity, radiogenic heating, and viscous dissipation of heat. It is written in a form that is amenable to finite-volume discretization. Since $\nabla \cdot \overline{\boldsymbol{v}} = 0$, it can be easily converted to Lagrangian form. Moreover, this equation can be rearranged to represent an equation that determines the melting rate Γ.

The term on the right-hand side containing $\alpha_\rho T \overline{\boldsymbol{v}} \cdot \mathbf{g}$, which also appeared in the enthalpy equation above, represents the exchange of sensible heat with density variation due to isentropic, volumetric work. It is this term that gives rise to the isentropic temperature gradient of the mantle (the *isentropic geotherm*). Inclusion of this physics is inconsistent with a strict application of the Boussinesq approximation, which states that density variations outside of body-force terms are neglected. However, there is no practical difficulty arising from its inclusion.

Consider a region of mantle far from the diffusive thermal boundary layers near the surface with $\Gamma = \mathbb{Q} = \Psi = 0$. Any changes in this region may be considered isentropic. Assuming and that a steady state holds, we expect an approximate balance of

$$\overline{\boldsymbol{v}} \cdot \nabla T \sim \alpha_\rho T \overline{\boldsymbol{v}} \cdot \mathbf{g}/c_P. \tag{8.30}$$

This equation tells us that vertical motions give rise to a vertical temperature variation described by

$$T \sim T_p \mathrm{e}^{-\alpha_\rho g z/c_P}, \tag{8.31}$$

where $\hat{z}$ points upward and T_p is a reference temperature known as the *potential temperature*. It represents the temperature that a parcel of mantle would have if it were transported isentropically from some position z where its temperature is T to $z = 0$. In the simple balance of equation (8.30), the potential temperature is a constant whereas in the mantle, potential temperature varies due to nonisentropic processes (see the

Literature Notes at the end of this chapter). Nonetheless, equation (8.31) clarifies the consequences of the reversible PV work term in (8.29).

Over the depth range of the asthenosphere (~300 km), the exponent in equation (8.31) remains small; it is appropriate to expand in a Taylor series and truncate at first order, giving $T \sim T_p\left(1-\alpha_\rho gz/c_P\right)$ with $z<0$ in the subsurface. Since $\alpha_\rho g/c_P \approx 10^{-4}$/km, the linear term plays a significant role only when the depth range is greater than about 100 km. For the purposes of analytical models, it is sometimes convenient to simplify the PV-work term in (8.29) by assuming a constant potential temperature. Then, at least in the context of one-dimensional models, $\alpha_\rho T\overline{\boldsymbol{v}}\cdot\mathbf{g} \rightarrow -\alpha_\rho T_0 W_0 g$, a constant.

8.6 Dissipation-Driven Melting and Compaction

Even a basic treatment of open-system mantle melting requires a model for mantle composition. However, to demonstrate some of the behavior of the energy equations above, it is helpful to consider the melting of a single-component, polycrystalline solid such as ice made from pure H_2O. Partially molten, polycrystalline ice has a similar pore structure to that of the olivine/basalt system.

We consider a layer of ice of thickness $\mathcal{H}(t)$ and initial, uniform porosity $\phi_0 \geq 0$ as shown in figure 8.1; such a configuration could be approximately realized in a laboratory experiment. The layer is isothermal and its temperature is equal to the melting point. It is held between flat, infinite plates that do not slip against the ice. The top plate is permeable and moves laterally at speed $\mathcal{S}$ in the x direction, imposing a shear on the ice below. It is also pushed downward (in the $-y$ direction) by an applied normal stress with magnitude $\mathcal{F}>0$. The bottom plate is rigid, impermeable, and unmoving. Hence, the ice layer is under a combination of shear and, if it is partially molten, compaction. We shall assume that:

- At the permeable boundary, $P^\ell = 0$ because the pores are open to the atmosphere. Also on this boundary we have

$$\boldsymbol{v}^s\cdot\hat{\boldsymbol{x}} = \mathcal{S} \text{ and } \hat{\boldsymbol{y}}\cdot\overline{\boldsymbol{\sigma}}\cdot\hat{\boldsymbol{y}} = -\mathcal{F} \text{ at } y=\mathcal{H}(t). \tag{8.32}$$

- The imposed stresses are much larger than body forces and therefore we neglect the latter.
- The compaction length δ is always much larger than the thickness of the layer $\mathcal{H}$ and hence we take $\delta \rightarrow \infty$. For finite solid viscosities, it must therefore be the case that $k_\phi/\mu \equiv K_\phi \rightarrow \infty$, which in turn means that there is no resistance to Darcy flow and the pressure gradient is zero.
- At the impermeable boundary,

$$\boldsymbol{v}^s = \mathbf{0} \text{ and } \frac{\mathrm{d}P^\ell}{\mathrm{d}z} = 0 \text{ at } y=0. \tag{8.33}$$

The problem as posed above is initially uniform in x and z. We shall assume that it remains so, despite the potential for mechanical instabilities (chapter 7). The solution must therefore be one-dimensional in y; all derivatives with respect to x and z vanish. We therefore propose the linear flow

$$\boldsymbol{v}^s = y\left[\mathcal{S}(t)/\mathcal{H}(t)\,\hat{\boldsymbol{x}} + \mathcal{C}(t)\,\hat{\boldsymbol{y}}\right] \tag{8.34}$$

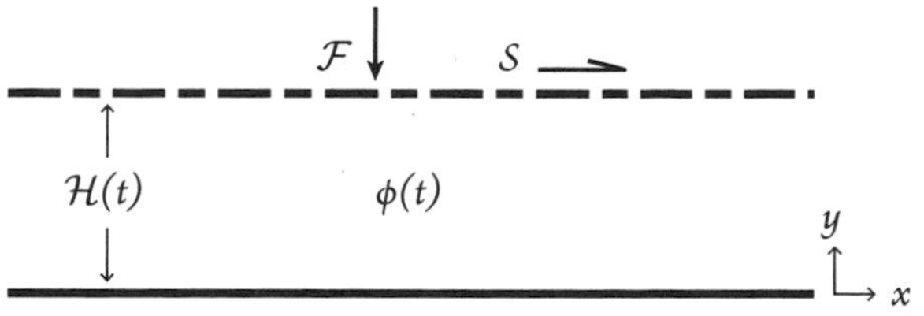

Figure 8.1. Schematic diagram of the compaction–dissipation problem. Bottom plate (thick solid line) is rigid, impermeable, and unmoving. The top plate (thick broken line) moves in the x direction with speed $\mathcal{S} > 0$ and moves in the $-y$-direction under normal stress $\mathcal{F} > 0$. The compaction length δ is much larger than $\mathcal{H}$ and hence taken to be infinite.

as a trial solution. This is the superposition of simple shear with uniform compaction ($\mathcal{C} < 0$), both of which were shown to be solutions to the governing equations under uniform viscosities and permeability.[4] Our trial solution has the properties

$$\nabla \cdot \boldsymbol{v}^s = \mathcal{C}, \tag{8.35a}$$

$$\dot{\boldsymbol{\varepsilon}}^s = \frac{1}{2}\begin{pmatrix} -2\mathcal{C}/3 & \mathcal{S}/\mathcal{H} & 0 \\ \mathcal{S}/\mathcal{H} & 4\mathcal{C}/3 & 0 \\ 0 & 0 & -2\mathcal{C}/3 \end{pmatrix}, \tag{8.35b}$$

that are uniform in space. $\mathcal{C}(t)$ and $\mathcal{H}(t)$ are unknowns to be determined. They are related by

$$\frac{\mathrm{d}\mathcal{H}}{\mathrm{d}t} = \int_0^{\mathcal{H}} \mathcal{C}\,\mathrm{d}y$$
$$= \mathcal{H}\mathcal{C}, \tag{8.36}$$

because $\mathcal{C}$ is uniform.

The solid velocity is obtained by using the y-component of the bulk stress balance from equation (4.46b). After dropping the pressure gradient (because $K_\phi \to \infty$) and the body force we have

$$\frac{\mathrm{d}(\xi_\phi \mathcal{C})}{\mathrm{d}y} = 0, \tag{8.37}$$

where $\xi_\phi \equiv \zeta_\phi + \frac{4}{3}\eta_\phi$. We can integrate (8.37) and apply the stress boundary condition (8.32) to obtain

$$\mathcal{C}(t) = -\mathcal{F}/\xi_\phi, \tag{8.38}$$

noting that $-\mathcal{F} < 0$ is a compressive stress. Combining this with equation (8.36) and the porosity evolution equation (4.46c),

$$\dot{\mathcal{H}} = -\mathcal{H}\mathcal{F}/\xi_\phi, \tag{8.39a}$$

[4]Uniformity of compaction rate in the y-direction is due to the effectively infinite compaction length, $\delta \gg \mathcal{H}$. See exercise 6.2.

$$\dot{\phi} = -(1-\phi)\mathcal{F}/\xi_\phi + \Gamma/\rho^s, \tag{8.39b}$$

where the dot indicates a derivative with respect to time.

To solve this system we require an additional equation for the volumetric melting rate Γ/ρ^s, which we obtain from the temperature equation (8.29). In fact, the temperature equation can be drastically simplified for present purposes. Because we are considering a thermodynamic system with only a single chemical component, the melting temperature is a function of pressure only. Thus, our partially molten system at fixed pressure has a uniform, constant temperature that is equal to the melting temperature. Again neglecting body forces, equation (8.29) becomes

$$\Gamma = \frac{\Psi}{L}. \tag{8.40}$$

Melting is thus driven by viscous dissipation at a rate that is reduced by the energetic cost associated with latent heat. We can use the strain rate tensor (8.35b) and the compaction rate (8.38) to rewrite the dissipation rate (8.7) as

$$\Psi = \xi_\phi \left(\frac{\mathcal{F}}{\xi_\phi}\right)^2 + \eta_\phi \left(\frac{\mathcal{S}}{\mathcal{H}}\right)^2. \tag{8.41}$$

where $K_\phi \to \infty$ eliminates the contribution from melt segregation $\boldsymbol{q}$. Which of the dissipative terms is larger? Multiplying (8.41) by η_ϕ and assuming that $\zeta_\phi/\eta_\phi \sim \phi^{-1}$, we can infer that shear dominates over compaction when $\mathcal{T}^2 \gg \phi\mathcal{F}^2$. Here we have defined $\mathcal{T} \equiv \eta_\phi \mathcal{S}/\mathcal{H}$ as the imposed shear stress required to maintain the upper plate at speed $\mathcal{S}$.

Assuming that simple shear is the dominant cause of viscous dissipation, we can rewrite the system (8.39) as

$$\dot{\mathcal{H}} = -\mathcal{H}\mathcal{F}/\xi_\phi, \tag{8.42a}$$

$$\dot{\phi} = \frac{\eta_\phi \mathcal{S}^2}{\rho^s L \mathcal{H}^2} - (1-\phi)\mathcal{F}/\xi_\phi. \tag{8.42b}$$

Melting in this system is driven by shear $\mathcal{S} > 0$, but the overall behavior depends on how that shear is imposed. There are three obvious cases:[5]

1. Constant displacement rate, $\mathcal{S} = \mathcal{S}_0$; as the layer thickness decreases, the strain rate increases.
2. Constant strain rate, $\mathcal{S}/\mathcal{H} = \mathcal{S}_0/\mathcal{H}_0$; as the layer thickness decreases, the displacement rate slows proportionately.
3. Constant shear stress, $\eta_\phi \mathcal{S}/\mathcal{H} = \mathcal{T}_0$; the strain rate increases when the viscosity decreases (and vice versa).

Below, we consider case 1 of constant strain rate. The other two cases are considered in exercises 8.4 and 8.5 at the end of this chapter.

We seek a solution $\mathcal{H}(t)$, $\phi(t)$ to the system of equations (8.42). $\mathcal{F}$, $\mathcal{S}/\mathcal{H}$, ρ^s, and L are constants; closure conditions are required for ξ_ϕ and η_ϕ. Assuming a Newtonian

[5] In these cases, a subscript $_0$ indicates the initial value of a quantity.

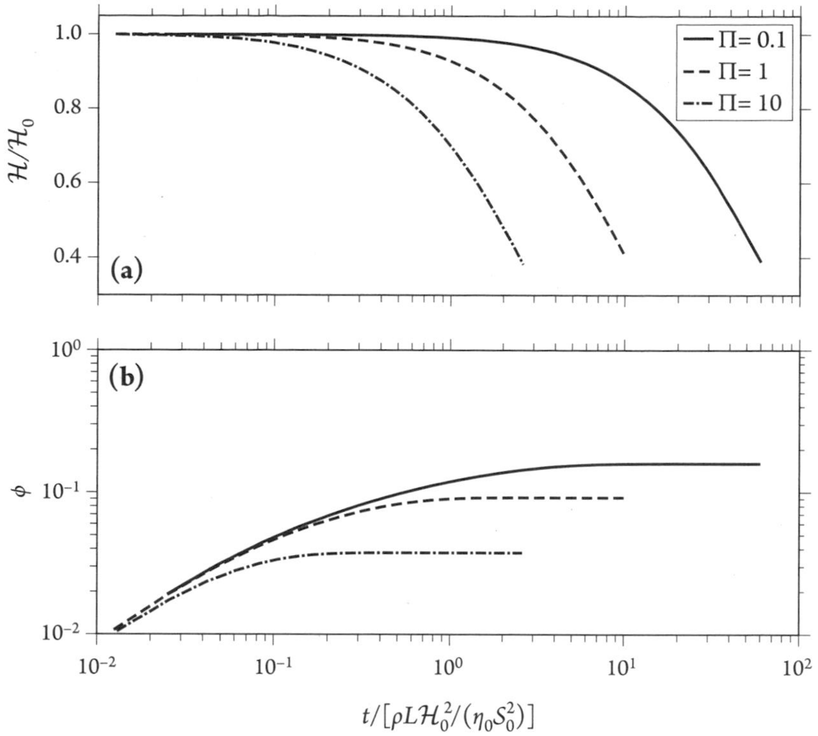

Figure 8.2. Numerical solutions of equations (8.45) for porosity and layer thickness with three distinct values of Π. **(a)** Layer thickness relative to initial versus time. **(b)** Porosity versus time. Note the logarithmic axes, both horizontal and vertical.

viscosity in which $\zeta_\phi \propto \eta_\phi/\phi \gg \eta_\phi$, the following are appropriate closures,

$$\eta_\phi = \eta_0 e^{-\lambda\phi}, \tag{8.43a}$$

$$\xi_\phi = \eta_0 \phi^{-1}. \tag{8.43b}$$

Substituting (8.43) into (8.42) and rescaling layer thickness and time, with

$$[\mathcal{H}] = \mathcal{H}_0, \quad [t] = \frac{\rho^s L \mathcal{H}_0^2}{\eta_0 S_0^2}, \tag{8.44}$$

we have

$$\dot{\mathcal{H}} = -\Pi \mathcal{H}\phi, \tag{8.45a}$$

$$\dot{\phi} = e^{-\lambda\phi} - \Pi(1-\phi)\phi, \tag{8.45b}$$

where all symbols are now dimensionless. We have introduced the dimensionless number

$$\Pi \equiv \frac{\rho^s L \mathcal{F} \mathcal{H}_0^2}{\eta_0^2 S_0^2}, \tag{8.46}$$

which we shall refer to as the *compaction–dissipation number*. It represents the ratio of compaction to shear-driven dissipative heating in the initial state. With this scaling, a dimensionless time of unity corresponds to the duration of shear required to fully melt the sample if the dissipation rate is fixed at its initial value, with zero porosity.

Numerical solutions of equations (8.45) are shown in figure 15.5 for three values of Π. Note the initial increase in porosity that eventually gives way to a pseudo–steady state in which melting balances compaction. This is not a true steady state because inevitably the layer thickness goes to zero. Although the porosity is steady, the layer thickness decreases at a constant rate. The solution with the largest value of Π has the lowest porosity, as expected, but also melts the fastest because low ϕ corresponds to high η_ϕ.

8.7 Decompression Melting

Another simplified application of conservation of energy involves decompression melting of the mantle. Decompression melting occurs where mantle upwells; a parcel undergoes decreasing pressure as it moves to the surface. The melting temperature of the mantle is an increasing function of pressure. Hence hot, upwelling mantle, even without any addition of heat, may eventually begin to melt as it ascends to lower-pressure areas.

To model this process simply, we can ignore compaction and melt segregation and consider only the temperature equation (8.29). We'll assume no radiogenic or dissipative heating, zero permeability (and, hence, $\boldsymbol{v}^s = \boldsymbol{v}^\ell \equiv \boldsymbol{v}$), no diffusive heat transfer, and densities that are constant and equal between phases (the Boussinesq approximation). Since the melting process under these conditions involves no addition or loss of heat to the surroundings and is (by further assumption) always in equilibrium, the process may be considered to be isentropic. We impose a constant upwelling rate $\boldsymbol{v} \cdot \mathbf{g} = -Wg$ for $W > 0$ and $g = |\mathbf{g}|$. Then the temperature equation becomes

$$\frac{\mathrm{D}T}{\mathrm{D}t} = -\left(\frac{L\Gamma}{\rho c_P} + \frac{\alpha_\rho TWg}{c_P}\right). \tag{8.47}$$

On the left-hand side, we can reformulate the derivative by applying the assumptions above to write $\mathrm{d}t = W^{-1}\,\mathrm{d}z$, where z increases upward (opposite gravity).

At depths greater than the onset of decompression melting, the mantle is entirely solid and $\Gamma = 0$. There we have

$$\frac{\mathrm{d}T}{\mathrm{d}z} = -\frac{\alpha_\rho gT}{c_P}, \tag{8.48}$$

leading to the familiar steady-state, isentropic geothermal temperature gradient from (8.31) above. The isentropic gradient T_z is rather small: taking $\alpha_\rho = 3 \times 10/\mathrm{K}$, $c_P = 1200$ J/kg/K and $T \sim 1650$ K, we have $T_z \sim -0.4$ K/km. Melting typically occurs within the upper 100 km of the mantle. Over this range, $-\alpha_\rho gz/c_P$ is positive but $\ll 1$. Hence it is appropriate to linearize the isentropic temperature profile (the *isentrope*) as

$$T = T_p(1 - \alpha_\rho gz/c_P). \tag{8.49}$$

The mantle cools as it rises because parcels of mantle do work by expanding. Despite this, the isentrope eventually crosses the mantle *solidus*, the temperature at

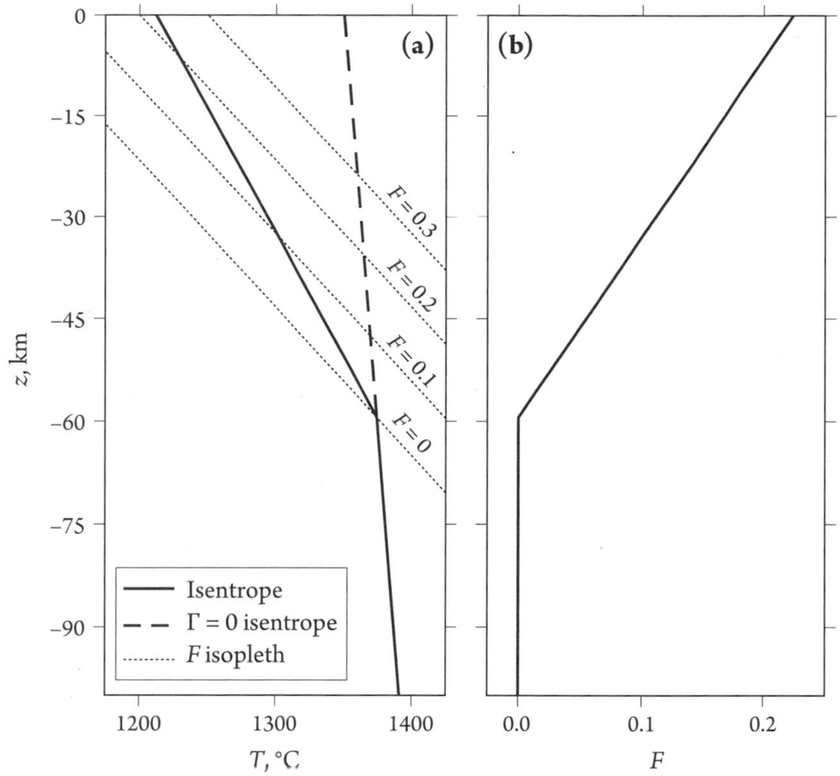

Figure 8.3. Decompression melting curves with no melt segregation. **(a)** Temperature as a function of z. The solid curve shows the isentropic temperature profile (the *isentrope*) computed according to (8.48) and (8.53). The dashed line shows the isentrope for no melting. Dotted lines are isopleths of the degree of melting F. **(b)** The degree of melting associated with the isentrope in (a). Parameter values used to compute the curves in the figure are given in table 8.1.

which the mantle begins to melt, which has an even steeper slope. This is shown in figure 8.3a. The slope of the mantle solidus in pressure–temperature space is given by the Clausius–Clapeyron relationship,[6] $\mathrm{d}P/\mathrm{d}T = \Delta s/\Delta(1/\rho)$, where Δs and $\Delta(1/\rho)$ are the difference in specific entropy and specific volume between the liquid and solid phases on the solidus. Above the solidus temperature, mantle melting takes place over a range of temperatures. This is in contrast to melting of a single-component, pure substance such as was considered in section 8.6. The mantle has $\sim$13 chemical components and melts over a broad range of temperature. The detailed chemical thermodynamics of mantle melting is beyond the scope of this book, but in the next two chapters we develop simplified theory to model conservation of component mass in a system of N components. We shall see that once melting begins, the solidus temperature changes with the evolving composition of the solid residue. Since composition is excluded from the current model, we avoid calculating this change. Hence, for the present purposes we

[6]More strictly, the Clausius–Clapeyron relationship holds only for single-component systems, which is certainly not what we are modeling here. However, the phenomenological nature of the model permits us to take a loose interpretation of the Clausius–Clapeyron slope.

Table 8.1. Parameters used to compute curves in figure 8.3.

Quantity	Value	Units
ρ	3000	kg/m^3
c_P	1200	J/kg/K
α_ρ	3×10^{-5}	K^{-1}
L	5×10^5	J/kg
$\mathcal{M}_F$	1/500	K^{-1}
ϑ	6.5×10^6	Pa/K
$T_0^{\mathcal{S}}$	1373	K
T_P	1623	K

consider the solidus to be the melting temperature of the initial mantle composition, before melting begins.

In that context, it is sufficient to prescribe a model for the *degree of melting* F as a function of temperature above the solidus,

$$F = \begin{cases} 0, & \text{if } T \leq T^{\mathcal{S}}(P), \\ \mathcal{M}_F\left[T - T^{\mathcal{S}}(P)\right], & \text{if } T^{\mathcal{S}}(P) < T < T^{\mathcal{L}}(P), \\ 1, & \text{if } T \geq T^{\mathcal{L}}(P). \end{cases} \tag{8.50}$$

This equation states that below the solidus temperature $T^{\mathcal{S}}$, the degree of melting is zero; above the solidus, F increases linearly with temperature (slope $\mathcal{M}_F$) at constant pressure. When the temperature reaches the *liquidus temperature* $T^{\mathcal{L}}$, melting stops because the solid is fully molten. $\mathcal{M}_F = \partial F/\partial T|_P$ is known as the *isobaric productivity*.

The solidus and liquidus are functions of pressure, according to the Clausius–Clapeyron relationship. We define $\vartheta \equiv \Delta s/\Delta(1/\rho)$ and assume that ϑ is a constant, independent of all variables. Then the solidus and liquidus are

$$T^{\mathcal{S}} = T^{\mathcal{S}}|_{P=0} + \vartheta^{-1}P, \tag{8.51a}$$

$$T^{\mathcal{L}} = T^{\mathcal{S}}|_{P=0} + \vartheta^{-1}P + \tfrac{1}{\mathcal{M}_F}. \tag{8.51b}$$

Here we have again assumed that for purposes of modeling melt production, the relevant pressure is equal between phases. In following sections, we take this pressure to be the lithostatic pressure.

It is important to note the crucial difference between F and ϕ. The degree of melting F is a property of the solid: how much has it previously been melted. The porosity ϕ is the instantaneous volume fraction of liquid. If liquid and solid can move independently, it is theoretically possible to have any combination of F and ϕ at a point in the two-phase continuum. This discussion highlights a shortcoming of our simple melting model: because it does not track chemical composition, it cannot account for the role of chemistry in controlling melting and freezing. Moreover, because it neglects melt segregation, it precludes the emergence of variations in bulk chemistry. We return to this in chapters 9 and 10.

The melting rate is obtained from the degree of melting by taking the Lagrangian derivative of F,

$$\Gamma = \rho\frac{\mathrm{D}F}{\mathrm{D}t} = \rho\mathcal{M}_F\left[\frac{\mathrm{D}T}{\mathrm{D}t} - \vartheta^{-1}\frac{\mathrm{D}P}{\mathrm{D}t}\right],$$

$$= \rho\mathcal{M}_F\left[\frac{\mathrm{d}T}{\mathrm{d}z} + \vartheta^{-1}\rho g\right]W, \quad \text{if } T^{\mathcal{S}}(P) < T < T^{\mathcal{L}}(P), \tag{8.52}$$

and otherwise, if the temperature is below the solidus temperature or above the liquidus temperature, $\Gamma = 0$. Here we have used $\mathrm{d}t = W^{-1}\mathrm{d}z$ and $\mathrm{d}P/\mathrm{d}z = -\rho g$.

Substituting equation (8.52) into (8.47) and rearranging gives the isentrope within the melting region:

$$\frac{\mathrm{d}T}{\mathrm{d}z} = -\frac{\alpha_\rho Tg/c_P + \rho g/\vartheta}{1 + L\mathcal{M}_F/c_P}. \tag{8.53}$$

Compare equations (8.53) and (8.48). Melting adds a term in the numerator that is proportional to the solidus slope; this term causes the temperature to drop more rapidly than it did below the solidus. Melting also adds a term in the denominator proportional to the latent heat of melting; this term moderates the rate of temperature reduction, keeping the temperature above the solidus.

To combine (8.53) and (8.48) for the full temperature profile, we need the depth at which melting begins z_F. Equating the subsolidus temperature profile (8.49) with the solidus temperature profile gives

$$z_F = -\frac{T_p - T^{\mathcal{S}}|_{P=0}}{\rho g/\vartheta - \alpha_\rho g/c_P} < 0. \tag{8.54}$$

Below this depth, $F = \phi = 0$, whereas above it, $F = \phi > 0$ (note that the degree of melting and porosity are equal in this case because we have assumed no segregation of melt and assumed equal density between phases). The overall isentropic gradient is plotted in figure 8.3.

Figure 8.3(a) shows that the isentrope changes slope where it intersects the solidus (the $F = 0$ isopleth) at z_F. Deeper than z_F, temperature decreases with z because sensible heat is lost to volumetric work. Shallower than z_F, temperature decreases more rapidly because sensible heat is lost to both volumetric work and latent heat. Figure 8.3(b) shows the degree of melting associated with the isentrope in figure 8.3(a).

8.8 Literature Notes

A conservation of energy equation was derived by McKenzie [1984] and analyzed in its application to one-dimensional melting columns by Ribe [1985]. An entropy formulation with compositional effects was developed by Iwamori et al. [1995]. More recently, the theory was generalized to incorporate damage generation, healing, and surface energy by Bercovici et al. [2001a] and Bercovici and Ricard [2003]. Boundaries between partially molten and unmolten regions were addressed in detail by Hewitt and Fowler [2008], where they also built on earlier column models.

Various authors have addressed question of the thermodynamic pressure in a compacting, partially molten system. The appendix of Jull and McKenzie [1996] contains an argument based on evaluating variations in the Gibbs free energy of the solid with respect to elastic stresses. It comes to the conclusion that the solid pressure ($P^\ell - \zeta_\phi \mathcal{C}$) is a close approximation of the thermodynamic pressure.[7] Rudge et al. [2011] discussed the more general two-phase formulation of Bercovici and Ricard [2003] (see also section 9.2.3 of Ribe [2018]). Perhaps the most satisfying approach, from the theoretical perspective is that of Ricard [2007], which considers that the pressure difference between phases contributes to the difference of chemical potential between phases. The chemical potential difference, in turn, drives mass transfer between phases (as we discuss following, in section 10.2). In practice, this makes the theory more complicated with more (poorly constrained) parameters. For partially molten rocks with a vast gap in viscosity between the phases, the approach presented here is adequate for most problems that are currently relevant.

Energy conservation equations that neglect melt segregation and transport of energy by melt are more abundant in the literature. These include the satisfying exchange between Waldbaum [1971] and Ramberg [1971]. The relationship of viscous dissipation and adiabatic PV work during mantle convection was demonstrated by Jarvis and McKenzie [1980], using the extended Boussinesq approximation. With regard to mantle melting, McKenzie [1984] made an important contribution, analyzing the melting rate associated with isentropic decompression. In appendix D of that work, McKenzie [1984] considered a one-component chemical system but allowed its temperature to be a function of pressure and degree of melting, as we have done in section 8.7. Asimow et al. [1997] extended this analysis of isentropic decompression to investigate the consequences of multicomponent thermochemistry that had previously been neglected. Isentropic models of this form are typically used as an approximation to adiabatic decompression melting, which is probably the dominant cause of melting in Earth. Ganguly [2005] has shown, in the context of assumptions similar to those of McKenzie [1984], that the difference between these two may be of an order of 30%. Moreover, for parcels of subsolidus rock that are less dense than the ambient mantle, the adiabatic geotherm may be smaller than the isentropic gradient (in an extreme case, even negative).

An outstanding challenge is to couple the energy equation to models of melting and melt segregation. Three pioneering papers to address this were by Asimow and Stolper [1999], Šrámek et al. [2007], and Rudge et al. [2011]. All of these incorporated chemical potentials into the conservation of energy, though Rudge et al. [2011] considered non-equilibrium thermodynamics of melting.

Dissipation-driven melting represents a fundamental connection between the mechanics and thermodynamics of a two-phase system. This coupling is important in the lateral margins of ice streams. There, localization of shear leads to dissipation and melting [Jacobson and Raymond, 1998]. Depending on how rapidly the meltwater drains to the bed, this process may be self-reinforcing [Haseloff et al., 2019] due to the viscosity reduction of the marginal ice by the grain-interstitial water fraction. Dissipation-driven melting is also important in the planetary context, where tidal variations in gravity of an orbiting body drive viscoelastic deformation and, hence, viscous dissipation [e.g., Beuthe, 2013]. If this dissipation is sufficiently intense, as on Jupiter's moon Io, it can drive silicate melting and volcanism [e.g., Peale et al., 1979; Spencer

[7] Jull and McKenzie [1996] also note that "[b]ecause of the contribution of the elastic strain to the Gibbs free energy, the concept of a solidus is not well defined on the scale of an individual crystal." This statement is consistent with the fact that macroscopic stresses lead to variation in elastic strain within a grain. These create differences in chemical potential (and hence in the Gibbs free energy) that drive diffusion of matter at the grain scale (which is manifest as diffusion creep at the macroscopic scale).

et al., 2020a]. Tidal heating is also important on "water worlds" such as Enceladus [Roberts and Nimmo, 2008], where tidal pumping of water through the unconsolidated, permeable core may make a substantial contribution to dissipation [Liao et al., 2020].

8.9 Exercises

8.1 In (8.6) we showed that the work done by surface and body forces on the system can be written in a more convenient form containing a term for energy dissipation. Verify one of the key results that we used in this simplification:

$$(1-\phi)\boldsymbol{\tau}^s : \nabla \boldsymbol{v}^s = 2\eta_\phi \left(\dot{\boldsymbol{\varepsilon}}^s\right)^2 .$$

8.2 Using the conservation of mass equations (4.11), show that for any scalar quantities q^ℓ, q^s with units of stuff per unit mass, conservation can be equivalently expressed in Eulerian (left-hand side) and Lagrangian (right-hand side) forms as

$$\frac{\partial \phi\rho^\ell q^\ell}{\partial t} + \nabla\cdot\left(\phi\rho^\ell q^\ell \boldsymbol{v}^\ell\right) = \phi\rho^\ell \frac{\mathrm{D}_\ell q^\ell}{\mathrm{D}t} + \Gamma q^\ell, \tag{8.55a}$$

$$\frac{\partial (1-\phi)\rho^s q^s}{\partial t} + \nabla\cdot\left((1-\phi)\rho^s q^s \boldsymbol{v}^s\right) = (1-\phi)\rho^s \frac{\mathrm{D}_s q^s}{\mathrm{D}t} - \Gamma q^s. \tag{8.55b}$$

Then show that conservation of the total mass of q per unit volume $\overline{\rho q}$ can be equivalently expressed in Eulerian and Lagrangian forms as

$$\frac{\partial \overline{\rho q}}{\partial t} + \nabla\cdot\left(\overline{\rho q \boldsymbol{v}}\right) = \phi\rho^\ell \frac{\mathrm{D}_\ell q^\ell}{\mathrm{D}t} + (1-\phi)\rho^s \frac{\mathrm{D}_s q^s}{\mathrm{D}t} - \Gamma \Delta q. \tag{8.56}$$

8.3 Find approximate analytical solutions to the steady porosity values shown in figure 15.5. *Hint: Use a Taylor expansion of the exponential function in (8.45).*

8.4 In this question we will look at the dissipation-driven melting and compaction under constant shear stress.

(a) In the case of constant shear stress, $\eta_\phi \mathcal{S}/\mathcal{H} = \mathcal{T}_0$, use (8.42) to obtain an analogous system of equations to (8.45).

(b) Using the evolution equation for the porosity from (a), find $\phi(t)$ under the assumptions $\lambda \sim \mathcal{O}(1)$ and $\Pi\phi \ll 1$. Furthermore, given that at $t=0$, $\phi=0$, find the time t^* at which disaggregation occurs ($\phi_\Xi \approx 0.3$).

(c) Adapt the code provided in the online supplement (the URL is provided in the preface) to solve the equations obtained in (a) numerically. Produce solutions for $\Pi = 0.1,\ 5,\ 20$, and $\lambda = 5$, and compare the values t^* to the values that you obtained in (b). *You will need to adjust t_{max} to get results.*

8.5 Consider the coupled equations (8.45) governing the layer thickness and porosity in the case of constant strain rate.

(a) Determine an ODE for $\mathcal{H}(\phi)$, independent of any time derivatives.

(b) By writing $e^{-\lambda\phi} \approx 1-\lambda\phi$ and $1-\phi \approx 1$, use the ODE from (a) to solve for $\mathcal{H}(\phi)$.

(c) Assuming that $\lambda \sim \mathcal{O}(1)$, use (8.45b) to find $\phi(t)$ in the limits $\Pi\phi \ll 1$ and $\Pi\phi \gg 1$.

8.6 In figure 15.5 we have noticed that at constant strain rate, the porosity tends to a pseudo–steady state. This behavior occurs for $\Pi > \Pi_c(\lambda)$, where the critical value of the compaction dissipation number depends on the parameter λ. Below the critical value $\Pi < \Pi_c$, the porosity does not tend to a steady state and rather increases continuously (illustrated by the asymptotic behaviour you obtained in exercise 8.5(c)).

(a) Starting from the porosity evolution equation for constant strain rate, equation (8.45b), for fixed λ, find Π_c. *Hint: At* $\Pi = \Pi_c$, *there is a unique porosity* ϕ_c *that gives a steady state in which* $\ddot{\phi}(\phi_c) = \dot{\phi}(\phi_c) = 0$.

(b) Starting from the porosity evolution equation for constant shear stress (exercise 8.4(a)) for fixed λ, find Π_c. *Hint: You can avoid doing the similar algebra again by noting the symmetry* ($\lambda \leftrightarrow -\lambda$) *of this governing equation and equation* (8.45b).

(c) For a given λ, is the critical value of the compaction–dissipation number Π_c greater for constant shear stress or for constant strain rate? Recalling that for systems dominated by shear dissipation, $\Psi = \eta_\phi \,(\mathcal{S}/\mathcal{H})^2 = \mathcal{T}^2/\eta_\phi$, and $\Gamma = \Psi/L$, explain mathematically and physically why you would expect one of them to be higher.

8.7 Consider a steady, one-dimensional model of decompression melting of the mantle where we assume no radiogenic heating, zero permeability, and no decompressional cooling, with equal densities between the two phases and a constant upwelling rate $\overline{\boldsymbol{v}} = W_0\hat{\boldsymbol{z}}$. The temperature equation in this case becomes

$$W_0 \frac{\mathrm{d}T}{\mathrm{d}z} = \kappa \frac{\mathrm{d}^2 T}{\mathrm{d}z^2} - \frac{L\Gamma}{c_P \rho}, \tag{8.57}$$

and therefore the temperature is a function of depth only: $T = T(z)$.

Beneath the region of partial melting we have $T \to T_0$ as $z \to -\infty$. Within the region of partial melting, the temperature is fixed by the melting process to remain on the solidus $T^S(z) = T_0^S - \rho g z/\vartheta < T_0$ (we shall see this in more detail in chapter 10). We assume that the melting rate is Γ_0, a constant. The depth of the bottom of the partially molten region is $z_0 < 0$, to be determined as part of the problem. Just below the depth z_0, thermal diffusion will cool the upwelling mantle. Our aim here is to quantify this diffusive boundary layer.

(a) Solve for the temperature at $z \geq z_0$ and, at the same time, obtain z_0. Note that the heat flux must be continuous across z_0.

(b) Assuming no diffusion ($\kappa = 0$), determine the depth z_0^* and compare this to z_0 from the previous part. Comment on the significance of diffusion in this region.

CHAPTER 9

Conservation of Chemical-Species Mass

Mantle rock and magma are composed of many chemical species. Some of these are considered *major elements* because they constitute a significant fraction of the total mass (usually >1%). Magnesium, iron, silicon, oxygen, calcium, and aluminum are all major elements. Major elements are all participants in the chemical thermodynamics of melting. Volatile species such as water and carbon typically have concentrations on the order of 100 ppm (by mass) in the asthenosphere (10^{-6}%) but, despite this, they too play an important role in the thermodynamics. Trace elements typically have mass concentrations in the range of parts per million. But unlike volatiles, trace elements have a negligible effect on the Gibbs free energies of the solid and liquid phases. Some trace elements are produced or destroyed by radiogenic decay; these are referred to as radiogenic elements. Fortunately, all elements obey similar conservation equations. Much of the theory developed here is adapted from Spiegelman and Elliott [1993].

The mass of a chemical species per unit mass within a phase (liquid or solid) is known as the *concentration* of that species. For species j in phase $i = \ell, s$, the volume-averaged, continuum concentration is defined in terms of microscopic quantities as

$$\phi^i \rho^i c_j^i = \frac{1}{\delta V} \int_{\text{RVE}} \Phi^i \breve{\rho} \breve{c}_j \, \mathrm{d}^3 \breve{\boldsymbol{x}}, \tag{9.1}$$

where $\breve{c}_j$ is the microscopic concentration of species j at $\breve{\boldsymbol{x}}$ within the RVE (notation used in equations (9.1)–(9.3) was introduced in sections 4.1–4.2).

While the integral in (9.1) is well defined, it raises an issue that we have not considered previously in this book: that variations within phases at the microscopic scale might have important effects at the macroscopic scale. Diffusion in the liquid phase is fast enough to homogenize concentration within pores rapidly. But diffusion in the solid is extremely slow, and hence grain-scale concentration gradients can persist over long periods. This *crystal zoning* means that the interior of grains (especially larger ones) can have a composition that is distinct from the rim and from the volume average. Equations based on the volume-averaged composition of the solid phase may introduce systematic bias into the model. We address this below in the context of trace elements, which have especially low diffusivities.

Putting this concern aside for the moment, a statement of conservation of species mass can then be written as

$$
\begin{aligned}
\frac{\mathrm{d}}{\mathrm{d}t}\int_{\mathrm{RVE}} \Phi^i \breve{\rho}\breve{c}_j \,\mathrm{d}^3\breve{\boldsymbol{x}} = &-\int_{\partial\mathrm{RVE}} \Phi^i \breve{\rho}\breve{c}_j \breve{\boldsymbol{v}}\cdot \mathrm{d}\boldsymbol{S} \\
&-\int_{\partial\mathrm{RVE}} \Phi^i \breve{\boldsymbol{J}}_j \cdot \mathrm{d}\boldsymbol{S} + \int_{\mathcal{I}_{\mathrm{RVE}}} \breve{\rho}^i \breve{c}_j^{\mathcal{I}} (\breve{\boldsymbol{v}}^{\mathcal{I}} - \breve{\boldsymbol{v}}^i)\cdot \breve{\boldsymbol{n}}^i \,\mathrm{d}S_{\mathcal{I}}.
\end{aligned} \tag{9.2}
$$

The rate of change of species mass within a RVE is on the left-hand side of this equation. On the right-hand side are terms representing, respectively, advective and diffusive transport through the boundaries of the RVE, and interphase mass transfer at the interface between phases. $\mathcal{I}_{RVE}$ denotes the interface between liquid and solid phases within the RVE; symbols evaluated on that interface are appended by $\mathcal{I}$. Also, $\Phi^i \breve{\boldsymbol{J}}_j$ represents the microscopic diffusive flux of component j within phase i.

The interphase mass-transfer term requires extra attention. The concentration of components in the mass that is transferred is not, in general, equal to either the source phase or the destination phase. Indeed, the concentration that is used in the integral along the interface is $\breve{c}_j^{\mathcal{I}}$; it is unconstrained by simple mass conservation statements. Bearing this in mind, the interface term can be written using the definition

$$
\begin{aligned}
\Gamma_j &\equiv \frac{1}{\delta V}\int_{\mathcal{I}_{\mathrm{RVE}}} \breve{\rho}^\ell \breve{c}_j^{\mathcal{I}} (\breve{\boldsymbol{v}}^{\mathcal{I}} - \breve{\boldsymbol{v}}^\ell)\cdot \breve{\boldsymbol{n}}^\ell \,\mathrm{d}S_{\mathcal{I}} \\
&= -\frac{1}{\delta V}\int_{\mathcal{I}_{\mathrm{RVE}}} \breve{\rho}^s \breve{c}_j^{\mathcal{I}} (\breve{\boldsymbol{v}}^{\mathcal{I}} - \breve{\boldsymbol{v}}^s)\cdot \breve{\boldsymbol{n}}^s \,\mathrm{d}S_{\mathcal{I}},
\end{aligned} \tag{9.3}
$$

where Γ_j is the rate of mass transfer of species j per unit volume of the two-phase continuum.

Applying procedures discussed in section 4.1, we obtain conservation equations for species j in each phase and, by summing them, in the bulk, obtain

$$
\frac{\partial}{\partial t}\left(\phi^\ell \rho^\ell c_j^\ell\right) + \boldsymbol{\nabla}\cdot\left(\phi^\ell \rho^\ell c_j^\ell \boldsymbol{v}^\ell\right) = \Gamma_j + \boldsymbol{\nabla}\cdot \phi^\ell \rho^\ell \mathcal{D}_j \boldsymbol{\nabla} c_j^\ell, \tag{9.4a}
$$

$$
\frac{\partial}{\partial t}\left(\phi^s \rho^s c_j^s\right) + \boldsymbol{\nabla}\cdot\left(\phi^s \rho^s c_j^s \boldsymbol{v}^s\right) = -\Gamma_j, \tag{9.4b}
$$

$$
\frac{\partial}{\partial t}\overline{\rho c_j} + \boldsymbol{\nabla}\cdot \overline{\rho c_j \boldsymbol{v}} = \boldsymbol{\nabla}\cdot \phi^\ell \rho^\ell \mathcal{D}_j \boldsymbol{\nabla} c_j^\ell, \tag{9.4c}
$$

where, using Fick's law, we have written the diffusive flux as $\boldsymbol{J}_j^\ell = -\phi^\ell \rho^\ell \mathcal{D}_j \boldsymbol{\nabla} c_j^\ell$, with $\mathcal{D}_j$ being the diffusivity of component j in the liquid phase. We have assumed that diffusion in the solid is so slow as to be negligible. Note that we have not assumed that diffusion within solid grains is negligible; in important cases it is not. However, our continuum model resolves only gradients in the mean species concentration of each phase. After averaging over the RVE, we cannot resolve grain-scale diffusion in the continuum model.

In equations (9.4), Γ_j is a portion of the total interphase mass-transfer rate Γ. We can express this fraction as c_j^Γ and understand it as representing the concentration of j

in an increment of mass that is transferred between phases. Its definition is

$$c_j^\Gamma \equiv \Gamma_j / \Gamma. \tag{9.5}$$

We can think of c_j^Γ as the interface concentration averaged over the area of the liquid–solid interface within the RVE. Hence, from $\breve{c}_j^{\mathcal{I}}$ it inherits a lack of constraint.

The conservation of species mass equations can be rewritten in a Lagrangian form using conservation of phase mass and the definition of c_j^Γ. Expanding the derivatives on the left-hand sides of equations (9.4), then substituting equations (4.11) and (9.5), gives

$$\phi^\ell \rho^\ell \frac{\mathrm{D}_\ell c_j^\ell}{\mathrm{D}t} = \left(c_j^\Gamma - c_j^\ell\right)\Gamma + \nabla\cdot\phi^\ell\rho^\ell\mathcal{D}\nabla c_j^\ell, \tag{9.6a}$$

$$\phi^s \rho^s \frac{\mathrm{D}_s c_j^s}{\mathrm{D}t} = -\left(c_j^\Gamma - c_j^s\right)\Gamma, \tag{9.6b}$$

$$\phi^\ell \rho^\ell \frac{\mathrm{D}_\ell c_j^\ell}{\mathrm{D}t} + \phi^s \rho^s \frac{\mathrm{D}_s c_j^s}{\mathrm{D}t} = \Delta c_j \Gamma + \nabla\cdot\phi^\ell\rho^\ell\mathcal{D}\nabla c_j^\ell. \tag{9.6c}$$

Above we have accounted for chemical diffusion even though, in most cases of interest, it transports chemistry slowly relative to melt segregation. However, chemical diffusivity is formally related to chemical *dispersivity*, which occurs during porous flow and may be more effective than diffusion. Dispersion arises because of the microscopic velocity gradients within pores and the tortuosity of the pore network. Indeed, the microscopic velocity of liquid is not, in general, aligned with its volume-averaged value. Fluctuations around this mean create a distribution of microscopic advection that leads to enhanced spreading of chemistry (especially in the direction parallel to the mean flow). At the continuum scale, this can be modeled with a dispersivity tensor $\tilde{\boldsymbol{\mathcal{D}}}$ that depends on the segregation rate $\Delta\boldsymbol{v}$ (and is hence invariant to changes in inertial reference frame). This tensor is then combined with diffusion by writing $\boldsymbol{\mathcal{D}} \equiv \tilde{\boldsymbol{\mathcal{D}}} + \mathcal{D}\boldsymbol{I}$ to give an effective diffusivity tensor $\boldsymbol{\mathcal{D}}$. We do not give further consideration to dispersivity in this book.

9.1 Thermodynamic Components

Thermodynamic components comprise the subset of chemical species that actively participate in the thermodynamics of phase change. This includes the major elements, by virtue of the fact that they are present in large quantities, as well as the volatile elements that, despite low concentrations, have an important effect on melting. The concentrations of mantle thermochemical components are generally not affected by radiogenic decay. Although there are a multitude of trace elements always present, it is convenient to treat the thermodynamic components as comprising the full mass of the material.

The thermodynamic components need not be defined in terms of the elements themselves, but rather can be defined as groups of elements that appear independently in the phases present. Oxides such as MgO and SiO_2 are sometimes used for mantle minerals, but in many cases we could more conveniently use end-member mineral formulas themselves (e.g., Mg_2SiO_4, forsterite olivine).

For a set of N thermochemical components that comprise the full mass of a real or model system, a few simple but important relations hold. Separately summing equations (9.4a) and (9.4b) over j from 1 to N and comparing with equations (4.11), we find that for each phase

$$\sum_{j=1}^{N} c_j^i = 1, \quad \sum_{j=1}^{N} \Gamma_j = \Gamma, \quad \sum_{j=1}^{N} \boldsymbol{J}_j^\ell = 0. \tag{9.7}$$

The first of these sums tells us that if we know the concentration of $N-1$ components, the Nth concentration is given as

$$c_N^i = 1 - \sum_{j=1}^{N-1} c_j^i. \tag{9.8}$$

This means that in the general case of N thermochemical components in chemical disequilibrium, we need to solve $2(N-1)$ conservation equations: one for each of $N-1$ components in each of the two phases. The second sum in (9.7) tells us that concentrations c_j^Γ in the melting (or freezing) reaction must also satisfy a unity sum. However, it does not mean that concentrations in the melting reaction are necessarily positive. As we shall see below, negative concentrations are associated with incongruent melting. The third sum in (9.7) tells us that not all of the N components in the liquid phase can be diffusive; at least one of them must be antidiffusive.

The interphase species-mass transfer rate Γ_j is associated with a set of melting reactions. This can be illustrated by consideration of the melting of forsterite (Fo) and magnesian enstatite[1] (En), which can be expressed as

$$Mg_2SiO_4(\text{Fo}, s) \longrightarrow 2MgO(\ell) + SiO_2(\ell), \tag{9.9a}$$

$$Mg_2Si_2O_6(\text{En}, s) \longrightarrow 2MgO(\ell) + 2SiO_2(\ell), \tag{9.9b}$$

where symbols in parentheses indicate the phase (ℓ or s) where the component exists and, where relevant, the mineral name. Taking component j to be SiO_2, both of these reactions would contribute to Γ_j. More generally, for a set of N_r reactions involving component j, we could construct Γ_j as

$$\Gamma_j = \sum_{k=1}^{N_r} \Gamma_{jk}. \tag{9.10}$$

Each reaction might progress at a different rate, depending on the thermodynamic conditions that are driving it (see chapter 10).

In two-phase models of magma/mantle dynamics, it has generally not been feasible to incorporate a set of thermochemical components that includes all of the thirteen (approximately) that are relevant for the mantle. Inclusion of such a large set would require a very robust and computationally fast thermodynamic model as well as the solution of a huge set of PDEs for species conservation. Instead, the approach that has been taken is to group the real components into sets of *effective components* that

[1] Also known as *orthoenstatite*.

behave in a way that is deemed to be qualitatively consistent with more detailed models. For example, a set of three effective components could be *refractory component*, *fertile component*, and *volatile component*, which loosely correspond to olivine, pyroxene, and H_2O+CO_2. We discuss this further at the end of the current chapter as well as in section 11.2.2.

9.1.1 CONGRUENT MELTING

We consider the two-component system comprising MgO and SiO_2. The solid can then be a mixture of grains of forsterite (olivine; Mg_2SiO_4) and enstatite (orthopyroxene; $Mg_2Si_2O_6$). We have the two melting reactions (9.9) and these may proceed at very different rates. Let's assume that both reaction rates are positive, with forsterite melting at a rate of a and enstatite melting at a rate of b, in moles per cubic meter–seconds. We consider how to model these rates in chapter 10. Using the molar mass M_j of each species that participates in the reaction, we can write the component concentrations in the instantaneous melt as

$$c^{\Gamma}_{\mathrm{MgO}} = \frac{2(a+b)M_{\mathrm{MgO}}}{aM_{\mathrm{Mg_2SiO_4}} + bM_{\mathrm{Mg_2Si_2O_6}}}, \qquad c^{\Gamma}_{\mathrm{SiO_2}} = \frac{(a+2b)M_{\mathrm{SiO_2}}}{aM_{\mathrm{Mg_2SiO_4}} + bM_{\mathrm{Mg_2Si_2O_6}}}. \tag{9.11}$$

The denominator of each of these expressions represents the total melting rate Γ. Since we expect pyroxene to melt much more readily than olivine, we assume $b \approx 10a$ mol/m^3-sec; then $c^{\Gamma}_{\mathrm{MgO}} \approx 0.41$ and $c^{\Gamma}_{\mathrm{SiO_2}} \approx 0.59$. Note that $\sum_j c^{\Gamma}_j = 1$.

It is important to note that although the melting reactions (9.9) involves two minerals, the two-phase flow formulation as developed in this book doesn't track multiple solid phases. However, the above illustrates the concept of components and their relationship to instantaneous melt composition.

9.1.2 INCONGRUENT MELTING

Let's next consider a melting reaction in which orthopyroxene (enstatite) decomposes into olivine (forsterite) and melt:

$$\mathrm{Mg_2Si_2O_6}(s) \longrightarrow \mathrm{Mg_2SiO_4}(s) + \mathrm{SiO_2}(\ell). \tag{9.12}$$

This is actually just the sum of reaction (9.9b) proceeding at a rate $+a$ and reaction (9.9a) proceeding at a rate $-a$ (i.e., going in reverse). Together, this is an *incongruent* melting reaction because a solid mineral decomposes into another solid mineral and some liquid.

The reaction concentration of SiO_2 is

$$c^{\Gamma}_{\mathrm{SiO_2}} = \frac{M_{\mathrm{SiO_2}}}{M_{\mathrm{Mg_2Si_2O_6}} - M_{\mathrm{Mg_2SiO_4}}}. \tag{9.13}$$

There is no other component in the liquid associated with this reaction and so $c^{\Gamma}_{\mathrm{SiO_2}} = 1$. As with congruent melting, $\sum_j c^{\Gamma}_j = 1$.

The incongruent melting reaction (9.12) is a linear combination of the reactions (9.9) in a -1:1 ratio. More generally, incongruent reaction could occur in a ratio of a:b such that we have

$$a\mathrm{Mg_2SiO_4}(s)+b\mathrm{Mg_2Si_2O_6}(s)\longrightarrow 2(a+b)\mathrm{MgO}(\ell)+(a+2b)\mathrm{SiO_2}(\ell). \qquad (9.14)$$

In this case, the component concentrations in the instantaneous melt are again given by (9.11). Note that when $a=-b$, $c^{\Gamma}_{\mathrm{SiO_2}}=1$ and $c^{\Gamma}_{\mathrm{MgO}}=0$.

In the development above we chose MgO and SiO_2 as our thermochemical components. But instead we could have chosen the formulas for forsterite and enstatite. In this case, the melt composition is a combination of Mg_2SiO_4 and $Mg_2Si_2O_6$ and the concentrations in the instantaneous melt are

$$c^{\Gamma}_{\mathrm{Mg_2SiO_4}}=\frac{aM_{\mathrm{Mg_2SiO_4}}}{bM_{\mathrm{Mg_2Si_2O_6}}+aM_{\mathrm{Mg_2SiO_4}}}, \qquad (9.15a)$$

$$c^{\Gamma}_{\mathrm{Mg_2Si_2O_6}}=\frac{bM_{\mathrm{Mg_2Si_2O_6}}}{bM_{\mathrm{Mg_2Si_2O_6}}+aM_{\mathrm{Mg_2SiO_4}}}. \qquad (9.15b)$$

To be considered incongruent, the reaction must consume enstatite ($b>0$) and produce forsterite ($a<0$). At the same time, to be considered melting (a net transfer of mass from the solid to the liquid phase), it must be true that $aM_{\mathrm{Mg_2SiO_4}}+bM_{\mathrm{Mg_2Si_2O_6}}>0$. Rearranging, inserting values for the molar masses, and using the constraints above tells us that $-1.43<a/b<0$; in this range we have $c^{\Gamma}_{\mathrm{Mg_2SiO_4}}<0$. For example, if we take $a/b=-1$, satisfying these requirements, we have $c^{\Gamma}_{\mathrm{Mg_2SiO_4}}\approx-2.33$ and $c^{\Gamma}_{\mathrm{Mg_2Si_2O_6}}\approx 3.33$. In the polybaric natural system of decompression melting beneath a mid-ocean ridge, the coefficients of the melting reaction vary substantially (see references in the Literature Notes at the end of this chapter).

9.2 Trace Elements

A model of trace elements must account for transfer between phases during melting and freezing, but it should also account for transfer that is independent of phase change. This is transfer that occurs during equilibration of the surface of a grain with the ambient melt. If the melt is oversaturated in a trace element with respect to the outside of the grain, the trace element concentration will increase in the grain and decrease in the melt as atoms diffuse from the melt into the grain. Of course this isn't really independent of phase change because the trace element itself has mass and, in such reactions, that mass is transferred between phases. However, since trace elements, by definition, comprise such a tiny fraction of the mass of either phase, we assume that exchange between phases can occur without affecting the porosity. This also means that a set of trace-element concentrations of different elements are not subject to a unity sum, distinct from the concentrations of major chemical component.

Adjusting the governing equations (9.4) to accommodate the consideration above leads to

$$\frac{\partial}{\partial t}\left(\phi^{\ell}\rho^{\ell}c_j^{\ell}\right)+\nabla\cdot\left(\phi^{\ell}\rho^{\ell}c_j^{\ell}\boldsymbol{v}^{\ell}\right)=+\left(\Gamma_j+\mathcal{X}_j\right)+\nabla\cdot\phi^{\ell}\rho^{\ell}\mathcal{D}\nabla c_j^{\ell}, \qquad (9.16a)$$

$$\frac{\partial}{\partial t}\left(\phi^{s}\rho^{s}c_j^{s}\right)+\nabla\cdot\left(\phi^{s}\rho^{s}c_j^{s}\boldsymbol{v}^{s}\right)=-\left(\Gamma_j+\mathcal{X}_j\right), \qquad (9.16b)$$

$$\frac{\partial}{\partial t}\overline{\rho c_j} + \nabla \cdot \overline{\rho c_j \boldsymbol{v}} = \nabla \cdot \phi^\ell \rho^\ell \mathcal{D} \nabla c_j^\ell. \tag{9.16c}$$

Here we have introduced $\mathcal{X}_j$ as the exchange reaction rate of trace element j. Note that its sign is opposite for the two phases and hence it cancels in the equation for the bulk concentration, which is identical to (9.4c). Exchange reactions are unlike the mass transfer due to melting Γ_j, which is driven by changes in pressure, temperature, and major element concentrations. In contrast, they are driven by the local concentration ratio of the trace elements themselves.

9.2.1 EQUILIBRIUM TRANSPORT MODEL

Trace-element equilibrium is defined by a concentration ratio between the two phases,

$$c_j^s = D_j c_j^\ell, \tag{9.17}$$

where D_j is known as the *partition coefficient* of the species j. When $D_j < 1$, species j is partitioned preferentially into the liquid; this species is then considered to be *incompatible*. A *compatible* species, in contrast, has $D_j > 1$ and, hence, partitions preferentially into the solid.

If we assume that trace elements are always and everywhere in equilibrium, then we need track only the concentration in one phase. Having obtained it, we can always use equation (9.17) to obtain the concentration in the other phase. Substituting the partitioning relationship (9.17) into equation (9.16c) for conservation of mass in the bulk mixture, making the extended Boussinesq approximation and rearranging gives

$$\frac{\partial c_j^\ell}{\partial t} + \boldsymbol{v}^{D_j} \cdot \nabla c_j^\ell = c_j^\ell \left[\beta_j \Gamma - \dot{D}_j\right] + \nabla \cdot \phi^\ell \mathcal{D} \nabla c_j^\ell. \tag{9.18}$$

Because this equation is derived from mass conservation in the bulk, the trace element exchange rates are excluded (moreover, since these reactions are driven by disequilibrium, they are identically zero in the equilibrium formulation). In equation (9.18) we have introduced

$$\beta_j \equiv \frac{1}{\rho}\left(\frac{D_j - 1}{\phi^\ell + D_j \phi^s}\right), \quad \dot{D}_j \equiv \frac{\phi^s}{\phi^\ell + D_j \phi^s}\left(\frac{\partial D_j}{\partial t} + \boldsymbol{v}^s \cdot \nabla D_j\right). \tag{9.19}$$

The latter is the rate of change of the partition coefficient in the reference frame of the solid. More importantly, we have defined

$$\boldsymbol{v}^{D_j} \equiv \frac{\phi^\ell \boldsymbol{v}^\ell + D_j \phi^s \boldsymbol{v}^s}{\phi^\ell + D_j \phi^s} \tag{9.20}$$

as the *chromatographic velocity* of species j. Note that $\boldsymbol{v}^{D_j}$ depends on the partition coefficient. Highly incompatible elements have $D_j \ll 1$ and move with $\boldsymbol{v}^{D_j} \sim \boldsymbol{v}^\ell$; highly compatible elements with $D_j \gg 1$ move with $\boldsymbol{v}^{D_j} \sim \boldsymbol{v}^s$. This dependence of transport velocity on elemental partitioning between liquid and solid is the basis for chromatography; it can also fractionate trace elements in the mantle.

Neglecting trace-element diffusion, we can rewrite equation (9.18) in a simple form that is convenient for numerical solution,

$$\frac{\partial C_j^\ell}{\partial t} + \boldsymbol{v}^{D_j} \cdot \nabla C_j^\ell = \beta_j \Gamma - \dot{D}_j, \tag{9.21}$$

where $C_j^\ell \equiv \ln c_j^\ell$. This logarithmic variable can more accurately accommodate the large dynamic range of the concentration; it has the added benefit of guaranteeing positivity of concentration.

To solve equations (9.18) or (9.21) one needs the phase velocities $\boldsymbol{v}^\ell$ and $\boldsymbol{v}^s$, a closure for the melting rate Γ, and models for the partition coefficients D_j. Partition coefficients are known to vary with pressure, temperature, and the solid minerals present. The assemblage of minerals is sensitive to the major element chemistry, and also depends on pressure and temperature. This can become complicated quickly; in published models it has generally been assumed that $D_j = D_j(P)$, where P is the lithostatic pressure.

It is worth returning to the definition of $\boldsymbol{v}^{D_j}$ to consider how two different but incompatible trace elements can be *fractionated*. For porosity small enough that $\phi^s = 1 - \phi^\ell \sim 1$,

$$\boldsymbol{v}^{D_j} \approx \frac{\boldsymbol{v}^\ell + D_j \boldsymbol{v}^s/\phi^\ell}{1 + D_j/\phi^\ell}. \tag{9.22}$$

This equation shows that an incompatible trace element moves with the liquid when $D_j \ll \phi^\ell$ and moves with the solid when $D_j \gg \phi^\ell$. Consider two different trace elements with partition coefficients $D_j < D_k < 1$. Under what conditions are these two elements separated by melt transport? In other words, under what conditions are $\boldsymbol{v}^{D_j}$ and $\boldsymbol{v}^{D_k}$ sufficiently different from each other?

To answer this, we simplify our model of the flow to be unidirectional with speeds $v^\ell > v^s$. Then we find that

$$\frac{v^{D_j} - v^{D_k}}{v^\ell - v^s} = \frac{\phi^\ell (D_k - D_j)}{(\phi^\ell + D_j)(\phi^\ell + D_k)}. \tag{9.23}$$

This relationship is plotted in figure 9.1 for two sets of D_j, D_k. Note that fractionation of the trace elements is most efficient when $D_j < \phi < D_k$ and drops off for porosity outside of this range. In physical terms, fractionation of trace elements under equilibrium partitioning occurs at a porosity such that one element moves with the melt whereas the other moves with the solid.

9.2.2 DISEQUILIBRIUM TRANSPORT MODEL

Liquid magma and solid mantle are in contact only at the liquid–solid interface. At this surface, trace element equilibrium is attained rapidly. However, the concentrations that we solve for in equations (9.16) are the volume-averaged concentrations, which include contributions from material that is adjacent to the interface as well as away from it. The equilibrium statement (9.17) is strictly valid for the interface concentrations only, but we have assumed that it applies to these averaged concentrations. More generally, even if equilibrium is established at the interface, the partitioning equation does not hold for the volume-averaged concentrations. Concentration gradients exist

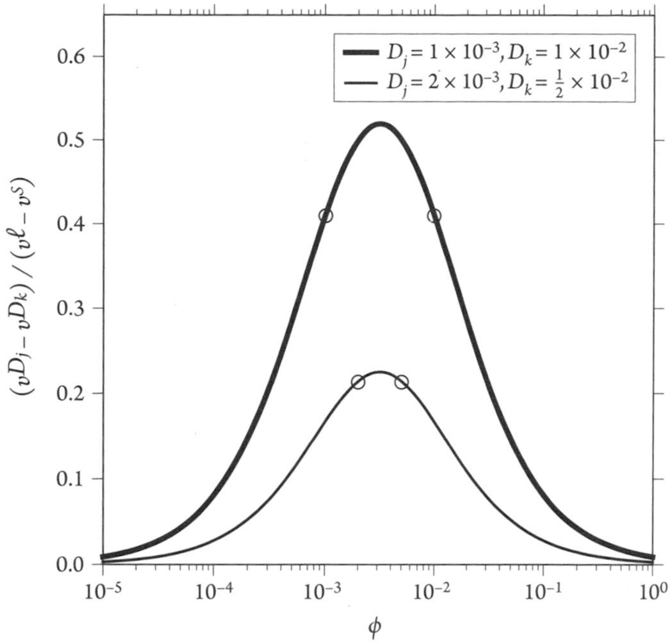

Figure 9.1. Segregation rate of two trace elements by equilibrium transport as a function of porosity, according to (9.23). The curve is plotted for two sets of D_j, D_k. Note the speed difference between the two trace elements is plotted as a fraction of the speed difference between liquid and solid.

within grains and/or pores, and material must diffuse through the solid or the liquid (at scales smaller than the RVE) to bring the concentrations into equilibrium. In the liquid, this happens rapidly because pores are small, concentration gradients are sharp, and the diffusivity is relatively large; in the solid, diffusion is very slow. Depending on the grain size, the equilibration times can be long (i.e., time scales that are much longer than typical melt-extraction times).

For a diffusion coefficient of $\mathcal{D}_j^s \approx 10^{-15}$ m^2/s and a grain size d of 1 cm, the characteristic diffusion time is $d^2/\mathcal{D}_j^s \approx 3000$ years. To achieve equilibrium with a melt, the adjacent melt should be either stationary or have constant concentration for at least this time period. If the concentration in the liquid is fluctuating with a period that is less than this time scale, disequilibrium will prevail. In that case, equation (9.18) is not a good model of the system. One approach to modeling the disequilibrium would be to add an additional, independent variable representing the radius of a typically sized grain at each point in space and solving for the concentration as a function of position, time, and radius. Here we take a simpler approach, already introduced, and use the exchange-rate term $\mathcal{X}_j$. We discuss this term in more detail below.

Returning to equations (9.16), writing the contribution of melting as $\Gamma_j = c_j^\Gamma \Gamma$, using conservation of mass, and neglecting diffusion gives

$$\phi^\ell \rho^\ell \frac{\mathrm{D}_\ell c_j^\ell}{\mathrm{D}t} = + \left(c_j^\Gamma - c_j^\ell\right)\Gamma + \mathcal{X}_j, \tag{9.24a}$$

$$\phi^s \rho^s \frac{\mathrm{D}_s c_j^s}{\mathrm{D}t} = -\left(c_j^\Gamma - c_j^s\right) \Gamma - \mathcal{X}_j. \tag{9.24b}$$

Several closures are needed to enable solution of this system. Of course the melting rate Γ remains an unknown, the treatment of which we defer to chapter 10. The trace-element concentration in the mass that is transferred between phases c_j^Γ can be modeled as follows. Let us assume that each instantaneous increment of melt is produced in equilibrium with the solid phase; let us further assume that instantaneous increments of solid that are produced by freezing are in equilibrium with the liquid. Hence, we write

$$c_j^\Gamma \equiv \begin{cases} c_j^s / D_j & \text{if } \Gamma > 0, \\ c_j^\ell D_j & \text{if } \Gamma < 0, \end{cases} \tag{9.25}$$

Because these instantaneous melts and solids are produced in equilibrium only with the source phase, the model in (9.25) represents *fractional melting* (see exercise 9.1) and *fractional crystallization*. However, these new melts or crystals do not necessarily move the local two-phase system toward trace-element equilibrium.

The process of equilibration is accommodated by $\mathcal{X}$. A simple (but nonunique) model for this rate is that it is proportional to the amount of trace-element disequilibrium. This linear kinetic rate law can be expressed as

$$\mathcal{X}_j \equiv r_j \left(c_j^s - D_j c_j^\ell\right), \tag{9.26}$$

where r_j is an exchange-rate coefficient with dimensions of mass per volume per time. The efficiency of the reaction (and hence r_j) is determined by the time required for the core of a typical crystal to diffusively equilibrate with its rim. Hence, we might expect the exchange-rate coefficient to be related to the grain size and shape, as well as the diffusivity of species j in the crystal, as a function of temperature and pressure. A dimensionally correct guess would be

$$r_j = r_0 \rho^s \mathcal{D}_j^s / d^2, \tag{9.27}$$

where r_0 is an $\mathcal{O}(1)$ dimensionless constant representing the grain shape, d is the grain size, and $\mathcal{D}_j^s$ is the rate of diffusion of species j in the solid grain.

We can substitute equation (9.26) into equation (9.24) and rewrite the result in terms of logarithmic concentrations by again using $C_j \equiv \ln c_j$,

$$\frac{\mathrm{D}_\ell C_j^\ell}{\mathrm{D}t} = +\frac{1}{\phi^\ell \rho^\ell} \left\{ \left[\mathrm{e}^{\left(C_j^\Gamma - C_j^\ell\right)} - 1 \right] \Gamma + r_j \left(\mathrm{e}^{\Delta C_j} - D_j\right) \right\}, \tag{9.28a}$$

$$\frac{\mathrm{D}_s C_j^s}{\mathrm{D}t} = -\frac{1}{\phi^s \rho^s} \left\{ \left[\mathrm{e}^{\left(C_j^\Gamma - C_j^s\right)} - 1 \right] \Gamma - r_j \left(\mathrm{e}^{\Delta C_j} - D_j\right) \mathrm{e}^{-\Delta C_j} \right\}, \tag{9.28b}$$

where $\Delta C_j = C_j^s - C_j^\ell$. If $r_j = 0$, these equations model fractional melting/freezing and transport of species j (see exercise 9.1); if $r_j \to \infty$, they model batch melting and freezing with equilibrium transport. However, equilibrium is more conveniently described by equation (9.21).

9.3 Radiogenic Trace Elements and Their Decay Chains

Some trace elements that are present in the mantle undergo radioactive decay, changing to a different isotope or different element by a release of nuclear mass. These decays occur along sequences of nuclides. The half-lives of nuclides vary widely, but some decay at time scales similar to melt-tranport scales of 10^3–10^5 years. For nuclides in this class, as they travel in the magma from the mantle source toward a volcano, their concentrations will evolve in a way that depends on the details of the melting and melt-transport processes. Nuclides with half-lives of 10^6–10^9 years can be modeled as nonradiogenic trace elements over magmatic time scales; their evolution is instead shaped by the time scales and pathways of mantle convection.

As an example of a radiogenic decay chain that is relevant to melt transport, consider

$$^{238}\mathrm{U} \xrightarrow{\lambda_1} \left(^{234}\mathrm{Th} \rightarrow {}^{234}\mathrm{Pa} \rightarrow\right) {}^{234}\mathrm{U} \xrightarrow{\lambda_2} {}^{230}\mathrm{Th} \xrightarrow{\lambda_3} {}^{226}\mathrm{Ra} \xrightarrow{\lambda_4} (\ldots \rightarrow)\, {}^{206}\mathrm{Pb},$$

where the λ_j are decay constants (units of inverse time) for each step in the decay chain. Superscripts preceding element names are the mass number that identifies a particular isotope. This is one of the two uranium series (or U-series) decay chains. Elements and decay steps in parentheses have relatively short half-lives ($\tau^{(1/2)}$) that can be considered to be instantaneous in this context. For example, ^{234}Th has a half-life of about 24 d, whereas its parent ^{238}U has $t_{1/2} \approx 4.5$ billion years. A summary of the decay chain with decay constants and half-lives is given in table 9.1.

Measurement of isotope ratios of U-series elements can potentially provide an observational constraint on the time scales of melt generation and extraction. If the measurements are made on fresh lavas erupted from, say, a mid-ocean ridge, these ratios may hold information about the time since elemental fractionation occurred. Fractionation is a marker of a particular stage in the melting process that can, hypothetically, be ascribed to a particular depth in the mantle beneath the ridge axis. However, the interpretation of measured U-series isotope ratios in terms of melting and melt transport is nontrivial. Models are therefore useful, though they require significant mathematical development. Here we set out a foundation; the framework for modeling mid-ocean ridge observations is described in section 11.5.

It is convenient to index the elements of a decay chain in order of their rank in the chain. With this choice, element $j-1$ decays to element j, which decays to $j+1$.

The rate of decay of a radioactive element is linearly proportional to its concentration:

$$\dot{c}_j^i = -\lambda_j c_j^i \quad \text{(decay)}, \tag{9.29}$$

where λ_j is the decay coefficient of the jth element. The rate of decay is also known as the *activity* a_j^i of the element,

$$a_j^i \equiv \lambda_j c_j^i. \tag{9.30}$$

For the jth element in a decay chain (excluding those at the top and bottom of the chain), decay of the *parent element* $j-1$ is a source and decay of the *daughter element* j is a sink. The net rate of change is

$$\dot{c}_j^i = \lambda_{j-1} c_{j-1}^i - \lambda_j c_j^i \quad \text{(net)}. \tag{9.31}$$

Table 9.1. Decay constants and half-lives of isotopes in the U-series decay chain that begins with ^{238}U. Another series, beginning with ^{235}U is not shown. These D_j numbers are effective values based on a typical mantle source mineralogy (without garnet) [Spiegelman and Elliott, 1993]. Elkins et al. [2008] measured residue–melt partitioning of U and Th for garnet and clinopyroxene. The presence of garnet in the residue reduces D_3 (Th) and increases D_4 (U) compared to a garnet-free residue that contains cpx.

j	Isotope	Decay constant, y	Half-life	Partition coefficient
1	^{238}U	$\lambda_1 = 1.5 \times 10^{-10}$	4.5×10^9 y	$D_1 = 0.0086$
	^{234}Th		24.1 d	
	^{234}Pa		1.17 min	
2	^{234}U	$\lambda_2 = 2.83 \times 10^{-6}$	2.45×10^5 y	$D_2 = 0.0086$
3	^{230}Th	$\lambda_3 = 9.19 \times 10^{-6}$	7.54×10^4 y	$D_3 = 0.0065$
4	^{226}Ra	$\lambda_4 = 4.33 \times 10^{-4}$	1.60×10^3 y	$D_4 = 0.0005$
	(various)		4 d [max.]	
5	^{210}Pb	$\lambda_5 = 3.12 \times 10^{-2}$	22.2 y	
	(various)		5 d [max.]	
6	^{206}Pb		stable	

This difference, scaled with $\phi^i \rho^i$, can be incorporated into the right-hand side of equations (9.16a) and (9.16b) to give phase-specific evolution equations for concentration of an element in a decay chain. Summing those two equations gives a bulk conservation equation analogous to equation (9.16c). In this system, equilibrium elemental partitioning between liquid and solid phases is no different from that for trace elements. Hence, we use equation (9.17), $c_j^s = D_j c_j^\ell$, to eliminate c_j^s from the bulk conservation equation. Neglecting diffusion, we obtain

$$\frac{\partial c_j^\ell}{\partial t} + \boldsymbol{v}^{D_j} \cdot \nabla c_j^\ell = c_j^\ell \left[\beta_j \Gamma - \dot{D}_j\right] + \lambda_{j-1} D_{j-1}^r c_{j-1}^\ell - \lambda_j c_j^\ell, \tag{9.32}$$

where β_j, $\dot{D}_j$, and $\boldsymbol{v}^{D_j}$ are as given in (9.19) and (9.20); we have introduced

$$D_{j-1}^r \equiv \frac{\phi^\ell + D_{j-1}\phi^s}{\phi^\ell + D_j \phi^s}. \tag{9.33}$$

Note that $D_{j-1}^r \to 1$ when the porosity ϕ^ℓ is much greater than either of the partition coefficients.

Observations of uranium series are typically presented in terms of *secular disequilibrium* of *activity ratios* in the magma. An activity ratio is the quotient of the activity of the daughter and the activity of the parent. In secular equilibrium, these are equal and hence their ratio is unity: each decay of the daughter is balanced by a decay of the parent, leading to steady-state concentration of the daughter. If the daughter is decaying more rapidly than the parent, $a_j^\ell / a_{j-1}^\ell > 1$, indicating secular disequilibrium. With time, the concentration of the daughter will decrease, its decay will slow, and this will restore the system back to secular equilibrium.

The mantle convects slowly and parcels remain isolated for hundreds of millions to billions of years—ample time to restore secular equilibrium. How then is secular disequilibrium created in the first place? It is created by chemical fractionation of parent and daughter elements—alteration of their concentration ratio—which occurs by melting and melt segregation. For equilibrium transport, this requires that the parent and daughter isotopes have different partition coefficients and that the porosity falls in a range that causes segregation (c.f., section 9.2.1 and fig. 9.1). To model the evolution of activity during melting and segregation, we can multiply equation (9.32) by λ_j and use the definition of activity, (9.30), to obtain

$$\frac{\partial a_j^\ell}{\partial t} + \boldsymbol{v}^{D_j} \cdot \nabla a_j^\ell = a_j^\ell \left[\beta_j \Gamma - \dot{D}_j\right] + \lambda_j \left(D_{j-1}^r a_{j-1}^\ell - a_j^\ell\right). \tag{9.34}$$

Equation (9.34) states that, at a fixed location, the activity of an element changes for three reasons. The first is advection (the second term on the left-hand side; see exercise 9.3). Elements move at their chromatographic velocities, (9.20), a weighted average of the solid and liquid velocities where the weight depends on the partition coefficient and the porosity. The second reason is interphase transfer by melting (or freezing; first group on the right-hand side of (9.34)). Melting concentrates or dilutes an element in the liquid depending on the partition coefficient and the porosity. Interphase transfer can also be driven by changes in the partition coefficient itself, which can happen when a mineral phase is exhausted from the solid residue (e.g., garnet). The last reason for secular changes in activity is *ingrowth* (represented by the second group on the right-hand side of (9.34)). Ingrowth arises from secular disequilibrium within the decay chain: if the parent is decaying more rapidly than the daughter, $a_{j-1}^\ell > a_j^\ell$ (and the porosity is sufficiently large that $D^r \sim 1$), then the activity of the daughter will increase.

Secular disequilibrium is measured by the ratio of parent to daughter activities (in particular, the deviation of this ratio from unity). To create secular disequilibrium, either or both of the activities must change. An important way to create secular disequilibrium is to fractionate parent and daughter concentrations. This occurs when parent and daughter have different partition coefficients such that equilibrium melting preferentially concentrates one of them in the liquid, and then they segregate because of their different chromatographic velocities. The disequilibrium created in this way is then modified by ingrowth. It is straightforward to show that ingrowth ultimately drives the decay chain back toward secular equilibrium. But ingrowth can also cause disequilibrium activity ratios to cascade down (or even propagate up) the chain (see exercise 9.5).

In the next section, we consider example solutions that illustrate these effects, though we exclude melt segregation and chromatographic fractionation (we will return to those in section 11.5).

9.4 Closed-System Evolution of a Decay Chain

In a closed-system model, we assume that there is no melt segregation and hence we discard the advection term in (9.34). Furthermore, for simplicity, we assume constant (but not equal) partition coefficients and prescribe the melting rate as $\mathcal{G} \equiv \Gamma/\rho = \mathrm{d}\phi/\mathrm{d}t$. Then

$$\phi(t) = \mathcal{G}t, \tag{9.35}$$

where it should be noted that in a closed system with no melt segregation and $\rho^\ell = \rho^s = \rho$ we have $\phi = F$ (recall that F is the degree of melting, the fraction of the initial solid mass that has been converted to liquid). Equation (9.34) becomes

$$\frac{\mathrm{d}a_j^\ell}{\mathrm{d}t} = -a_j^\ell \frac{(1-D_j)\mathcal{G}}{D_j + (1-D_j)\mathcal{G}t} + \lambda_j \left[D_{j-1}^r a_{j-1}^\ell - a_j^\ell\right]. \tag{9.36}$$

In this closed system, changes in activity are due to interphase transfer and ingrowth only. We consider each of these separately, below, then analyse them jointly in section 9.4.3.

9.4.1 EVOLUTION WITH MELTING ONLY

Setting $\lambda_j = 0$ for all j reduces the system to a set of inert trace elements with partition coefficients D_j. Returning to equation (9.32), neglecting advection, assuming constant partition coefficients, taking $\lambda_j = 0$ for all j, and recasting in terms of the natural logarithm of concentration, gives

$$\frac{\mathrm{d}}{\mathrm{d}t}(\ln c_j^\ell) = -\frac{(1-D_j)\mathcal{G}}{D_j + (1-D_j)\mathcal{G}t}. \tag{9.37}$$

Recall that the liquid–solid system is in chemical equilibrium and hence $c_j^s = D_j c_j^\ell$. This equation can be integrated directly with the initial condition

$$c_j^\ell = c_j^s|_0 / D_j \quad \text{at} \quad t = 0 \tag{9.38}$$

to give

$$c_j^\ell(t) = \frac{c_j^s|_0}{D_j + (1-D_j)\mathcal{G}t}. \tag{9.39}$$

This is the batch melting equation, parameterized as a function of time. Curves for the elements of the uranium series (see table 9.1) are plotted in figure 9.2(a). For $\mathcal{G}t = F \lesssim D_j$, an element j (say, uranium) concentrates in the liquid by a roughly constant factor $1/D_j$. Once the melt fraction exceeds this level, additional melting only dilutes the liquid and the concentration of j (in our example, uranium) declines. Because melt fractions cannot exceed unity, our simple model makes sense only for $t < \mathcal{G}^{-1}$.

9.4.2 EVOLUTION DUE TO INGROWTH ONLY

Without melting ($\mathcal{G} = 0$) but with some constant porosity $\phi \gg \max(D_j)$, equation (9.33) tells us that $D_j^r \sim 1$, and equation (9.36) becomes

$$\frac{\mathrm{d}a_j^\ell}{\mathrm{d}t} = \lambda_j \left[a_{j-1}^\ell - a_j^\ell\right]. \tag{9.40}$$

For a decay chain with N elements, this equation can be expressed in matrix–vector format as

$$\begin{pmatrix} -\lambda_1 & 0 & 0 & \cdots & 0 \\ \lambda_1 & -\lambda_2 & 0 & & \vdots \\ 0 & \lambda_2 & -\lambda_3 & & \\ \vdots & & & \ddots & 0 \\ 0 & \cdots & 0 & \lambda_{N-1} & -\lambda_N \end{pmatrix} \begin{pmatrix} a_1^\ell \\ a_2^\ell \\ a_3^\ell \\ \vdots \\ a_N^\ell \end{pmatrix} = \frac{\mathrm{d}}{\mathrm{d}t} \begin{pmatrix} a_1^\ell \\ a_2^\ell \\ a_3^\ell \\ \vdots \\ a_N^\ell \end{pmatrix} \tag{9.41}$$

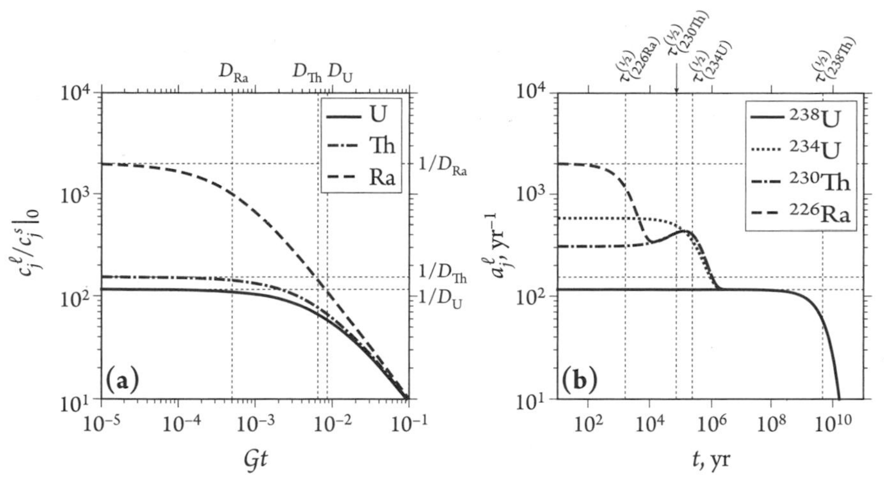

Figure 9.2. Closed system evolution of the uranium-series decay chain. **(a)** Under melting only, as per equation (9.39). **(b)** Under ingrowth only, as per equation (9.45). The initial activities are $a_j^s|_0/D_j$, where $a_j^s|_0 = 1, 5, 2$, and 1 per year for ^{238}U, ^{234}U, ^{230}Th, and ^{226}Ra, respectively, are chosen to create an interesting evolution. Vertical dotted lines mark the isotopic half-lives.

or, in terms of a decay matrix $\boldsymbol{\mathcal{L}}$ and activity vector $\boldsymbol{a}^\ell$,

$$\boldsymbol{\mathcal{L}}\boldsymbol{a}^\ell = \dot{\boldsymbol{a}}^\ell. \tag{9.42}$$

Performing an eigenvalue decomposition on $\boldsymbol{\mathcal{L}}$ gives

$$\boldsymbol{V}\grave{\boldsymbol{\mathcal{L}}}\boldsymbol{V}^{-1}\boldsymbol{a}^\ell = \dot{\boldsymbol{a}}^\ell, \tag{9.43}$$

where $\grave{\boldsymbol{\mathcal{L}}} = \text{diag}(-\lambda_1, -\lambda_2, ..., -\lambda_N)$ is a diagonal matrix of eigenvalues and $\boldsymbol{V}$ is a matrix of eigenvectors. Both $\boldsymbol{V}$ and its inverse can be formed analytically as

$$\left.\begin{aligned} V_{ij} &= \prod_{m=i+1}^{N} \frac{\lambda_m - \lambda_j}{\lambda_m}, \\ V_{ij}^{-1} &= \left(\prod_{m=j+1}^{N} \lambda_m\right)\left(\prod_{m=j}^{N,\, m\neq i} \frac{1}{\lambda_m - \lambda_i}\right) \end{aligned}\right\} \text{ for } j \leq i. \tag{9.44}$$

The solution of this equation in terms of the vector of activities $\boldsymbol{a}^\ell|_0$ at $t = 0$ is

$$\boldsymbol{a}^\ell(t) = \boldsymbol{V}\mathrm{e}^{\grave{\boldsymbol{\mathcal{L}}}t}\boldsymbol{V}^{-1}\boldsymbol{a}^\ell|_0. \tag{9.45}$$

This solution has a surprising property: the evolution of any activity can be computed independently of the evolution of all the others, and is based only on the initial condition.[2]

Curves obtained from (9.45) for the elements of the uranium series (table 9.1) are plotted in figure 9.2(b). The activities are initialized in secular *dis*equilibrium as noted in the figure caption. Radium has the shortest half-life and hence evolves most rapidly, reaching secular equilibrium with its parent thorium after about 10^4 years. Thorium then evolves toward its parent ^{234}U. However, because the half-lives of this parent–daughter pair are similar, the parent and daughter are both evolving at the same time scale; the activity ratio of unity becomes a moving target. The pair then coevolves to secular equilibrium with the ultimate parent of the chain, ^{238}U. This is a long-lived isotope but it is eventually removed by decay (and its activity then goes asymptotically to zero).

9.4.3 EVOLUTION BY BOTH MELTING AND INGROWTH

In the previous two sections, we illustrated the consequences of interphase transfer and radiogenic ingrowth separately. But, according to equation (9.36), these processes occur simultaneously, as part of the decay-chain system. The structure of equation (9.36) is such that we can decompose the general solution into a product of the melting effect (9.39) and the ingrowth effect (9.45):

$$a_j^\ell = \left(\frac{a_j^s|_0}{D_j + (1 - D_j)\mathcal{G}t} \right) \tilde{a}_j^\ell. \tag{9.46}$$

Substituting this trial solution into equation (9.36), requiring that the decay chain is initially in secular equilibrium and rearranging gives

$$\frac{\mathrm{d}\tilde{a}_j^\ell}{\mathrm{d}t} = \lambda_j \left[\tilde{a}_{j-1}^\ell - \tilde{a}_j^\ell \right], \tag{9.47}$$

which is formally equivalent to (9.40) for ingrowth in the absence of melting, which we solved in section 9.4.2. Hence, we can write a general solution that is the combination of (9.46) and (9.45) as

$$\boldsymbol{a}^\ell(t) = \boldsymbol{V} \mathrm{e}^{\grave{\mathcal{L}} t} \boldsymbol{V}^{-1} \left[\frac{\boldsymbol{a}^s|_0}{\boldsymbol{D} + (1 - \boldsymbol{D})\mathcal{G}t} \right], \tag{9.48}$$

where the division is taken entrywise, without any summation.

Activity ratios computed with equation (9.48) for the uranium-series decay chain (table 9.1) are plotted in figure 9.3 for three values of the melting rate $\mathcal{G}$. In each case, the initial condition is a solid composition in secular equilibrium ($\left[a_{j-1}^s / a_j^s \right]_0 = 1$). However, the first infinitesimal increment of melting causes fractionation of each element,

[2] This solution is only valid when all of the decay coefficients in the decay chain are distinct. For repeated decay coefficients we have repeated eigenvalues and the inverse of the eigenvector matrix is undefined. Fortunately, in this universe, there are no decay chains with repeated coefficients.

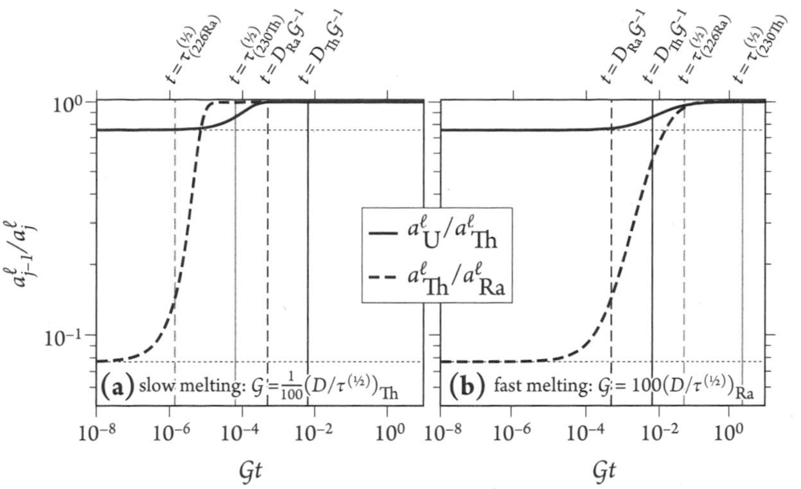

Figure 9.3. Simultaneous melting and ingrowth according to equation (9.48). The activities are plotted in terms of parent/daughter isotope ratios as a function of $\mathcal{G}t$, which corresponds to F for values up to unity (at which F is capped). Vertical lines mark key time scales for ingrowth and melting. Horizontal lines mark the D ratios of parent–daughter pairs that control the initial elemental fractionation (for small F). **(a)** Evolution for a slow melting rate. Activity ratios go to secular equilibrium by ingrowth. **(b)** Evolution for a fast melting rate. Activity ratios go to secular equilibrium by dilution back to the activities of the solid (which were in secular equilibrium initially).

giving an activity in the liquid of λ_j/D_j. These liquid activities are not in secular equilibrium. Note, however, that ^{238}U and ^{234}U have the same partition coefficient and hence are not fractionated by melting; they remain in secular equilibrium.

There are two extreme, end-member scenarios that elucidate the behavior of the system. If melting is extremely slow (i.e., $\mathcal{G} \to 0$), the activity ratios in the liquid evolve toward secular equilibrium by ingrowth only. This is what took place in figure 9.2(b). On the other hand, if melting is extremely fast (i.e., $\mathcal{G} \to \infty$), the system becomes fully molten before any ingrowth can occur (as illustrated in fig. 9.2(a)). Since the initial, unmelted solid was in secular equilibrium, the liquid produced by instantaneous, complete melting is also in secular equilibrium. Thus, we see that melting can equilibrate (at large F) and disequilibrate (at small F) the isotope system. Ingrowth can only drive the overall system toward secular equilibrium.

Figure 9.3(a) shows a slow-melting scenario. Recall that $F \sim D$ is the point at which further melting leads to dilution of the element in the liquid. The melting rate $\mathcal{G}$ is small enough that the initial disequilibrium of the liquid has decayed away before $F \sim D_j$ for any j. In particular, the melting rate is set to be $\mathcal{G} = \frac{1}{100} D_{\text{Th}}/\tau_{\text{Th}}^{(1/2)}$, meaning that the time required to reach $F = D_{\text{Th}}$ is one hundred times greater than the half-life of ^{230}Th.

Figure 9.3(b) shows a fast-melting scenario. Melting proceeds to $F = D_{\text{Ra}}$ in 1/100th the time of the half-life of ^{226}Ra. Hence, the disequilibrium of the liquid is relaxed by dilution rather than by ingrowth. This is evident by inspection of the shape of the radium curve, which differs between figures 9(a) and (b). In an intermediate melting-rate case (not shown), dilution and ingrowth occur on similar time scales such that both contribute to restoring secular equilibrium.

9.5 Literature Notes

The major-element petrology of mantle melting has been studied extensively from geological, experimental, and theoretical perspectives. The literature on this topic is immense and continues to evolve. The theory in the present chapter creates a framework for the coupling of magma segregation with chemical reactions. This is part of a class of problems referred to as *reactive flow*. Chapter 12 considers an instance of reactive flow relevant to magma extraction from the mantle. The reaction chemistry of segregating basalt and mantle peridotite was studied by Kelemen [1990]. These reactions provide a crucial background for understanding magmatic channelization [e.g., Kelemen et al., 1995a].

Conservation of mass for chemical species was derived by M^c^Kenzie [1984] and used to model trace-element transport, including an analysis of chromatographic segregation in their appendix E. Subsequent one-dimensional models of trace-element transport were published by Richter [1986] and Navon and Stolper [1987], the latter focusing on trace-element chromatography, which we discuss further in chapter 11 below. The linear kinetic re-equilibration rate introduced here for the disequilibrium transport model was adapted from Glueckauf [1955] by Navon and Stolper [1987]. End-member cases of equilibrium and fractional transport theory were applied to trace-element transport in two dimensions beneath mid-ocean ridges by Spiegelman [1996].

Kenyon [1990] quantified the limits of the validity of models formulated around trace-element equilibrium. She proposed a model to evaluate the importance of the equilibration time arising from diffusion through the solid phase. That model was elaborated and analyzed by Spiegelman and Kenyon [1992]. Theory for disequilibrium trace-element transport was developed by Qin [1992], Iwamori [1992], and Iwamori [1993a] on the basis of radial zonation of crystals with crystal/melt equilibrium at the rim only. This was applied by Iwamori [1993b] to a model of transport by porous and channel flow.

Diffusive trace-element equilibration of crystals with the surrounding melt was studied by Liang [2003]. Some of the concepts developed there were applied by Liang and Liu [2016, 2018] and Liu and Liang [2017] to model transport of isolated trace-element and isotopic anomalies. An important aspect of these models is transport through both high-porosity channels and low-porosity regions between them. Bo et al. [2018] considered trace-element transport in a true one-dimensional column, but modeled heterogeneity in terms of a Fourier decomposition rather than isolated heterogeneities. This allowed them to calculate the variability of "filtration" properties of the column.

M^c^Kenzie [1985] extended the theory of equilibrium trace element transport to radiogenic trace elements. One-dimensional models were thoroughly analyzed by Spiegelman and Elliott [1993], which is the basis for the treatment here. A web-based calculator of uranium-series column models is described by Spiegelman [2000]. Disequilibrium transport models with radial diffusion through solid grains were considered by Iwamori [1994]. Recent measurements of uranium and thorium partitioning during mantle melting are presented in Elkins et al. [2008]. A review by Elkins et al. [2019]

of U-series in the context of a lithologically heterogeneous mantle provides further references and a discussion of constraints and open questions.

9.6 Exercises

9.1 We consider the disequilibrium trace-element model with $\mathcal{X}=0$ and $\boldsymbol{v}^\ell=\boldsymbol{v}^s=\mathbf{0}$. We take the degree of melting increasing linearly with time, $F=\phi=\mathcal{G}t$. Then, using the relationship between F and Γ from (8.52), the melting rate becomes $\Gamma=\rho\mathcal{G}$. Making the Boussinesq approximation, equations (9.24) become

$$\phi^\ell\frac{\mathrm{d}c_j^\ell}{\mathrm{d}t}=+\left(c_j^s/D_j-c_j^\ell\right)\mathcal{G}, \tag{9.49a}$$

$$\phi^s\frac{\mathrm{d}c_j^s}{\mathrm{d}t}=-\left(c_j^s/D_j-c_j^s\right)\mathcal{G}. \tag{9.49b}$$

Show that the solution to this coupled system is equivalent to the canonical fractional melting model, which is

$$c_j^s=c_0^s\,(1-F)^{\frac{1}{D_j}-1},$$

$$c_j^\ell=c_0^s\left(\frac{1-(1-F)^{\frac{1}{D_j}}}{F}\right).$$

9.2 By modifying equations (9.16a) and (9.16b) to account for radioactive decay (9.31), derive equation (9.32).

9.3 In section 9.2.1 we noted that two different trace elements can be fractionated due to their difference in transport velocity. Here we consider a simple one-dimensional model of transport through a channel of length L. Two species with partition coefficients D_1 and D_2 travel through the channel. Assume that the solid velocity $\boldsymbol{v}^s=0$ and that the two species travel as wave packets with compact support with lengths d_1 and d_2, respectively. Both species enter the channel at the same time. Find the minimal length L for the two species to separate, i.e., to exit the channel disjointly. (You may assume that $D_1>D_2$).

9.4 Consider the disequilibrium transport model with only variations in concentration (everything else is constant).

$$\phi\rho^\ell\left(\frac{\partial c_j^\ell}{\partial t}+\boldsymbol{v}^\ell\cdot\nabla c_j^\ell\right)=+r_j\left(c_j^s-D_jc_j^\ell\right),$$

$$(1-\phi)\rho^s\left(\frac{\partial c_j^s}{\partial t}+\boldsymbol{v}^s\cdot\nabla c_j^s\right)=-r_j\left(c_j^s-D_jc_j^\ell\right).$$

(a) Assuming a traveling wave solution, i.e., $c_j^\ell=f(z-ut)$ and $c_j^s=g(z-ut)$, simplify the equations and obtain the form

$$\frac{\mathrm{d}f}{\mathrm{d}\zeta} = \alpha(g - D_j f), \tag{9.50a}$$

$$\frac{\mathrm{d}g}{\mathrm{d}\zeta} = \beta(g - D_j f), \tag{9.50b}$$

where $\zeta = z - ut$. Find expressions for α and β.

(b) Find the condition on α, β for there to exist a solution of the form

$$f = A_1 \sin(\zeta) + A_2 \cos(\zeta),$$
$$g = B_1 \sin(\zeta) + B_2 \cos(\zeta).$$

(c) For general $f(\zeta), g(\zeta)$ in (9.50), obtain a differential equation in $f(\zeta)$.

9.5 Using Python code provided in the online supplement (URL in the preface) that calculates the effects of both melting and decay, experiment with four-element decay chains defined by λ_i and D_i. Look for conditions under which secular disequilibrium

(a) cascades down the decay chain and
(b) propagates up the decay chain.

Discuss the key parametric control on these behaviors.

CHAPTER 10

Petrological Thermodynamics of Liquid and Solid Phases

The general melting relations of the mantle are highly complex, involving up to about thirteen components in various phases,[1] over a large range of pressure and temperature. Thermochemical equilibrium, which determines the fraction of melt present, is obtained when a system finds its minimum Gibbs free energy. Much of the complexity arises from the thermodynamic interaction between chemical components in each phase. For example, the effect of one chemical component in the liquid phase on the Gibbs free energy of the system depends, in general, on the concentrations of other components in that phase. These interactions do not obey any simple, consistent thermodynamic or chemical principle; they must therefore be calibrated against a database of experimental results. Furthermore, the problem of finding the minimum Gibbs free energy state (i.e., the thermochemical equilibrium) is highly nonlinear. It is thus well beyond the present scope to address this problem in any generality.

Instead we will consider a thermodynamic framework in which chemical potentials for both liquid and solid phases behave according to *ideal solution* theory. According to this theory, the change in Gibbs free energy associated with chemical solution is entirely due to the *entropy of mixing* (i.e., the *enthalpy of mixing* is assumed to be zero). The entropy of binary mixing is the configurational entropy of dissolving (mixing at the molecular level) one component into the other. Since this change in entropy is always positive (it creates disorder), the associated change in free energy is always negative. Hence there is a decrease in the Gibbs free energy associated with dissolution. This means that dissolution is always a spontaneous process under ideal solution theory.

The change in Gibbs free energy with chemical composition is measured in terms of a quantity called *chemical potential.* As we demonstrate in this chapter, the chemical potential of a species in an ideal solution simply depends on pressure, temperature, the material properties of the pure component and, crucially, on its own molar concentration in the solution. Hence, its chemical potential is independent of the concentrations (or, indeed, the presence) of other components. Chemical potentials, whether obtained through ideal solution theory or otherwise, give us a means to access the equilibrium

[1] In the asthenosphere, these phases can include minerals (e.g., olivine, pyroxene, plagioclase, garnet, spinel), melts (e.g., basaltic, carbonatitic), vapor (e.g., water, CO_2), or supercritical fluids. Immiscible liquids can coexist with solid and vapor, though such situations are uncommon. It is generally appropriate to simplify our analysis to that of a coexisting liquid and (composite) solid, as we have done throughout this book.

state of a thermochemical system. We can express that state in terms of partition coefficients that relate the concentration of a component in the solid to that of the liquid.

In this context, we are restricted to a representation of equilibrium in which all components have complete *solid solution*. This means that the solid phase grades continuously, in compositional space, between the pure-component end members. We also assume complete liquid solution, and hence that liquids with any composition are miscible. In this context, we have excluded the ability to model cotectic and eutectic systems, though these can be parameterized, as we shall see below.

The resulting thermochemical model is not a good description of the detailed petrology of the mantle. It cannot, for example, be used to accurately predict the concentration of aluminum in the magma during melting or crystallization. This is unfortunate but acceptable for our present purposes. Instead of modeling the true chemical components that comprise the mantle, we will consider *effective* (also referred to as *fictitious* or *pseudo-*) *components* that represent categories of chemical constituents. For example, one effective component could be "refractory constituents" representing a spectrum of olivine compositions. Because the components are fictitious and not described by a specific chemical formula, their properties cannot be calibrated by recourse to an experimental database. However, those properties can be assigned by semiquantitative calibration to a variety of more general constraints.

10.1 The Equilibrium State

Consider a closed system with a specified thermodynamic pressure P, temperature T, and bulk chemical composition $\bar{c}_j$. The equilibrium state of this system can be defined in terms of the phases present when the system reaches a state of minimum energy. In particular, it is defined by the composition and mass fraction of each phase. The phases may include vapors as well as immiscible liquids and solids. In the present discussion, we will assume that there is, at most, one solid phase and one liquid phase present (no gas). In this context, f^ℓ and f^s are the mass fractions of liquid and solid present. Of course as with volume fractions, the mass fractions are subject to a unity sum, $f^\ell + f^s = 1$; hence, we use f to refer to the liquid fraction and $(1-f)$ to refer to the solid fraction. Under the extended Boussinesq approximation, which is the assumption that the liquid and solid densities are equal except in terms representing the gravitational body force, the melt fraction and the porosity are equal. Otherwise, the two are related by

$$f = \phi\rho^\ell/\bar{\rho}. \tag{10.1}$$

In exercise 10.1 we obtain the porosity in terms of the melt fraction and phase densities.

For a system with N chemical components and two phases at a given pressure and temperature, the equilibrium state is defined by $2N+1$ numbers: f, c_j^ℓ, and c_j^s, where j is in the set of integers from unity to N. We will therefore require $2N+1$ equations to obtain a unique solution to the equilibrium state. We can immediately write down the following $N+1$ equations

$$\bar{c}_j = fc_j^\ell + (1-f)c_j^s, \tag{10.2a}$$

$$\sum_{j=1}^{N} c_j^s = 1 \quad \text{or} \quad \sum_{j=1}^{N} c_j^\ell = 1. \tag{10.2b}$$

The first of these equations, (10.2a), is known as the *lever rule*; it is simply a balance of component mass between the two phases ($\bar{c}_j$ is the bulk concentration of component j and satisfies a unity sum over j). There are N such equations, one for each component; any of them can be rearranged to give the melt fraction if the phase compositions are known,

$$f = \frac{c_j^s - \bar{c}_j}{\Delta c_j}, \tag{10.3}$$

where $\Delta c_j = c_j^s - c_j^\ell$. More generally, however, both the melt fraction and the phase compositions are unknown. The second equation, (10.2b), uses the unity sum of concentration in the solid or the liquid phases to provide an additional constraint on the overall composition. For partially molten rock with intermediate melt fraction f, either of these two equations can be used. However, for a single-phase system of pure solid (or pure melt) that is approaching the solidus (or liquidus) it is essential to choose the relevant one. Below we suggest retaining both at all times.

The $N+1$ equations (10.2) can be combined by solving (10.2a) for the phase concentrations and then substituting into (10.2b) to give

$$\sum_{j=1}^{N} \frac{\bar{c}_j}{f + (1-f)\left(c_j^s/c_j^\ell\right)} = 1 \quad \text{or} \quad \sum_{j=1}^{N} \frac{\bar{c}_j}{(1-f) + f/\left(c_j^s/c_j^\ell\right)} = 1. \tag{10.4}$$

In this equation set, the phase concentrations only appear in the ratio of the solid concentration to the liquid concentration. This ratio appeared in reference to trace elements in section 9.2, where we defined a partition coefficient D_j that we treated as a constant. For the thermochemical components that determine the melt fraction we define an equivalent partition coefficient

$$\check{K}_j \equiv \check{c}_j^s / \check{c}_j^\ell, \tag{10.5}$$

where a check ($\check{\cdot}$) above a symbol indicates that the quantity represented takes the value appropriate for thermodynamic equilibrium. In general, the partition coefficients of thermochemical components vary with temperature, pressure, and the composition of the phases present. As noted above, however, we will assume that both the solid and the liquid are ideal solutions and hence that $\check{K}_j$ depends on P and T only.[2] The next section provides an overview of the thermodynamic theory of ideal solutions. We will use it to obtain N equations for the coefficients $\check{K}_j$; these will enable us to eliminate the concentration ratios from equations (10.4) and solve for f.

10.1.1 PARTITION COEFFICIENTS FROM IDEAL SOLUTION THEORY

At fixed pressure and temperature, thermochemical equilibrium between coexisting liquid and solid phases is attained when the Gibbs free energy G (units of joules; an extensive quantity) of the system is at a minimum. This minimum is achieved by exchange of components between the phases, where the required amount of molar exchange of the jth component is quantified by the chemical potential

[2]This statement implicitly refers to equation (10.18). In a strict adherence to ideal solution theory, only the *molar* partition coefficients $\check{K}_j^*$ are independent of each other. A detailed examination and explicit approximation is provided below in section 10.1.1.

$\mu_j = (\partial G/\partial n_j)\,|_{P,T,n_{i\neq j}}$ with units of joules per mole. Here, n_j represents the moles of the component that are exchanged; the subscripts P, T, and $n_{i\neq j}$ indicate that the partial derivative is taken at fixed pressure, temperature, and all other chemical components. At the minimum Gibbs free energy, the chemical potential is equal between phases:

$$\mu_j^\ell = \mu_j^s. \tag{10.6}$$

If the chemical potential was imbalanced, the difference would drive chemical transfer between phases to minimize the Gibbs free energy. The Literature Notes at the end of this chapter cite textbooks that provide a more detailed exposition of the material in this section.

Equation (10.6) represents the minimum energy, equilibrium state. This is precisely the state that interests us and hence this equation is a very powerful (if somewhat opaque) constraint. To make (10.6) useful, we must express the chemical potential in terms of the quantities of interest: concentrations, temperature, and pressure. In general, the chemical potentials of each component in each phase depends on the concentration of all components in that phase.

Solutions are mixtures of chemical components within a material phase, with homogeneity at all scales down to the molecular. The thermodynamic properties (e.g., entropy and enthalpy) of the impure phase are, in general, not a linear combination of those of the components. Indeed the relationships are nonlinear and, for complex materials such as silicate melts, must typically be obtained empirically. Under ideal solution theory,[3] the variations of enthalpy, density, and heat capacity are neglected. Entropy, however, increases in an ideal solution relative to the pure-phase end members. Therefore the Gibbs free energy of mixing is always negative. This means that the formation of solutions is a spontaneous process.

These considerations lead to a formulation[4] in which the chemical potential of each component in a solution is a function of temperature, pressure, and the concentration of that component only. In particular,

$$\mu_j^i = \mathring{\mu}_j^i + \nu RT \ln X_j^i, \tag{10.7}$$

where $\mathring{\mu}_j^i(P,T)$ is the chemical potential (J/mol) in a phase i that is composed of *pure* component j. The symbol X_j^i represents the molar concentration of component j in phase i. The factor ν counts the number of lattice sites per formula unit of the solid in which substitutions associated with dissolution can occur (e.g., $\nu = 2$ for olivine $(MgFe)_2SiO_4$). Because our approach is intended for application to fictitious, effective chemical components (for which exact chemical formulas cannot be written, in general), we take $\nu = 1$ in what follows.

Equation (10.7) states that changes in the Gibbs free energy of phase i due to changes in the mole fraction of component j arise from two sources. The first is the direct effect of component j due to its intrinsic chemical potential. The second is the configurational entropy associated with dissolution of component j into phase i. This is captured by the second term on the right-hand side of equation (10.7). The contribution of the second

---

[3]We refer here to a Roultian ideal solution in which activity is equal to the mole fraction and the mixing properties of enthalpy, volume, and heat capacity are all equal to zero.

[4]For details of the derivation, see a textbook such as Anderson [2017].

term is always negative because $0 \leq X_j^i \leq 1$. Note that the second term can be computed without further manipulations whereas the first term requires a model for chemical potential in pure phases.

The difference in chemical potential of component j between solid and liquid phases that are in thermal equilibrium but not chemical equilibrium is

$$\begin{aligned}\Delta\mu_j &= \Delta\mathring{\mu}_j + RT\Delta\left(\ln X_j\right),\\ &= \Delta\mathring{\mu}_j + RT\ln\left(X_j^s/X_j^\ell\right).\end{aligned} \tag{10.8}$$

This equation relates the difference in chemical potential between phases (left-hand side) to the difference at a reference state (pure phases, °) and the actual composition of the solid and liquid phases (second term on the right-hand side). At chemical equilibrium, we have $\Delta\mu_j = 0$ from equation (10.6) and hence that

$$\ln\left(\check{X}_j^s/\check{X}_j^\ell\right) = -\frac{\Delta\mathring{\mu}_j}{RT}. \tag{10.9}$$

On the left-hand side of this equation is the equilibrium solid-to-liquid ratio of mole fractions,

$$\check{K}_j^* \equiv \frac{\check{X}_j^s}{\check{X}_j^\ell}, \tag{10.10}$$

where the * indicates a ratio of molar quantities. The ratio in equation (10.10) is closely related to the object of interest, the ratio of mass fractions $\check{K}_j$. On the right-hand side of equation (10.9) is the difference in chemical potential of component j when it forms a pure solid and liquid. The molar partition coefficient $\check{K}_j^*$ is obtained by developing a model for this difference.

For solid and liquid phases composed purely of component j, the difference $\Delta\mathring{\mu}_j$ at temperature T is equal to the difference in Gibbs free energy

$$\begin{aligned}\Delta\mathring{\mu}_j &= \Delta g_j M_j\\ &= \left(\Delta h_j - T\Delta s_j\right)M_j,\end{aligned} \tag{10.11}$$

where Δg_j is the free-energy difference per unit mass and M_j is the molar mass of component j. Both the mass-specific enthalpy difference Δh_j and entropy difference Δs_j are negative for materials of interest here (recalling that $\Delta q = q^s - q^\ell$).

A solid of pure component j melts at a univariate melting point $T_j^m(P)$. Since the pure liquid and solid coexist in equilibrium at $T = T_j^m$, we have $\Delta g_j(T_j^m) = 0$. Using this in equation (10.11) tells us that

$$\Delta s_j^m = \Delta h_j^m / T_j^m. \tag{10.12}$$

This is exactly true for equilibrium between liquid and solid composed of pure component j. For reasons that we shall not detail here, Δh_j and Δs_j remain nearly constant over the temperature interval associated with melting of the impure solid. Hence, we approximate them as remaining constant at their melting-point values (equation (10.12)) and rewrite equation (10.11) as

$$\Delta\mathring{\mu}_j = \Delta h_j^m \left(1 - T/T_j^m\right) M_j. \tag{10.13}$$

Substituting this result into equation (10.9) we obtain

$$\ln \check{K}_j^* = \frac{-\Delta h_j^m}{R/M_j}\left(\frac{1}{T} - \frac{1}{T_j^m}\right). \tag{10.14}$$

This equation provides a closed-form expression for the partitioning of moles of component j between the solid and liquid phases. Recall, however, that to solve equation (10.4) for f, we require $\check{K}_j$, the partition coefficient in terms of mass fractions.

Therefore, to make equation (10.14) useful, we must relate the mole-fraction ratio X_j^s/X_j^ℓ to mass partition coefficient $\check{K}_j$. Mass fraction and mole fraction are related by

$$\mathcal{M}^i c_j^i = M_j X_j^i \quad \text{(no implied summation)}, \tag{10.15}$$

where

$$\mathcal{M}^i \equiv \sum_j M_j X_j^i = \left(\sum_j \frac{c_j^i}{M_j}\right)^{-1}. \tag{10.16}$$

We use equation (10.15) and the definition of the equilibrium partition coefficient (10.5) to rewrite (10.14) as

$$\begin{aligned}\check{K}_j &= \frac{\mathcal{M}^\ell}{\mathcal{M}^s}\exp\left[\frac{-\Delta h_j^m}{R/M_j}\left(\frac{1}{T} - \frac{1}{T_j^m}\right)\right], \qquad (10.17)\\ &= \left[\frac{\sum_k (\bar{c}_k/M_k)\,\check{K}_k}{\sum_k (\bar{c}_k/M_k)}\right]\exp\left[\frac{\Delta h_j^m}{R/M_j}\left(\frac{1}{T} - \frac{1}{T_j^m}\right)\right].\end{aligned}$$

This is a nonlinear equation that couples together all the N equilibrium constants of the N-component system. Hence to obtain $\check{K}_j$, all N equations must be solved numerically as a coupled system.

In cases where, for example, we use a small set of fictitious "effective" components to approximate the full thermochemical system, the error that we make in taking $\mathcal{M}^\ell/\mathcal{M}^s \approx 1$ is acceptably small. With this approach, we arrive at the simple result

$$\check{K}_j(P,T) \approx \exp\left[\frac{L_j}{R_j}\left(\frac{1}{T} - \frac{1}{T_j^m(P)}\right)\right], \tag{10.18}$$

where $L_j \equiv -\Delta h_j^m$ is the latent heat of melting for a solid composed of pure component j and $R_j \equiv R/M_j$ is the modified gas constant. Below, in sections 10.1.3 and 10.1.4, we treat the ratio L_j/R_j as a fitting parameter for a system of fictitious components. Equation (10.18) can be employed to compute the equilibrium state using the approach discussed in the next section.

10.1.2 COMPUTING THE EQUILIBRIUM STATE

Rewriting equation (10.4) with partition coefficients $\check{K}_j$ gives two implicit equations for the equilibrium melt fraction $\check{f}$ as

$$\sum_{j=1}^{N} \frac{\bar{c}_j}{\check{f} + (1-\check{f})\check{K}_j(T,P)} = 1 \quad \text{or} \quad \sum_{j=1}^{N} \frac{\bar{c}_j}{\check{f}/\check{K}_j(T,P) + (1-\check{f})} = 1. \tag{10.19}$$

Alternatively, one of these equations can be subtracted from the other to give

$$\sum_{j=1}^{N} \frac{\bar{c}_j\left[1-\check{K}_j(T,P)\right]}{\check{f} + (1-\check{f})\check{K}_j(T,P)} = 0, \tag{10.20}$$

which applies for all $\check{f}$ between zero and unity. This equation can be solved for $\check{f}$ using a numerical root-finder based on, for example, Newton's method. Once a value of $\check{f}$ has been obtained, it can be used to determine the phase composition using equations (10.2a) and (10.5),

$$\check{c}_j^{\ell} = \frac{\bar{c}_j}{\check{f} + (1-\check{f})\check{K}_j}, \tag{10.21a}$$

$$\check{c}_j^{s} = \frac{\bar{c}_j}{\check{f}/\check{K}_j + (1-\check{f})}. \tag{10.21b}$$

This method will work for any temperature T, even those above the liquidus or below the solidus, provided we limit the equilibrium melt fraction to $0 \leq \check{f} \leq 1$ before substituting into (10.21).

It is instructive, nonetheless, to compute the solidus and liquidus temperatures as functions of composition and pressure. This is achieved by solving equation (10.20) for T with either $\check{f} = 0$ (solidus) or $\check{f} = 1$ (liquidus). An example of a two-component system in the next section illustrates this method. Thermodynamic data for the olivine solid solution (forsterite–fayalite) is given in the exercises at the end of the chapter.

10.1.3 APPLICATION TO A TWO-PSEUDO-COMPONENT SYSTEM

The simplest system for which ideal solution theory can be applied is a system with two chemical components. However, as discussed above, it is impossible to identify two real chemical components of the mantle that reasonably approximate the thermodynamic behavior of the full chemical system. Hence we rely the concept of *pseudo-components*. These are behavioral end-members that may, themselves, be chemically complex and variable. To avoid this complexity, we don't attempt to specify their chemical formulas, but rather treat the parameters associated with their ideal solution behavior as fitting parameters.

A set of pseudo-components that arguably provides a basic description of mantle petrology are "olivine" and "basalt." These are placed in quotation marks to emphasize that they do not refer literally to the mineral olivine and the rock basalt. Instead they provide an expectation of the two components' properties, which we assign without quantitative constraints. Figure 10.1 contains plots of an ideal solution model that will illustrate these ideas. In what follows, we drop the quotation marks and accept that the meaning of component names will be context-specific; the reader should pay careful attention in parsing the following sections.

Our petrological model is roughly described as follows. We assume that the mantle is mostly composed of olivine. The olivine has a melting temperature that is very high. The rest of the solid mantle has the composition of basalt, which is, in contrast, a readily fusible material. The pure-component melting temperatures of pseudo-components olivine and basalt are shown in figure 10.1(a). Recall that a single-component material held at a fixed pressure melts at a fixed temperature. This fixed temperature is the $T_j^m(P)$ that appears in, for example, equation (10.18). The change of this melting temperature with pressure is given by the Clausius–Clapeyron equation. For simplicity, we assume a constant value of $\vartheta = \Delta s/\Delta(1/\rho)$, which gives

$$T_j^m(P) = T_j^m\Big|_{P=0} + P/\vartheta. \tag{10.22}$$

This produces the straight lines in figure 10.1(a). Other approaches for parameterizing the pure-component melting temperatures include polynomials or the Simon–Glatzel equation (exercise 10.3); these are not considered here.

The thermochemical behavior is modeled by combining the component solidi (10.22) with the partition coefficients (10.18) and the equilibrium constraints (10.20) and (10.21), with parameters as given in table 10.1.

By treating both the liquid and solid as ideal solutions of all components, we capture only one solid phase and one liquid phase (i.e., no distinct minerals or immiscible liquids). Both of these phases have smooth compositional gradation, which is a consequence of the smoothness of the partition coefficients $\check{K}_j$, shown in figure 10.1(b), that are computed with equation (10.5) (parameter values are given in table 10.1). The partition coefficients give rise to the solidus and liquids curves shown in figure 10.1(c) at $P = 1$ GPa. This panel is a *phase diagram*: the solidus curve bounds the region of liquid stability and the liquidus curve bounds the region of solid stability. The region between these two curves is where both solid and liquid are stable. In other words, for temperatures between the solidus and liquidus, the system is partially molten, with $0 < \check{f} < 1$.

Figure 10.1(d) shows an isobaric (1 GPa) melting curve computed in the two-component system. The bulk composition is initially 25% basalt component and remains constant because the system is closed. Starting from 1050°C and increasing the temperature, the system reaches the solidus and melts progressively until it reaches the liquidus, where it is fully molten. We can follow the evolution of the phase compositions during progressive melting in (c). Starting from low temperature, the solid composition (solid black line) is constant until it reaches the solidus at about 1250°C. There, melting begins; the liquid appears with a concentration given by the liquidus curve at the same temperature. This initial liquid is enriched in the basaltic (fusible) component. As the temperature increases further, the solid composition changes along the solidus curve. It is progressively depleted of the basaltic component, becoming less and less fusible and requiring higher and higher temperatures for additional melting. Simultaneously, the liquid composition evolves along the liquidus curve toward the bulk composition. As the melt fraction approaches unity, the liquid concentration approaches the bulk

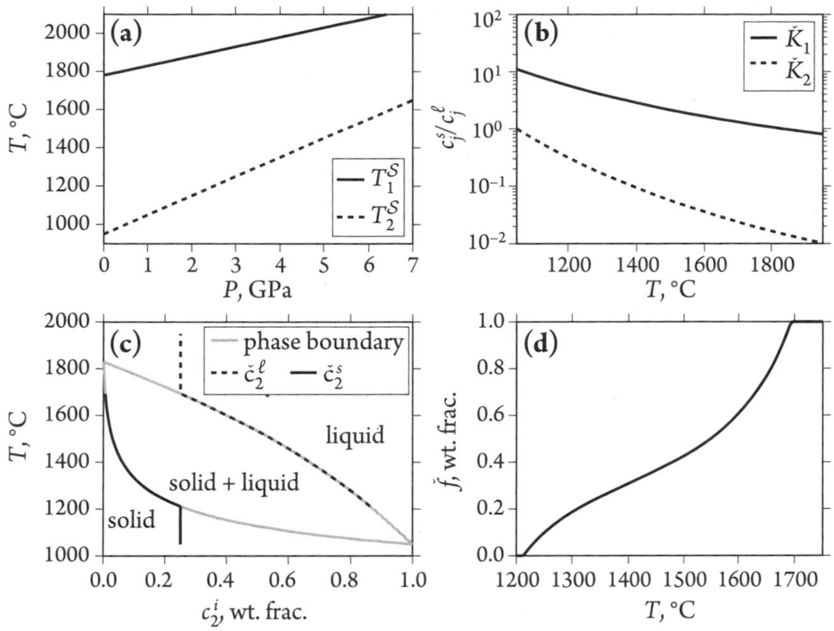

Figure 10.1. Ideal-solution model of a two-component system of "olivine" ($j = 1$) and "basalt" ($j = 2$). Parameter values are given in table 10.1. **(a)** Pure-component melting temperatures T_j^m as a linear function of pressure. **(b)** Partition coefficients $\check{K}_j$ as a function of temperature at a constant pressure of 1 GPa. **(c)** Gray lines are the solidus and liquidus temperatures as functions of the basalt fraction at a pressure of 1 GPa. The composition along the solidus curve refers to the solid phase; that along the liquidus curve refers to the liquid phase. Black lines indicate the compositional evolution of a closed system with bulk composition of 25% basalt component. **(d)** Equilibrium melt fraction as a function of temperature for a bulk composition of 25% basalt component at 1 GPa.

concentration and, at $\check{f} = 1$ where the solid phase disappears, the liquid concentration is equal to the bulk concentration. The liquid then remains at constant composition with increasing temperature.

10.1.4 APPLICATION TO A THREE-PSEUDO-COMPONENT SYSTEM

Extending the two-component system from the previous section to a three-component system allows us to incorporate a component that represents volatiles. In particular, we will use the third component to include the effect of water. Pure water has a melting temperature of 0°C at atmospheric pressure that drops to −20°C at 200 MPa. At higher pressures the melting temperature of water increases slowly, reaching 0°C again at 600 MPa and exceeding 20°C at 900 MPa. None of this is relevant for the mantle, however, where silicate melting temperatures are two orders of magnitude larger. Even the lowest temperatures at depth in subduction zones are significantly outside this range. Deep beneath mid-ocean ridges, the bulk concentration of water is of order 100 parts per million (ppm) by mass ($c^s_{H_2O} \sim 1 \times 10^{-4}$). This is small enough that it can be fully dissolved in the unmolten solid. Hence it is inconvenient to include pure water as a thermochemical component.

Table 10.1. Parameter values for the two- and three-component ideal-solution systems shown in figures 10.1 and 10.2, respectively. These values are useful for the demonstration of ideal-solution phase diagrams but should not be taken as an optimal calibration for the mantle.

Name	j	$T_j^m(P=0)$ [°C]	ϑ_j [GPa/K]	L_j [kJ/kg]	R_j [J/kg/K]
olivine	1	1780	1/50	600	70
basalt	2	950	1/100	450	30
hydrous basalt	3	410	1/50	330	30

Instead of considering pure water, we include a thermochemical component representing basalt hydrated with 5 wt% water, as specified in table 10.1. In this system, a bulk composition that includes 0.2 wt% of the hydrated basalt component has 100-ppm water by mass. The melting points of the pure-component solid phase are shown in figure 10.2(a). The hydrated basalt has a moderately lower melting point—sufficiently low to model the solidus depression of water at mid-ocean ridges but much higher than the melting point of pure H_2O-ice. Figure 10.2(b) shows the partition coefficients for the three components as a function of temperature. The lines in figures 10.2(a) and (b) for the anhydrous components are unchanged from the two-component system; therefore, the phase loop on the left side of figure 10.2(c), where $\check{c}_3^i = 0$, is also unchanged from figure 10.1. The modification comes from the temperature drop in the solidus surface with addition of $\check{c}_3^i$: water sharply reduces the melting temperature of mantle rock, even at very low concentration. This behavior is evident in figure 10.2, where the solid black line is an isobaric (1 GPa) melting curve for a system with $\bar{c}_3 = 0.002$ wt. frac. (i.e., 100-ppm (by mass) water). Solidus depression gives a low-$\check{f}$ tail to the water-present "wet" curve. However, the difference between the wet and dry isobaric melting curves vanishes for $\check{f} \geq 0.2$.

There are clear limitations of the three-component system as formulated, with the choice of end-member compositions considered here. First, the maximum bulk content of water that can be achieved is 5 wt% water and this requires $\bar{c}_3 = 1$. Second, it is impossible to model liquids that contain ~50 wt% water, such as those that percolate off relatively cold subducting slabs and saturate the cold mantle rock adjacent to the slab. Modifications to the choice of fictitious components and to the partition coefficients would be required to model that case.

10.1.5 APPROACHING THE EUTECTIC PHASE DIAGRAM

Ideal solution theory provides thermodynamically consistent formulas to construct phase diagrams based on complete solution between components in the solid phase. These phase diagrams take the form of loops, as shown in figure 10.1(c). However, it isn't necessary to employ a rigorous thermodynamic theory to obtain equations that describe a phase diagram. Those equations can be parameterized based on empirical information or pure intuition, as long as they are well posed.

Furthermore, some binary systems do not behave as solid solutions. Instead they behave according to a eutectic phase diagram in which two separate but intermingled solid phases melt at a temperature beneath the pure-phase melting points of either of the two. Water plus salt (e.g., NaCl or NH_4Cl) is an example; at temperatures below the

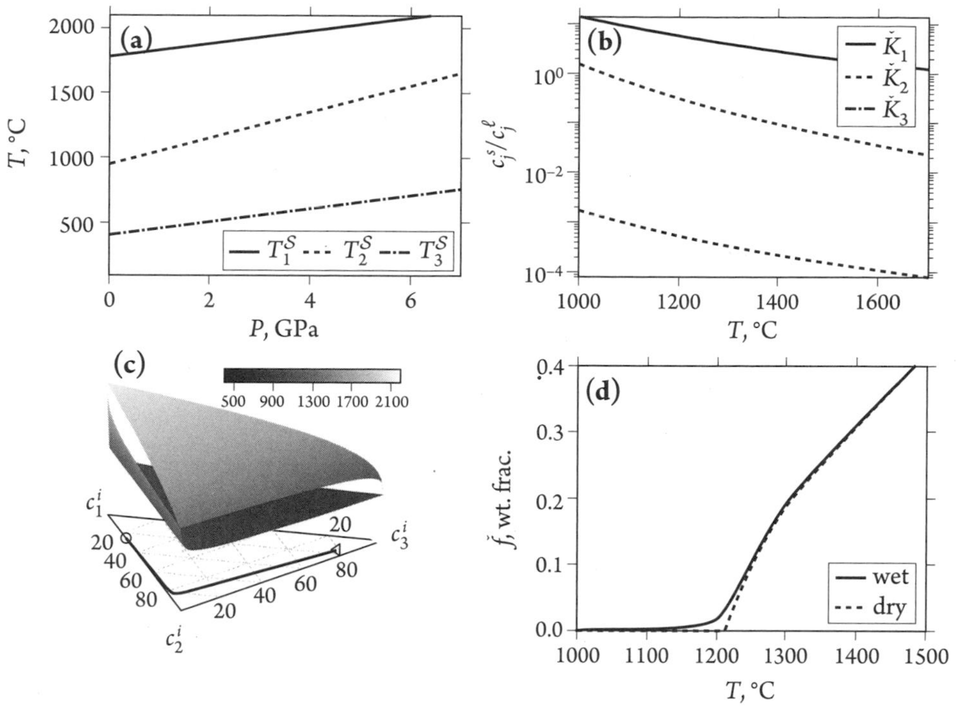

Figure 10.2. Ideal-solution model of a three-component system of "olivine" ($j=1$), "basalt" ($j=2$), and "hydrous basalt" ($j=3$). Parameter values are given in table 10.1. **(a)** Pure-component melting temperatures T_j^m as a linear function of pressure. **(b)** Partition coefficients $\check{K}_j$ as a function of temperature at a constant pressure of 1 GPa. **(c)** Surfaces are the solidus and liquidus temperature through the full, three-component space at a pressure of 1 GPa. The composition along the solidus surface refers to the solid phase; that along the liquidus surface refers to the liquid phase. The compositional evolution of the melt for a closed system with 75 wt% olivine and 0.2 wt% hydrous basalt is shown by the solid line. It starts at the triangle and progresses toward the circle with increasing T and $\check{f}$. **(d)** Equilibrium melt fraction as a function of temperature for a bulk composition of 75 wt% olivine component with and without 0.2 wt% of hydrated basalt component at 1 GPa.

eutectic, crystals of pure water coexist with crystals of pure salt. When energy is added to a system at the eutectic temperature, melts of the eutectic composition are produced while the temperature remains fixed. Temperature begins to rise only when one of the two solid phases has been exhausted from the residue.

Mantle lherzolite, which comprises about five solid mineral phases, has a melting behavior that is "eutectic-like" in that the melting temperature of the polycrystalline aggregate is lower than any of the individual minerals alone.

The key feature of the eutectic diagram is that over some range of compositions terminating at the eutectic point, the solidus is a line of fixed temperature, as shown by the solid line at 1000 °C in figure 10.3a. Bulk compositions in this range begin to form a melt at the eutectic temperature T_E with the eutectic composition c_j^E. The system remains at this temperature, producing these melts, until the solid composition has

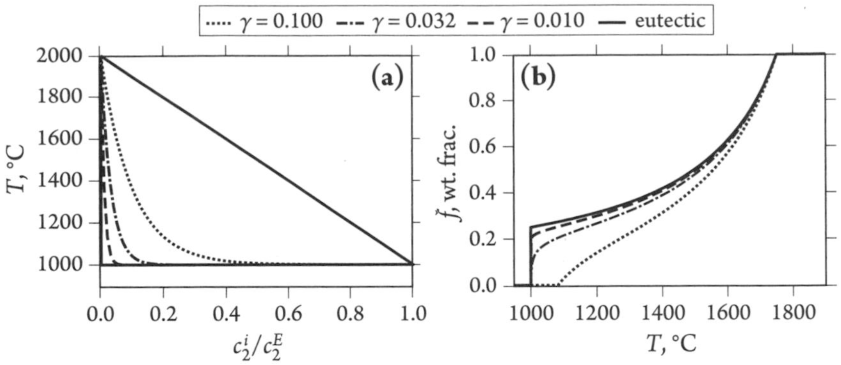

Figure 10.3. A eutectic phase diagram modeled as a smoothly varying solidus curve. **(a)** Solidus and liquidus curves computed with the parametrically smoothed model (10.23) and compared with a sharp eutectic. As the smoothing parameter $\gamma \to 0$, the smoothed solidus approaches the sharp solidus (black line). **(b)** Isobaric melting curves computed based on the smoothed and sharp phase diagrams.

traversed along the horizontal line away from the eutectic. Further melting occurs as solid of pure component 1 becomes liquid, while temperature increases along the liquidus curve. For simplicity, we have assumed the liquidus is a straight line connecting the eutectic temperature to the pure-component melting temperature.

Such eutectic systems can be parameterized directly, using conditional statements in code that implement the calculations. However, these conditionals can make derivatives of the thermodynamic system discontinuous, which can lead to difficulties in convergence of nonlinear numerical algorithms such as Newton's method. A discontinuity can be removed with a different parameterization of the phase diagram. For example, we can write the solidus and liquidus curves as

$$T^{\mathcal{S}} = T_0 + (T_E - T_0)\,\frac{1 - \exp\left(-c_2^s/c_2^E\gamma\right)}{1 - \exp\left(-1/\gamma\right)}, \tag{10.23a}$$

$$T^{\mathcal{L}} = T_0 + (T_E - T_0)\,\frac{c_2^\ell}{c_2^E}, \tag{10.23b}$$

where a sub- or superscript E indicates a quantity evaluated at the eutectic point and a subscript 0 indicates a quantity evaluated at zero concentration. Concentrations given without subscript refer to component 2. We have neglected the pressure-dependence of the phase boundaries, although this can be reincorporated with minor changes.

Figure 10.3a shows that as we take $\gamma \to 0$, the smooth solidus curve (10.23a) approaches the solidus, which is horizontal except at $c_2^i = 0$, where it is vertical. Moreover, figure 10.3b shows that the isobaric melting path for $\gamma \to 0$ converges to that of the sharp solidus.

This convergence doesn't prove anything, except maybe that there exist parameterized representations of the eutectic phase diagram that are more convenient for numerical models. Moreover, some eutectic phase diagrams include a limited amount of solid solution, which moves the solidus off the vertical axis near $c_2^i = 0$. This brings the sharp eutectic closer to the smoothed eutectic model.

10.1.6 LINEARIZING THE TWO-COMPONENT PHASE DIAGRAM

Phase diagrams constructed with ideal-solution theory provide a useful basis for developing the melting relations required for magma/mantle dynamics. Their nonlinearity, however, precludes a use in analytical calculations; instead they must be handled numerically. It is therefore important to formulate an approximation in which the equilibrium state is computed analytically.

This can be achieved by linearizing the solidus and liquidus surfaces about a reference composition. The result remains strictly valid only for compositions near the reference composition. In fact, a phase diagram with linear solidus and liquidus surfaces (e.g., dashed lines in fig. 10.4(a)) necessarily violates a basic thermodynamic property of pure substances: that melting of a pure solid, composed of only one component, occurs at a single, fixed temperature for any given pressure. However, as long as our solutions remain compositionally distant from the pure end-members, this violation is inconsequential and the linearized form behaves similarly to the ideal-solution theory. For linear variations in pressure and composition, we can write

$$T^{\mathcal{S}} = T^{\mathcal{S}}_{\text{ref}} + (P - P_{\text{ref}})/\vartheta + M^{\mathcal{S}}\left(\check{c}^s - \check{c}^s_{\text{ref}}\right), \tag{10.24a}$$

$$\check{c}^\ell = \check{c}^s - \Delta\check{c}, \tag{10.24b}$$

where $M^{\mathcal{S}}$ is the constant slope of the solidus with concentration in the two component space and $T^{\mathcal{S}}_{\text{ref}}$ is a reference temperature at $P = P_{\text{ref}}$ and $\check{c}^s = \check{c}^s_{\text{ref}}$. Also, $\Delta\check{c} \equiv \check{c}^s - \check{c}^\ell$ is the concentration difference between the solidus and the liquidus. If $\Delta\check{c}$ is taken to be a constant then the liquidus slope $M^{\mathcal{L}}$ and the solidus slope $M^{\mathcal{S}}$ are equal. Equations (10.24) can be combined to give an analytical expression for the liquidus temperature as a function of liquid concentration.

Figure 10.4(a) shows a comparison of an ideal-solution phase loop with a solidus and liquidus that are linearized about the reference composition. The compositional difference $\Delta\check{c}$ is taken to be a function of temperature such that $M^{\mathcal{S}} \neq M^{\mathcal{L}}$. Note that by considering c^i to be the fertile component c^i_2, we have $\Delta\check{c} < 0$; this requires that $M^{\mathcal{S}}, M^{\mathcal{L}} < 0$ too. Figure 10.4(b) shows isobaric melting curves computed from the loop and linearized phase diagrams. The match is excellent close to the solidus temperature, where the linearized phase boundaries are tangent to the loop. At higher temperatures the curves diverge; above about 1350°C, the solid composition becomes negative. This behavior is sensitive to the values of $M^{\mathcal{S}}$, $M^{\mathcal{L}}$, and $\Delta\check{c}$. Nonetheless, the comparison illustrates that a linearized phase diagram is useful for compositions near the reference composition and should be applied with care outside of that neighborhood.

10.1.7 DEGREE OF MELTING

It will be beneficial to prepare for chapter 11 by recognizing that there is a simple relationship between the composition of the solid and the equilibrium *degree of melting F*, which is a measure of the total amount of previous melting that the solid has undergone (see also section 8.7). It is a form of the lever rule (10.2a), but is referenced to the "initial" concentration $c^s|_0$ of the solid—the concentration it had before it started melting

$$c^s|_0 = F\check{c}^\ell + (1 - F)\check{c}^s. \tag{10.25}$$

The initial concentration $c^s|_0$ cannot, by definition, be altered by subsequent melting–it is a "history property" of the solid phase, and is therefore transported with that phase.

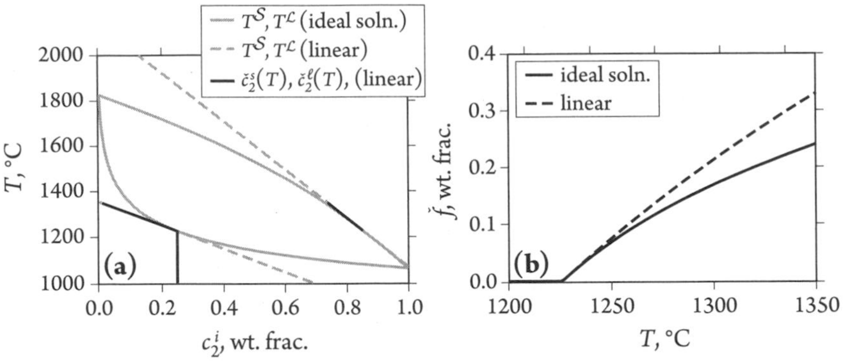

Figure 10.4. Comparison of a two-component, ideal-solution phase loop with a linearized solidus and liquidus at 1 GPa. The phase loop uses the parameters from table 10.1. **(a)** Solidus and liquidus curves. Evolution of an initially unmolten rock with 25-wt% basaltic component for increasing temperature, along the linearized phase boundaries. **(b)** Isobaric melting curves.

This can be rearranged for F:

$$F = \frac{\check{c}^s - c^s|_0}{\Delta\check{c}}. \tag{10.26}$$

Compare this with equation (10.3) for the melt fraction. Both the melt fraction $\check{f}$ and the degree of melting F depend on phase concentrations, but they are referenced to different bulk concentrations: the instantaneous and the initial concentrations, respectively. If the initial composition (i.e., before melting begins) of a parcel of the solid is known, then F can be computed after some melt has been extracted, based on concentration in the residual solid. In a closed system where melt and solid do not segregate, the initial, unmolten solid concentration is equal to the bulk concentration; in that case, the degree of melting and the melt fraction are equal.

The degree of melting defined in this section is qualitatively identical to that of section 8.7. However in that case, prior to the development of the compositional model, we specified F using the temperature relative to the solidus. Here we see the deficiency of that specification; the degree of melting of a solid that has been raised above the solidus temperature, melted, segregated from the melt and then cooled below the solidus temperature should have nonzero F. According to equation (10.26) it does; according to equation (8.50) it does not.

It is also worth noting that the equilibrium treatment developed above enables us to compute $\check{f}$ and F, but it doesn't explicitly constrain the melting rate Γ. In the next section, we develop a simple model of disequilibrium melting, which yields an explicit formulation of the melting rate. After that, in section 10.3, we return to the equilibrium model and reconsider this issue.

10.2 Thermochemical Disequilibrium and the Rate of Interphase Mass Transfer

Thermochemical equilibrium is rarely, if ever, fully realized in a dynamic, multiphase system. Dynamics of the liquid and solid phases are associated with variations in pressure, temperature, and composition that require constant chemical adjustment. The kinetics of silicate systems, even at high temperature, can fail to keep pace with this forcing, which drives the system out of equilibrium. It may be that this disequilibrium is small enough that it is well approximated by the equilibrium model. But even so, there is another reason to consider disequilibrium models: they result in an explicit description of the melting rate Γ that is the key closure required of the thermodynamics.

Here we develop the simplest, physically based model of disequilibrium melting. It is formulated in terms of a quantity called the *affinity*. The affinity is defined for a given chemical component (the jth component of N, say) that exists in two phases (one solid and one liquid, say) that are in physical contact. At equilibrium, the chemical potential of that component in each of the two phases is equal. In disequilibrium, there is a difference in the chemical potential $\Delta\mu_j$ of component j between the phases. This difference in chemical potential is the affinity, and it drives a transfer of component mass between the phases. Hence, a theory for disequilibrium melting must be built from two parts: first, it requires a means for computing the affinity of each component, given the composition of each phase; and second, it requires a kinetic law relating the affinity to the mass-transfer rate. We develop a simple version of these below; for more detail the reader is referred to De Groot and Mazur [1984] and references in the Literature Notes at the end of this chapter.

10.2.1 AFFINITY AS THE THERMODYNAMIC FORCE FOR LINEAR KINETICS

For simplicity and internal consistency, we will use the ideal-solution theory that was summarized in the previous section as our model for chemical affinity. Recalling equation (10.8),

$$\Delta\mu_j = \Delta\mathring{\mu}_j + RT\ln\left(X_j^s/X_j^\ell\right). \tag{10.27}$$

The first term on the right-hand side represents the chemical potential difference when the liquid and solid phases are composed purely of component j; the second term represents the difference of the entropy of mixing component j into each phase. Equations (10.9) and (10.10) allow us to rewrite this as

$$\Delta\mu_j = RT\ln\left(\frac{X_j^s/X_j^\ell}{\check{K}_j^*}\right) = RT\ln\left(\frac{c_j^s/c_j^\ell}{\check{K}_j}\right). \tag{10.28}$$

This equation is our recipe for computing the affinity that drives reactions.

Next we consider the kinetic law relating affinity to the mass-transfer rate Γ_j of component j from the solid to the liquid. This disequilibrium mass transfer is an irreversible process, and hence is associated with an irreversible increase in entropy. The rate of entropy increase is the product of a thermodynamic force (proportional to the affinity $\Delta\mu_j$) and its consequent thermodynamic flow (the mass-transfer rate Γ_j). Hence, the entropy production rate (dissipation rate; units of energy per volume per time) arising from all reactions is

$$\Psi_\mu \equiv \overline{\rho}\sum_{j=1}^{N}\frac{\Delta\mu_j}{M_jT}\Gamma_j. \tag{10.29}$$

Here we see that the thermodynamic force is actually $\Delta\mu_j/M_jT$ for each component.

Close to equilibrium, we expect that the thermodynamic flow is linearly proportional to the thermodynamic force. Hence, we write

$$\Gamma_j = \sum_{k=1}^{N} E_{jk} \frac{\Delta\mu_k}{M_k T}, \tag{10.30}$$

where E_{jk} is a matrix of coefficients that describes the reaction rates. Onsager's reciprocal relations, derived from the second law of thermodynamics, require that this matrix be symmetric and positive semidefinite.[5] Off-diagonal terms couple the affinity of one component with mass transfers of another component. There is no general theory to determine the entries in this matrix for magmatic systems. Furthermore, far from equilibrium, a linear form for the mass-transfer rate may be a poor approximation; see the Literature Notes at the end of this chapter for more information.

The simplest possible case of the reaction-rate matrix has a single rate coefficient,

$$E_{jk} = E_0 \delta_{jk}, \tag{10.31}$$

where δ_{jk} is the Kronecker delta (the identity matrix) and $E_0 > 0$ is the rate constant. We will not consider more complex forms of E_{jk} in this book. Combining equations (10.31), (10.30), and (10.29) gives a rate of entropy production

$$\begin{aligned}\Psi_\mu &= \overline{\rho} E_0 \sum_{j=1}^{N} \left(\frac{\Delta\mu_j}{M_j T}\right)^2, \\ &= \overline{\rho} E_0 \sum_{j=1}^{N} \left[R_j \ln\left(\frac{c_j^s / c_j^\ell}{\check{K}_j}\right)\right]^2, \end{aligned} \tag{10.32}$$

From this equation it is obvious that the dissipation rate due to spontaneous reactions is always positive. This dissipation rate could be added to the mechanical dissipation rate Ψ that appears in chapter 8, although it is generally considered to be negligible.

More importantly, we have an explicit expression for the mass transfer rate of each thermodynamic component,

$$\Gamma_j = \mathcal{R}_j \ln\left(\frac{c_j^s / c_j^\ell}{\check{K}_j}\right), \tag{10.33}$$

where $\mathcal{R}_j \equiv E_0 R / M_j$ is a lumped reaction-rate constant with units of mass per volume per time. Some authors have taken this rate to be a function of the specific interfacial surface area $\mathcal{A}$ and the concentration of component j in the source phase; it might also have an Arrhenius factor if it occurs over a broad range of temperatures.

The total melting rate Γ must be equal to a sum over all N thermochemical components of the component mass transfer rates Γ_j, as in equations (9.7).

[5] A positive-definite matrix A_{ij} has the property $u_i A_{ij} u_j > 0$ for all vectors $\boldsymbol{u}$ of real numbers. A positive-semidefinite matrix has the same property except that $u_i A_{ij} u_j \geq 0$.

10.2.2 LINEARIZED MELTING RATES

If the deviation from equilibrium is sufficiently small, we can linearize equation (10.33). We express the deviation in terms of a small parameter, ϵ,

$$\frac{c_j^s/c_j^\ell}{\check{K}_j} = 1+\epsilon. \tag{10.34}$$

Then, by expanding the natural logarithm in a Taylor series about unity and truncating at first order in ϵ, equation (10.33) becomes

$$\Gamma_j \approx \mathcal{R}_j\left(\frac{c_j^s/c_j^\ell}{\check{K}_j} - 1\right) \quad \text{for} \quad \epsilon \ll 1. \tag{10.35}$$

From this expression it is clear that if the solid contains a higher concentration of component j than it should (according to $\check{K}_j$), the right-hand side is positive and hence component j is transferred from the solid to the liquid ($\Gamma_j > 0$).

In some cases it is useful to simplify the reaction rate further. This is typically achieved by assuming that variations in the liquid concentration are large relative to those of the solid (which is reasonable at small f and small $\check{K}$), and that these variations in the liquid concentration drive the reaction. In particular, starting from equation (10.35), we assume that the solid concentration is near equilibrium and hence that $c_j^s/\check{K}_j \approx \check{c}_j^\ell$. Then, we define $\mathcal{R}_j' \equiv \mathcal{R}_j/c_j^\ell$ and hence write

$$\Gamma_j = \mathcal{R}_j'\left(\check{c}_j^\ell - c_j^\ell\right). \tag{10.36}$$

This equation states that if the liquid is oversaturated in component j, the mass of that component is transferred from the liquid to the solid phase ($\Gamma_j < 0$). Equation (10.36) is used in chapter 12 on reactive flow.

10.3 Computing the Melting Rate at Equilibrium

A comparison of sections 10.1 and 10.2 indicates that whereas the melting rate Γ is absent from the equilibrium formulation, it naturally appears in the disequilibrium formulation as a sum over the component transfer rates Γ_j. This doesn't mean that the melting rate is zero at equilibrium! Rather, it indicates that the melting rate is implicitly determined by the requirement that the system remain in equilibrium as the local physical conditions (pressure, temperature, and bulk composition) change in space and time.

However, the melting rate appears in the system of equations (4.46), governing the fluid dynamics, as well as in the temperature evolution equation (8.29). Hence it is useful to have a means to compute it explicitly in the case of thermodynamic equilibrium.

There are two approaches to obtaining the melting rate. The first of these simply uses a conservation of mass equation; e.g., (4.11b) for the solid phase,

$$\Gamma = \rho^s\left[\frac{\mathrm{D}_s\phi}{\mathrm{D}t} - (1-\phi)\mathcal{C}\right], \tag{10.37}$$

to compute the melting rate *diagnostically*: it is the rate of interphase mass transfer that is consistent with the Lagrangian derivative of porosity after accounting for the rate of compaction. Of course this approach can only be applied when both terms on the right-hand side of equation (10.37) are known, which in general is not the case.[6]

The second approach uses conservation of energy to solve for Γ, rather than for the temperature. The temperature is known by inversion of the equilibrium conditions. Consequently, there are two requirements for this approach to be valid. First, the temperature must be on the solidus. In subsolidus or supersolidus regions, the temperature cannot be obtained by inversion of the solidus relations. Hence it is necessary to know the boundaries of the partially molten region and the point of evaluation. Second, the equilibrium thermochemical constraints must be uniquely invertible for temperature. This is generally the case for ideal solutions because the solidus and liquidus surfaces are continuous, smooth, single-valued functions. It is also fine at a eutectic or cotectic, although those have nonunique mappings from temperature onto melt fraction. The inverse mapping fails when the solidus and liquidus temperatures are multivalued (e.g., are parallel to the temperature axis), which would seem uncommon in silicate melting systems but is theoretically possible.

Assuming that these two conditions are met, and that the phase compositions are known, the temperature is determined via the equilibrium relationship,

$$T = T^{\mathcal{S}}(P, c_j^s) = T^{\mathcal{L}}(P, c_j^\ell). \tag{10.38}$$

If equilibrium is defined in terms of a partition coefficient as in equation (10.18), the temperature is

$$T = \left[\frac{R_j}{L_j}\ln\frac{c_j^s}{c_j^\ell} + \frac{1}{T_m(P)}\right]^{-1}. \tag{10.39}$$

Then, returning to the Boussinesq temperature equation (8.29) and rearranging it to solve for the melting rate, we have

$$\Gamma = \frac{\rho}{L}\left[\overline{\mathbb{Q}} + \Psi/\rho + \alpha_\rho T^{\mathcal{S}}\overline{\boldsymbol{v}}\cdot\mathbf{g} - c_P\left(\frac{\partial T^{\mathcal{S}}}{\partial t} + \overline{\boldsymbol{v}}\cdot\nabla T^{\mathcal{S}} - \kappa\nabla^2 T^{\mathcal{S}}\right)\right], \tag{10.40}$$

where we have replaced the temperature with the solidus temperature (which is valid if both liquid and solid are locally present).

Evaluation of the first two terms on the right-hand side of (10.40), for radiogenic heating and dissipation, is straightforward. The third term, which represents the rate of pressure–volume work, requires an evaluation of the solidus temperature, which was discussed above. The last three terms all depend on derivatives of the solidus temperature. Evaluation of these terms requires the use of its total differential,

$$\mathrm{d}T^{\mathcal{S}} = \frac{\partial T^{\mathcal{S}}}{\partial P}\mathrm{d}P + \sum_{j=1}^{N-1}\frac{\partial T^{\mathcal{S}}}{\partial c_j^s}\mathrm{d}c_j^s. \tag{10.41}$$

[6]But this is indeed the approach that becomes useful under the enthalpy method, which is introduced below in section 13.4.

Partial derivatives in this equation can be evaluated analytically for a simple solidus relationship; they can also be evaluated using finite differences, if that is more convenient. Using this result, the gradient of the solidus temperature becomes

$$\nabla T^{\mathcal{S}} = -\frac{\partial T^{\mathcal{S}}}{\partial P}\rho\mathbf{g} + \sum_{j=1}^{N-1}\frac{\partial T^{\mathcal{S}}}{\partial c_j^s}\nabla c_j^s, \tag{10.42}$$

where we have taken the pressure gradient to be lithostatic. This equation and those arising from a similar treatment of $\partial T^{\mathcal{S}}/\partial t$ and $\nabla^2 T^{\mathcal{S}}$ are substituted into equation (10.40) to obtain the melting rate.

It is important to note that equation (10.40) could equivalently be written in terms of the liquidus temperature $T^{\mathcal{L}}$. For systems with highly incompatible components where the solid concentration is extremely small, this choice may make numerical evaluation more accurate and stable.

Each of the above two methods are useful in different situations. The approach that uses mass conservation is incorporated into the enthalpy method, as described in section 13.4, below. This is a proven approach for numerical models of the evolution of the coupled thermal, chemical, and mechanical systems, and has been applied for tectonic-scale modeling. However, is not useful for analytical treatment of simpler problems. For those problems, the approach using energy conservation provides a means to obtain explicit expressions for Γ. This is illustrated in section 11.2.2 of the next chapter.

10.4 Remarks about Mantle Thermochemistry

The theory developed in chapter 9 is general enough to model the full suite of major-element oxides that participate in mantle melting. Including them could help to avoid a bias in our calculations by hewing as close as possible to the natural system. And yet in the present chapter, we advocate an approach that is, at best, a decent approximation of mantle petrology. Going beyond these low-dimensional models of mantle composition remains a research challenge at the cutting edge. Here are a few reasons to exercise caution when moving in that direction.

Model complexity. It is often simpler to build a realistic model than it is to understand it. In fact, a key goal of models is to be simpler than the natural system that they seek to represent. Inclusion of the full mantle thermochemical system will likely lead to unexpected behavior that is difficult or impossible to interpret in terms of natural processes.

Computational cost. Modeling the full suite of N major-element oxides requires solution of $N-1$ transport equations in equilibrium (double that in disequilibrium). Furthermore, realistic, nonideal solution models make the thermochemical calculations nonlinear. Large, complex free-energy minimization calculations are required to compute the equilibrium state. At the time of writing, the computational cost of both the transport and equilibrium calculations will exceed the cost of solving for the fluid dynamics.

Lack of constraints. Despite the large number of petrological experiments that have been conducted on realistic and end-member compositions, solution models for silicate phases remain poorly constrained over the full parameter space of mantle melting.

A broadly accepted, physics-based theoretical framework for silicate liquids, in particular, is currently lacking. The state of affairs for disequilibrium theory (and empirical constraints) in the context of full mantle petrology is even more skeletal.

Lack of appropriate software. At present there are a variety of sophisticated codes available to compute mantle thermochemical equilibria. All of these codes suffer from at least one of the following problems: stable only in a restricted pressure–temperature–composition space, significantly offset in temperature from the natural system, unable to handle key volatile species, prone to convergence failure for reasons that are not always understood, and not implemented in such a way that they can be incorporated into tectonic-scale computational simulations of two-phase flow. Even when all of these obstacles are overcome, simpler chemical models will still be needed to interpret the results of more complicated ones.

10.5 Literature Notes

An introduction to equilibrium chemical thermodynamics for petrology can be found in the book by Philpotts and Ague [2009], which thoroughly presents the development of ideal-solution theory outlined above. Anderson [2017] also provides an introduction, including some concepts associated with disequilibrium models. Ganguly [2008] is comprehensive but provides a helpful focus on the solid Earth, including a treatment of adiabatic processes. Denbigh [1981] is an advanced reference on equilibrium chemical thermodynamics.

Ideal solution theory was applied to two-component igneous rocks with full solid solution by Bradley [1962a,b]. The same approach was adopted and clarified by Rudge et al. [2011]; it was applied to three- and four-component systems by Keller and Katz [2016]. Ideal-solution theory was employed by Boukaré et al. [2015] in developing an empirically constrained phase diagram of MgO–FeO–SiO_2 up to 140 GPa of pressure.

Ribe [1985] was probably the first to incorporate phase diagrams quantitatively into calculations of magma transport. His approach was adapted by Katz [2008]. Linearized, two-component phase diagrams were employed by Katz [2010] and Hewitt [2010] to model reactive melt transport. This use was extended by Katz and Weatherley [2012], Weatherley and Katz [2012], and Weatherley and Katz [2016] to facilitate simulation of chemically heterogeneous mantle. The smoothed eutectic phase diagram discussed in section 10.1.5 was proposed and employed by Spencer et al. [2020b] to model Jupiter's moon Io. A sharp eutectic was used by Katz and Worster [2008] to model solidification of aqueous NH_4Cl.

There are studies that have coupled magma/mantle dynamics with a thermochemical model based on Gibbs free energy minimization. An early example of this is by Tirone et al. [2009] using up to six oxides as chemical components [see also Baker Hebert et al., 2009]. A framework for this coupling was developed by Oliveira et al. [2018] and Oliveira et al. [2020]. Of course there is an entire field of petrological thermodynamics in which multicomponent free-energy minimization calculations are used to predict equilibrium phase assemblages during melting. See, for example, pMELTS [Ghiorso et al., 2002] or PERPLE_X [Connolly, 2009]. The MELTS algorithm was coupled to a melting column model by Asimow and Stolper [1999].

The classic reference on nonequilibrium thermodynamics is De Groot and Mazur [1984]. This theory was first adapted to model mantle melting by Šrámek et al. [2007] and elaborated by Rudge et al. [2011]. The approach of Rudge et al. [2011] was

Table 10.2. Parameters and values for theforsterite–fayalite binary system after Rudge et al. [2011].

Quantity	Symbol	Forsterite	Fayalite	Units
Chemical formula		Mg_2SiO_4	Mg_2SiO_4	
Latent heat	L_j	8.71×10^4	5.17×10^5	J/kg
Molar mass	M_j	0.1406934	0.2037736	kg/mol
Specific gas constant	R_j	59.096	40.803	J-K/kg
Reference melting temperature	T_j^{m0}	2163	1478	K (at 0.1 MPa)
Simon–Glatzel equation coefficient	a_j	10.83	15.78	GPa
	b_j	3.70	1.59	

modified by Keller and Katz [2016] for use in two-dimensional models incorporating three chemical components. A broader application of nonequilibrium thermodynamics and rational thermodynamics to constraining closures in multiphase geodynamic applications was developed in Keller and Suckale [2019].

10.6 Exercises

10.1 Obtain a formula for $\phi(f)$ when $\rho^\ell \neq \rho^s$.

10.2 Develop, code and test an algorithm for computing the equilibrium melt fraction $\check{f}$ as a function of pressure, temperature, and bulk composition for the partitioning coeffficient defined in equation (10.18). Include functionality to compute the solidus and liquidus temperatures as a function of pressure and bulk composition. Do this for

(a) a two-component system;
(b) a three-component system.

10.3 Apply your code from exercise 10.2 to the forsterite–fayalite solid solution. Parameters for this system are given in table 10.2. You will need the Simon–Glatzel equation for the pure-component melting temperatures:

$$T_j^m(P) = T_j^{m0}\left(1 - \frac{P - P_{\text{ref}}}{a_j}\right)^{1/b_j}. \qquad (10.43)$$

(a) Plot the binary loops for $P = 1, 2,$ and 3 GPa on a c–T diagram.
(b) Compute and plot $\check{f}(T)$ at $P = 1, 2,$ and 3 GPa for a bulk concentration that is 50 wt% fayalite.

CHAPTER 11

Melting Column Models

Melting columns are one-dimensional approximations of mantle melting in which the domain is aligned with the gravity vector. Hence, column models provide a context in which to explore the interplay of two processes in particular: (decompression) melting and gravity-driven segregation of magma. These are the main focuses of the chapter, but other topics are also considered, including the lithospheric barrier to melt segregation and vertical transport of radiogenic nucleids.

We define the z direction as pointing upward, opposite to gravity, in the direction of melt segregation, such that $\mathbf{g} = -g\hat{z}$. Furthermore, we seek only steady, time-independent solutions. The stability of these solutions will not concern us, although we have seen above in section 6.5 that certain perturbations can lead to growth and propagation of porosity waves.

In this context, we require that all variables depend only on z. The phase velocities reduce to

$$\boldsymbol{v}^{\ell} = w\hat{z}, \quad \boldsymbol{v}^{s} = W\hat{z}. \tag{11.1}$$

We place the origin of the z-axis at the bottom of the melting column, where melting initiates and the porosity becomes nonzero. The column extends vertically to $z = z_0$, the surface of the Earth. However, since our interest is in processes that occur in the asthenosphere, we ignore the cold temperatures in the thermal boundary layer just beneath $z = z_0$ and treat the mantle as rising quasi-isentropically all the way to the surface (in particular, we neglect thermal diffusion). Melting in the column is driven by decompression. We consider the model to correspond to the mantle directly beneath the axis of a mid-ocean ridge. For a variety of reasons, this is an approximate correspondence (e.g., the column does not accommodate laterally convergent melt).

Our assumptions of steady state and one-dimensionality lead to important simplifications of the system of governing equations that make them more amenable to analysis. This becomes evident below.

11.1 Fluid Mechanics

At steady state, in one dimension, and with the Boussinesq approximation, the conservation of mass equations (4.11) and (4.12) become

$$\frac{d}{dz}(\phi w) = \Gamma/\rho, \tag{11.2a}$$

$$\frac{d}{dz}[(1-\phi)W] = -\Gamma/\rho, \tag{11.2b}$$

$$\frac{d}{dz}[\phi w + (1-\phi)W] = 0. \tag{11.2c}$$

These can be integrated from 0 to z using boundary condition $W(0) = W_0$ at the bottom of the column, where $\phi = 0$, to give

$$\phi w = W_0 F(z), \tag{11.3a}$$

$$(1-\phi)W = W_0[1 - F(z)], \tag{11.3b}$$

$$\phi w + (1-\phi)W = W_0. \tag{11.3c}$$

In these equations, $F(z)$ is the degree of melting that the solid has undergone when it has reached height z above the depth at which melting begins. F is defined as

$$F(z) \equiv \frac{\int_0^z \Gamma dz}{\rho W_0}, \tag{11.4}$$

which is the ratio of the amount of melt produced up to height z to the flux of material (pure solid) into the bottom of the column. The melting rate (and hence $F(z)$) can be specified under simplifying assumptions or can be computed in the context of a thermochemical model.

Equation (11.3c) states that, at any depth within the partially molten region, the total vertical mass flux is equal to that at the base of the column. The flux at the base of the column is given by the boundary condition W_0 on the solid upwelling rate. The value of this parameter could be inferred from the spreading rate of a mid-ocean ridge using the corner-flow solution of section 3.3. This would produce only an approximation, however, because the plate spreading at a mid-ocean ridge imposes the solid flow at the *top* of the column whereas here the upwelling rate is imposed at the bottom.

Combining the conservation of momentum equations for the two-phase aggregate (equation (4.46b)) with that for the liquid phase (equation (4.46a)) in one dimension gives

$$\phi\,(w - W) = K_\phi \left[(1-\phi)\Delta\rho g - \frac{d}{dz}\xi_\phi \frac{dW}{dz} \right], \tag{11.5}$$

where

$$K_\phi \equiv k_\phi/\mu^\ell = K_1 \phi^n \tag{11.6}$$

is the mobility and $\xi_\phi \equiv \zeta_\phi + \frac{4}{3}\eta_\phi$ is the augmented compaction viscosity. We avoid normalizing the porosity in the mobility relationship because we don't have an estimate of a characteristic porosity in the column—indeed we shall see that it is sensitive to the permeability itself. Hence, note that the scale factor K_1 differs from the factor K_0 used in other chapters.

There are two terms on the right-hand side that represent buoyancy of the magma and gradients in compaction pressure, respectively. In most of the column, melt segregation (on the left-hand side) balances the buoyancy term on the right-hand side. To see this, we rescale variables according to characteristic scales,

$$[z] = |z_0|, \quad [W] = W_0, \quad [\xi_\phi] = \xi_0, \tag{11.7}$$

where $|z_0|$ is the depth at the onset of melting. The rescaled equation is

$$\phi\,(w - W) = \phi^n \left[(1-\phi)\mathcal{Q} - \delta^2 \frac{\mathrm{d}}{\mathrm{d}z} \xi_\phi \frac{\mathrm{d}W}{\mathrm{d}z} \right], \tag{11.8}$$

where all symbols are dimensionless and we have defined

$$\delta \equiv \frac{\sqrt{K_1 \xi_0}}{z_0} \quad \text{and} \quad \mathcal{Q} \equiv \frac{K_1 \Delta\rho g}{W_0}. \tag{11.9}$$

The first of these two dimensionless quantities is the ratio of the compaction length to the column height. For a compaction length that is of order 1 km (see section 6.6) and a column height of about 60 km, $\delta \ll 1$. The second is a parameter that relates the magma buoyancy to the segregation flux. Using values $\Delta\rho \approx 500$ kg/m^3, $g \approx 10$ m/s^2, $W_0 \approx 3$ cm/yr, and $K_1 = 10^{-8}$ m^2/Pa-s (for $n = 2$) gives $\mathcal{Q} \approx 5 \times 10^4$. The ratio of $\mathcal{Q}$ to δ^2 tells us that gradients in compaction pressure are negligible over most of the column and we can consider the balance of buoyancy and Darcy drag.

Then, making the zero-compaction-length approximation and using equation (11.3c) (with velocities rescaled by W_0), we can rewrite (11.8) as

$$\phi w = \mathcal{Q}\phi^n (1-\phi)^2 + \phi. \tag{11.10}$$

And using equation (11.3a) (after rescaling) to eliminate the liquid flux, we obtain

$$\mathcal{Q}\phi^n (1-\phi)^2 + \phi - F = 0. \tag{11.11}$$

Assuming small porosities such that $(1-\phi)^2 \sim 1$, equation (11.11) simplifies further to

$$\mathcal{Q}\phi^n + \phi - F = 0. \tag{11.12}$$

Analytical solutions to this polynomial exist when $n = 2, 3$. In the former case,

$$\phi(z) \approx \frac{1}{2\mathcal{Q}} \left(\sqrt{4\mathcal{Q}F + 1} - 1 \right). \tag{11.13}$$

This solution is shown in figure 11.1c for a $F(z)$ that increases linearly with z. The liquid and solid upwelling rates, $w(z)$ and $W(z)$, are reconstructed using equations (11.3a) and (11.3b) (after nondimensionalizing).

It is useful to further approximate (11.12) by recognizing that over most of the column, $\mathcal{Q} \gg \phi^{1-n}$ and hence that the first term balances the degree of melting. This balance gives

$$\phi(z) \sim [F(z)/\mathcal{Q}]^{1/n}, \tag{11.14}$$

which is the asymptotic solution in the limit of large $\mathcal{Q}$. Combining this result with equation (11.3a) gives an asymptotic solution for the liquid speed,

$$w(z) \sim \mathcal{Q}^{1/n} F(z)^{1-1/n}. \tag{11.15}$$

These asymptotic results are plotted in figure 11.1c. From the above we see that the largest porosity and segregation speed are attained at the top of the column, where $F = F_{\text{max}}$. We can write, in dimensional terms,

$$w_{\text{max}}/W_0 \sim F_{\text{max}}/\phi_{\text{max}} \quad \text{and} \quad \mathcal{Q} \sim (w_{\text{max}}/W_0)^n F_{\text{max}}^{1-n}, \tag{11.16}$$

which enables an estimate of $\mathcal{Q}$, given constraints on the melt and solid velocities and the maximum degree of melting.

We could also view (11.16) as a means to estimate the maximum melt ascent rate by diffuse porous flow, given observational constraints on mantle upwelling rate, maximum degree of melting, and maximum porosity. For a mid-ocean ridge, the spreading rate combined with the corner-flow solution provide a good estimate for W_0; spreading rate varies widely but a characteristic value for upwelling rate might be 10 cm/yr. The maximum degree of melting can be estimated on the basis of petrological analysis of basalt chemistry to be about 0.2. The maximum porosity is much more difficult, but estimates from seismic tomography have typically been about 0.01. Together these estimates give a maximum melt ascent rate of 2 m/yr. This is small in comparison to estimates based on the response to Iceland deglaciation, which are tens of meters per year. This difference may be reconciled by channelized (rather than diffuse) melt transport, which we consider in chapter 12.

The asymptotic, nondimensional results (11.14) and (11.15) are incorrect at the bottom of the melting region, where ϕ is small and $\mathcal{Q} \ll \phi^{1-n}$. Here we have $\phi \sim F$, indicating no melt segregation and hence that $w \sim W - 1$ at $z \to 0$. This is captured by the more complete solution (11.13), which nonetheless neglects the compaction pressure gradient. However, as we saw in the filter-press problem (section 6.1) and the permeability-step problem (section 6.2), gradients of compaction pressure become important within a narrow region near obstacles to flow. Hence we expect a boundary layer near the bottom of the melting column, where the porosity goes to zero at $z = 0$. In this narrow region, the buoyancy of the liquid is balanced by a gradient in the compaction pressure, $\Delta\rho g \sim \text{d}\mathcal{P}/\text{d}z$. Since $\mathcal{P} = \xi_\phi \mathcal{C}$, we expect very slow compaction if the augmented compaction viscosity ξ_ϕ is large. This should be true, since the porosity will be vanishingly small in this layer. This is sometimes called the *visco-gravitational regime*. It is analyzed in section 11.3, where we show that it forms a very narrow boundary layer with insignificant effect on the rest of the domain.

Melting driven by mantle decompression causes the porosity to grow with height in the column. The augmented compaction viscosity decreases and, outside the visco-gravitational boundary layer, the compaction pressure gradient is no longer sufficient to balance the buoyancy of the liquid. Buoyancy is then balanced by Darcy drag; the matrix compacts freely as melt segregates. In this *Darcian regime*, the gradient in porosity is small, solid upwelling is slow, and hence advection of porosity by solid upwelling is negligible. Since this regime occupies most of the melting column, the solutions to the ZCL model above are broadly useful.

Solutions to the full equation (11.5), retaining gradients in compaction pressure, capture both the visco-gravitational and Darcian regimes. These can be obtained numerically or analytically by the method of matched asymptotics (see the Literature Notes at the end of this chapter). In this chapter we generally neglect the visco-gravitational regime and hence the compaction boundary layer.

11.2 Melting-Rate Closures

To solve equation (11.12) for the porosity, and hence to obtain $w(z)$, we require a specification of the melting rate Γ (or the degree of melting $F(z)$). In this section we consider several approaches to obtaining this closure. The first approach simply prescribes the melting rate (as a constant) throughout the column. After that we consider thermodynamically consistent models of melting, simplified from our approach in chapter 10.

11.2.1 PRESCRIBED MELTING RATE

We can close the system by prescribing the maximum degree of melting F_{max} that is attained by the solid when it has reached the top of the melting column at $z = z_0$. A second, independent assumption is that the column reaches this maximum degree with constant decompressional productivity, $\text{d}F/\text{d}z = F_{\text{max}}/z_0$. We then require that the melting rate is proportional to the vertical mass flux which, in a steady-state, extended Boussinesq melting column, is $\rho\overline{\boldsymbol{v}} = \rho W_0\hat{\boldsymbol{z}}$. In dimensional terms we have

$$\Gamma = \rho W_0 F_{\text{max}}/z_0. \tag{11.17}$$

Consistent with the assumptions above, equation (11.4) tells us that $F(z) = F_{\text{max}}z/z_0$, again in dimensional terms. This can be used in equation (11.13) to obtain porosity for $n = 2$. For other values of n, $F(z)$ can be used in equation (11.11) or (11.12).

The small-porosity Darcy solution for $n = 2$ is plotted in figure 11.1 for two values of the liquid flux parameter $\mathcal{Q}$. The solid upwelling rate in (b) decreases linearly with height due to melting and melt segregation. This melting leads to an upward increase in porosity in figure 11.1(c). The sharp curvature of porosity near $z = 0$ is associated with the nonlinear relationship between porosity and permeability. Figure 11.1(d) shows that larger $\mathcal{Q}$ is associated with smaller porosity and faster liquid upwelling and segregation. This difference in segregation rates is, in fact, the cause of the smaller porosity.

In figure 11.2, the column model is plotted on logarithmic axes to highlight the structure near the base of the column. Figure 11.2(a) shows the approximate porosity solution (11.13) versus height in the column for both values of $\mathcal{Q}$. As noted above, there are two asymptotic regimes, a deeper one where $\phi \sim F$ and a shallower one where $\phi \sim (F/\mathcal{Q})^{1/n}$. The transition between these regimes occurs at $z \sim \mathcal{Q}^{1/(1-n)}/F_{\text{max}}$.

By the assumptions used to obtain the solution (11.13), the buoyancy of the melt is balanced by Darcy drag. Nonetheless, there is compaction associated with the balance of melting and segregation of the melt; we can compute its attendant compaction pressure. This is shown in figure 11.2b, for two hypothetical forms for the augmented compaction viscosity ξ_ϕ. In both cases we assume that the compaction viscosity ζ_ϕ dominates as $\phi \to 0$. When $\xi_\phi \propto \phi^{-1}$ (black lines), there is a strong,[1] negative singularity as $z \to 0$.

[1]Here, strength is a measure of with the rate at which a function goes to $\pm\infty$ as the singularity is approached.

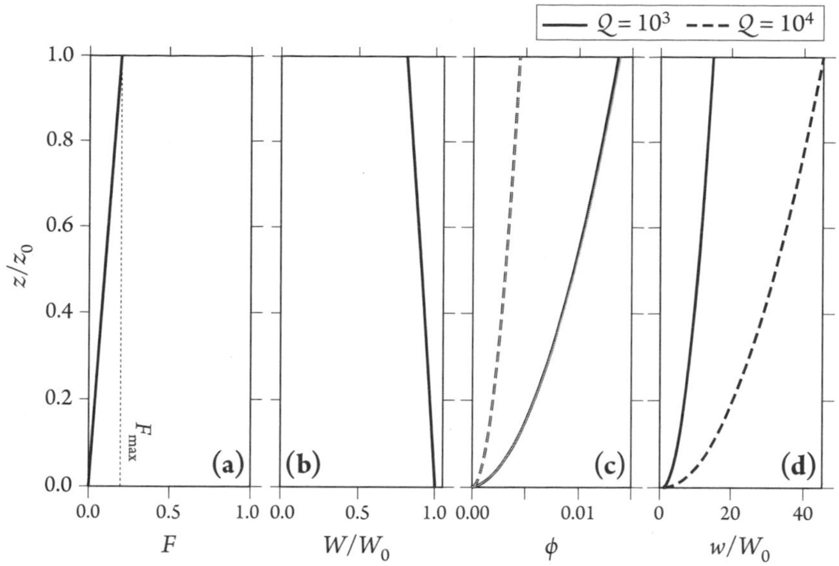

Figure 11.1. Solutions of the melting column model with prescribed melting rate (11.17), under the assumption that Darcy drag balances buoyancy of the liquid phase. Black lines show the analytical solution (11.13) and $w(z)/W_0 = F(z)/\phi(z)$. Parameters are $F_{max} = 0.2$, $n = 2$, and $\mathcal{Q}$ as given in the legend. **(a)** Degree of melting. **(b)** Scaled solid upwelling rate. **(c)** Porosity. Thin gray lines show the numerical solution of the implicit equation (11.11). **(d)** Liquid upwelling rate scaled with the inflow solid upwelling rate.

In contrast, with $\xi_\phi \propto -\ln\phi$, the singularity is weak. The strength of this singularity controls the compaction length as $\phi \to 0$.

For both cases in figure 11.2(b), compaction pressure is negative and hence $P^\ell < P^s$; this is as expected given the sign of $\mathrm{d}W/\mathrm{d}z$. The liquid underpressure arises from its buoyancy relative to the solid. Near $z = 0$, this buoyancy acts on sharply diminished porosity and hence sharply increased compaction viscosity. The steep gradient in ΔP here, especially for $\xi_\phi \propto \phi^{-1}$, is inconsistent with our assumption that the gradient in compaction pressure can be neglected (however, the existence of a singularity is physically inconsistent). This indicates that a boundary layer, which we had neglected by making the ZCL approximation, should appear at the bottom of the column. We calculate the properties of this boundary layer in section 11.3, below.

11.2.2 THERMODYNAMICALLY CONSISTENT MELTING RATE

In the previous section we prescribed the melting rate to close the equations and allow us to obtain a solution. This is not entirely satisfactory, however, because the melting rate should be consistent with the (approximated) thermochemistry of the mantle system. In this section we introduce the conservation of energy and derive a consistent formulation of the melting rate. We first do this for a single-component system and then for a two-component system. For simplicity and according to our assumption of quasi-isentropic melting, we neglect thermal diffusion. This is a good approximation away from the lithosphere, which is the conductive boundary layer near the surface (though also see exercise 8.7).

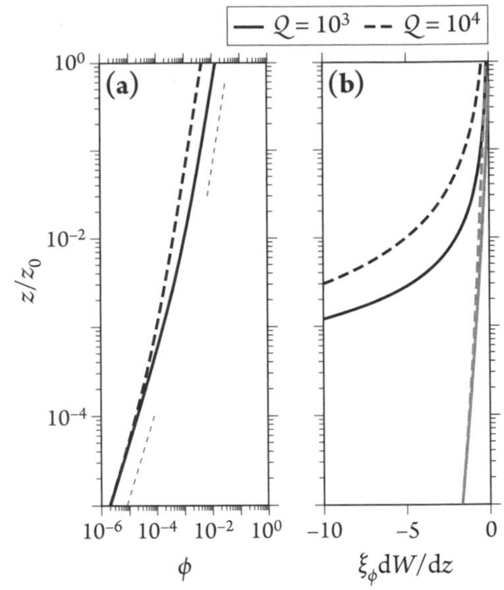

Figure 11.2. Solutions of the melting column model with prescribed melting rate (11.17). Parameters are as in figure 11.1. **(a)** The scaled porosity from equation (11.13) as a function of column height with logarithmic axes. Narrow, dashed lines show scalings $\phi \propto z$ and $\phi \propto z^{1/n}$. **(b)** The a posteriori, nondimensional compaction pressure, computed from the solution for $W(z), \phi(z)$ from (11.13) and (11.3a). Two forms for the augmented compaction viscosity are considered: the black lines represent $\xi_\phi = \phi^{-1}$; the gray lines represent $\xi_\phi = -\ln \phi$.

In one dimension and at steady state, the temperature equation can be simplified using the dimensional mass conservation equation (11.3c), which tells us that $\overline{\boldsymbol{v}} = W_0\hat{\boldsymbol{z}}$ (making the Boussinesq approximation). Equation (8.29) then reads

$$\rho c_P W_0 \frac{\mathrm{d}T}{\mathrm{d}z} = -\left(L\Gamma + \rho\alpha_\rho W_0 g T\right). \tag{11.18}$$

Here, for simplicity, we have taken $\mathbb{Q} = \Psi = 0$ and $\rho^s = \rho^\ell = \rho$. With this equation, we will follow the approach outlined in section 10.3: assume that temperature is equal to the solidus temperature and use the equilibrium constraints of the phase diagram to obtain an explicit expression of the melting rate. Moreover, we treat thermodynamic parameters as constant and equal between phases. These include thermal expansivity α_ρ, specific heat capacity c_P, mass-specific latent heat L, and density ρ.

One-Component Mantle Column

If the mantle were composed of only a single component, Earth would be very different, but models of Earth would be easier to understand! In particular, for a one-component mantle, the melting temperature is a function of pressure only,

$$\begin{aligned} T^m(P) &= T^m|_{P=0} + \int_0^P \vartheta^{-1}(P')\mathrm{d}P', \\ &\approx T^m|_{P=0} + P/\vartheta, \end{aligned} \tag{11.19}$$

where P' is a dummy variable of integration and $\vartheta(P) \equiv \mathrm{d}P/\mathrm{d}T^m$ is the Clausius-Clapeyron slope of the melting point. In the approximation on the second line, we have assumed that $\vartheta = $ const. and have integrated accordingly.

Recall that for a single-component system in thermodynamic equilibrium, the temperature must remain exactly equal to the melting temperature wherever both phases are present. Hence we can use (11.19) to eliminate the temperature gradient from (11.18) if we assume a lithostatic pressure gradient (i.e., $\mathrm{d}P = -\rho g \mathrm{d}z$). Then, where $\phi > 0$,

$$\frac{\mathrm{d}T}{\mathrm{d}z} = \frac{1}{\vartheta}\frac{\mathrm{d}P}{\mathrm{d}z} = -\rho g/\vartheta \tag{11.20}$$

and (11.18) can be rearranged for the melting rate to give

$$\Gamma = \frac{\rho g W_0}{L}\left[\frac{\rho c_P}{\vartheta} - \alpha_\rho T^m\right], \tag{11.21a}$$

$$= \frac{\rho g W_0}{L}\left[\frac{\rho c_P}{\vartheta} - \alpha_\rho\left(T_0^m - \frac{\rho g z}{\vartheta}\right)\right], \tag{11.21b}$$

where $T_0^m = T^m|_{P=0} + \rho g z_0/\vartheta$ is the melting point at the bottom of the melting column. We can compare the contribution of the three terms within the square brackets of equation (11.21b) by taking $T_0^m \sim 1600$ K, $z_0 \sim 60$ km, and other parameters from table 8.1. Relative to the size of the first term, the second term is $\mathcal{O}(10^{-1})$ and the third term is $\mathcal{O}(10^{-2})$. We therefore neglect the third term and define the one-component maximum degree of melting

$$F_{\text{max}}^{1c} \equiv \frac{\rho g/\vartheta - \alpha_\rho g T_0^m/c_P}{L/c_P} z_0 \tag{11.22}$$

and approximate Γ as

$$\Gamma \approx \rho W_0 F_{\text{max}}^{1c}/z_0. \tag{11.23}$$

This equation has been written in a form that highlights its similarity to equation (11.17). The one-component model with constant thermodynamic coefficients leads to an approximately constant decompressional productivity (the deviation is of order 10^{-2} and is associated with the variation in the PV-work term in (11.21) that we neglected). Constant productivity is not obtained in one-component models with more detailed thermodynamics, where the Clapeyron slope will vary substantially over the pressure range associated with decompressional melting of the asthenosphere.

To understand the melting rate for this one-component model, it is helpful to think about the different groups of variables in (11.21a). The quantity $\rho g W_0$ is the rate of depressurization of the two-phase aggregate (also the rate of change of potential energy), which we expect to drive melting. In the square brackets, the first, dimensionless term is the drop in melting temperature with decompression multiplied by the volumetric heat capacity. This factor converts the decompression rate into an available power density. Some power density is lost to volumetric work of expansion, represented by the second, dimensionless term in the square brackets. The net power density is reduced by the energetic cost (per unit mass) of melting, the latent heat, to give the melting rate.

As noted above, the melting rate for the one-component model is well approximated by a constant productivity F_{max}^{1c}/z_0. Therefore, the nondimensional governing equations for the one-component model are identical to those for the prescribed melting rate model, except for replacement of F_{max} by F_{max}^{1c}. The solution is also identical; it is given by (11.13) with the same replacement. Plots of the solution for $\mathcal{Q} = 10^3$, 10^4 are shown in figure 11.3.

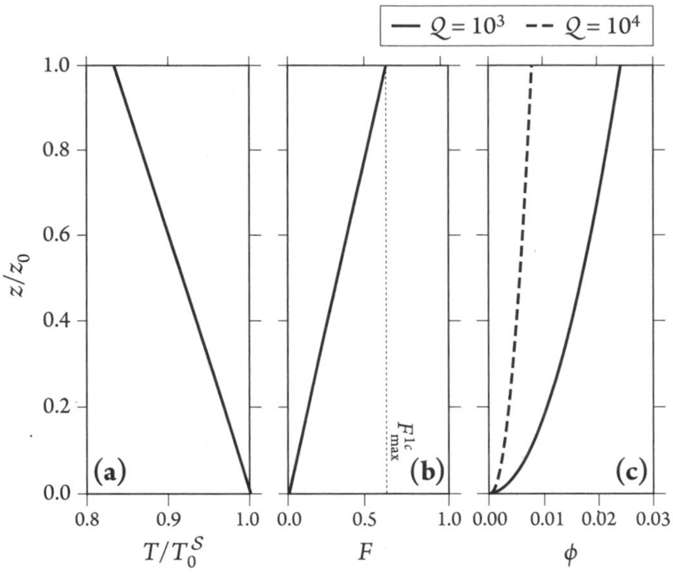

Figure 11.3. Solutions of the one-component melting column model with melting rate (11.23) under the assumption that Darcy drag balances buoyancy of the liquid phase. Parameters are $z_0 = 60$ km, $T_0^m = 1650$ K, $n=2$, $\mathcal{Q}$ as given in the legend, and other parameters as given in table 8.1. **(a)** Temperature. The slope of this curve is $-\rho g/\vartheta$. **(b)** Degree of melting with F^{1c}_{max} as given in (11.22). **(c)** Scaled porosity from (11.13).

Comparison of figure 11.3 for the one-component column model and figure 11.1 for the prescribed melting rate model reveals that they are qualitatively identical. In both cases, melting proceeds linearly with height in the column, leading to porosity increase, melt segregation, compaction, and decreasing solid upwelling speed. The quantitive difference is in the amount of melt produced. The one-component column produces much more melt; it has a maximum F of about 0.6 whereas F_{max} was prescribed to be 0.2 in figure 11.1. The reason for this factor-of-three difference stems from the use of thermodynamic parameters to compute Γ in figure 11.3, rather than imposing the rate as in figure 11.1.

The maximum degree of melting in the one-component column of figure 11.3 is also greater than the column model without segregation in section 8.7. How do we account for this? In the latter case, we approximated melting of real, multicomponent mantle rock by having distinct solidus and liquidus temperatures for the mantle. With each increment of melting, the residual solid is depleted of the fusible components and becomes more refractory; the melting rate decreases with height in the column. This is a realistic and desirable feature of a simple model that is lacking in the present, one-component model. However, an issue with the formulation in section 8.7 is that it is awkward to define or track the "composition" of the liquid and solid phases. Hence there is no natural way to model freezing, metasomatism, heterogeneity, or reactive melting. A better approach is to extend the one-component model to handle multiple chemical components. In the next section we consider a two-component version.

Two-Component Mantle Column

A two-component model is a step in complexity, requiring at least one additional PDE for conservation of species mass. However, this additional complexity enriches the model: it enables us to describe progressive depletion by tracking the concentration of a conserved chemical component. In the context of a steady-state melting column model, this gives us better control of the maximum extent of melting. In the context of non-steady, two-dimensional models at the tectonic scale, a two-component system provides the simplest formulation that allows for mantle heterogeneity.

What should we choose for our two mantle components? By mass, olivine is the dominant mantle mineral, but it is much less fusible than pyroxene. Melting of pyroxene is most important for magma genesis, but this mineral is sometimes completely removed from the residue. The difficulty of capturing mantle melting in a chemically simple system was considered in chapter 10. With a two-component system, the best we can do is to introduce fictional components that represent *fertile* and *refractory end-members*. Presence of the fertile component lowers the melting temperature and enhances melting. The liquid has a higher concentration of the fertile component relative to the solid residue from which it is derived. The refractory component is concentrated in the solid residue with progressive melting, making that solid increasingly difficult to melt. In this context, it makes some sense to consider the fertile component to be "basalt" and the refractory component to be "olivine." These terms are placed in quotes to remind us that they should not be taken literally, for reasons noted in chapter 10.

For analytical tractability, we apply the linearized solidus and liquidus curves of section 10.1.6. We make the further simplification of equal slopes $M^{\mathcal{L}} = M^{\mathcal{S}}$ and hence that $c^s - c^\ell = \Delta c$, a constant. This will simplify the calculations below.

Consistent with our previous simplifications for column models, we consider composition in steady state, along one dimension, with constant and equal phase densities, and without diffusion. We assume thermochemical equilibrium holds throughout the domain but drop the $\check{}$ from concentrations for simplicity of expression. These assumptions mean that our open-system model produces melts that are identical to those produced by closed-system, *batch melting* to the same degree of melting, F (see the Literature Notes at the end of this chapter).

Under these assumptions, equation (9.4c) for conservation of the phase-averaged, bulk species mass becomes

$$\frac{\mathrm{d}}{\mathrm{d}z}\left[\phi w c^\ell + (1-\phi) W c^s\right] = 0, \tag{11.24}$$

where c^i is the concentration of one of the two chemical components in phase i. The concentration of the other component can always be recovered as $1 - c^i$. Equation (11.24) can be integrated from $z = 0$, where the porosity is zero, to any height z, as we did for the mass conservation equations (11.2). Then the dimensional governing equations for energy, mass, and composition are

$$\rho c_P W_0 \frac{\mathrm{d}T}{\mathrm{d}z} = -\left(L\Gamma + \rho\alpha_\rho W_0 g T\right), \tag{11.25a}$$

$$W_0 = \phi w + (1-\phi) W, \tag{11.25b}$$

$$W_0 c_0^s = \phi w c^\ell + (1-\phi) W c^s, \tag{11.25c}$$

where c_0^s is the concentration in the unmolten mantle that flows into the domain from below. It will be useful to combine equations (11.25b) and (11.25c) to eliminate $(1-\phi)W$ and obtain

$$\phi w = W_0 \frac{c^s - c_0^s}{\Delta c}, \tag{11.26}$$

an equation relating the fluid flux to the depletion of the solid phase. Note that the quotient on the right-hand side was recognized in equation (10.26) as the degree of melting F, and this makes (11.26) consistent with (11.3a).

At steady state, conservation of liquid mass (4.11a) relates the divergence of the liquid flux to the melting rate. Differentiating (11.26), and using this relationship, gives

$$\frac{\mathrm{d}}{\mathrm{d}z}(\phi w) = \frac{\Gamma}{\rho} = \frac{W_0}{\Delta c}\frac{\mathrm{d}c^s}{\mathrm{d}z}. \tag{11.27}$$

The melting rate Γ must be consistent with conservation of energy. Furthermore, our assumption of thermochemical equilibrium implies that the temperature and phase compositions in the partially molten region are constrained by the phase diagram. In particular, they must satisfy the solidus and liquidus relations in (10.24). We can re-express the linearized solidus equation (10.24a) in terms of its vertical derivative,

$$\frac{\mathrm{d}T^S}{\mathrm{d}z} = -\frac{\rho g}{\vartheta} + M^S \frac{\mathrm{d}c^s}{\mathrm{d}z}, \tag{11.28}$$

where we have taken the thermodynamic pressure to be equal to the lithostatic pressure. This relationship, which is a two-component analogue of (11.20), must hold throughout the partially molten region.

Equations (11.27) and (11.28) can be combined to eliminate the compositional gradient and give an equation relating the temperature gradient to the melting rate. Substituting this into the conservation of energy equation (11.25a) to eliminate $\mathrm{d}T/\mathrm{d}z$, and rearranging for Γ, gives

$$\Gamma \approx \rho W_0 \frac{\rho g/\vartheta - \alpha_\rho g T_0^S/c_P}{M^S \Delta c + L/c_P}, \tag{11.29}$$

where we have approximated the PV-work term by fixing the temperature to the melting temperature at the bottom of the column (as we did for the one-component solution above). Once again we find that the predicted melting rate is approximately constant in the column.

Using equation (11.29) to form the degree of melting via equation (11.4) gives

$$F(z) \approx \left(\frac{\rho g/\vartheta - \alpha_\rho g T_0^S/c_P}{M^S \Delta c + L/c_P}\right) z, \tag{11.30}$$

where we have used $F=0$ at $z=0$. By evaluating this at the top of the column where $z=z_0$, we can define

$$F_{\text{max}}^{2c} \equiv \left(\frac{\rho g/\vartheta - \alpha_\rho g T_0^S/c_P}{M^S \Delta c + L/c_P}\right) z_0. \tag{11.31}$$

Compare this with the equivalent result for the one-component column in equation (11.22). The numerator is identical between the two cases, but for the two-component melting column, the denominator is augmented by $M^S \Delta c$. This quantity represents the increase of solidus temperature as the solid residue becomes increasingly refractory with increasing height in the column. This diminishes the maximum degree of melting.

Hence, in the context of our two-component column model with constant thermodynamic coefficients, the decompressional productivity is again found to be approximately constant with height in the column. This result arises from the simplicity of the phase diagram that we impose, which has constant Clapeyron slope, a linearized solidus, and constant Δc. None of these assumptions hold more generally and hence decompressional productivity is, again, not anticipated to be constant in more detailed models. This is illustrated below in our model of a two-component system where one component represents volatile elements (and we relax the assumption of constant Δc).

This caveat aside, the porosity solution (11.13) from the prescribed-melting-rate case applies, after replacing F_{max} with F^{2c}_{max}. To plot the solution, however, we need a value for F^{2c}_{max}. This is not readily available, because our two-component system is fictitious: it cannot be subjected to laboratory experiments to determine M^S or Δc. We can therefore understand the equation for F^{2c}_{max} as an opportunity to calibrate our simple, two-component model to the natural system. However, we can constrain only the product $M^S \Delta c$, which represents the difference between the liquidus and solidus temperatures at a fixed bulk composition—the *melting interval*.

Various lines of geochemical and geophysical evidence suggest that the maximum degree of melting beneath mid-ocean ridges is between 20% and 25%. Setting $F^{2c}_{\text{max}} = 0.225$, $z_0 = 60$ km, $T^S_0 = 1650$ K, and other parameters as in table 8.1, gives a melting interval of $M^S \Delta c = 700$ K.

Figure 11.4 shows the temperature, porosity, and bulk composition of the column. Other fields are exactly as one would expect on the basis of figures 11.3 and 11.1 and are hence not shown. Temperature in (a) is computed using the solidus equation (10.24a). The difference between the solidus curve and the dotted line showing the Clausius–Clapeyron slope arises from the progressive removal of the fertile component with height in the column, which leads to lower melt productivity. Porosity is computed using the Darcy solution under the ZCL approximation (equation (11.13) for $n = 2$), replacing F_{max} with F^{2c}_{max}. The bulk composition is defined as

$$\begin{aligned} \bar{c} &= \phi c^\ell + (1-\phi)c^s, \\ &= c^s_0 + (F - \phi)\Delta c, \end{aligned} \tag{11.32}$$

where the second equation is obtained using a rearranged lever rule $c^s = F\Delta c + c^s_0$. Since we have not yet specified c^s_0 or Δc, we avoid doing so now and plot $(\bar{c} - c^s_0)/\Delta c$ instead. This represents the change in the bulk composition due to liquid segregation. Figure 11.4(c) shows that the bulk composition increases with height in the column. This is because the liquid, for which $c^\ell < c^s$, is being removed by segregation, leaving the bulk composition more refractory. More rapid melt segregation (larger $\mathcal{Q}$) leads to smaller porosity and a bulk composition that is closer to the solid composition.

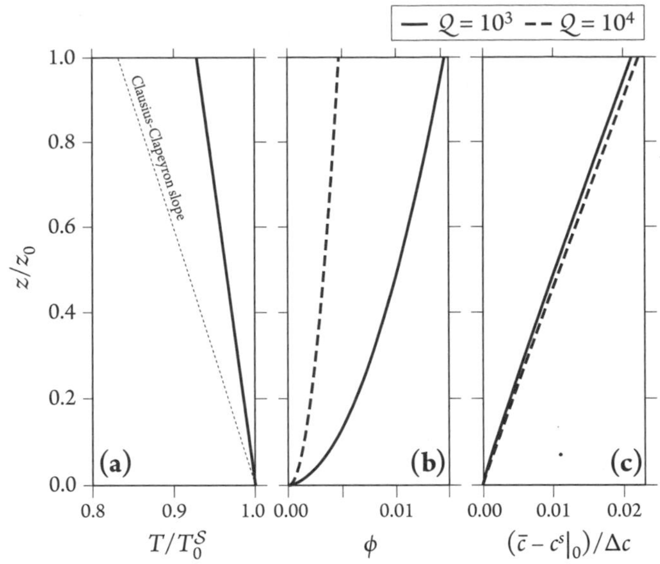

Figure 11.4. Solutions of the two-component, equilibrium-melting column model with melting rate (11.29), under the assumption that Darcy drag balances buoyancy of the liquid phase. Parameters are $M^S\Delta c = 700$ K, $z_0 = 60$ km, $T_0^S = 1650$ K, $n = 2$, $\mathcal{Q}$ as given in the legend, and other parameters as given in table 8.1. **(a)** Temperature. Dotted line shows the Clausius–Clapeyron slope $-\rho g/\vartheta$. **(b)** Scaled porosity. **(c)** Scaled bulk composition. The degree of melting F is not shown; it increases linearly to $F_{\text{max}}^{2c} = 0.225$, as discussed in the main text.

Two-Component Column with an Incompatible Component

The two-component melting column described above can be modified to model the effect of an incompatible component, including volatiles such as water and carbon. These are present in small concentrations but drastically lower the solidus temperature at a given pressure. Melting thus begins at higher pressure. However, it does not proceed linearly, as was the case above. Incompatible elements partition strongly into the liquid phase, and hence melting rapidly depletes the solid. In terms of the two-component phase diagrams of chapter 10, it is still advantageous to linearize the solidus and liquidus curves, but a difference in slope between them is now essential. Indeed the compositional distance between these curves Δc is now a function of the concentration itself: we adopt a partitioning coefficient and write $c^s = Dc^\ell$ and hence we have

$$\Delta c = c^s\,(1 - 1/D)\,. \tag{11.33}$$

Combining this relationship with the integrated mass conservation equations (11.25b) and (11.25c), rearranging for the liquid flux and differentiating with respect to z gives

$$\frac{\mathrm{d}}{\mathrm{d}z}(\phi w) = \frac{\Gamma}{\rho} = \frac{W_0 c_0^s}{1 - 1/D} \cdot \frac{1}{(c^s)^2}\frac{\mathrm{d}c^s}{\mathrm{d}z}. \tag{11.34}$$

This is analogous to equation (11.27) from the two-component model with constant Δc, but here we obtain a nonlinear dependence on the volatile concentration in the solid.

Taking a different strategy from the previous subsection, we seek a differential equation in c^s alone. We substitute the solidus gradient (11.28) into the energy equation (11.25a), rearrange for the melting rate and linearize the PV work term to give

$$\frac{\Gamma}{\rho} = -\frac{c_P W_0}{L}\left[M^S \frac{\mathrm{d}c^s}{\mathrm{d}z} - \frac{\rho g}{\vartheta} + \frac{\alpha_\rho g T_0^S}{c_P}\right]. \tag{11.35}$$

Substituting this into equation (11.34) to eliminate Γ/ρ, and rearranging, gives

$$\left[1 + \frac{A_1}{(c^s)^2}\right]\frac{\mathrm{d}c^s}{\mathrm{d}z} = A_2, \tag{11.36}$$

where

$$A_1 \equiv \frac{L c_0^s}{c_P M^S (1 - 1/D)}, \qquad A_2 \equiv \frac{g}{M^S}\left(\frac{\rho}{\vartheta} - \frac{\alpha_\rho T_0^S}{c_P}\right). \tag{11.37}$$

With the boundary condition $c^s(0) = c_0^s$, equation (11.36) has the solution

$$c^s(z) = \frac{1}{2}\left[c_0^s - \frac{A_1}{c_0^s} + A_2 z + \sqrt{4A_1 + \left(A_2 z + c_0^s - \frac{A_1}{c_0^s}\right)^2}\,\right]. \tag{11.38}$$

This is plotted in figure 11.5(a). Two regimes are evident, corresponding to the two terms on the left-hand side of equation (11.36). From the bottom of the column, the solid concentration is linearly depleted until $c^s \lesssim \sqrt{A_1}$. In the second regime, the solid concentration is buffered near (but above) zero.

The solution for c^s can be used in equation (11.35) to compute the melting rate. The degree of melting is then obtained as

$$\begin{aligned} F &= \int_0^z \frac{\Gamma}{\rho W_0}\mathrm{d}z, \\ &= \frac{c_P}{L}\left[\left(\frac{\rho g}{\vartheta} - \frac{\alpha_\rho g T_0^S}{c_P}\right) z + M^S\left(c_0^s - c^s\right)\right]. \end{aligned} \tag{11.39}$$

This result is plotted in figure 11.5(b). The solution for $c^s(z)$ is also used in the solidus relation (10.24a) to compute the temperature structure $T(z)$, which is plotted in (c). The other column-model variables (ϕ, W, w) are computed using equation (11.13) for the porosity and $w(z) = F(z)/\phi(z)$. The porosity structure is plotted in (d).

Figure 11.5 shows that the temperature drops continuously with upwelling, but the absolute rate of change is larger in the top quarter of the column. There the temperature slope is approximately the Clausius–Clapeyron slope. This upper regime corresponds to a transition from low-productivity to high-productivity upwelling. The transition is evident in (b): in the bottom three quarters of the column, F remains close to zero; in the top quarter of the column, F increases rapidly.

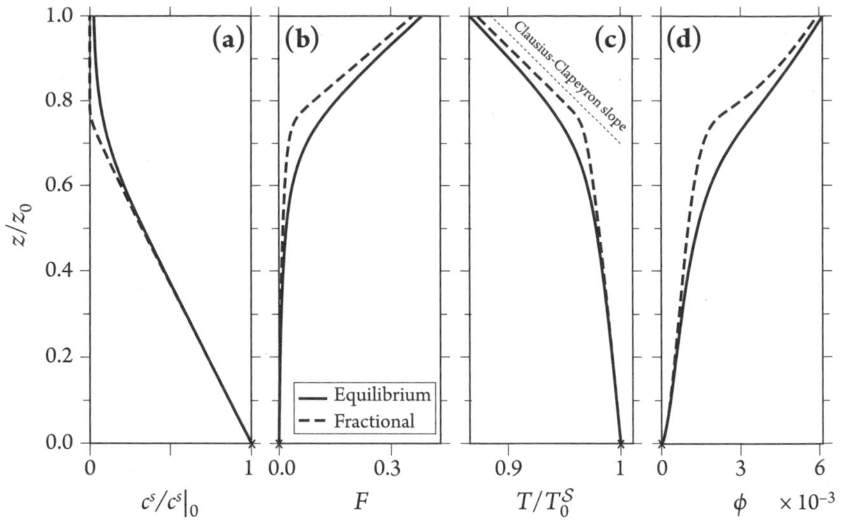

Figure 11.5. A solution of the two-component melting column model with a volatile component. Parameters as in table 8.1 and also $M^S = -4$ K/ppm, $c_0^s = 100$ ppm, and $\mathcal{Q} = 10^4$. The onset of melting is marked by a "×" in each panel. The bottom of the column $z = 0$ corresponds to a depth of 130 km and $T_0^S = 1300\,°\mathrm{C}$. The solid line represents the solution developed in this section for equilibrium between solid and liquid phases; the dashed line represents disequilibrium (fractional) melting, developed in exercise 11.8 using equations (9.24). **(a)** The normalized volatile concentration in the solid from equation (11.38). **(b)** Degree of melting as a function of height, computed according to equation (11.39). **(c)** Temperature as a function of height in the melting column, obtained from the solidus relation (10.24a). **(d)** Porosity computed according to the approximate Darcy solution (11.13) for $n = 2$.

Evidently the two-component system with a solidus-lowering incompatible component at low concentration captures the low-productivity tail at depths beneath that of the dry melting point. But at shallower depth than this, the solution behaves like the one-component system from section 11.2.2: temperature follows the Clausius–Clapeyron slope and the melt production is too large. A three-component system where the first component is a refractory silicate, the second component is a fertile silicate (constant Δc_2), and the third component is an incompatible element (Δc_3 as a function of D) can ameliorate this problem. Then the melting curves include a low-productivity tail controlled by the incompatible component at depth and a shallow productivity that is moderated by progressive removal of the fertile component. This extension of the present treatment is straightforward; see the Literature Notes at the end of this chapter for more information.

11.3 The Visco-Gravitational Boundary Layer

Here we develop a model of the narrow boundary layer above the onset of melting (at $z = 0$), where the compaction pressure gradient is important. We assume constant decompressional productivity and hence that $F(z) = F_{\max} z$, in terms of the nondimensional height in the column. Returning to the compaction equation (11.8), making the small-porosity approximation and balancing the liquid buoyancy against the gradient in compaction pressure gives

$$\frac{d\mathcal{P}}{dz} \sim \frac{\mathcal{Q}}{\delta^2}, \tag{11.40}$$

where

$$\mathcal{P} = \frac{1}{\phi}\frac{dW}{dz} \tag{11.41}$$

is the dimensionless compaction pressure. In (11.41), we have assumed that $\xi_\phi = \phi^{-1}$. Equation (11.40) states that the compaction pressure is approximately linear with height in the visco-gravitational boundary layer. Integrating (11.40) gives

$$\mathcal{P} \sim \frac{\mathcal{Q}}{\delta^2} z + C, \tag{11.42}$$

where C is a constant of integration. We determine C by requiring that this boundary-layer compaction pressure match that of the Darcy solution at the top of the boundary layer.

In the Darcy region above the boundary layer, where buoyancy balances Darcy drag, the porosity is small and hence $(1-\phi) \sim 1$ and $W \sim 1 - F$. Therefore we have $dW/dz \sim -F_{\text{max}}$ (see fig. 11.1b). Moreover, as we approach the boundary layer, equation (11.12) reduces to $\phi \sim F_{\text{max}} z$. Together this implies that $\mathcal{P} \sim -1/z$ in the Darcy region at the top of the visco-gravitational boundary layer. To match this with the boundary-layer solution, we substitute into (11.40) and solve for $z = z_b$, the height of the top of the boundary layer. We obtain

$$z_b = \delta/\sqrt{\mathcal{Q}}. \tag{11.43}$$

Hence the boundary layer thickness is smaller than the compaction length by a factor $1/\sqrt{\mathcal{Q}}$ (e.g., 1% of the compaction length if $\mathcal{Q} = 10^4$). This confirms that the zero-compaction-length approximation leading to Darcy flow is valid over most of the melting column. Hence the boundary condition on (11.42) is

$$\mathcal{P} = -1/z_b = -\sqrt{\mathcal{Q}}/\delta \quad \text{at} \quad z = z_b. \tag{11.44}$$

This tells us that $C = -2\sqrt{\mathcal{Q}}/\delta$ in (11.42), giving

$$\mathcal{P} \sim -\frac{1}{z_b}\left(2 - \frac{z}{z_b}\right). \tag{11.45}$$

This boundary-layer solution is shown in figure 11.6 as a function of z/z_b for $\mathcal{Q} = 10^4$, $F_{\text{max}} = 0.22$, $\delta = 0.1$, and $n = 2$.

We next use the result (11.45) to obtain boundary-layer solutions for W and ϕ. Combining dimensionless mass-conservation equations to eliminate w gives

$$\phi = \frac{F_{\text{max}} z - V}{1 - V} \sim F_{\text{max}} z - V, \tag{11.46}$$

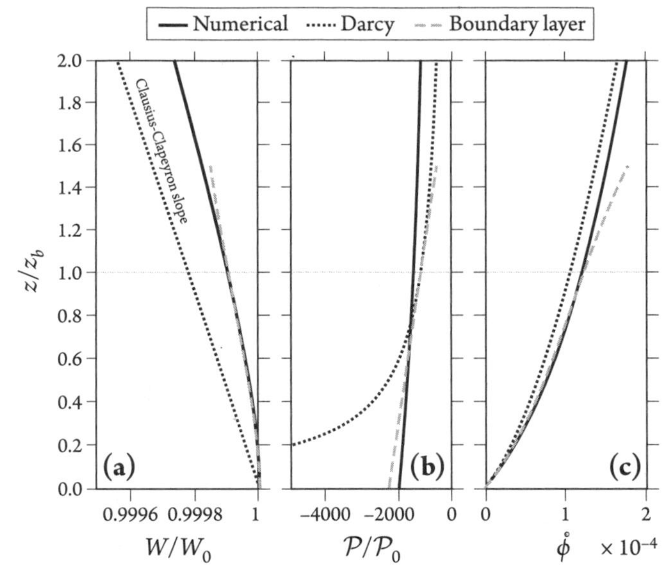

Figure 11.6. Numerical, Darcy, and boundary-layer solutions to the one-dimensional melting column with constant decompressional melting productivity. The numerical solution of the system (11.50) is discretized by finite differences and run to steady state. The domain is limited to $0 \leq z \leq 2z_b$. Nondimensional parameters used to compute these solutions are $\delta = 0.1$, $\mathcal{Q} = 10^4$, $n = 2$, and $F_{\max} = 0.22$. For these parameters, $z_b = z_0/100$. The nondimensional augmented compaction viscosity is $\xi_\phi = \phi^{-1}$. **(a)** Solid upwelling rate. **(b)** Compaction pressure nondimensionalized by $\mathcal{P}_0 = \xi_0 W_0/z_0$. **(c)** Porosity.

where we have defined

$$V \equiv 1 - W \sim \mathcal{O}(F_{\max} z) \ll 1. \tag{11.47}$$

The asymptotic result in (11.46) comes from the fact that $1 - W$ is very close to unity throughout the boundary layer. Using equations (11.46) and (11.45) and the definition of V in (11.41) gives, after rearranging,

$$\frac{\mathrm{d}V}{\mathrm{d}z} = \frac{1}{z_b}\left(F_{\max} z - V\right)\left(2 - \frac{z}{z_b}\right). \tag{11.48}$$

V must satisfy the boundary condition $V = 0$ (i.e., $W = 1$) at $z = 0$. This has the solution

$$V(z) = F_{\max}\left\{z - z_b\sqrt{\frac{\pi}{2}}\exp\left(\frac{(z/z_b - 2)^2}{2}\right)\left[\operatorname{erf}\sqrt{2} + \operatorname{erf}\left(\sqrt{2}\frac{(z/z_b - 2)}{2}\right)\right]\right\}. \tag{11.49}$$

Recall that these variables, V, z, and z_b, are dimensionless (using scales from (11.7)). We can reconstruct the porosity profile in the boundary layer using equation (11.46). Results are shown in figure 11.6.

Since both the Darcy solution and the boundary-layer solution are approximations, it is helpful to compare them to a numerical solution of the full equations. The dimensionless, time-dependent governing equations are

$$(1-W)=\phi^n\left[\mathcal{Q}-\delta^2\frac{\partial}{\partial z}\frac{1}{\phi}\frac{\partial W}{\partial z}\right], \tag{11.50a}$$

$$\frac{\partial\phi}{\partial t}+W\frac{\partial\phi}{\partial z}=F_{\max}+\frac{\partial W}{\partial z}. \tag{11.50b}$$

Figure 11.6 shows a numerical solution of these equations, run to a steady state; (a) shows that at $z\sim 0$, the solid velocity $W(z)$ of the boundary-layer solution is closely matched with the numerical solution. Distinct from the Darcy solution, the boundary layer and numerical solutions have $\mathrm{d}W/\mathrm{d}z\sim 0$ as $z\to 0$. This shows that the large compaction viscosity inhibits compaction (and melt segregation) in the boundary layer, but doesn't entirely prevent it. Figure 11.6(b) shows that the gradient of the boundary-layer compaction pressure is qualitatively consistent with the numerical solution. The large, negative compaction pressure in the Darcy solution arises from its spurious compaction as $z\to 0$. In (c), the boundary-layer porosity is in excellent agreement with the numerical solution. The consistency of the boundary-layer solution with the full, numerical solution validates our assumption, used to obtain (11.40), that gradients in compaction pressure balance buoyancy near $z=0$.

Figure 11.6 illustrates how the physical balance shifts within a thin boundary layer near the bottom of the melting column. It also shows that this boundary layer is certainly no thicker than $\sqrt{\xi_0 W_0/\Delta\rho g}$. Outside of this boundary layer, the numerical solution is parallel to the Darcy solution, with a small offset. The offset arises from the presence of the visco-gravitational boundary layer. Nonetheless, the Darcy solution (and the ZCL approximation) are valuable tools for modeling diffuse porous flow in upwelling mantle.

11.4 The Decompaction Boundary Layer

In the chapter thus far, we have ignored the fact that the top of a melting column is the base of the lithosphere. In the lithosphere, the temperature drops rapidly with height along a conductive geotherm. Melting in the asthenosphere transitions to freezing in the lithosphere. In the current section we analyze this transition, in which the porosity and permeability drop to zero with height. We again consider a one-dimensional domain, aligned with gravity and in steady state. This time, however, we take $z=0$ to be the depth at which all the melt produced below has frozen; at this depth the porosity and the magma flux are zero. To keep our model simple, we neglect the asthenospheric melting process and consider only the dynamics associated with the zone of freezing. Hence we assume that melt enters the domain, driven by its buoyancy, at $z=-\infty$ where the porosity is ϕ_0.

The freezing rate is a consequence of reduced temperatures at the base of the lithosphere. These, in turn, result from diffusive loss of heat to the surface of the solid Earth. It would therefore be appropriate to model the freezing rate on the basis of conservation of energy and component mass, as we have done for melting in this chapter. However, to demonstrate the principles of the decompaction boundary layer, we instead prescribe the form of the freezing function analytically, choosing with foresight a function that has the required properties. These properties are both subtle and stringent; 11.6 explores a different choice of function that appears valid but fails to produce a decompaction boundary layer.

Mass and momentum conservation in steady state can be written

$$\frac{\mathrm{d}}{\mathrm{d}z}(\phi w) = \Gamma(z)/\rho, \tag{11.51a}$$

$$\frac{\mathrm{d}W}{\mathrm{d}z} = -\Gamma(z)/\rho, \tag{11.51b}$$

$$\phi w = -K_\phi \left(\xi_0 \frac{\mathrm{d}\mathcal{C}}{\mathrm{d}z} - \Delta\rho g\right), \tag{11.51c}$$

where we have made the extended Boussinesq approximation, taken $\xi_\phi = \xi_0$ to be constant and, assuming small porosity, taken $(1-\phi) \sim 1$. Also we have neglected W in the modified Darcy's law (11.51c) under the assumption that it is much smaller than w. The melting rate $\Gamma(z) \leq 0$ is unknown but represents freezing. The liquid mobility is given by $K_\phi = K_0(\phi/\phi_0)^n$ and hence, sufficiently far from the decompaction boundary layer, the melt flux is

$$q_\infty \equiv K_0 \Delta\rho g. \tag{11.52}$$

Integration of equation (11.51a) yields

$$\phi w = q_\infty + \int_{-\infty}^{z} \frac{\Gamma(z)}{\rho} \mathrm{d}z. \tag{11.53}$$

We will use this equation to constrain the properties of $\Gamma(z)$ on the basis of the boundary conditions

$$\phi = 0 \qquad \text{at } z = 0, \tag{11.54a}$$

$$\phi w = q_\infty \qquad \text{at } z \to -\infty. \tag{11.54b}$$

Equation (11.53) can be rewritten by defining a shape function $G(\zeta)$,

$$\phi w \equiv q_\infty \left[1 - G(z/\delta_f)\right], \tag{11.55}$$

where δ_f is a length scale over which freezing occurs. The function $G(\zeta)$ is related to Γ by

$$\Gamma(z) = -\frac{\rho q_\infty}{\delta_f} G'(z/\delta_f), \tag{11.56}$$

where the prime symbol on G represents a derivative with respect to its argument ζ.

Recalling that $\mathcal{C} \equiv \mathrm{d}W/\mathrm{d}z$ and combining equations (11.51) and (11.55) with the mobility law allows us to solve for the porosity profile as

$$\frac{\phi}{\phi_0} = \left[\frac{1 - G(z/\delta_f)}{1 - R_\delta^2 G''(z/\delta_f)}\right]^{1/n}, \tag{11.57}$$

where $R_\delta \equiv \delta_0/\delta_f$ is the ratio of the compaction length to the length scale over which freezing occurs.

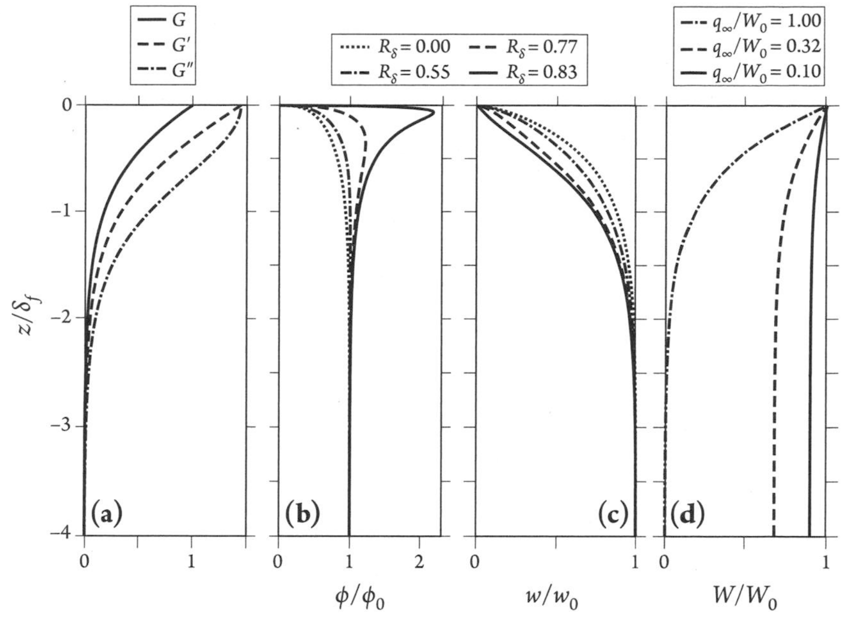

Figure 11.7. The decompaction column. **(a)** Quantities G, G', and G'' as given by (11.58). **(b)** The normalized porosity for various values of $R_\delta \equiv \delta_0/\delta_f$, computed with equation (11.57). **(c)** Normalized liquid upwelling speed for various values of R_δ. **(d)** Normalized solid upwelling speed for various flux ratios q_∞/W_0.

To investigate the behavior of the solution for different values of R_δ, we must furnish the shape function $G(\zeta)$ that can be used to evaluate equation (11.57). The boundary conditions (11.54) require that $G(0) = 1$ and $G(-\infty) = 0$. The hyperbolic tangent (tanh) function has a shape with the desired properties.[2] Taking $G'(\zeta) = A(1 + \tanh\zeta)$, then integrating once, gives $G(\zeta) = A\,[\ln 2 + \zeta + \ln(\cosh\zeta)]$. This is suitable if $A = 1/\ln(2)$. Hence we have

$$G(\zeta) = \frac{1}{\ln 2}\,[\zeta + \ln(\cosh\zeta)] + 1, \tag{11.58a}$$

$$G'(\zeta) = \frac{1}{\ln 2}(1 + \tanh\zeta), \tag{11.58b}$$

$$G''(\zeta) = \frac{1}{\ln 2}\operatorname{sech}^2\zeta. \tag{11.58c}$$

These functions are plotted in figure 11.7(a). Note that G' is nonzero at $z < -\delta_f$, meaning that some freezing occurs outside the depth range defined by δ_f. But the G curve indicates that this is only about 20% of the total freezing in the column.

[2]Not all functions that satisfy conditions (11.54) via equation (11.55) give rise to decompaction boundary layers (exercise 11.6). However, numerical solutions with self-consistent determination of the freezing rate do produce such layers. The relevant properties shared by shape functions that capture this behavior are complicated and not relevant to the current discussion.

The normalized porosity ϕ/ϕ_0 is plotted in figure 11.7(b) for various values of the ratio R_δ of compaction length to freezing length. In the case of zero-compaction-length ($R_\delta = 0$), the porosity decreases monotonically upward and attains the lowest value at any height, compared to curves for larger R_δ. With increasing R_δ, the porosity increases until a local maximum in porosity appears and becomes sharp at the largest value of R_δ.

The speed of melt upwelling, shown in figure 11.7(c), decreases as the porosity increases. w/w_0 is obtained by combining equations (11.55) and (11.57) to eliminate porosity. At any depth, melt moves upward more slowly with increasing R_δ.

Both of these trends can be explained by consideration of the compaction pressure gradient. Recalling that $\mathcal{P} = \xi_0\mathcal{C}$, and using equations (11.51b) and (11.56), this gradient is

$$\frac{\mathrm{d}\mathcal{P}}{\mathrm{d}z} = \Delta\rho g R_\delta^2 G''. \tag{11.59}$$

Since G'' is always positive (fig. 11.7(a)), this gradient is always positive; the compaction pressure increases as melt approaches the impermeable boundary at $z = 0$ from below. Equation (11.51c) tells us that the compaction pressure gradient balances some portion of the melt buoyancy. Indeed when R_δ is sufficiently large that $R_\delta^2 \max\left(G''\right) = 1$, the compaction-pressure gradient neutralizes the melt buoyancy at some height in the column. At this value of R_δ, our simple model becomes invalid because a steady state cannot be attained (and the porosity solution is singular). But for R_δ below this the back pressure slows magmatic ascent, reducing w/w_0. The porosity must increase where magma ascends more slowly because the total magmatic flux is independent of R_δ (equation (11.55)).

The scaling of the compaction pressure gradient with $\Delta\rho g$ in equation (11.59) indicates that this gradient is a consequence of the melt buoyancy. Indeed, the compaction pressure gradient is analogous to the normal force experienced by a mass that rests on a rigid, horizontal surface in a gravitational field. Moreover, a larger compaction length gives a larger back pressure from the impermeable boundary. This analogy reinforces the point that the downward pressure gradient is the cause of the layer of elevated porosity.

To understand the motion of the solid through the decompaction layer, we sum the mass-conservation equations (11.51a) and (11.51b) and integrate from $-\infty$ to zero. The result is $\phi w + W = W_0$; here, W_0 is the solid upwelling speed at $z = 0$, where the porosity is zero. Evaluating this equation at $z \to -\infty$ we obtain $W_0 = W(-\infty) + q_\infty$, which states that the solid and liquid fluxes into the bottom of the domain must balance the solid flux out of the top. Since the change in solid upwelling occurs over a length scale of $\sim\delta_f$, we can estimate that $\mathrm{d}W/\mathrm{d}z = \mathcal{C} \sim q_\infty/\delta_f > 0$. The compaction rate is positive and hence we refer to the region of freezing as a *decompaction layer*. Interestingly, however, the compaction rate is independent of the compaction length δ_0.

The normalized solid upwelling speed $W(z)/W_0$ is obtained by combining the integrated total mass equation with equation (11.55) to obtain

$$\frac{W}{W_0} = 1 - \frac{q_\infty}{W_0}\left[1 - G(z/\delta_f)\right]. \tag{11.60}$$

This result is plotted in figure 11.7(d) for several values of the flux ratio q_∞/W_0.

Our one-dimensional model of the decompaction layer assumes that flow of both liquid and solid is purely vertical. A more realistic model would account for lateral motion too. In a mid-ocean ridge setting, the plate moves away from the ridge axis, dragging the solid flow with it and creating a corner-flow pattern. The lithosphere thickens with distance from the ridge axis, which means that the freezing front (and hence the decompaction layer) are tilted downward away from the axis. Hence there is a component of gravity that points along the decompaction layer. A hypothesis prevalent in the literature states that the decompaction layer acts as a channel that transports liquid laterally toward the ridge axis. We develop models for this lateral melt focusing in section 13.3.

11.5 Isotopic Decay-Chain Disequilibria in a Melting Column

Column models provide a context in which to consider how melt segregation creates secular disequilibrium in the uranium-series decay chain—they give us insight into the physical controls and hence inform interpretations of geochemical measurements. The calculations in this section are based on Spiegelman and Elliott [1993]; they build on the theory presented in sections 9.3 and 9.4, which the reader should understand before proceeding. We adopt the ZCL Darcy solution (11.13) for imposed melting rate and $n=2$. We assume that the solid that enters the column at $z=0$ is in secular equilibrium. In particular,

$$\frac{a^s_{j-1}|_0}{a^s_j|_0} = 1 \tag{11.61}$$

for all elements j in the decay chain (except, of course, for the element at the top of the chain). Recall that the activity a^i_j of element j in phase i is the product of the decay coefficient λ_j and the concentration c^i_j. Hence, equation (11.61) states that the decay rate of each element in the chain provides exactly enough atoms to maintain a constant concentration (and hence constant decay rate) of the subsequent elements. It applies only at $z=0$.

The secular equilibrium of the incoming mantle can be disturbed by melting, which fractionates elements in a way that depends on their partition coefficients D_j. We assume that these partition coefficients are constant and uniform, but have values that differ for each element j. Furthermore, we assume steady state and rewrite equation (9.34) as

$$w^{D_j}\frac{\mathrm{d}a^\ell_j}{\mathrm{d}z} = -a^\ell_j \frac{W_0}{z_0} F_{\max}\left(\frac{1-D_j}{\phi+(1-\phi)D_j}\right) + \lambda_j\left(D^r_{j-1}a^\ell_{j-1} - a^\ell_j\right), \tag{11.62}$$

where we have used $\Gamma = \rho W_0 F_{\max}/z_0$. The maximum degree of melting $F_{\max}$ will be considered to be a parameter, but it could also be calculated according to conservation principles, as we showed above. Using vertically integrated conservation of mass equations (11.3), the effective transport speed becomes

$$w^{D_j} = W_0 \frac{(1-F)+F/D_j}{(1-\phi)+\phi/D_j}, \tag{11.63}$$

where $F = F_{\max} z/z_0$. Rescaling height with $[z]=z_0$ and using (11.63) allows us to write equation (11.62) as

$$\frac{1}{a_j^\ell}\frac{\mathrm{d}a_j^\ell}{\mathrm{d}z} = -\frac{(1-D_j)F_{\max}}{D_j+(1-D_j)F_{\max}z} + \frac{\lambda_j z_0}{w^{D_j}}\left[\left(\frac{\phi+(1-\phi)D_{j-1}}{\phi+(1-\phi)D_j}\right)\frac{a_{j-1}^\ell}{a_j^\ell} - 1\right], \quad (11.64)$$

where $0 \leq z \leq 1$ is now dimensionless. To derive this equation, recall that we have assumed the solid is in chemical equilibrium with the liquid. Therefore, having solved (11.64), we can obtain the concentration of elements in the liquid and solid as $c_j^\ell = a_j^\ell/\lambda_j$ and $c_j^s = D_j c_j^\ell$, respectively.

Equation (11.64) states that vertical advection of activity in the liquid is balanced by processes associated with the two terms on the right-hand side. The first term represents dilution due to melting (for small D_j), which reduces concentration and activity with height in the column. The second term represents ingrowth. It depends on the difference of the activity ratio of parent and daughter elements from unity and, for small porosity, the ratio of the partition coefficients. If the parent's activity is larger (and the porosity is sufficiently larger than the partition coefficients D_{j-1}, D_j), the daughter element is produced faster than it decays. In that case, the daughter's activity increases by ingrowth with height in the column.

It is helpful to reintroduce the decomposition that we developed in section 9.4.3, writing the activity as a product of a batch-melting factor and an ingrowth factor,

$$a_j^\ell = \left(\frac{a_j^s|_0}{D_j+(1-D_j)F_{\max}z}\right)\tilde{a}_j^\ell. \quad (11.65)$$

Substituting (11.65) into (11.64) and assuming that the unmolten solid is in secular equilibrium (i.e., as per (11.61)) gives an equation governing the ingrowth factor $\tilde{a}_j$,

$$\frac{\mathrm{d}\tilde{a}_j^\ell}{\mathrm{d}z} = \lambda_j z_0\left[\frac{\tilde{a}_{j-1}^\ell}{w^{D_{j-1}}} - \frac{\tilde{a}_j^\ell}{w^{D_j}}\right]. \quad (11.66)$$

The boundary condition (11.61) at $z=0$ is satisfied by the batch melting factor and hence we must take $\tilde{a}_j^\ell|_0 = 1$.

If the decay constants λ_j are all zero then the concentrations of isotopes in the column are described by the batch-melting equation. This is the case for nonradiogenic trace elements: their concentration at any height in the column is computed as a batch melt produced by the total degree of melting at that height.[3] Equation (11.66) then represents the evolution (by ingrowth) of departures from secular equilibrium caused by melting and melt segregation.

If the partition coefficients are all much smaller than the maximum degree of melting, that is, $D_j \ll F_{\max}$ (expected for the U-series in a column with $F_{\max} \approx 0.2$), then at the top of the column we have

$$a_j^\ell|_1 \sim \left(\frac{a_j^s|_0}{F_{\max}}\right)\tilde{a}_j^\ell|_1. \quad (11.67)$$

[3]This depends, of course, on the assumption of instantaneous chemical equilibrium between phases.

In parent–daughter activity ratios at the column top, the factor of $F_{\max}$ cancels and

$$\left.\frac{a_{j-1}^{\ell}}{a_j^{\ell}}\right|_1 = \left.\frac{\tilde{a}_{j-1}^{\ell}}{\tilde{a}_j^{\ell}}\right|_1, \tag{11.68}$$

where we have used the assumption (11.61) of secular equilibrium at the bottom of the column. This means that because the melting column achieves a relatively large degree of melting, the problem is reduced to computing the ingrowth factors $\tilde{a}_j^{\ell}$ at the top of the column. Furthermore, with the additional assumption of $\phi_{\max} \ll 1$ (which is reasonable for the melting column), the effective transport rates at the top of the column reduce to

$$w_0^{D_j}|_1 \sim \frac{W_0 F_{\max}}{\phi_{\max} + D_j}. \tag{11.69}$$

The maximum porosity $\phi_{\max}$ is attained at the top of the column; it can be computed using, for example, the Darcy solution (11.13), or prescribed as a problem parameter (that scales the porosity throughout the column).

11.5.1 CONSTANT TRANSPORT RATES

The effective transport rates for each isotope w^{D_j} depend on the porosity that is, in general, a function of z (as we have seen in the sections above). Hence, these rates are functions of z that differ from each other due to the differences in D_{j-1} and D_j. To explore the behavior of equation (11.66) without resorting to numerical solutions we neglect the variation of the effective transport rates with z. We take

$$w^{D_j} \sim w_0^{D_j} \neq w_0^{D_{j-1}}, \tag{11.70}$$

where $w_0^{D_j}$ is the imposed, uniform, effective transport rate. This is not consistent with our derivation of (11.66), which assumed progressive melting in z, but it provides a basis for understanding the results of consistent calculations (presented below). Motivated by the fact that ^{238}U has a half-life that is approximately equal to the age of the Earth, we shall assume that the activity of the isotope at the top of the chain ($j = 1$) remains constant through the column. Without loss of generality, we can choose $a_1^s|_0 = 1$; then the assumption of secular equilibrium in the solid at $z = 0$ (equation (11.61)) gives us $a_j^s|_0 = 1$.

With these assumptions, equation (11.66) can be rewritten for the activity of the daughters ($j = 2, 3$) as

$$\frac{d\tilde{a}_2^{\ell}}{dz} = \lambda_2 \left(\tau_1^{\text{col}} - \tilde{a}_2^{\ell}\tau_2^{\text{col}}\right), \tag{11.71a}$$

$$\frac{d\tilde{a}_3^{\ell}}{dz} = \lambda_3 \left(\tilde{a}_2^{\ell}\tau_2^{\text{col}} - \tilde{a}_3^{\ell}\tau_3^{\text{col}}\right), \tag{11.71b}$$

where

$$\tau_j^{\text{col}} \equiv \frac{z_0}{w_0^{D_j}} \tag{11.72}$$

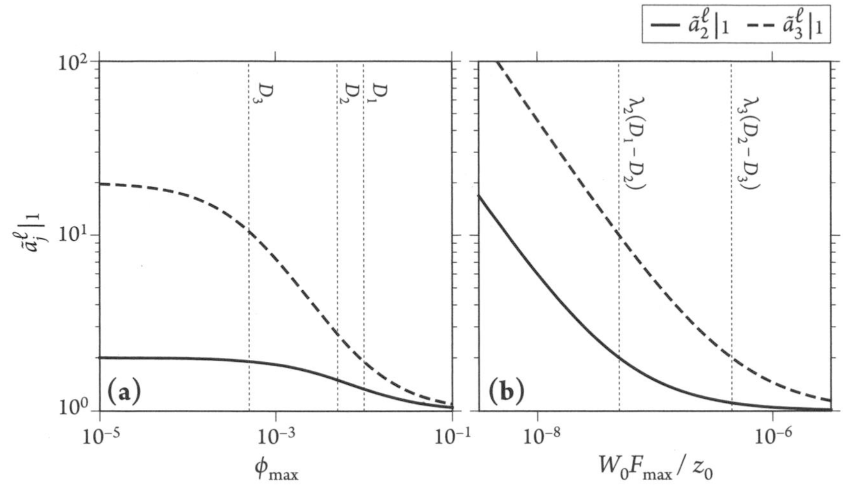

Figure 11.8. Ingrowth factors at the top of the column. Parameters used to produce this figure are $F_{max}=0.2$, $z_0=60$ km, $D=[10^{-2}, 5\times10^{-3}, 5\times10^{-4}]$, $\lambda_2=10^{-5}$/yr, $\lambda_3=10^{-4}$/yr, which are inspired by but not equal to the values for the U-series decay chain in table 9.1. **(a)** Slow-transport regime, computed using equations (11.75). **(b)** Fast-transport regime, computed using equations (11.77).

is the residence time of isotope j in the column. The solution of (11.71) with ingrowth factor $\tilde{a}_j^\ell|_0=1$ is

$$\tilde{a}_2^\ell(z)=\frac{\tau_1^{col}}{\tau_2^{col}}+\left(1-\frac{\tau_1^{col}}{\tau_2^{col}}\right)e^{-\lambda_2\tau_2^{col}z}, \tag{11.73a}$$

$$\tilde{a}_3^\ell(z)=\frac{\tau_1^{col}}{\tau_3^{col}}+\frac{\lambda_3(\tau_1^{col}-\tau_2^{col})}{\lambda_2\tau_2^{col}-\lambda_3\tau_3^{col}}e^{-\lambda_2\tau_2^{col}z}+\left[1-\frac{\tau_2^{col}}{\tau_3^{col}}\left(\frac{\lambda_2\tau_1^{col}-\lambda_3\tau_3^{col}}{\lambda_2\tau_2^{col}-\lambda_3\tau_3^{col}}\right)\right]e^{-\lambda_3\tau_3^{col}z}. \tag{11.73b}$$

Before considering the systematics of this solution when each nucleide has a distinct residence time in the column, it is worth noting that when the effective transport rates are equal (i.e., $D_{j-1}=D_j$), the ingrowth factor $\tilde{a}_2^\ell$ is unity. And, since the partition coefficients are equal, the fractionation due to batch melting is identical for the two isotopes. Hence they remain in secular equilibrium. This is, in fact, the case for ^{238}U and ^{234}U. This means that we can neglect ^{234}U and consider the second element in the decay chain to be ^{230}Th.

There are two end-member behavioral regimes of the solution (11.73), shown in figure 11.8. In both regimes, chromatographic fractionation between isotopes is required to cause deviations from secular equilibrium. However, the mechanisms differ between the two. They can be understood by considering the top of the column ($z=1$).

The first regime is associated with slow transport: $\tau_2^{col}\gg\tau_2^{(1/2)}$ and $\tau_3^{col}\gg\tau_3^{(1/2)}$, which means that each daughter resides in the column for much longer than its half-life and so ingrowth dominates the behavior. In this regime we take $\lambda_2\tau_2^{col}, \lambda_3\tau_3^{col}\gg1$

and hence, from equations (11.73), the activity at the top of the column is

$$\tilde{a}_2^\ell|_1 \sim \tau_1^{\text{col}}/\tau_2^{\text{col}} = w_0^{D_2}/w_0^{D_1}, \tag{11.74a}$$

$$\tilde{a}_3^\ell|_1 \sim \tau_1^{\text{col}}/\tau_3^{\text{col}} = w_0^{D_3}/w_0^{D_1}. \tag{11.74b}$$

In the limit of $F_{\text{max}} \gg D_j$ and $\phi_{\text{max}} \ll 1$ (which is reasonable for the melting column and uranium series), we can use equation (11.69) for the transport rates and hence (11.74) becomes

$$\tilde{a}_2^\ell|_1 \sim \frac{\phi_{\text{max}} + D_1}{\phi_{\text{max}} + D_2}, \quad \tilde{a}_3^\ell|_1 \sim \frac{\phi_{\text{max}} + D_1}{\phi_{\text{max}} + D_3}. \tag{11.75}$$

So, in this regime the excesses are independent of degree of melting and mantle upwelling rate, and depend exclusively on the maximum porosity relative to the partition coefficients. Figure 11.8(a) shows that if ϕ_{max} is large compared with $D_{1,2,3}$ then there is no chromatographic fractionation of the elements; they travel at the same rate and approach secular equilibrium long before they reach the top of the column. Secular disequilibrium at the top of the column depends on the chromatographic fractionation and is hence a function of ϕ_{max}. In particular, strong chromatographic fractionation requires that ϕ_{max} be less than the smallest partition coefficient.

The second regime is fast transport, where $\tau_2^{\text{col}} \ll \tau_2^{(1/2)}$ and $\tau_3^{\text{col}} \ll \tau_3^{(1/2)}$, which means that each daughter resides in the column for much less time than its half-life and undergoes little decay. In this case, Taylor expanding equations (11.73) about zero to first order in z and evaluating at $z = 1$, we have

$$\tilde{a}_2^\ell|_1 \sim 1 + \lambda_2 \left(\tau_1^{\text{col}} - \tau_2^{\text{col}}\right), \tag{11.76a}$$

$$\tilde{a}_3^\ell|_1 \sim 1 + \lambda_3 \left(\tau_2^{\text{col}} - \tau_3^{\text{col}}\right). \tag{11.76b}$$

And using the definition (11.72) of the column residence time and the approximate transport rate from (11.69), these column-top activity estimates become

$$\tilde{a}_2^\ell|_1 \sim 1 + \frac{\lambda_2 z_0}{W_0 F_{\text{max}}}(D_1 - D_2), \quad \tilde{a}_3^\ell|_1 \sim 1 + \frac{\lambda_3 z_0}{W_0 F_{\text{max}}}(D_2 - D_3). \tag{11.77}$$

In this regime of fast transport, shown in figure 11.8(b), the excesses are independent of the porosity. Instead they depend on the difference in column residence times, which is a function of the column height z_0, the bulk upwelling rate W_0, the maximum degree of melting F_{max}, and the partition coefficients for the parent and the daughter. If $D_{j-1} = D_j$, the effective transport rates are equal, there is no chromatographic fractionation, and the ingrowth factor is unity. Small differences in the small partition coefficients can lead to preferential retention of parent or daughter and a consequent deviation of the ingrowth factor from unity.

In the next section we employ these lessons to interpret numerical solutions to equation (11.62).

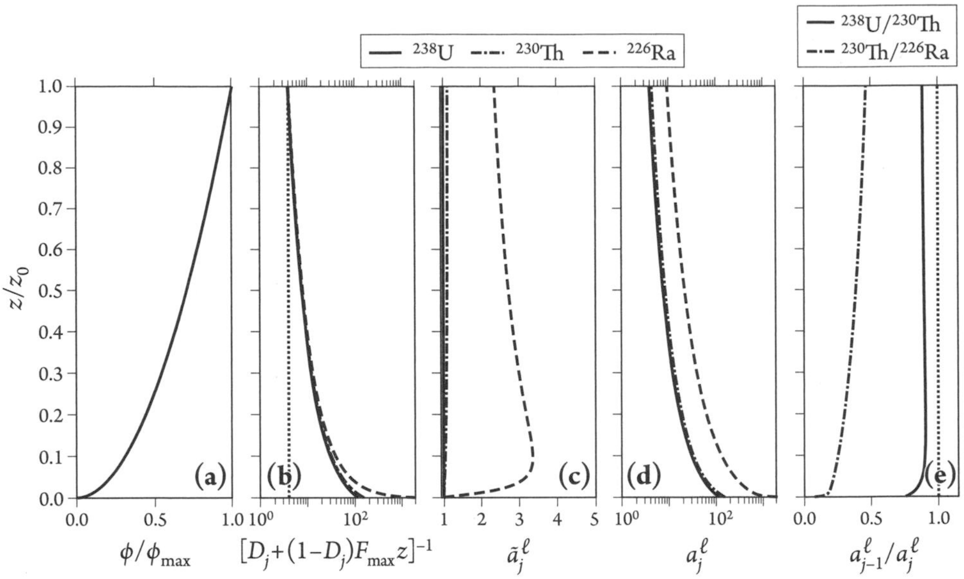

Figure 11.9. A numerical solution of equation (11.62) for parameters from table 9.1 and $z_0 =$ 50 km, $W_0 = 5$ cm/yr, $F_{\max} = 0.25$, $\phi_{\max} = 0.005$, $\mathcal{Q} = 10^3$, and $n = 2$ (these are chosen to approximately match figure 3 of Spiegelman and Elliott [1993]). **(a)** Porosity from (11.13). **(b)** The batch melting factor of (11.65) showing enrichment in the liquid due to the small partition coefficients. The dotted line has a value of $1/F_{\max}$. **(c)** The ingrowth factor of (11.65). **(d)** The total activity. **(e)** Parent–daughter activity ratios. In secular equilibrium, these ratios are unity (dotted line).

11.5.2 VARIABLE TRANSPORT RATES

To attain physical consistency, the uranium-series activity equations (11.62) can be coupled to the column model for porosity and upwelling rates. Here we use the Darcy solution for an upwelling column with permeability exponent $n = 2$. Parameters for the uranium series are listed in table 9.1 with values chosen to reproduce the results of Spiegelman and Elliott [1993]. Equation (11.62) must be solved numerically. An example result is shown in figure 11.9.

We learned above that, because ^{238}U and ^{234}U enter the column in secular equilibrium and are not fractionated by melting or chromatographic transport, their activity ratio doesn't deviate from unity. Hence, we exclude ^{234}U from our calculation without approximation. Assuming permeability exponent $n = 2$, the porosity in the column can be written $\phi = \sqrt{\phi_{\max}^2 z + (2\mathcal{Q})^{-2}} - (2\mathcal{Q})^{-1}$, where z has been nondimensionalized with the total column height z_0; this is plotted in figure 11.8(a). The approximate maximum porosity in the column is $\phi_{\max}$ (it is exact for $\mathcal{Q} \to \infty$), attained at $z = 1$. We will treat $\phi_{\max}$ as a problem parameter, but it could also be interpreted in terms of equation (11.13) as $\phi_{\max} \equiv (F_{\max}\mathcal{Q})^{1/n}$.

Figure 11.9(b) shows the batch-melting factor, given in parentheses in equation (11.65). All three elements have small partition coefficients and are enriched in the liquid at the bottom of the column, then diluted by melting with height in the column.

There is little fractionation between uranium and thorium because their partition coefficients are extremely close together: $\lambda_1 = 0.0086$ and $\lambda_2 = 0.0065$, respectively. There is significant fractionation of thorium and radium at the bottom of the column, however; radium has a partition coefficient of $\lambda_3 = 0.0005$.

Figure 11.9(c) shows the ingrowth factor $\tilde{a}_j^\ell$. For uranium it is almost exactly unity throughout the column, consistent with the very long half-life of this nuclide. For thorium, the ingrowth factor is also close to unity. The small, positive difference from unity means that decay of uranium enriches thorium and gives it an activity above what it would have due to batch melting alone. The ingrowth-factor curve for radium is far greater than unity, showing that decay of thorium to radium has a large effect.

Figure 11.9(d) shows the total activity in the liquid for the three nuclides, where each is normalized to its value at $z = 0$. Because of the strong chemical fractionation that occurs at small porosity (fig. 11.9(b)), the liquid has its largest secular disequilibrium at the bottom of the column. With height in the column, all the activities are affected by dilution; only radium is significantly affected by ingrowth (fig. 11.9(c)). Nonetheless, its activity has an overall decrease upward.

The parent–daughter activity ratios in figure 11.9(e) show secular disequilibrium is maintained through the height of the column for both pairs, ^{230}Th/^{238}U and ^{226}Ra/^{230}Th. The latter has larger disequilibrium, consistent with the larger ingrowth factor. Both curves show an adjustment to the initial fractionation for $F < D_j$ over a short length scale.

Except in this narrow boundary layer at the bottom of the column, secular disequilibrium for both parent–daughter pairs is caused by chromatographic fractionation of parent and daughter. As we saw in the previous section (fig. 11.8), there are two regimes associated with this fractionation. In the slow-transport regime, the residence time of the daughter in the column is long compared to the half-life; chromatographic fractionation is controlled by the porosity. In the fast-transport regime, the residence time of the daughter is short compared to the half-life; chromatographic fractionation is controlled by the amount of decay that occurs over the difference between parent and daughter residence times.

Which of these regimes applies to the disequilibria exhibited in fig. 11.9(e)? The answer can be discerned from its position in the parameter space shown in figure 11.10. Figures 11.10(a) and (b) plot column-top activity-ratio isopleths for ^{230}Th/^{238}U and ^{226}Ra/^{230}Th, respectively. The example shown in figure 11.9 is indicated in figure 11.10 by a star. Both subfigures of figure 11.10 show the two behavioral regimes: one in which the activity ratio depends on $\phi_{\max}$ (slow transport) and one in which it depends on W_0 (fast transport). The star marking column-top ^{226}Ra/^{230}Th in figure 11.10(b) is clearly in the slow-transport regime. This is unsurprising given the short half-life of radium. The star for ^{230}Ra/^{238}Th in figure 11.10(a) is near the transition between regimes but more closely associated with the fast-transport regime.

Observations of uranium-series disequilibrium thus potentially constrain the porosity of the partially molten mantle, which is related to its permeability, as well as the solid upwelling rate. The difficulty is that the effective partition coefficients are challenging to constrain in the natural system because they depend on the lithology over the depth-range of melting. The garnet content of the mantle, in particular, has a significant influence on partition coefficients of the uranium-series elements.

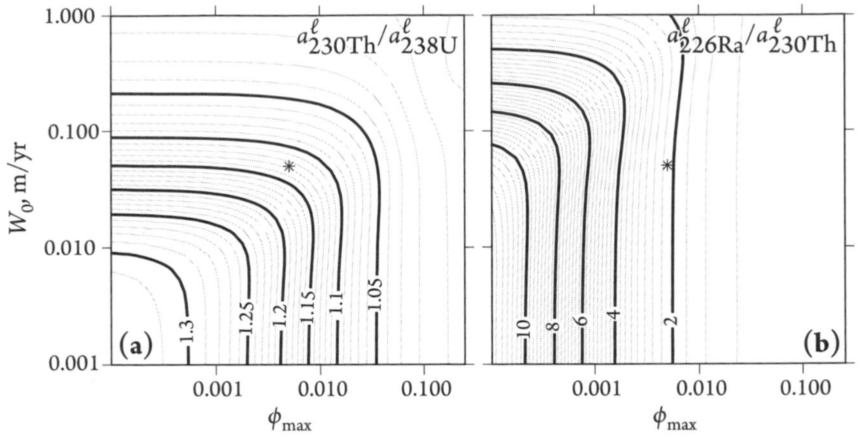

Figure 11.10. Parameter space diagrams for the column-top daughter–parent activity ratios. Lines are activity-ratio isopleths. Stars represent column-top values from the column shown in fig. 11.9. **(a)** ^{230}Th/^{238}U. **(b)** ^{226}Ra/^{230}Th. Intended to reproduce fig. 5 of Spiegelman and Elliott [1993].

11.6 Literature Notes

One-dimensional, gravity-aligned models of melt generation and segregation were part of the earliest work on the magma/mantle dynamics. The first such model, largely conceptual, is that by Frank [1968]. Detailed theory by Turcotte and Ahern [1978] and Ahern and Turcotte [1979] followed Frank [1968] in assuming magmatic segregation through the pore space between grains, driven by buoyancy. The next significant work on such models was by Ribe [1985], who incorporated the use of binary phase diagrams. The formalism used above is based on that of Cerpa et al. [2019].

There are many studies that consider the detailed thermodynamics of isentropic upwelling, including McKenzie [1984]. The most thorough is probably Asimow et al. [1997]. That work, however, does not model the dynamics of melt segregation, but was incorporated into Asimow and Stolper [1999]. The latter demonstrated the conditions under which a one-dimensional, steady-state column produces melts that are identical to batch melts from a closed system of the same initial composition.

Several papers have considered the physical regimes in the boundary layers at the top and bottom of melting columns. Fowler [1989] was the first to apply matched asymptotics to the boundary layers of the melting column. Šrámek et al. [2007] extended the analysis of the various physical balances that are possible near the bottom of the column. Hewitt and Fowler [2008] derived matched asymptotic solutions connecting the boundary layers at the top and bottom of the column with the Darcy region in the middle.

Katz [2008] used numerical solutions of column models with two thermochemical components to benchmark a code for more complex, two-dimensional models of mid-ocean ridges. The binary-loop phase diagram was linearized in Hewitt [2010], who extended the analysis of Ribe [1985] and considered the linear stability of the one-dimensional solution to lateral perturbations. A three-component melting-column model was developed by Rees Jones et al. [2018] to describe melt transport in subduction zones; the two-component column with volatiles described here is a simplification of that model. Cerpa et al. [2019] built on the two-component (silicate and volatile) column to investigate the consequences of sea-level variations on magmatism. They noted that the forced, time-dependent deviations from the steady state are small and hence derived linearized equations to model the perturbations. Rees Jones and Rudge [2020] used a nonlinear, time-dependent column model to analyze the magmatic response to Icelandic deglaciation.

The decompaction layer and its implications for melt focusing were first discussed by Sparks and Parmentier [1991]. The case of magma rising into a permeability barrier, *not* freezing, but rather being extracted into a volcanic plumbing system was analyzed by Spencer et al. [2020a] (in their appendix B2). The precursor to the decompaction model presented here was developed by Spiegelman [1993c], which was adapted for analytical treatment by John Rudge [personal communication, 2020]. Numerical models that resolve the layer include Ghods and Arkani-Hamed [2000], Katz [2008], and Katz [2010]. A parameterized analysis is discussed by Hebert and Montési [2010].

Column models incorporating uranium-series disequilibria were initially developed by McKenzie [1985] McKenzie [1984, appendix E] and Navon and Stolper [1987] applied the idea of chromatography to the mantle column, informing ideas about trace-element transport, though not directly considering decay chains. The theory introduced by these three papers was elaborated and thoroughly analyzed by Spiegelman and Elliott [1993], which is the source for much of the material presented here. Later studies have built on these foundations to consider two-dimensional [e.g., Elliott and Spiegelman, 2003] and parameterized [e.g., Jull et al., 2002] models of melt transport and uranium-series. However, as the focus of this chapter is column models, that literature is not considered further here.

11.7 Exercises

11.1 In this basic exercise we will examine the significance of the column-model boundary conditions.

(a) Verify that for no melting and constant solid and liquid velocities, any constant porosity ϕ_0 is a solution to (11.2) for boundary condition $\phi = \phi_0$ at $z = 0$.

(b) Assuming no melting ($\Gamma = 0$), we can trivially determine the degree of melting F. Do this, and then use (11.12) to find the porosity ϕ (assuming small ϕ and ignoring the boundary condition). How many possible values of ϕ are allowed? (Compare to the first part of this question.) Explain why this is the case.

11.2 We consider the disequilibrium trace-element transport column model with $\mathcal{X} = 0$. Making the Boussinesq approximation and assuming a constant, uniform melting rate $\Gamma = \rho W_0 \left.\frac{\mathrm{d}F}{\mathrm{d}z}\right|_\mathrm{s}$, equations (9.24) become

$$\phi w \frac{\mathrm{d}c_j^\ell}{\mathrm{d}z} = +\left(c_j^s/D_j - c_j^\ell\right) W_0 \left.\frac{\mathrm{d}F}{\mathrm{d}z}\right|_\mathrm{s}, \tag{11.78a}$$

$$(1-\phi) W \frac{\mathrm{d}c_j^s}{\mathrm{d}z} = -\left(c_j^s/D_j - c_j^s\right) W_0 \left.\frac{\mathrm{d}F}{\mathrm{d}z}\right|_\mathrm{s}, \tag{11.78b}$$

where W_0 is the upwelling rate at the bottom of the melting column. Show that the solution to these equations is equivalent to the canonical fractional melting model.

11.3 For a one-component mantle column, use table 8.1 and $F^{1c}_{\text{max}} = 0.6$ to estimate the depth of the partially molten mantle z_0.

11.4 For a two-component mantle column, use table 8.1, $F^{2c}_{\text{max}} = 0.225$ and $M^S \Delta c = 700$ K to estimate the depth of the partially molten mantle z_0.

11.5 Recall that in the decompaction boundary-layer problem, we have defined the shape function $G(\zeta)$ to write the integral form of (11.53) in the more convenient form

$$\phi w \equiv q_\infty \left[1 - G(z/\delta_f)\right].$$

Use this simplification and (11.51c) to obtain (11.57).

11.6 Our analysis of the porosity profile for the decompaction boundary layer relied on the exact form of the shape function $G(\zeta)$.

(a) Show that the function $\exp(\zeta)$ satisfies all the conditions in (11.54) required of the shape function $G(\zeta)$.

(b) For the shape function used in the chapter, we saw in figure 11.7 that, for a certain value R_δ, there is a local maximum in the normalized porosity. Using the shape function $G(\zeta) = \exp(\zeta)$, show that, independent of the value of $R_\delta > 0$, there is no local maximum.

11.7 Starting with (11.64), obtain (11.66) by substituting the decomposition into the batch-melting factor and the ingrowth factor (11.65).

11.8 Consider the trace-element transport column model based on equations (9.24) with $\mathcal{X} = 0$.

$$\rho^\ell \phi w \frac{\mathrm{d}c_j^\ell}{\mathrm{d}z} = +\left(c_j^s/D_j - c_j^\ell\right)\Gamma, \tag{11.79a}$$

$$\rho^s(1-\phi) W \frac{\mathrm{d}c_j^s}{\mathrm{d}z} = -\left(c_j^s/D_j - c_j^s\right)\Gamma. \tag{11.79b}$$

This system represents fractional melting with disequilibrium transport.

In this exercise, we will solve (11.79b) for the solid concentration, assuming a nonuniform melting rate Γ caused by a volatile component. Assuming no diffusion, no dissipation, and a steady state, the temperature equation in this case can be simplified to

$$W_0 \frac{\mathrm{d}T}{\mathrm{d}z} = -\frac{L}{c_e \rho}\Gamma - \frac{\alpha_\rho W_0 g T_0^S}{c_e}. \tag{11.80}$$

To close the system, recall that the solidus temperature is governed by

$$T^S = T_0^S + (P - P_{\text{ref}})/\vartheta + M^S\left(c^s - c_0^s\right),$$

with lithostatic pressure $P = -g\rho z$. In the following two parts, we will take different approaches to solve for the solid concentration.

Table 11.1. Parameters to compute curves for 11.9.

Quantity	Value	Units
ρ	3000	kg/m^3
c_P	1200	J/kg/K
α_ρ	3×10^{-5}	K^{-1}
L	5×10^5	J/kg
M^S	-4	K^{-1}
ϑ	6.5×10^6	Pa/K
T_0^S	1373	K
c^s	100	-
D	0.01	-

(a) Recall that the vertical solid velocity satisfies

$$(1-\phi)W(z) = W_0(1-F).$$

Use this and the definition (11.4) of the degree of melting F, to obtain an implicit expression for c^s.

(b) For mid-ocean ridges, the maximum degree of melting $F_{\max} \approx 0.2$. Knowing this, we can approximate

$$(1-\phi)W(z) \approx W_0$$

and hence find an explicit expression for c^s. *Hint: The general solution to the equation*

$$z\left(1-f(z)\right)\frac{df}{dz} = f(z)$$

is $f = \mathcal{W}(c_0 z)$, *where* $\mathcal{W}$ *is the first branch of the Lambert W function and* c_0 *is an arbitrary constant. To obtain this form, you might find it useful to substitute* $y = e^{\alpha z}$, *for a suitable value of* α.

11.9 Write code to plot the implicit and explicit forms for solid concentration that were obtained in 11.8. Also plot the expression (11.38) derived earlier in this chapter. Discuss the differences between the implicit and explicit solutions and between fractional and batch melting. Use the following parameter values for your plots.

CHAPTER 12

Reactive Flow and the Emergence of Melt Channels

In chapter 11, where we explored melting-column models, we derived expressions for the melting rate in approximately isentropic upwelling mantle (i.e., with no thermal diffusion). Those expressions show that the melting rate is proportional to the bulk vertical mass flux, ρW_0 (no subscript on density because of the extended Boussinesq approximation). The bulk mass flux has contributions from both the vertical solid flux $-\rho(1-\phi)\boldsymbol{v}^s \cdot \mathbf{g}/|\mathbf{g}|$ and the vertical liquid flux $-\rho\phi\boldsymbol{v}^\ell \cdot \mathbf{g}/|\mathbf{g}|$. Melting driven by the liquid flux is sometimes referred to as *reactive melting* or *flux melting*. The present chapter will demonstrate the dynamic effect of reactive melting.

Geological observations, geochemical measurements and models show that this dynamic effect may be to localize the magmatic flux into a network of channels. Most importantly, observations of ophiolites have been interpreted as field evidence for reactive magmatic channelization. These exhumed sections of asthenosphere contain, in some cases, an abundance of tabular-shaped bodies of dunite (rock that is nearly pure olivine). This dunite appears to be *replacive* in origin—it replaces the harzburgite (olivine plus orthopyroxene) that was previously there. As we discuss below, the prevalent hypothesis is that replacement occurs by a flux-driven melting reaction and, furthermore, that that reaction is the cause of magmatic channelization and that dunite bodies are the consequence. The reader is referred to the Literature Notes at the end of this chapter for further information and references.

In a one-dimensional upwelling column, the bulk mass flux is uniformly ρW_0. Melt cannot flow laterally, so there is no possibility of forming high-flux channels that alternate with low-flux interchannel regions. To allow for the possibility of horizontal variations, our analysis in this chapter must be two-dimensional (at least). However, we seek to illustrate the physical processes while retaining the possibility of analytical solutions. Hence we consider the linearized stability of one-dimensional column solutions in two dimensions. This corresponds, conceptually, to modeling the emergence of channels from an unchannelized system.

This analysis was first developed by Aharonov et al. [1995], based on earlier work on reactive-flow fronts (see the Literature Notes at the end of this chapter). The approach is highly simplified in its thermochemistry, dynamics, and domain/boundary conditions. It is a topic of current research to correct some of these deficiencies; a key step is to incorporate empircal constraints into a rigorous treatment of nonequilibrium thermodynamics. Below, however, we draw on the analysis of Rees Jones and Katz [2018].

12.1 Governing Equations

We begin with equations (4.46), representing the two-phase fluid dynamics under the extended Boussinesq approximation. Our interest is the reactive flow of the liquid phase; it is appropriate to simplify the equations by neglecting large-scale shear flow of the solid. Defining a coordinate system in which $\hat{z} = -\mathbf{g}/|\mathbf{g}|$, we express the solid velocity as

$$\boldsymbol{v}^s \equiv W_0\hat{z} + \boldsymbol{v}^{\mathcal{C}}, \tag{12.1}$$

where $\nabla \cdot \boldsymbol{v}^{\mathcal{C}} = \mathcal{C}$ and, by assumption, $\nabla \times \boldsymbol{v}^{\mathcal{C}} = 0$; hence we are neglecting any shear contribution to the solid flow. We express conservation of mass and momentum as

$$\frac{\mathrm{D}_s\phi}{\mathrm{D}t} - (1-\phi)\mathcal{P}/\zeta_\phi - \Gamma/\rho = 0, \tag{12.2a}$$

$$\frac{\partial\phi}{\partial t} + \nabla\cdot\phi\boldsymbol{v}^\ell - \Gamma/\rho = 0, \tag{12.2b}$$

$$\phi\left(\boldsymbol{v}^\ell - \boldsymbol{v}^s\right) + K_\phi\left[\nabla\mathcal{P} - (1-\phi)\Delta\rho g\hat{z}\right] = 0, \tag{12.2c}$$

where $\mathcal{P} \equiv \zeta_\phi\mathcal{C}$ is the compaction pressure, $K_\phi \equiv k_\phi/\mu$, and for simplicity we have taken $\zeta_\phi + \frac{4}{3}\eta_\phi \sim \zeta_\phi$. We further assume that the compaction viscosity ζ_ϕ is constant and give it subscript 0 (the effect of nonconstant compaction viscosity is considered in exercise 12.3). For the mobility we take the usual form, $K_\phi = K_0(\phi/\phi_0)^n$.

Reactive flow, by definition, involves thermochemical reaction in a multicomponent system. Hence, to the mechanical system above, we add equations for conservation of chemical species mass. We take the simplest possible approach and adopt a two-component chemical system. Since concentrations are subject to a unity sum, we need write equations for only one of the components in each phase. Recalling equations (9.6), we have

$$\phi\frac{\mathrm{D}_\ell c^\ell}{\mathrm{D}t} = \left(c^\Gamma - c^\ell\right)\Gamma/\rho + \mathcal{D}\nabla\cdot\phi\nabla c^\ell, \tag{12.3a}$$

$$(1-\phi)\frac{\mathrm{D}_s c^s}{\mathrm{D}t} = -\left(c^\Gamma - c^s\right)\Gamma/\rho. \tag{12.3b}$$

In this system, c^Γ represents the concentration of the relevant component in the mass that is transferred between phases during melting.

To close the system (12.2)–(12.3), we require expressions for Γ and c^Γ.

12.2 The Melting-Rate Closure

We express the melting rate as a sum of two terms that represent decompression melting and reactive melting,

$$\Gamma \equiv \Gamma_D + \Gamma_R, \tag{12.4}$$

and consider the decompression melting part first. We write Γ_D as the vertical mass flux of solid times the isentropic productivity $\left.\frac{\mathrm{d}F}{\mathrm{d}z}\right|_s$. For simplicity, we will assume that the solid mass flux can be approximated by the mean, background flux

$$\rho(1-\phi)\boldsymbol{v}^s \cdot \hat{\boldsymbol{z}} \approx \rho W_0.$$

Hence the decompression melting rate is written $\Gamma_D = \rho W_0 \left.\frac{\mathrm{d}F}{\mathrm{d}z}\right|_{\mathrm{s}}$. We consider the reactive part next.

The essence of reactive melting is that as liquid moves upward, it becomes undersaturated in a soluble component. This undersaturation drives a melting reaction whereby the soluble component is transferred from the solid to the liquid. Of course, this simple picture becomes more complicated when we look at the petrological details of the mantle. However, it is not incorrect to say that as magma rises to lower pressures, it becomes undersaturated in SiO_2. The undersaturation drives an incongruent melting reaction in which pyroxene (Px) is dissolved from the solid while olivine (Ol) is precipitated:

$$\mathrm{Liq}_1 + \mathrm{Px} \rightarrow \mathrm{Liq}_2 + \mathrm{Ol},$$

where Liq_1 and Liq_2 represent silicate melts of different compositions. The net result for the magma is an increase in both its mass fraction and its silica content. For the residual solid, the net result is a conjugate decrease in mass fraction and the replacement of pyroxene with olivine. The latter leads to a lithological change from harzburgite to dunite.

To model reactive melting, we assume that there is an equilibrium concentration of the soluble component in the melt. This equilibrium concentration $\check{c}^\ell$ depends only on depth (i.e., $\check{c}^\ell = \check{c}^\ell(z)$). The rate of melting is then related to the compositional difference from equilibrium. In particular, we assume a linear proportionality to this difference:

$$\Gamma_R = -\rho\mathcal{R}\left(c^\ell - \check{c}^\ell\right), \tag{12.5}$$

where $\mathcal{R}$ is the constant of proportionality. This is a statement of linear kinetics. Note that if the liquid is *undersaturated* in the soluble component (i.e., $c^\ell < \check{c}^\ell$) then the reactive melting rate is positive. The reaction rate coefficient $\mathcal{R}$ has the same units as Γ (mass/volume/time). In a more detailed kinetic model, $\mathcal{R}$ would depend on the specific surface area of the interface between liquid and solid (related to the grain size), the concentration of the soluble component in the solid phase, and the difference in chemical potentials of the component between phases (the chemical affinity; see section 10.2). For simplicity, we neglect these dependencies and treat $\mathcal{R}$ as constant.

The equilibrium solubility could be computed on the basis of a thermodynamic model such as that discussed in chapter 10. However, for our present purposes, we impose a linear variation with depth

$$\check{c}^\ell = \beta z + \check{c}^\ell_0, \tag{12.6}$$

where β and $\check{c}^\ell_0$ are constants. The latter is taken to be zero without loss of generality; the former can be interpreted as the vertical gradient of SiO_2 solubility in silicate melts in the shallow asthenosphere, with a value that is $\mathcal{O}\left(10^{-6}/\mathrm{m}\right)$.

Assembling the melting model according to equation (12.4), we have

$$\Gamma/\rho = W_0 \left.\frac{\mathrm{d}F}{\mathrm{d}z}\right|_{\mathrm{s}} - \mathcal{R}\left(c^\ell - \beta z\right). \tag{12.7}$$

This expression will be substituted into the system (12.2)–(12.3).

To complete the melting closure, we must specify the composition of the melt produced by decompression melting c^{Γ_D} and reactive melting c^{Γ_R}. These can be distinct and, indeed, logic suggests that they should be. Reactive melting is a process of chemical equilibration between the phases, whereas decompression melting is the breakdown of solid to form liquid. The latter should produce melts that are in equilibrium with their parent solid, $c^{\Gamma_D}=\check{c}^\ell$. Reactive melting, in the case considered here, involves an incongruent reaction where pyroxene is dissolved and olivine is precipitated with a net effect of increasing the mass of melt. Incongruent melting was discussed in section 9.1.2, where it was shown that c^{Γ_R} is not required to be between zero and unity. We will assume that it is given by

$$c^{\Gamma_R}=\check{c}^\ell(z)+\alpha, \tag{12.8}$$

where α is a compositional offset from equilibrium. In general, α should depend, at the least, on the sign of $c^\ell-\check{c}^\ell$ to ensure that the reaction moves the liquid toward equilibrium. However in the case considered below, the liquid will only be undersaturated and hence α will be taken to be a positive constant.

12.3 Problem Specification

We now combine the governing equations (12.2)–(12.3) with the melting model (12.7)–(12.8) and $\check{c}^\ell=\beta z$ to give the system

$$\frac{\partial\phi}{\partial t}+\boldsymbol{v}^s\cdot\nabla\phi=(1-\phi)\mathcal{P}/\zeta_\phi+W_0\left.\tfrac{\mathrm{d}F}{\mathrm{d}z}\right|_s-\mathcal{R}\left(c^\ell-\beta z\right), \tag{12.9a}$$

$$\frac{\partial\phi}{\partial t}+\nabla\cdot\phi\boldsymbol{v}^\ell=W_0\left.\tfrac{\mathrm{d}F}{\mathrm{d}z}\right|_s-\mathcal{R}\left(c^\ell-\beta z\right), \tag{12.9b}$$

$$\phi\frac{\partial c^\ell}{\partial t}+\phi\boldsymbol{v}^\ell\cdot\nabla c^\ell=\mathcal{D}\nabla\cdot\phi\nabla c^\ell+\left(\beta z-c^\ell\right)W_0\left.\tfrac{\mathrm{d}F}{\mathrm{d}z}\right|_s-\left(\beta z+\alpha-c^\ell\right)\mathcal{R}\left(c^\ell-\beta z\right), \tag{12.9c}$$

$$\phi\left(\boldsymbol{v}^\ell-\boldsymbol{v}^s\right)=-K_\phi\left[\nabla\mathcal{P}-(1-\phi)\Delta\rho g\hat{\boldsymbol{z}}\right]. \tag{12.9d}$$

Here we have dropped the compositional equation (12.3b) for the solid. This is because it is decoupled from the other equations (i.e., c^s does not appear in the system (12.9)). In a more detailed treatment, we would expect a nonconstant $\mathcal{R}=\mathcal{R}(\phi,c^s)$ to model the availability of the reactant in the solid phase. For present purposes however, where we are considering only the stability of a uniform, nonchannelized state, we expect only negligibly small variations in c^s. Note also that for (12.9a) and (12.9b), we substituted equation (12.7) for Γ/ρ. For the governing equation (12.9c), we modified the equation to accommodate the different reaction products c^{Γ_D} and c^{Γ_R} by splitting the melting term into two.

We use the system (12.9) to analyze a two-dimensional region extending to $\pm\infty$ in the x-direction and from 0 to H in the z-direction, shown schematically in figure 12.1. The bottom of the domain is located at some distance above the onset of melting, such that a Darcy flow regime is well established; gradients in compaction pressure are small (see section 11.3). The porosity at this depth is uniform. The top of the domain is located beneath the bottom of the zone of freezing at the base of the lithosphere.

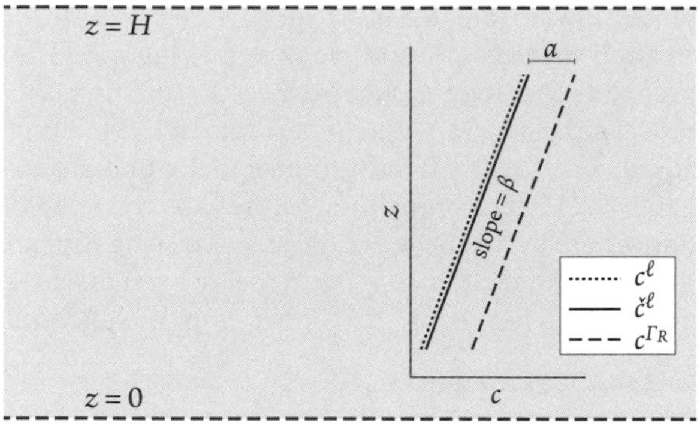

Figure 12.1. Schematic diagram of the reactive flow domain. The gray area is the region in which the equations are solved; it extends to $\pm\infty$ in the x-direction but is bounded between 0 and H in the z-direction. The inset plot shows the equilibrium solubility $\check{c}^\ell$, the reactive product concentration c^{Γ_R}, and an illustration of the expected melt concentration c^ℓ (although this latter quantity is to be solved for).

Hence, it is not affected by the decompaction boundary layer there, which contains a sharp gradient in the compaction pressure (section 11.4). Mantle is assumed to upwell uniformly into the domain at a rate W_0 that is very small compared to the speed of melt migration. The liquid flux into the bottom of the domain is fixed as a boundary condition.

The first part of our analysis is to rescale the equations, define nondimensional groups of parameters that highlight the key physics, and drop small terms. We then decompose variables into a steady, one-dimensional base-state solution and time-dependent perturbations to that state. Next we solve for the leading-order base-state solution and assess its stability to lateral perturbations in the form of channels. In particular, we seek an expression for the growth (or decay) rate of channel perturbations as function of problem parameters.

12.4 Scaling and Simplification

We apply scaling analysis to the system (12.9) to simplify the equations. Based on previous analysis of melting columns in chapter 11, we expect that the porosity remains near the inflow value, suggesting that $[\phi] = \phi_0 \ll 1$ and hence that $(1-\phi) \sim 1$. The mobility at this porosity is K_0 and hence a typical speed of vertical melt transport w_0 is given by balancing equation (12.9d) as $\phi_0 w_0 \sim K_0 \Delta\rho g$. We therefore take

$$\left[\boldsymbol{v}^\ell\right] = w_0 \equiv K_0 \Delta\rho g/\phi_0, \tag{12.10a}$$

$$\left[\boldsymbol{v}^s\right] = \phi_0 w_0. \tag{12.10b}$$

Based on our experience with porosity-banding instabilities in chapter 7, we expect that localization (such as channels) will occur on a horizontal scale that is no larger than the compaction length $\delta \equiv \sqrt{K_0 \zeta_0}$. The vertical structure of solutions, however, is likely to be determined by the equilibrium solubility forcing, which has a characteristic size of the domain height. Hence the length scale of solutions is more conveniently referenced to the domain size than to the compaction length and we scale $[\boldsymbol{x}] = H$. The change in the equilibrium liquid concentration of the soluble component over the domain height is βH. This suggests a scaling for concentrations as

$$\left[c^{\ell}\right] = \beta H. \tag{12.11}$$

If the reaction-rate coefficient $\mathcal{R}$ and/or the incongruent concentration offset α are large, then the system should be close to equilibrium everywhere. This suggest that $|c^{\ell} - \check{c}^{\ell}| \ll \alpha$ and that $\nabla c^{\ell} \sim \beta\hat{\boldsymbol{z}}$. At steady state in equation (12.9c), we expect vertical advection of concentration to balance reactive melting, i.e., $\phi_0 w_0 \beta \sim \alpha\Gamma_R$, motivating

$$[\Gamma_R] = \phi_0 w_0 \beta/\alpha. \tag{12.12}$$

Furthermore, in equation (12.9a), we expect a steady balance between compaction and reactive melting, $-\mathcal{C} \sim \Gamma_R$. Combining this balance with (12.12) we can assume $[\mathcal{C}] = \phi_0 w_0 \beta/\alpha$ as a scaling for the compaction length and hence

$$[\mathcal{P}] = \zeta_0 \phi_0 w_0 \beta/\alpha \tag{12.13}$$

as a scaling for the compaction pressure. Finally, we scale time according to a characteristic reactive time,

$$[t] = \frac{\alpha}{w_0 \beta}. \tag{12.14}$$

Here we note the existence of a reactive length scale α/β, which represents the vertical distance over which a substantial change in equilibrium solubility occurs.

Applying these scales to the symbols in (12.9) and dropping terms of order ϕ_0, we find that

$$\frac{\partial \phi}{\partial t} = \mathcal{P} + \mathcal{G} - \text{Da}(c^{\ell} - z), \tag{12.15a}$$

$$\mathcal{M}\frac{\partial \phi}{\partial t} + \nabla\cdot\phi\boldsymbol{v}^{\ell} = \mathcal{M}\left[\mathcal{G} - \text{Da}(c^{\ell} - z)\right], \tag{12.15b}$$

$$\mathcal{M}\phi\frac{\partial c^{\ell}}{\partial t} + \phi\boldsymbol{v}^{\ell}\cdot\nabla c^{\ell} = \frac{1}{\text{Pe}}\nabla\cdot\phi\nabla c^{\ell} - \mathcal{M}\mathcal{G}(c^{\ell} - z) - \text{Da}\left[1 - \mathcal{M}(c^{\ell} - z)\right](c^{\ell} - z), \tag{12.15c}$$

$$\phi\boldsymbol{v}^{\ell} = \phi^{n}\left(\hat{\boldsymbol{z}} - \mathcal{S}\nabla\mathcal{P}\right), \tag{12.15d}$$

where various quantities, introduced in this system of equations, are defined below, and all symbols are nondimensional. We have defined

$$\mathcal{G} \equiv \frac{\alpha W_0 \left.\frac{\text{d}F}{\text{d}z}\right|_{\text{s}}}{\phi_0 w_0 \beta}, \tag{12.16}$$

a dimensionless ratio of the rates of melt production by decompression and by reaction. According to mantle reference values given below in table 12.1,[1] we expect that $\mathcal{G} \sim \mathcal{O}(10)$.

We have defined

$$\text{Da} \equiv \frac{\alpha \mathcal{R} H}{\phi_0 w_0}, \tag{12.17}$$

a Damköhler number that represents the ratio of a time scale for melt transport over the domain height (H/w_0) to the time scale for equilibration of melt at the characteristic porosity ($\phi_0/(\alpha\mathcal{R})$). A large Damköhler number means that the system is close to chemical equilibrium. We expect that this is true of the mantle and hence that $\text{Da} \gg 1$. However, the dynamics depend on a term that is of order $1/\text{Da}$ and so we retain it in the system to be analyzed.

We have also defined

$$\mathcal{M} \equiv \beta H/\alpha. \tag{12.18}$$

This parameter represents the volume of melt produced by reactive melting, per unit of volume, per unit of melt flux, per unit of chemical disequilibrium, due to upwelling of magma over the domain height. It can loosely be considered to be the *reactive potential* of the system. A large value of $\mathcal{M}$ means that reactive melting produces a large volume of melt. We expect that $\mathcal{M} < 1$ and is perhaps even $\ll 1$.

We have defined

$$\mathcal{S} \equiv \mathcal{M}\frac{\delta^2}{H^2} = \frac{\beta\delta}{\alpha}\frac{\delta}{H}, \tag{12.19}$$

a stiffness number that is a combination of the reactive potential and rigidity. For small reactivity $\mathcal{M}$, the ratio of the compaction length to the domain height must be large to achieve a stiffness of $\mathcal{O}(1)$ or greater.

Finally, we have defined the Péclet number

$$\text{Pe} \equiv w_0 H/\mathcal{D}, \tag{12.20}$$

which is a ratio of diffusive to advective time scales. Diffusion is slow in silicate melts and hence $\text{Pe} \gg 1$.

Because the Damköhler number is large, we expect that the melt is close to equilibrium with the solid and hence that, in dimensionless units, $|c^\ell - z| \ll 1$. This motivates a change of variables where instead of solving for the liquid concentration of the soluble component, we solve for the scaled *undersaturation* χ. The two are related by

$$\chi \equiv \text{Da}(z - c^\ell). \tag{12.21}$$

Multiplication of the undersaturation by Da ensures that $\chi \sim \mathcal{O}(1)$. Substitution of (12.21) into (12.15) and dropping terms that are $\mathcal{O}(\mathcal{M}/\text{Da}) \ll 1$ gives

$$\frac{\partial \phi}{\partial t} = \mathcal{P} + \mathcal{G} + \chi, \tag{12.22a}$$

$$\mathcal{M}\frac{\partial \phi}{\partial t} + \nabla \cdot \phi \boldsymbol{v}^\ell = \mathcal{M}(\mathcal{G} + \chi), \tag{12.22b}$$

[1]We use those values to roughly constrain nondimensional constants here, deferring a more detailed discussion to section 12.7 below.

$$\phi \boldsymbol{v}^{\ell} \cdot \left(\nabla \chi / \mathrm{Da} - \hat{z}\right) = \frac{1}{\mathrm{Da}\,\mathrm{Pe}} \frac{\partial}{\partial x} \phi \frac{\partial \chi}{\partial x} - \chi, \tag{12.22c}$$

$$\phi \boldsymbol{v}^{\ell} = \phi^{n} \left(\hat{z} - \mathcal{S} \nabla \mathcal{P}\right). \tag{12.22d}$$

Regarding the diffusion of undersaturation (liquid concentration): it is certainly true that $1/(\mathrm{Da}\,\mathrm{Pe}) \ll 1$. However, we have retained in (12.22c) the term representing horizontal diffusion because we anticipate that closely spaced channels will have short-wavelength fluctuations in the undersaturation and hence a second derivative proportional to the square of the horizontal wavenumber (which may be very large, especially given our choice of length scale H).

The simplified system of equations (12.22) is a minimal representation of the physics to be analyzed.

12.5 Linearized Stability Analysis

To formally represent a steady, one-dimensional base state and a (potentially) growing perturbation to that state, we expand the variables in powers of the small parameter ϵ,

$$\phi = \phi^{(0)}(z) + \epsilon \phi^{(1)}(x, z, t), \tag{12.23a}$$

$$\chi = \chi^{(0)}(z) + \epsilon \chi^{(1)}(x, z, t), \tag{12.23b}$$

$$\mathcal{P} = \mathcal{P}^{(0)}(z) + \epsilon \mathcal{P}^{(1)}(x, z, t), \tag{12.23c}$$

$$\boldsymbol{v}^{\ell} = w^{(0)}(z)\hat{z} + \epsilon \boldsymbol{v}^{(1)}(x, z, t). \tag{12.23d}$$

and truncate the series at $\mathcal{O}(\epsilon)$. The first-order velocity $\boldsymbol{v}^{(1)}$ is written, for concision, without the ℓ superscript.

Substituting the expansion (12.23) into the system (12.22), and dropping terms of $\mathcal{O}(\epsilon)$ and $\mathcal{O}(1/\mathrm{Da})$, gives the leading-order balance

$$-\mathcal{P}^{(0)} = \mathcal{G} + \chi^{(0)}, \tag{12.24a}$$

$$\frac{\mathrm{d}}{\mathrm{d}z} \phi^{(0)} w^{(0)} = \mathcal{M}\left(\mathcal{G} + \chi^{(0)}\right), \tag{12.24b}$$

$$\phi^{(0)} w^{(0)} = \chi^{(0)}, \tag{12.24c}$$

$$\phi^{(0)} w^{(0)} = \left(\phi^{(0)}\right)^{n} \left(1 - \mathcal{S} \frac{\mathrm{d}\mathcal{P}^{(0)}}{\mathrm{d}z}\right). \tag{12.24d}$$

The governing equations at $\mathcal{O}(\epsilon)$ are

$$\phi_t^{(1)} = \mathcal{P}^{(1)} + \chi^{(1)}, \tag{12.25a}$$

$$\mathcal{M}\phi_t^{(1)} + w^{(0)}\phi_z^{(1)} + \phi^{(0)} \nabla \cdot \boldsymbol{v}^{(1)} = \mathcal{M}\chi^{(1)}, \tag{12.25b}$$

$$\left(\phi^{(0)} w^{(1)} + \phi^{(1)} w^{(0)}\right)\left(1 - \frac{\chi_z^{(0)}}{\mathrm{Da}}\right) = \phi^{(0)} w^{(0)} \frac{\chi_z^{(1)}}{\mathrm{Da}} - \frac{\phi^{(0)}}{\mathrm{Da}\,\mathrm{Pe}} \chi_{xx}^{(1)} + \chi^{(1)} \tag{12.25c}$$

$$\phi^{(0)}\boldsymbol{v}^{(1)} + \phi^{(1)}w^{(0)}\hat{\boldsymbol{z}} = -\left(\phi^{(0)}\right)^n \mathcal{S}\nabla\mathcal{P}^{(1)} + n\left(\phi^{(0)}\right)^{n-1}\phi^{(1)}\left(1 - \mathcal{S}\mathcal{P}_z^{(0)}\right)\hat{\boldsymbol{z}}. \quad (12.25\text{d})$$

Here we have expressed some partial derivatives as subscripts. The perturbation to the vertical velocity $\boldsymbol{v}^{(1)} \cdot \hat{\boldsymbol{z}}$ is written as $w^{(1)}$.

12.5.1 THE BASE STATE

We now seek a solution to the leading-order system (12.24) that satisfies the boundary conditions

$$\chi^{(0)} = \phi^{(0)}w^{(0)} = 1 \quad \text{at } z = 0. \quad (12.26)$$

Note the use of a nonzero undersaturation $\chi^{(0)}$ at the inflow boundary. This prevents formation of a chemical boundary layer in the domain immediately above the inflow boundary.

Combining equations (12.24b) and (12.24c) to form an equation for $\chi^{(0)}$ and using the boundary conditions tells us that

$$\chi^{(0)} = (1 + \mathcal{G})\,\mathrm{e}^{\mathcal{M}z} - \mathcal{G}. \quad (12.27)$$

From this solution, we can immediately obtain $\mathcal{P}^{(0)}$ using (12.24a). The base-state porosity and liquid upwelling rate are determined by (12.24c) and (12.24d). The assembled, base-state solution is then

$$\chi^{(0)} = (1 + \mathcal{G})\,\mathrm{e}^{\mathcal{M}z} - \mathcal{G} \qquad \approx 1 + (1 + \mathcal{G})\,\mathcal{M}z, \quad (12.28\text{a})$$

$$-\mathcal{P}^{(0)} = \chi^{(0)} + \mathcal{G} \qquad \approx (1 + \mathcal{G})\,(1 + \mathcal{M}z)\,, \quad (12.28\text{b})$$

$$\phi^{(0)} = \left[\frac{\chi^{(0)}}{1 + \mathcal{S}\mathcal{M}\left(\chi^{(0)} + \mathcal{G}\right)}\right]^{1/n} \qquad \approx \mathcal{F}^{-1}\left[1 + (1 + \mathcal{G})\,\mathcal{M}z/n\right], \quad (12.28\text{c})$$

$$w^{(0)} = \left[\frac{1 + \mathcal{S}\mathcal{M}\left(\chi^{(0)} + \mathcal{G}\right)}{\left(\chi^{(0)}\right)^{1-n}}\right]^{1/n} \qquad \approx \mathcal{F}\left[1 + (1 + \mathcal{G})\,(1 - 1/n)\mathcal{M}z\right]. \quad (12.28\text{d})$$

The approximation that follows each exact solution is a Taylor series expansion for small $\mathcal{M}$, truncated after one term. We have introduced the constant

$$\mathcal{F} \equiv [1 + \mathcal{S}\mathcal{M}(1 + \mathcal{G})]^{1/n}, \quad (12.29)$$

which is an approximation that is valid when $\mathcal{M}^2 \ll 1$. When $\mathcal{S}\mathcal{M} \ll 1$ we have $\mathcal{F} \sim 1$.

Both the full solution and the linear approximation are plotted in figure 12.2 for two values of $\mathcal{M}$. In both cases, $\mathcal{S} = \mathcal{G} = 1$; this value of $\mathcal{G}$ corresponds to decompression melting occurring at the same rate as reactive melting. Note that, for $\mathcal{M} \ll 1$, the undersaturation, compaction pressure, porosity, and upwelling rate are approximately uniform. This is because the total amount of melt produced over the column is small compared to the flux into the bottom of the column. For larger $\mathcal{M}$ or $\mathcal{G}$, the slope of the

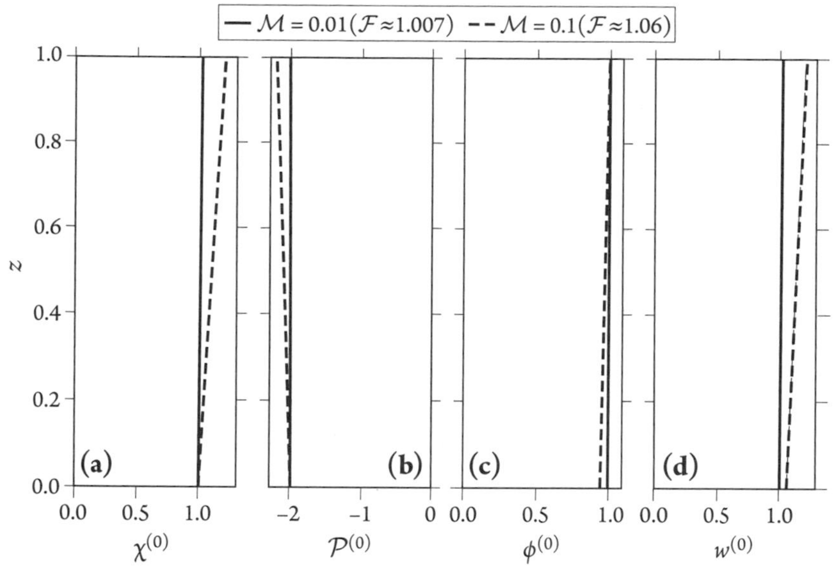

Figure 12.2. The base-state solution (12.30), plotted for two values of $\mathcal{M}$. Thick lines are the full solution and narrow lines are the linear approximation, which are almost indistinguishable from each other. In each case, $\mathcal{S}=1$, $\mathcal{G}=1$, and $n=3$. The values at $z=0$ of (c) $\phi^{(0)}$ and (d) $w^{(0)}$ are given by $\mathcal{F}^{-1}$ and $\mathcal{F}$, respectively.

base-state set of solutions is greater. The case of most immediate relevance, however, is reasonably represented by

$$\chi^{(0)} = -\mathcal{P}^{(0)} = \phi^{(0)} = w^{(0)} = 1. \tag{12.30}$$

The simplicity of this background state facilitates our analysis of the perturbations.

12.5.2 THE GROWTH RATE OF PERTURBATIONS

Using the constant base-state solution (12.30) and the asymptotic limit of $\mathcal{M} \ll 1$, the system of equations (12.25) for the perturbation quantities becomes

$$\phi_t^{(1)} = \mathcal{P}^{(1)} + \chi^{(1)}, \tag{12.31a}$$

$$\phi_z^{(1)} + \nabla\cdot \boldsymbol{v}^{(1)} = 0, \tag{12.31b}$$

$$w^{(1)} + \phi^{(1)} = \frac{\chi_z^{(1)}}{\text{Da}} + \chi^{(1)} - \frac{1}{\text{DaPe}}\chi_{xx}^{(1)} \tag{12.31c}$$

$$\boldsymbol{v}^{(1)} = (n-1)\,\phi^{(1)}\hat{z} - \mathcal{S}\nabla\mathcal{P}^{(1)}. \tag{12.31d}$$

Equation (12.31a) can then be used to eliminate $\chi^{(1)}$ and equation (12.31d) can be used to eliminate the velocity. This leads to the system

$$n\partial_z\phi^{(1)} = \mathcal{S}\nabla^2\mathcal{P}^{(1)}, \tag{12.32a}$$

$$\left[\partial_t + \frac{\partial_{tz}}{\text{Da}} - n - \frac{\partial_{txx}}{\text{DaPe}}\right]\phi^{(1)} = \left[1 + \left(\frac{1}{\text{Da}} - \mathcal{S}\right)\partial_z - \frac{\partial_{xx}}{\text{DaPe}}\right]\mathcal{P}^{(1)}, \tag{12.32b}$$

where we have introduced an operator notation for partial derivatives: subscripts on the ∂ symbol indicate a set of partial derivatives. Equations (12.32) can be combined to give a single equation for $\mathcal{P}^{(1)}$,

$$\left[\partial_t + \frac{\partial_{tz}}{\text{Da}} - n - \frac{\partial_{txx}}{\text{DaPe}}\right]\mathcal{S}\nabla^2\mathcal{P}^{(1)} = \left[1 + \left(\frac{1}{\text{Da}} - \mathcal{S}\right)\partial_z - \frac{\partial_{xx}}{\text{DaPe}}\right]n\partial_z\mathcal{P}^{(1)}. \tag{12.33}$$

This is a linear PDE that is third order in the z-direction. An appropriate ansatz, constructed from eigenfunctions of the linear differential operators, is

$$\mathcal{P}^{(1)}(x, z, t) = \Re\sum_{j=1}^{3} A_j \exp\left(ikx + m_j z + \sigma t\right), \tag{12.34}$$

where $\Re$ means taking only the real part of the complex expression. This trial solution with unknown k, m_j, and σ represents the perturbation to the base-state compaction pressure. Note that σ and the m_j may be complex numbers. If the real part of the eigenvalue σ is positive for some mode (k, m) then that perturbation mode grows; this exponential growth is termed here *instability*. Substituting (12.34) into (12.33), we obtain the characteristic polynomial

$$\frac{\sigma}{\text{Da}}m^3 + \left(\sigma\mathcal{K} - \frac{n}{\text{Da}\mathcal{S}}\right)m^2 - \left(\frac{n\mathcal{K}}{\mathcal{S}} + \frac{\sigma}{\text{Da}}k^2\right)m + (n - \sigma\mathcal{K})\,k^2 = 0, \tag{12.35}$$

where $\mathcal{K} = 1 + k^2/\,(\text{Da Pe})$. The three roots m_j of this cubic equation can be obtained given values of horizontal wavenumber k and growth rate σ. We consider k to be an independent variable and σ an unknown. The only combinations of k, σ, and roots m_j that are acceptable are those that, when used in equation (12.34), describe a pressure perturbation that satisfies a set of boundary conditions. We use this requirement to determine σ.

The boundary conditions impose zero perturbation of the base state along the bottom boundary. Hence we take $\chi^{(1)} = \phi^{(1)} = w^{(1)} = 0$ at $z = 0$. Moreover, we require that perturbations to the flow are unimpeded (and unforced) by gradients in the compaction pressure at the top of the domain, $z = 1$. Using these boundary conditions and equations (12.31) we find that

$$\mathcal{P}^{(1)} = 0 \quad \text{at } z = 0, \tag{12.36a}$$

$$\partial_z\mathcal{P}^{(1)} = 0 \quad \text{at } z = 0, \tag{12.36b}$$

$$\partial_z\mathcal{P}^{(1)} = 0 \quad \text{at } z = 1. \tag{12.36c}$$

This set of conditions must hold for all x and $t > 0$. Combining these conditions with our ansatz (12.34) and expressing in terms of a matrix–vector multiplication gives

$$\begin{pmatrix} 1 & 1 & 1 \\ m_1 & m_2 & m_3 \\ m_1 e^{m_1} & m_2 e^{m_2} & m_3 e^{m_3} \end{pmatrix}\begin{pmatrix} A_1 \\ A_2 \\ A_3 \end{pmatrix} = \begin{pmatrix} 0 \\ 0 \\ 0 \end{pmatrix}. \tag{12.37}$$

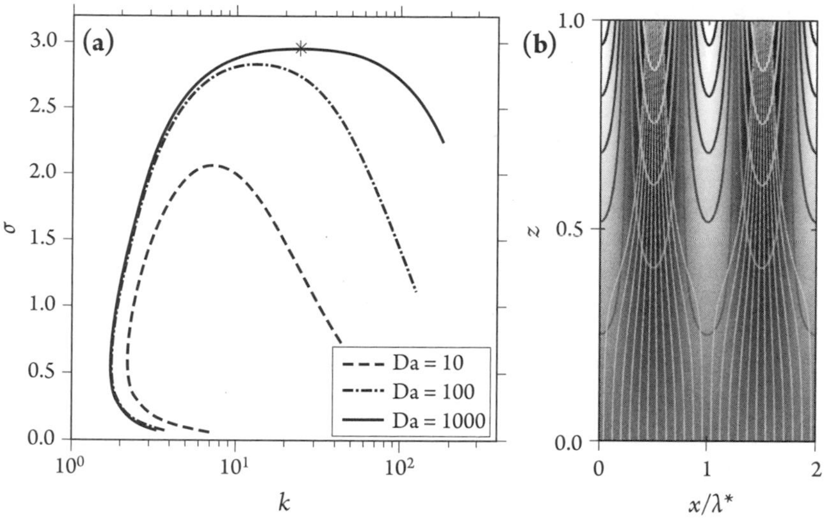

Figure 12.3. **(a)** Dispersion curves: growth rate σ as a function of wavenumber k from numerical solutions for σ, m_j, A_j for three different values of Da. All three curves use the parameters $\mathrm{Pe}=100$, $\mathcal{M}=0.01$, $\mathcal{G}=\mathcal{S}=1$, and $n=3$ (corresponding to the parameters used to compute the base state in fig. 12.2). The star symbol marks the maximum growth rate for the reference curve. **(b)** The eigenmode with maximum growth rate $\sigma^* \approx 2.96$ at $k^* \approx 24.6$ ($\lambda^* \approx 0.26$) for the curve with $\mathrm{Da}=1000$ corresponding to the star marker in (a). The perturbation to the compaction pressure $\mathcal{P}^{(1)}$ is shown in the gray-scale background image. The narrow lines are contours of the porosity perturbation $\phi^{(1)}$, which has maxima where the compaction pressure has minima. The white curves are streamlines of the flow $\boldsymbol{v}^\ell = \hat{\boldsymbol{z}} + \epsilon \boldsymbol{v}^{(1)}$, with ϵ chosen to be 3×10^{-5}. The velocity perturbation $\boldsymbol{v}^{(1)}$ is computed with equation (12.31d).

A nontrivial solution that satisfies the boundary conditions will exist if and only if the determinant of the matrix (which we will define as **B**) is zero. Therefore a solution in terms of values for σ, m_j is found when both (12.37) and

$$\det(\mathbf{B}) = 0 \tag{12.38}$$

are simultaneously satisfied. Numerically, for given values of n, $\mathcal{S}$, Da, Pe, and a chosen horizontal wavenumber k, we search for the value of σ such that the roots satisfy $|\det(\mathbf{B})| < tol$, where tol is the tolerance, for some suitably small numerical value of tol. Solutions to (12.33) are then represented by values of σ and m_j.

Figure 12.3 illustrates the results of numerical solutions of the stability problem. The growth rate of perturbations $\sigma(k)$ for a reference set of parameters and three values of Da are plotted in figure 12.3(a). The curve for $\mathrm{Da}=1000$ is taken to be the reference case. Its perturbation growth rate has a maximum at $k^* \approx 24.5$, which corresponds to a perturbation wavelength $2\pi/k^*$ of about 0.26, roughly one-quarter the height of the domain (characteristics of the fastest growing mode will be denoted with the superscript $*$). The growth rate σ^* at this Damköhler number is approximately equal to the permeability exponent n. Moving toward small $k < k^*$ (large wavelength), the growth rate declines rapidly. There is a wavenumber of $\mathcal{O}(1)$ below which there are no solutions with positive σ. There is a similar drop-off of growth rate for $k > k^*$ (small wavelength). Hence the system exhibits wavelength selection: there exists a unique, finite wavelength with the fastest exponential growth rate; this is the wavelength that might be expected to emerge in the physical system. A discussion of the physical mechanisms of instability and geological predictions of the selected wavelength are given below in section 12.6.

The fastest-growing solution $\mathcal{P}^{(1)}$ for the reference parameters is illustrated in figure 12.3(b). This is obtained by setting $A_1 = 1$ and solving (12.37) for A_2 and A_3. For the solution in (b), this is done at conditions that maximize σ. The structure of the compaction pressure is shown by the gray scale; white lines are streamlines of the melt. The streamlines converge into regions of low compaction pressure, which represent the emergent channels. The porosity field is shown as contours; porosity is larger where the streamlines converge. For smaller Damköhler number than the reference value of 1000, the dispersion curves have maxima at smaller values of k. The fastest-growing solution in those cases is qualitatively the same as that of figure 12.3(b), except that it is stretched in the horizontal direction, giving broader channels.

12.5.3 THE LARGE–DAMKÖHLER NUMBER LIMIT

Petrological understanding of magma transport in the asthenosphere, both from experimental studies and observations, suggests that magma is close to chemical equilibrium with its host matrix. Hence it is reasonable to consider the asymptotic limit of large Damköhler number (see table 11.1).

Inspection of roots m_j of the cubic polynomial (12.35) obtained numerically shows that one root is real, negative and very large, and that the other two are complex conjugates and of $\mathcal{O}(1)$. In the limit of $\text{Da} \gg 1$, the large root m_1 must cancel the first two terms of (12.35), so that

$$\frac{\sigma}{\text{Da}} m_1^3 + \sigma m_1^2 \sim 0, \tag{12.39}$$

where we have assumed that σ is $\mathcal{O}(1)$, $\mathcal{K} \sim 1$, and $n/\text{Da}\mathcal{S} \ll \sigma$. Equation (12.39) then gives $m_1 = -\text{Da}$. This result is readily confirmed by comparison with the full model considered in the previous section, solved numerically. The m_1 eigenmode in $\exp(-\text{Da}z)$ decays sharply with z and hence does not affect $\mathcal{P}^{(1)}(z)$ outside of a narrow boundary layer that accomodates the boundary condition in (12.36b).

The large-Da solution is therefore controlled by the two remaining roots $m_{2,3}$. When Da is large and (12.39) is approximately satisfied, the remaining roots are given by

$$\left(\sigma\mathcal{K} - \frac{n}{\text{Da}\mathcal{S}}\right) m^2 - \left(\frac{n\mathcal{K}}{\mathcal{S}} + \frac{\sigma}{\text{Da}} k^2\right) m + (n - \sigma\mathcal{K})\, k^2 = 0. \tag{12.40}$$

We retain the term in k^2/Da because k^2 could be large. The two roots of this pure-real equation are complex conjugates and hence we can represent them as

$$m_{2,3} = a \pm ib, \tag{12.41}$$

with a the real part and b the imaginary part. The solution is still proportional to a sum over eigenfunctions $\exp(m_j z)$, but now we sum over only two of these, multiplied by coefficients A_2 and A_3.

Boundary conditions (12.36a) and (12.36c) remain to be satisfied. They must apply at all x and $t > 0$ and therefore, for simplicity of exposition, we don't write the x and t

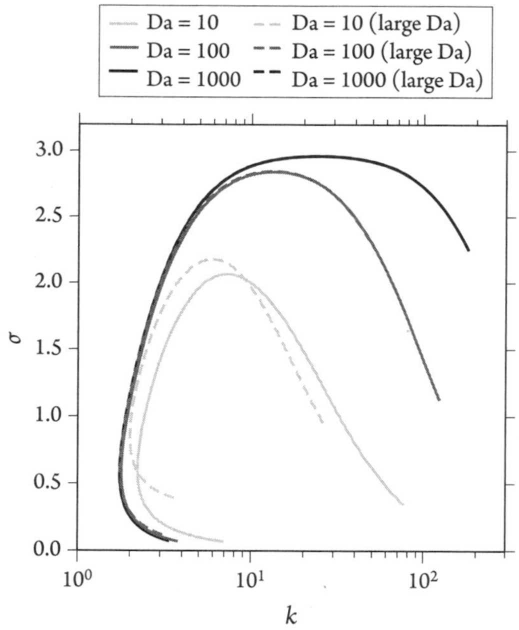

Figure 12.4. Dispersion curves for growth rate σ as a function of wavenumber k. Curves come from numerical solutions to the full problem (the cubic polynomial (12.35), solid lines) and the large-Da problem (the quadratic (12.40), dashed lines). The agreement for Da ≥ 100 suggests that the large-Damköhler approximation is very good for geologically relevant conditions.

dependencies of the eigenfunctions. The boundary condition $\mathcal{P}^{(1)}=0$ at $z=0$ tells us that $A_2=-A_3$, and hence we have $\mathcal{P}^{(1)}=\Re\left\{A_2\left[\exp(m_2z)-\exp(m_3z)\right]\right\}$. This can be rewritten as

$$\mathcal{P}^{(1)}=e^{az}\sin(bz), \tag{12.42}$$

where we have taken $A_2=-i/2$ to isolate the sine part that satisfies the boundary condition. This eigenfunction is monotonic in z for $0<b<\pi$ and becomes increasingly oscillatory in z for larger $b>\pi$.

The second boundary condition, $\partial_z\mathcal{P}^{(1)}=0$ at $z=1$, applied to this eigenmode, gives the constraint

$$a\sin(b)+b\cos(b)=0. \tag{12.43}$$

This equation can be rearranged as $\tan b=-b/a$, which has a trivial solution $b=0$ and an infinite set of roots at larger b. There is only one root $0<b<\pi$; this represents a channelized flow and turns out to be the fastest-growing mode.

For given parameters and at fixed k, the problem can be solved with numerical root finding. The solution is a value σ for which the real and imaginary parts of the roots of (12.40) satisfy (12.43); the latter equation replaces the condition $|\det(\mathbf{B})|=0$ for $\mathbf{B}$ given by (12.37). Dispersion curves are shown in figure 12.4; note the agreement of the large-Da solution with the full solution at Da ≥ 100.

12.5.4 A MODIFIED PROBLEM AND ITS ANALYTICAL SOLUTION

An analytical solution for the dispersion relationship can be obtained for a problem that is slightly modified from the one considered above. Although the problem is different, the analytical results will facilitate an understanding of how the solution scales with problem parameters. These scalings turn out to be identical to those of the stability analyses treated above.

To make the problem analytically tractable, we modify the boundary condition at $z = 1$. Instead of requiring $\partial_z \mathcal{P}^{(1)} = 0$, we require that the perturbation to the compaction pressure vanish on the upper boundary:

$$\mathcal{P}^{(1)} = 0 \text{ at } z = 1. \tag{12.44}$$

Applying this condition to the expression for the compaction pressure in equation (12.42) gives

$$b = l\pi \text{ for } l = 1, 2, 3... \tag{12.45}$$

For $l = 1$ we obtain the fundamental mode; higher values of l give zero-crossings $0 < z < 1$.

To solve for $m_{2,3}$ and hence for a and σ, we apply a variant of the quadratic formula to equation (12.40). The solution to a quadratic polynomial of the form $pm^2 + qm + r = 0$ is

$$m_{2,3} = \frac{-q}{2p} \pm i\sqrt{\frac{r}{p} - \left(\frac{q}{2p}\right)^2}. \tag{12.46}$$

Recalling that $m_{2,3} = a \pm ib$, we can identify a and b in this solution as

$$a = \frac{-q}{2p} = \frac{1}{2}\left(\frac{n\mathcal{K}/\mathcal{S} + \sigma k^2/\text{Da}}{\sigma\mathcal{K} - n/(\text{Da}\mathcal{S})}\right), \tag{12.47a}$$

$$b^2 + a^2 = \frac{r}{p} = \frac{(n - \sigma\mathcal{K})\,k^2}{\sigma\mathcal{K} - n/(\text{Da}\mathcal{S})}. \tag{12.47b}$$

We eliminate a and b by combining equations (12.47) with (12.45), taking $l = 1$ to select the fundamental (and fastest growing) mode that corresponds to channels. This gives an algebraic equation that can be solved for the growth rate σ as a function of wavenumber k and parameters n, Da, Pe, and $\mathcal{S}$ (recall that $\mathcal{K} = 1 + k^2/(\text{DaPe})$). Without further approximations, the growth rate of $l = 1$ perturbations is given by

$$\sigma = \pm\frac{n}{\mathcal{K}}\left\{\left[k^4\left(1 - \frac{\pi^2}{\text{Da}^4\mathcal{S}^2\mathcal{K}^2} - \frac{3}{\text{Da}\mathcal{S}}\right) - \frac{k^6}{\text{Da}^3\mathcal{S}\mathcal{K}^2} - k^2\left(\frac{\mathcal{K}^2}{\mathcal{S}^2} + \frac{2\pi^2}{\text{Da}^2\mathcal{S}^2}\right)\right.\right.$$
$$\left.\left. - \frac{\pi^2\mathcal{K}^2}{\mathcal{S}^2}\right]^{1/2} \pm \left(k^2 + \frac{2\pi^2 + k^2}{\text{Da}\mathcal{S}}\right)\right\}\left[2\left(\pi^2 + k^2\right) + \frac{k^4}{2\text{Da}^2\mathcal{K}^2}\right]^{-1}. \tag{12.48}$$

The $\pm$ symbols represent the upper and lower solution branches. Both branches are plotted in figure 12.5(a) for four values of the matrix stiffness parameter $\mathcal{S}$. The upper and lower branches connect where the dispersion curves are vertical. The maximum growth rate for each curve is marked with a star. An exploration of the parameter space of the dispersion relationship (12.48) reveals that, when the product $\text{Da}\mathcal{S} \gg 1$, the maximum growth rate $\sigma^* \lesssim n$. We exploit this observation below.

Eigenfunctions for the fastest-growing wavenumbers k^* are shown for $\mathcal{S} = 1, 0.1, 0.01$ in figure 12.5(b)–(d). For $\mathcal{S} \geq 1$, channels span the full height of the domain, whereas for $\mathcal{S} < 1$, the height of the unstable region decreases and emergent channels are confined to a layer near the top of the domain.

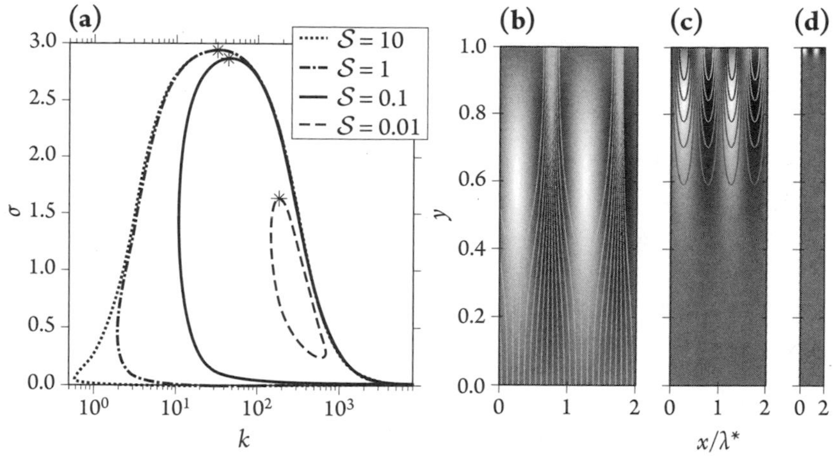

Figure 12.5. Results for the modified problem with $\mathcal{P}^{(1)} = 0$ at $z = 1$. **(a)** Dispersion curves for $n = 3$, $\mathrm{Da} = 1000$, $\mathrm{Pe} = 100$, and four values of $\mathcal{S}$. Maximum values of the growth rate for each curve are marked by stars. The eigenfunctions for each of these maxima are plotted in subsequent panels. **(b)** The $\mathcal{P}^{(1)}$ eigenfunction for $\mathcal{S} = 1$. White curves are streamlines of the flow $\mathbf{v}^{\ell} = \hat{\mathbf{z}} + \epsilon \mathbf{v}^{(1)}$, with ϵ chosen to be 3×10^{-3}. **(c)** $\mathcal{P}^{(1)}$ for $\mathcal{S} = 0.1$ with superimposed contours of the porosity perturbation $\phi^{(1)}$. Porosity is larger in the low-pressure channels. **(d)** $\mathcal{P}^{(1)}$ for $\mathcal{S} = 0.01$.

Equation (12.48) is exact but difficult to analyze. It would be helpful to have an approximate expression that could be differentiated with respect to k to obtain an expression for the peak growth rate σ^*. This would enable us to study how the character of the dominant mode scales with model parameters. To simplify (12.48), we consider the regime in which $\mathrm{Da}\mathcal{S} \gg 1$. It is reasonable to expect that the mantle is in this regime, even if $\mathcal{S} < 1$, because the Damköhler number is very large.

We therefore return to the system (12.47) with an interest in the behavior of the dispersion curve for $\sigma \sim \sigma^* \lesssim n$ in the limit of $\mathrm{Da}\mathcal{S} \gg 1$. Hence we (re)define the small parameter ϵ as

$$\sigma \sim n(1 - \epsilon), \quad \epsilon \ll 1 \tag{12.49}$$

and substitute, dropping the $n/\mathrm{Da}\mathcal{S}$ terms. Rewriting $(1-\epsilon)^{-1} \sim 1 + \epsilon + \mathcal{O}(\epsilon^2)$, substituting, simplifying, and truncating at the lowest possible order, we find that

$$a \sim \frac{1}{2\mathcal{S}} + \frac{k^2}{2\mathrm{Da}}, \tag{12.50a}$$

$$\epsilon \sim \frac{a^2 + b^2}{k^2} + \frac{k^2}{\mathrm{DaPe}}. \tag{12.50b}$$

Here we have made the additional approximation that near the maximum growth rate (i.e., for $k \approx k^*$ at which $\epsilon \ll 1$), $k^2/\mathrm{DaPe} \ll 1$ and hence that $\mathcal{K} \sim 1$. The expression for ϵ tells us that it is larger (and hence σ^* is smaller) for larger $b = l\pi$. This demonstrates that the mode with $l = 1$ grows the fastest. A comparison between the exact result (12.48) for growth rate σ (solid lines) and the approximation $\sigma \sim n(1 - \epsilon)$

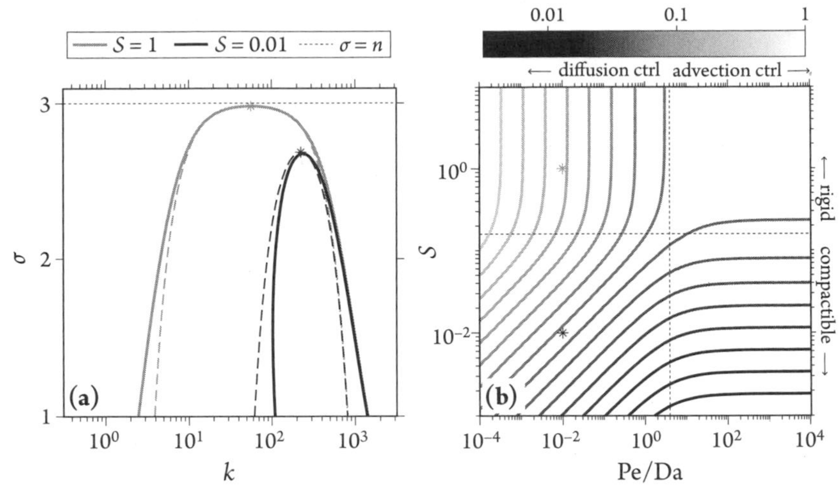

Figure 12.6. Properties of the dispersion curve near its maximum, with $n=3$, $\mathrm{Da}=10^4$, and $b=\pi$. **(a)** Comparison of the exact dispersion relation (12.48) (solid lines) with the asymptotic relations (12.49)–(12.50) (dashed lines) for $\mathrm{Pe}/\mathrm{Da}=10^{-2}$ and two values of $\mathcal{S}$, as given in the legend. **(b)** Contours of fastest-growing (nondimensional) wavelength λ^* for a range of $\mathcal{S}$ and Pe from equation (12.51). Dotted lines are at $\mathrm{Pe}/\mathrm{Da}=4$ and $\mathcal{S}=1/2\pi$. In both panels, stars indicate the maximum (k^*, σ^*) of the asymptotic curves, computed with (12.51) and (12.52).

(dashed lines) is shown in figure 12.6(a). Note that the approximation is designed to be valid only near the peak growth rate.

The maximum growth rate $\sigma \sim \sigma^*$ occurs where ϵ is at a minimum with respect to k. In particular, we seek the stationary point where $\mathrm{d}\epsilon/\mathrm{d}\left(k^2\right)=0$. Using equations (12.50) we find that

$$k^* \sim \left(\frac{4\mathrm{DaPe}\mathcal{B}}{4+\mathrm{Pe}/\mathrm{Da}}\right)^{1/4}, \tag{12.51}$$

where $\mathcal{B} \equiv b^2 + (2\mathcal{S})^{-2}$. The maximum growth rate σ^* of the channel instability is estimated by substituting k^* into equations (12.47), solving for ϵ^*, and forming $\sigma^* = n(1-\epsilon^*)$. The result is

$$\sigma^* \sim n\left[1 - 2\sqrt{\frac{\mathcal{B}(4+\mathrm{Pe}/\mathrm{Da})}{\mathrm{DaPe}}}\right]. \tag{12.52}$$

Recall that equations (12.51) and (12.52) are valid only in the limit $\mathrm{Da}\mathcal{S} \gg 1$. These results confirm that to leading order, the growth rate is controlled by the permeability, represented by n. They also represent how advection, diffusion, and compaction modify the dominant wavelength of instability and its growth rate via Da, Pe, and $\mathcal{B}(\mathcal{S})$, respectively.

It is instructive to consider the asymptotic regimes associated with Pe/Da and $\mathcal{S}$.

Diffusion-controlled regime. In the limit of finite Péclet number ($\mathrm{Pe} \ll \mathrm{Da}$), diffusion controls the wavelength selection and reduces the growth rate. Equation (12.51), written in terms of the most unstable wavelength λ^* and equation (12.52) contain

$$\lambda^* \propto (\mathcal{B}\mathrm{DaPe})^{-1/4}, \tag{12.53a}$$

$$\epsilon^* \propto (\mathcal{B}/\mathrm{DaPe})^{1/2}. \tag{12.53b}$$

Smaller Pe corresponds to faster diffusion. Equations (12.53) show that peak growth rate is reduced and the wavelength of the most unstable mode is increased by diffusion, although this latter sensitivity is very weak.

Figure 12.6(b) shows contours of the fastest-growing wavelength as a function of Pe/Da and stiffness parameter $\mathcal{S}$. The Damköhler number is fixed at 10^4 in this figure. The diffusion-controlled regime is the left half-plane.

When the matrix is rigid ($\mathcal{S} \gg 1$), $\mathcal{B} \sim b^2 = \pi^2$ and equations (12.51)–(12.52) become independent of the stiffness. On the other hand, in the limit of an easily compactible matrix ($\mathcal{S} \ll 1$), $\mathcal{B} \sim (2\mathcal{S})^{-2}$; in this case we find that $\lambda^* \propto \mathcal{S}^{1/2}$ and $\epsilon^* \propto \mathcal{S}^{-1}$. Hence a compactible system would produce more narrowly spaced channels with slower growth.

Advection-controlled regime. In the limit of infinite Péclet number ($\mathrm{Pe} \gg \mathrm{Da}$), advection controls the wavelength selection and equations (12.51)–(12.52) tell us

$$\lambda^* \propto \mathcal{B}^{-1/4}\mathrm{Da}^{-1/2}, \tag{12.54a}$$

$$\epsilon^* \propto \mathcal{B}^{1/2}\mathrm{Da}^{-1} \tag{12.54b}$$

The same rigid and compactible matrix regimes apply in this case. The advection-controlled regime is shown in the right half plane of figure 12.6(b) for fixed $\mathrm{Da} = 10^4$. Note that the fastest growing wavelength is independent of Pe/Da in this regime.

12.6 Physical Mechanisms

With the quantitative results gained by the stability analysis, it is instructive to return to a consideration of the equations governing perturbation growth to develop intuition for the physics. The dominant balances in these equations help to elucidate the mechanisms of growth. In discussing those balances, we first assume $\mathcal{S} \to \infty$, the rigid medium limit, whereby compaction is negligible relative to melting. Then the dominant balance of the modified Darcy's equation (12.25d) is

$$\phi^{(0)}w^{(1)} + \phi^{(1)}w^{(0)} = n\phi^{(1)}, \tag{12.55}$$

where we have simplified by using the base-state solution to replace $\left(\phi^{(0)}\right)^{n-1}$ with unity. From (12.55) we understand that it is the permeability, associated with the porosity exponent $n \geq 2$, that causes increased magmatic flux (left-hand side) where there is increased porosity (right-hand side). The dominant balance of conservation of species mass equation (12.25c) relates the flux to the undersaturation,

$$\phi^{(0)}w^{(1)} + \phi^{(1)}w^{(0)} = \chi^{(1)}, \tag{12.56}$$

where the nondimensional undersaturation is also the rate of reactive melting. From this we understand that increased vertical flux against the constant background solubility gradient (which has been scaled out of the equation) causes increased undersaturation and reactive melting. We can think of this as the advection of a *corrosive* liquid from below.

Combining (12.55) and (12.56) gives an expression for the undersaturation (equivalent to the nondimensional reactive melting rate) in terms of the porosity, $\chi^{(1)} = n\phi^{(1)}$. Substituting this into conservation of mass for the solid phase (12.25) with $\partial_t \to \sigma$ gives $\sigma\phi^{(1)} = n\phi^{(1)}$, where we have again assumed a rigid matrix and hence neglected compaction. This equation gives us the key result

$$\sigma = n \qquad \text{for } \mathcal{S} \to \infty. \tag{12.57}$$

This is a dimensionless equation; the dimensional growth rate has units of inverse time. Hence, using (12.14), the dimensional growth rate is $\sigma = nw_0\beta/\alpha$. This tells us that the emergence of channelized flow requires both of the following destabilizing factors: (*i*) permeability that increases with porosity and (*ii*) a background solubility gradient such that buoyantly upwelling liquid causes reactive melting. We compute a numerical value for the growth rate in section 12.7 below.

Further examination of equation (12.25c) reveals two stabilizing mechanisms that reduce the undersaturation and hence the propensity for channelization. The first is lateral diffusion of undersaturation ($\chi_{xx}^{(1)}/\text{DaPe}$), which spreads corrosivity from the channels to the interchannel regions. Smaller Péclet number favors diffusion. Moreover, this term scales with k^2 and hence reduces the growth rate most effectively at smaller wavelengths.

The second is advection of undersaturation by the base-state flow. In channels, the undersaturation increases upward ($\chi_z^{(1)}/\text{Da} > 0$); vertical advection thus tends to reduce undersaturation by transporting it out of the domain. Smaller Damköhler number magnifies this effect. It can be shown that this term is also wavelength dependent and that the mechanism is most effective at small wavelengths. Hence both base-state advection and diffusion of chemical perturbations decrease the growth rate at large wavenumber and contribute to wavelength selection.

Compaction is a third stabilizing mechanism, affecting the growth rate through mass conservation equation (12.31a). Compaction is driven by a liquid underpressure, $\mathcal{P} \sim (P^\ell - P^s) < 0$. Underpressure is associated with convergence of liquid flow, which is a characteristic of channels. Indeed, the vertical increase of porosity within channels is associated with convergent liquid flow, $-\nabla \cdot \boldsymbol{v}^{(1)}$, according to liquid mass conservation equation (12.31b). Combining that equation with the divergence of Darcy's law (12.31d) gives

$$\nabla^2 \mathcal{P}^{(1)} = \frac{n}{\mathcal{S}} \left(-\nabla \cdot \boldsymbol{v}^{(1)} \right). \tag{12.58}$$

This equation states that convergence of liquid flow must be driven by a positive curvature in the compaction pressure field, and hence by a region of negative compaction pressure. It also states that the amplitude of fluctuations in compaction pressure is large when the matrix stiffness $\mathcal{S}$ is small. Returning to equation (12.31a), negative compaction pressure suppresses porosity growth in the channels, in competition with undersaturation, which promotes growth. Hence it is the relative importance of compaction versus reactive melting that determines the stability of the system, and $\mathcal{S}$ is a control on this balance. Indeed there is a critical stiffness (not derived here),

$$\mathcal{S}_{\text{crit}} \sim \max\left(1/\text{Da},\ 2/\sqrt{\text{DaPe}}\right) \tag{12.59}$$

below which the reaction–infiltration instability is suppressed at all wavelengths.

12.7 Application to the Mantle

This discussion is simplified from Rees Jones and Katz [2018], which followed from Aharonov et al. [1995]. Based on results above, we can assess the conditions under which channels might emerge and the dimensions that they would have. However, this is an area of active research; the discussion below should therefore not be considered definitive but rather illustrative.

Table 12.1 shows the parameter values and ranges that are appropriate for decompression melting beneath mid-ocean ridges. We can immediately note that $\mathcal{M}$ and $\mathcal{S}$ are both small (and hence $\mathcal{F} \approx 1$). It is therefore reasonable to assume a uniform base state, as we did in section 12.5.1. Furthermore, since $\mathcal{S} \ll 1$, channels will grow in the compaction-limited regime, rather than in the rigid regime. We also note that although we have $\text{Pe} \gg \text{Da}$ in our reference values, the ranges of these parameters have a large overlap.

We therefore consider both of the growth-limiting mechanisms: diffusion and advection. We can introduce a reactive length scale that covers each of these cases

$$L_{\text{eq}} = \begin{cases} L_w \equiv \frac{\phi_0 w_0}{\alpha \mathcal{R}}, & \text{Pe} \gg \text{Da (advection-controlled)}, \\ L_{\mathcal{D}} \equiv 2\left(\frac{\phi_0 \mathcal{D}}{\alpha \mathcal{R}}\right)^{1/2}, & \text{Pe} \ll \text{Da (diffusion-controlled)}, \end{cases} \tag{12.60}$$

Table 12.1. Parameters and their values (with ranges in parentheses) for application to the mantle beneath mid-ocean ridges. These follow Rees Jones and Katz [2018] (who largely followed Aharonov et al. [1995]).

Variable (unit)	Symbol	Estimate (range)
Permeability exponent	n	3 (2 to 3)
Solubility gradient (m^{-1})	β	2×10^{-6} (10^{-6} to 4×10^{-6})
Compositional offset	α	1
Melting region depth (m)	H	8×10^4
Compaction length (m)	δ	10^3 (3×10^2 to 10^4)
Melt flux (m s^{-1})	$\phi_0 w_0$	3×10^{-11} (5×10^{-12} to 2×10^{-10})
Diffusivity (m^2s^{-1})	$\phi_0 \mathcal{D}$	3×10^{-14} (10^{-15} to 10^{-12})
Reaction rate (s^{-1})	$\mathcal{R}$	3×10^{-8} (10^{-11} to 10^{-4})
Decompression melting rate (s^{-1})	$W_0 \left.\frac{dF}{dz}\right\|_s$	3×10^{-15} (10^{-15} to 10^{-14})
Melt productivity ratio	$\mathcal{G}$	45 (1 to 200)
Reactive melt volume	$\mathcal{M}$	0.14 (0.07 to 0.28)
Péclet number	Pe	7×10^9 (3.5×10^7 to 1.4×10^{12})
Damköhler number	Da	7×10^7 (3.5×10^3 to 1.4×10^{12})
Stiffness number	$\mathcal{S}$	3×10^{-5} (1×10^{-6} to 6×10^{-3})

where L_w is the distance that the soluble component is transported by base-state advection in the liquid over the reaction time scale and $L_{\mathcal{D}}$ is the distance that the soluble component can diffuse over the reaction time scale. Since Pe $\gg$ Da corresponds to $L_w \gg L_{\mathcal{D}}$, we can write (12.60) as $L_{\text{eq}} \sim \max(L_w, L_{\mathcal{D}})$. For small $\mathcal{S}$, the most unstable wavelength is

$$\lambda^* = 2\pi\delta \left(L_{\text{eq}}\beta/\alpha\right)^{1/2} \tag{12.61}$$

in dimensional terms. This equation states that the fastest growing wavelength is smaller than the compaction length by a factor $2\pi \left(L_{\text{eq}}\beta/\alpha\right)^{1/2}$. Using the values from table 12.1 and L_{eq} from equation (12.60), this factor falls in an interval between 10^{-5} and 10^{-1}. Therefore, for a compaction length of 1 km, we expect emergence of channels with a wavelength of between 1 cm and 100 m.

Comparison of these wavelengths with the geological record is not straightforward. The tabular dunite bodies observed in ophiolites and interpreted as the residue of reactive melting may not represent the dimensions of the emergent instability. Instead, they may mark the zone through which the reactive flux has swept, dissolving away the pyroxene and precipitating olivine. Finite-time solutions to the nonlinear theory are required, as is a detailed understanding of the geological observations.

The dimensional growth rate of the fastest-growing mode is

$$\sigma^* \sim n\frac{w_0\beta}{\alpha}\left(1 - \frac{L_{\text{eq}}\alpha/\beta}{\delta^2}\right) \quad \text{for } \mathcal{S}_{\text{crit}} < \mathcal{S} \ll 1. \tag{12.62}$$

Comparing this with the dimensional version of equation (12.57), we see that the effect of compaction is expressed by the second term in parentheses. Using values from table 12.1, we compute a characteristic time scale of channel growth as $1/\sigma^*$ of about 2 million years. For this process to be viable as an explanation for the dunites observed in ophiolites, the time scale for growth should be similar to or shorter than the time to advect mantle rock through the melting region. Assuming a melting column height of 80 km beneath a mid-ocean ridge axis and an upwelling rate of 4 cm/y gives an advection time of 2 million years.

Figure 12.7 shows the time scale for channel growth as a function of horizontal channel wavelength. It was computed using the set of preferred parameter values in table 12.1. The shape of the curve suggests that channels will emerge having widths in a finite range. For the chosen set of parameters, wavelengths between 3 cm and 3 m all have a characteristic growth time within 2% of the minimum time.

12.8 Literature Notes

Quick [1982] was the earliest paper proposing a reactive origin for tabular dunite bodies observed in ophiolites. The case was supported by detailed geological, petrological, and geochemical observations by Kelemen et al. [1992] and Kelemen et al. [1995a] of the Oman ophiolite. Characteristics of the tabular dunites of Oman were measured and discussed by Kelemen et al. [2000] and Braun and Kelemen [2002]. The igneous petrology of reactive melting was explored in detail by Kelemen [1990] and later by Longhi [2002].

There is substantial literature on the theory of reaction–infiltration instabilities. The book by Ortoleva [1994] reviews much of it. Early work mainly focused on infiltration

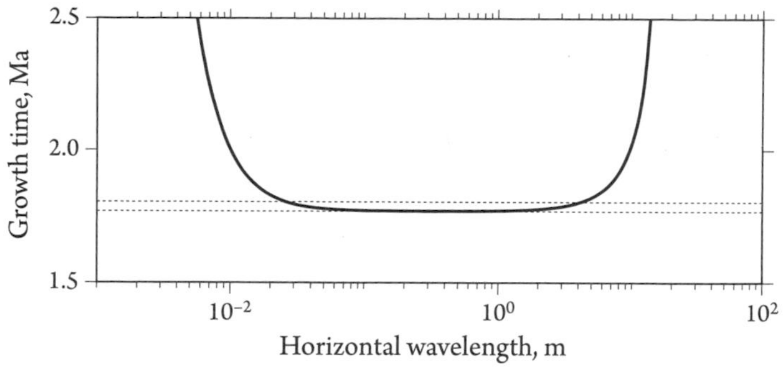

Figure 12.7. The time scale of channel growth $1/\sigma$ as a function of the horizontal wavelength of channels. This curve is computed using the full dispersion relation (12.48) with preferred parameter values from table 12.1. Horizontal dotted lines mark the minimum growth time ($\sigma = 1/n$, in nondimensional terms) and this value plus 2%.

instabilities at a dissolution front [e.g., Chadam et al., 1986; Hinch and Bhatt, 1990]; this remains a topic of interest [e.g., Szymczak and Ladd, 2013, 2014].

Aharonov et al. [1995] were the first to adapt reaction–infiltration theory to study the problem of magma segregation and obtained many of the key results discussed here. Spiegelman et al. [2001] built on their work with a revised stability analysis and numerical simulations. Whereas in stability analysis coalescence of channels can only be considered to be speculative, the numerical models demonstrate the relevance of coalescence [Spiegelman et al., 2001]. These same models indicate that the dynamic liquid underpressure in channels, if allowed to feed back through the equilibrium composition onto the melting rate, can promote channelization and channel coalescence. This intriguing result has not been reproduced. High-resolution models were developed [Schiemenz et al., 2011] and analyzed by Liang et al. [2010] with a focus on how dissolution can create dunite bodies by fully removing pyroxene. Hesse et al. [2011] returned to stability analyses but considered a system in chemical equilibrium. The focus of that paper was on compaction–dissolution waves, but channelization was also considered. Compaction–dissolution waves were first studied by Aharonov et al. [1995].

The development presented in this chapter closely follows Rees Jones and Katz [2018], who revisit the stability analysis of Aharonov et al. [1995]. Rees Jones and Katz [2018] extend the results of earlier work by using asymptotic analysis to explore the full parameter space. In particular, they investigate the small-$\mathcal{S}$ regime that is relevant for the mantle but difficult to access with numerical methods. Rees Jones et al. [2020] build on this analysis by incorporating a background, pure-shear flow of the solid matrix. This formulation combines the reaction–infiltration instability with the shear-driven instability of chapter 7 into a combined analysis. The results suggest that even a minor influence of matrix shear can promote the formation of tabular (rather than tubular) channels. Rees Jones et al. [2020] also embed the stability analysis into the context of corner-flow deformation associated with a mid-ocean ridge.

Laboratory experiments have played a role in our qualitative understanding of the reaction–infiltration instability, although they have not yet led to quantitative constraints. Hoefner and Fogler [1988] conducted experiments on limestone dissolution by acid. Kelemen et al. [1995b] constructed a reactive flow of water into a pack of glass balls and salt. Experiments have also been conducted on mantle materials at high temperature and pressure. Daines and Kohlstedt [1994] observed reaction between undersaturated basalts and pyroxene-bearing peridotites in experiments where melt flow was driven by surface energy. The reaction front showed a fingering instability, but because of the minimal melt flow in these experiments, it is unclear whether this represents channelization by reactive flow. Pec et al. [2015, 2017, 2020] created a pressure gradient to force undersaturated basaltic melt through peridotite under various conditions and documented reactive dissolution of pyroxenes, precipitation of olivine, and channel formation. The interaction of reactive infiltration with matrix deformation was investigated in experiments by King et al. [2011b].

A few theoretical and computational studies have extended the analysis conducted by Rees Jones and Katz [2018] and others to include additional physics. Hewitt [2010] introduced energy conservation and found that latent heat tends to reduce the reaction rate and stabilize the instability.[2] Weatherley and Katz [2012] and Katz and Weatherley [2012] looked at the role of finite perturbations in the form of mantle heterogeneity. Keller and Katz [2016] extended the petrological model to include volatile-enriched melting and showed that channelization can occur at the transition from volatile-induced to silicate melting.

Geochemical consequences of channelized melt transport were anticipated by the earliest workers in this area. These were clearly demonstrated and explored in numerical models by Spiegelman and Kelemen [2003]. Jull et al. [2002] and Elliott and Spiegelman [2003] presented models of uranium-series disequilibrium under channelized melt transport. All of this work reaches the fundamental conclusion that channelized melt transport can create large geochemical variations in the magmas delivered to the base of the crust. These variations hypothetically explain much of the geochemical variance in mid-ocean ridge and oceanic island lavas.

12.9 Exercises

12.1 Recall that in subsection 12.5.4 we showed that $m_{2,3} = a \pm ib$, where

$$a = \frac{-q}{2p} = \frac{1}{2}\left(\frac{n\mathcal{K}/\mathcal{S} + \sigma k^2/\mathrm{Da}}{\sigma\mathcal{K} - n/(\mathrm{Da}\mathcal{S})}\right), \tag{12.63}$$

$$b^2 + a^2 = \frac{r}{p} = \frac{(n - \sigma\mathcal{K})\,k^2}{\sigma\mathcal{K} - n/(\mathrm{Da}\mathcal{S})}. \tag{12.64}$$

Consider the regime in which $\mathrm{Da}\mathcal{S} \gg 1$, $\sigma \sim n(1-\epsilon)$, $\epsilon \ll 1$, and $\frac{k^2}{\mathrm{DaPe}} \sim \mathcal{O}(\epsilon)$. Assuming these approximations, derive (12.50).

12.2 Considering the algebraic expression for the growth rate (12.48) in the limit Da, Pe $\to \infty$ and $k \gg \pi$, derive an approximation for σ/n. Deduce that in this limit there is no high wavenumber cutoff.

12.3 In this question we examine the effect of nonconstant compaction viscosity on the stability of the system. Consider a dimensional compaction viscosity of the form $\zeta_\phi = \zeta_0 \frac{\phi_0}{\phi}$.

[2] Appendix C of Rees Jones and Katz [2018] explains that this stabilization is mostly a consequence of the use of a compaction viscosity $\zeta_\phi \propto \phi^{-1}$ (see exercise 12.3). Energy conservation also plays a role.

(a) Derive the equations for the base state and the first-order perturbation.
(b) Assuming constant base state (12.30), find the new growth rate σ in terms of the growth rate σ_0 of the system with constant compaction viscosity.
(c) What is the condition for instability to occur?
(d) Does the variable compaction viscosity reduce or increase the occurrence of instabilities?
(e) How does the system behave (to the first order) if $\zeta_\phi = \zeta_0 \left(1 - \log \frac{\phi}{\phi_0}\right)$?

CHAPTER 13

Tectonic-Scale Models and Modeling Tools

Partial melting of the mantle is closely linked to mantle flow at plate tectonic boundaries. The volcanos that are always associated with mid-ocean ridges and subduction zones attest to this fact. The melting region beneath these volcanos is of order 100 km in depth and width, and extends almost continuously along the plate boundary. This 100-km scale of partial melting is also probably characteristic of the diameter of mantle plumes, although some plumes may have a structure that is more complicated and broadly distributed. In this chapter our focus is on mid-ocean ridges. There is more published research on magma/mantle dynamics at mid-ocean ridges than on plumes or subduction zones. Hence their dynamics are better understood, although major open questions remain.

The theory developed in this book should apply to the partially molten mantle beneath tectonic boundaries, although its application remains a challenge. The main difficulties stem from the disparate temporal and spatial scales associated with mantle and magmatic processes. The mantle moves at speeds of a few centimeters per year; it therefore moves 100 km in about one million years. Magmatic rise due to buoyancy may be at speeds of several to tens of *meters* per year; hence it can travel vertically more than 100 km in thousands to tens of thousands of years. Furthermore, as we have noted in the discussions above, the characteristic length scale of emergent magmatic features such as channels is on the order of the compaction length, which is expected to be $\sim$1 km in the mantle. Models must therefore resolve processes that range over a factor of at least 100 in both time scale and length scale.

There are multiple ways to address this separation of scales. One is to develop judicious simplifications that enable approximate analytical solutions. Another is to volume-average features at the compaction length scale to derive an upscaled representation of their properties. A third and complementary approach is to develop numerical methods and codes that enable high-resolution computer simulation of magma/mantle dynamics at the tectonic scale. In this chapter we consider examples of the analytical approach and theoretical tools for the numerical approach. We analyze simplified models of magma focusing at mid-ocean ridges. We develop equations for magma dynamics in the limit of small porosity, where the bulk momentum equation reduces to Stokes equations for incompressible flow. And we outline the enthalpy method, which facilitates coupled solution of the dynamics and thermochemistry by using simplifications that come from the assumption of thermochemical equilibrium. Thus, our focus remains on gaining insight into magma/mantle dynamics through mathematical analysis; the reader is referred to the literature for detailed discussion of tectonic-scale numerical models. An introductory consideration of numerical methods is the subject of chapter 14.

13.1 Governing Equations in the Small-Porosity Approximation

To analyze the governing equations for problems at the tectonic scale, it is helpful to introduce decompositions that break the problem into smaller parts. These parts can then be treated as approximately independent of each other, reducing the nonlinear coupling between equations. We return to the system of equations representing conservation of mass and momentum (4.46) and decompose two of the variables:

$$\nabla P^\ell \equiv \nabla P + \nabla \mathcal{P} + \rho^s \mathbf{g}, \tag{13.1a}$$

$$\boldsymbol{v}^s \equiv \boldsymbol{v} + \boldsymbol{v}^{\mathcal{C}}. \tag{13.1b}$$

In the pressure decomposition, expressed in terms of gradients to avoid specifying a coordinate system, we have written the liquid pressure gradient as the sum of a gradient of dynamic pressure P, a gradient of compaction pressure[1] $\mathcal{P} \equiv \left(\zeta_\phi - \frac{2}{3}\eta_\phi\right)\mathcal{C}$ and an approximate lithostatic pressure gradient $\rho^s\mathbf{g}$. The true static pressure gradient will deviate from the lithostatic pressure gradient due to porosity and variations in phase density, but this is a good approximation.

The solid velocity is decomposed into two parts, corresponding to the Helmholtz decomposition of equation (4.51). The incompressible part is $\boldsymbol{v}$ and the irrotational part is $\boldsymbol{v}^{\mathcal{C}}$. Thus,

$$\nabla\cdot \boldsymbol{v}^s = \nabla\cdot \boldsymbol{v}^{\mathcal{C}} \equiv \mathcal{C}, \tag{13.2a}$$

$$\nabla\times \boldsymbol{v}^s = \nabla\times \boldsymbol{v} \equiv \boldsymbol{\omega}. \tag{13.2b}$$

In the latter equation, we have defined the vorticity, although we do not use it below.

Substituting the decompositions (13.1) into the system (4.46) we have

$$-\nabla P + \nabla\cdot \eta_\phi\left[\nabla\boldsymbol{v} + (\nabla\boldsymbol{v})^\mathsf{T}\right] - \phi\Delta\rho\mathbf{g} = \mathbf{0}, \tag{13.3a}$$

$$\phi\left(\boldsymbol{v}^\ell - \boldsymbol{v} - \boldsymbol{v}^{\mathcal{C}}\right) + \frac{k_\phi}{\mu^\ell}\left(\nabla P + \nabla\mathcal{P} + \Delta\rho\mathbf{g}\right) = \mathbf{0}, \tag{13.3b}$$

$$\frac{\mathrm{D}_s\phi}{\mathrm{D}t} - \Gamma/\rho^s - (1-\phi)\mathcal{C} = 0, \tag{13.3c}$$

$$\nabla\cdot\left[\phi\boldsymbol{v}^\ell + (1-\phi)\left(\boldsymbol{v} + \boldsymbol{v}^{\mathcal{C}}\right)\right] = 0. \tag{13.3d}$$

We rescale variables to highlight the dominant balances in these equations. The choice of scales is guided by the form of the equations and constitutive laws, by the intuition gained from the analyses developed earlier in this book and by foresight of the analysis to be introduced below.

[1] Note that this definition of the compaction pressure is different from that introduced in section 4.3.3 but, as we shall find below, this difference is quantitatively small.

$$[\phi] = \phi_0, \quad \left[\boldsymbol{v}^\ell, \boldsymbol{v}\right] = U_0, \quad \left[\boldsymbol{v}^C\right] = \phi_0 U_0, \quad [\mathcal{C}] = \phi_0 U_0/L,$$
$$[\boldsymbol{x}] = L, \quad [t] = U_0/L, \quad \left[\eta_\phi\right] = \eta_0, \quad \left[\zeta_\phi\right] = \eta_0/\phi_0, \tag{13.4}$$
$$[P] = \eta_0 U_0/L, \quad [\mathcal{P}] = \Delta\rho g L, \quad [\Gamma] = \phi_0 \rho U_0/L,$$

where the length scale L is given by

$$L \equiv \left(\frac{\eta_0 U_0}{\Delta\rho g}\right)^{1/2}. \tag{13.5}$$

We discuss the meaning of this length scale below in section 13.2.

The nondimensional governing equations are

$$-\nabla P + \nabla \cdot \eta_\phi \left[\nabla \boldsymbol{v} + (\nabla \boldsymbol{v})^\intercal\right] = \phi_0 \, \phi \hat{\mathbf{g}} \text{ (with } \nabla \cdot \boldsymbol{v} = 0), \tag{13.6a}$$

$$\phi \left(\boldsymbol{v}^\ell - \boldsymbol{v}\right) + \phi^n \left(\nabla P + \nabla \mathcal{P} + \hat{\mathbf{g}}\right) = \phi_0 \, \phi \boldsymbol{v}^C, \tag{13.6b}$$

$$\frac{\partial \phi}{\partial t} + \boldsymbol{v} \cdot \nabla \phi - \Gamma - \mathcal{C} = -\phi_0 \left(\phi \mathcal{C} + \boldsymbol{v}^C \cdot \nabla \phi\right), \tag{13.6c}$$

$$\mathcal{C} + \nabla \cdot \phi \left(\boldsymbol{v}^\ell - \boldsymbol{v}\right) = \phi_0 \nabla \cdot \phi \boldsymbol{v}^C, \tag{13.6d}$$

$$\mathcal{P} - \zeta_\phi \mathcal{C} = -\phi_0 \, \tfrac{2}{3} \eta_\phi \mathcal{C}, \tag{13.6e}$$

where we have used the dimensional permeability $k_\phi = k_0(\phi/\phi_0)^n$. On the left-hand side of each equation of (13.6) are the respective leading-order terms; all symbols are dimensionless and $\mathcal{O}(1)$ except for $\boldsymbol{v}^\ell$, which is $\mathcal{O}(1/v_r)$, where

$$v_r \equiv \frac{U_0}{w_0} = \frac{\phi_0 \mu^\ell U_0}{k_0 \Delta\rho g}. \tag{13.7}$$

This velocity ratio is typically much less than unity.

On the right-hand side of each equation of (13.6), all terms are $\mathcal{O}(\phi_0) \ll 1$. The physics represented by each of these equations is adequately captured on the left-hand sides except in (13.6a), where the right-hand side represents the buoyancy of the two-phase aggregate relative to the lithostatic pressure gradient $\rho\mathbf{g}$. For all the other equations of the system (13.6), it is reasonable to drop the terms on the right-hand side. We retain the buoyancy term in (13.6a) if porosity-driven convection is potentially relevant.

Equations (13.6) at leading order have a significant feature that should be noted. The dynamic pressure P and solenoidal solid velocity $\boldsymbol{v}$ are obtained by solving (13.6a), given a spatial distribution of shear viscosity η_ϕ. This is the solution to the incompressible Stokes equations, a problem for which numerical software is readily available. Dropping terms of $\mathcal{O}(\phi_0)$, (13.6a) can be treated as decoupled from the rest of the system (except via the viscosity). Then equations (13.6b)–(13.6e) must be solved for the compaction rate, compaction pressure, liquid velocity, and evolution of porosity. If the porosity-dependence of the incompressible flow is not negligible, but varies slowly relative to the compaction-related variables (or if it can be neglected entirely), equation (13.6a) can be solved at much larger time intervals than the system (13.6b)–(13.6e). If the porosity dependence in (13.6a) requires synchronous time-stepping with the compaction problem, the solution of the coupled system can be approximated by iterating between the solution of (13.6a) and of (13.6b)–(13.6e).

13.2 Corner-Flow with Magmatic Segregation

Far from transform faults, ridges are relatively uniform along the strike of their central axis and relatively symmetric across it. These properties suggest that two-dimensional models can capture some of the fundamental features of mid-ocean ridge processes. We have already introduced a single-phase model of mantle flow beneath mid-ocean ridges in section 3.3. In this section we extend that model to include the liquid phase and investigate how buoyancy and dynamic pressure gradients drive magmatic segregation in the context of the kinematic corner-flow model. The analysis follows Spiegelman and M^c^Kenzie [1987] and Morgan [1987].

Our approach is to employ the small-porosity approximation from section 13.1 above. This allows us to use the single-phase, kinematic corner-flow solution from section 3.3 to compute the dynamic pressure gradient that is among the drivers of liquid segregation. To obtain an analytical expression for the liquid velocity, we simplify even further by assuming uniform porosity $\phi = 1$ (rescaled by the reference value ϕ_0) and no melting $\Gamma = 0$. Furthermore, we consider a semi-infinite, two-dimensional half-space. Under this highly simplified scenario, there are no impermeable boundaries or variations in porosity or permeability; it is therefore reasonable to neglect the gradients in compaction pressure that would localize near such boundaries. Finally, we consider only the instantaneous solution at $t = 0$.

This simplified problem is governed by the system

$$\left.\begin{aligned} \nabla P &= \nabla^2 \boldsymbol{v} \\ \nabla \cdot \boldsymbol{v} &= 0 \end{aligned}\right\} \quad \text{Corner-flow problem of section 3.3,} \tag{13.8a}$$

$$\boldsymbol{v}^\ell = \boldsymbol{v} - \frac{1}{\mathcal{V}_r}\left(\nabla P + \hat{\mathbf{g}}\right). \tag{13.8b}$$

We have already obtained a solution to (13.8a) in section 3.3. That solution, put into nondimensional form using scales (13.4), is

$$f(\theta) = A \sin\theta - B\theta \cos\theta, \tag{13.9a}$$

$$\boldsymbol{v} = \left[f'(\theta)\hat{\boldsymbol{r}} - f(\theta)\hat{\boldsymbol{\theta}}\right], \tag{13.9b}$$

$$P = -\frac{1}{r}\left[f'''(\theta) + f'(\theta)\right] = -\frac{2B}{r}\cos\theta, \tag{13.9c}$$

where

$$A = \frac{2\sin^2\theta_p}{\pi - 2\theta_p - \sin 2\theta_2} \quad \text{and} \quad B = \frac{2}{\pi - 2\theta_p - \sin 2\theta_2}. \tag{13.10}$$

Recall that θ_p is the angle between the horizontal and the base of the rigid wedge that represents the lithosphere, as shown in figure 3.1. Using the cylindrical coordinate system defined in that figure,

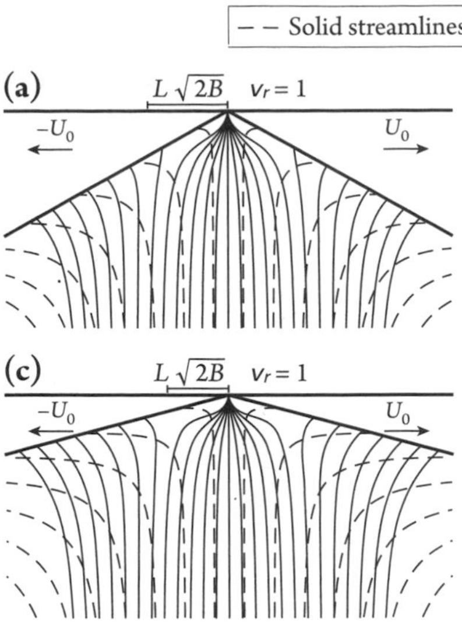

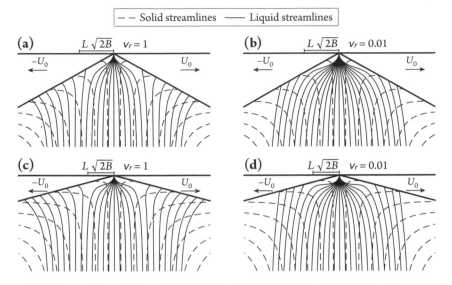

Figure 13.1. Streamlines of the solid (13.9b) and liquid (13.13) velocities for the corner flow solution with uniform viscosity and porosity. The domain has a fixed size in dimensional units. The top row of panels has a larger lithospheric wedge angle ($\theta_p = 30°$) than the bottom row ($\theta_p = 15°$). The left column has 100× slower melt segregation ($v_r = 1$) than the right column ($v_r = 0.01$).

$$\hat{\mathbf{g}} = \cos\theta\hat{\boldsymbol{r}} - \sin\theta\hat{\boldsymbol{\theta}}, \tag{13.11}$$

$$\nabla P = \frac{\partial P}{\partial r}\hat{\boldsymbol{r}} + \frac{1}{r}\frac{\partial P}{\partial \theta}\hat{\boldsymbol{\theta}}. \tag{13.12}$$

The liquid velocity is then

$$\boldsymbol{v}^\ell = \left(f'(\theta)\hat{\boldsymbol{r}} - f(\theta)\hat{\boldsymbol{\theta}}\right) - \frac{1}{v_r}\left\{\frac{1}{r^2}\left[\left(f''' + f'\right)\hat{\boldsymbol{r}} - \left(f'''' + f''\right)\hat{\boldsymbol{\theta}}\right] + \cos\theta\hat{\boldsymbol{r}} + \sin\theta\hat{\boldsymbol{\theta}}\right\}. \tag{13.13}$$

The streamlines of the solid velocity $\boldsymbol{v}$ and liquid velocity $\boldsymbol{v}^\ell$ are shown in figure 13.1. The basic observation is that liquid streamlines rise vertically until they are deflected either away from or toward the ridge axis. The initially vertical rise is due to the buoyancy of melt. Lateral deflection toward the ridge axis is caused by the strong suction (low pressures) at the axis where the solid deformation of corner flow is most intense; lateral deflection away from the ridge axis is a consequence of transport by the solid flow-field. It is evident from the figure that L represents a length scale over which liquid focusing occurs. This length scale can also be inferred from equations (13.13) and (13.9c), which together show that the radial component of the ∇P decreases as $2L^2B/r^2$ in dimensional terms; hence the characteristic decay length is $L\sqrt{2B}$.

The panels of figure 13.1 show four combinations of v_r and θ_p. Larger wedge angle θ_p corresponds to a stronger suction at the ridge axis and a larger focusing distance. Larger $v_r = U_0/w_0$ corresponds to slower melt segregation and hence stronger deflection by the solid flow. However, even for $v_r = 1$, deflection by the solid flow is negligible and liquid flow is focused toward the ridge axis.

How large is L in the mantle? Over what distance might this focusing operate? We can evaluate L using characteristic asthenospheric properties of $\eta_0 \approx 10^{19}$–10^{20} Pa-sec, $U_0 \approx 10$ cm/yr, and $\Delta\rho \approx 500$ kg/m^3. This gives $L \approx 2.5$–8 km, scarcely wider than the volcanic zone of most mid-ocean ridges. This mechanism is therefore not considered to be an effective means of melt focusing toward the ridge axis. In the next section we consider what may be a more effective mechanism.

13.3 Melt Focusing through a Sublithospheric Channel

In this section we take two approaches to modeling sublithospheric transport in a decompaction channel. The first of these is an extension of the mechanical decompaction column model of section 11.4, while the second one builds on a thermal basis.

13.3.1 LATERAL TRANSPORT IN SEMI-INFINITE HALF-SPACE

Before developing a model of a mid-ocean ridge, it is useful to consider the extension of the column model from section 11.3 that produced a decompaction boundary layer.[2] Here we consider that boundary layer as the top of a semi-infinite half-space shown in figure 13.2. Our coordinate system is aligned with the upper boundary such that x is parallel to the boundary. The gravity vector is tilted at an angle α to the negative z-direction. This means that the decompaction layer is inclined by α. We therefore refer to the direction parallel to the boundary as the "lateral" direction to make an important distinction from the horizontal direction.

Writing $\boldsymbol{v}^\ell = u\hat{\boldsymbol{x}} + w\hat{\boldsymbol{z}}$ and referring back to the system of equations (11.51), the tilted coordinate system compels us to write the momentum balance of the liquid phase as

$$\phi u = -K_\phi \left[\xi_0 \frac{\partial C}{\partial x} - \Delta\rho g \sin\alpha \right], \tag{13.14a}$$

$$\phi w = -K_\phi \left[\xi_0 \frac{\partial C}{\partial z} - \Delta\rho g \cos\alpha \right]. \tag{13.14b}$$

These equations can be simplified by our construction of the problem. Use of the semi-infinite half-space requires translational invariance in the x and y directions and hence that $\partial C/\partial x = 0$. Indeed, the derivative of any quantity with respect to x or y vanishes in this context. Furthermore, by the same logic, the mass conservation equations again reduce to their one-dimensional form given by (11.51a) and (11.51b). The solution for porosity obtained in section 11.3 still holds, but with the gravity reduced by a factor of $\cos\alpha$.

Then, using the standard expression for mobility K_ϕ, equation (13.14a) becomes

$$\phi u(z/\delta_f) = q_\infty \sin\alpha \, (\phi/\phi_0)^n. \tag{13.15}$$

[2] Understanding the current section requires a clear recollection of the material in section 11.3.

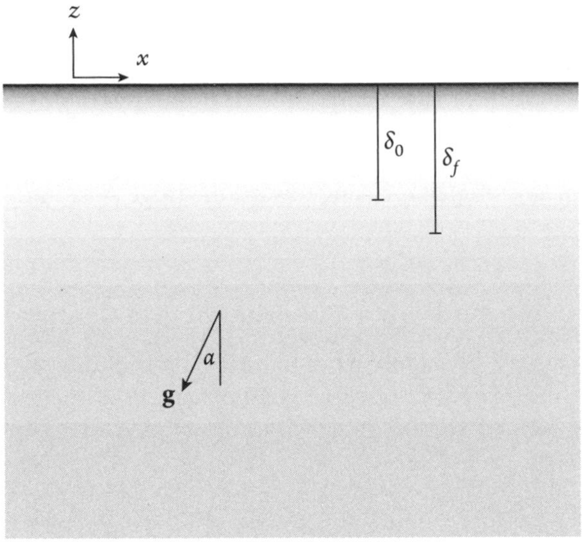

Figure 13.2. Schematic diagram of the infinite decompaction channel in a semi-infinite half space. Gray scale shows the normalized porosity from equation (11.57) for $R_\delta = 0.77$. The coordinate system is aligned with the boundary of the half-space and with the channel; gravitational acceleration makes an angle α to the z axis.

Recall that δ_f is a length scale over which freezing occurs near the impermeable boundary (fig. 13.2), and that $q_\infty \equiv K_0 \Delta\rho g$ is the buoyancy-driven liquid flux far from the boundary.

Using the solution for porosity, (11.57), we obtain the lateral component of the melt velocity as

$$\phi u(z/\delta_f) = q_\infty \sin\alpha \left[\frac{1 - G(z/\delta_f)}{1 - R_\delta^2 G''(z/\delta_f)}\right]. \tag{13.16}$$

Recall that G is a shape function that models the vertical variation in liquid flux associated with freezing beneath an impermeable boundary (see section 11.3 and equations (11.55) and (11.58)). $R_\delta \equiv \delta_0/\delta_f$ is the ratio of the compaction length to the freezing length. The vertical flux differs from that derived earlier as (11.55) by a factor of $\cos\alpha$,

$$\phi w(z/\delta_f) = q_\infty \cos\alpha \left[1 - G(z/\delta_f)\right]. \tag{13.17}$$

The lateral flux in the absence of a decompaction channel is $q_\infty \sin\alpha$. Therefore, the enhancement to this flux by the channel is

$$\frac{1}{\delta_f}\int_{-\infty}^{0} [\phi u(z) - q_\infty \sin\alpha]\mathrm{d}z. \tag{13.18}$$

Following from this, a measure of the efficiency of focusing is the lateral flux enhancement compared to the vertical flux far from the barrier, $q_\infty \cos\alpha$. This efficiency can be written as

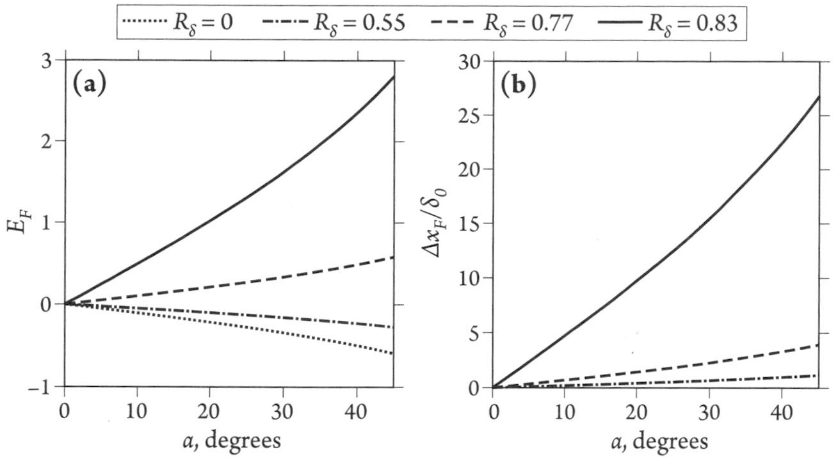

Figure 13.3. Metrics of focusing through an inclined decompaction channel atop a semi-infinite half-space. **(a)** Focusing efficiency calculated according to (13.19). **(b)** Focusing distance calculated according to (13.22), normalized by the reference compaction length δ_0.

$$E_F \equiv \tan\alpha \int_{-\infty}^{0} \left[\frac{R_\delta^2 G''(\zeta) - G(\zeta)}{1 - R_\delta^2 G''(\zeta)} \right] d\zeta, \tag{13.19}$$

where we have used equation (13.15) for the lateral flux and substituted $\zeta \equiv z/\delta_f$.

The focusing efficiency E_F is plotted as a function of the inclination α of the decompaction layer in figure 13.3(a). The efficiency is zero for $\alpha = 0$ with any value of R_δ because there is no component of the buoyancy force in the lateral direction. At increasing α, the lateral driving force increases. If a decompaction channel exists (larger R_δ, cf. fig. 11.7), the permeability is larger near the boundary and buoyancy drives an enhanced lateral flow. This increases the focusing efficiency. But if the porosity and permeability decrease toward the boundary (smaller R_δ), then the lateral flux is diminished relative to the case of uniform porosity.

Another measure of focusing efficiency is the lateral distance traveled by melt as it traverses the full height of the domain, before freezing. We can express this as

$$\int_{-\infty}^{0} \frac{\phi u(z)}{\phi} dt, \tag{13.20}$$

where $t = 0$ is taken to be the present, when the last increment of melt is frozen into the boundary at $z = 0$. We wish to evaluate this integral along streamlines of the melt, which depend on z only. Hence we write $dt = dz/w(z/\delta_f) = \delta_f d\zeta/w(\zeta)$. Furthermore, we subtract off the lateral transport that, in the absence of a decompaction channel, is associated with the inclination of the domain. This background transport is given by

$$\int_{-\infty}^{0} \frac{q_\infty \sin\alpha}{q_\infty \cos\alpha} d\zeta. \tag{13.21}$$

Subtracting the expression (13.21) from the quantity (13.20), we define the focusing distance

$$\Delta x_F \equiv \int_{-\infty}^{0} \left[\frac{\phi u(z/\delta_f)}{\phi w(z/\delta_f)} - \tan\alpha \right] \mathrm{d}z,$$

$$= \delta_f \tan\alpha \int_{-\infty}^{0} \frac{R_\delta^2 G''(\zeta)}{1 - R_\delta^2 G''(\zeta)} \mathrm{d}\zeta,$$

$$= \delta_0 R_\delta \tan\alpha \int_{-\infty}^{0} \frac{G''(\zeta)}{1 - R_\delta^2 G''(\zeta)} \mathrm{d}\zeta, \quad (13.22)$$

where we have used the change of variables from dt to dζ. This focusing distance is plotted as $\Delta x_F/\delta_0$ in figure 13.3(b). For values of R_δ that give rise to a high-porosity decompaction channel, the focusing distance can be a large multiple of the reference compaction length δ_0. To convert this to real distances requires estimates of δ_0 and δ_f. However, the fundamental control is now clear: the combination of a large compaction length with a small freezing length is conducive to lateral melt focusing along an impermeable barrier below the lithosphere.

13.3.2 LATERAL TRANSPORT TO A MID-OCEAN RIDGE

The previous section considered lateral focusing of melt from a mechanical perspective. It developed a model for how gradients in compaction pressure divert flow from a vertical path and force it through a decompaction channel along a sloping, impermeable boundary. The thermal balance controlling freezing was replaced by the prescription of a shape function for Γ.

The present section, in contrast, places the emphasis on the thermal balance. It models the flow by mass conservation combined with simplifying assumptions. By neglecting the mechanical details of the decompaction channel, we are able to develop a slightly less idealized model than what we considered in section 13.3.1. Indeed, here we consider a simple model of melt focusing beneath a mid-ocean ridge. The model is meant to explore the thermal physics in a transparent manner. It is too simplified to provide quantitatively correct results. Moreover, whether this type of melt focusing actually occurs in the Earth is debated on the basis of observations. The aim here is to build a foundation to understand the hypothesized physics.

We assume that a partially molten region of triangular shape is associated with the mid-ocean ridge. It is confined to the zone labeled $\mathcal{Z}$ in figure 13.4 that is between depths $z_b(x)$ and $z_h(x)$. The z-direction is upward with its origin at the surface. In this context, z_h represents the top of the region and z_b the bottom. Mantle comes into the bottom of the triangle at temperature T_m and the surface of the solid Earth has temperature 273 K. We emphasize that the melting region is prescribed; its boundaries are used to constrain other parameters.

The temperature within the melting region $\mathcal{Z}$ is assumed to be the solidus temperature. The temperature field above $\mathcal{Z}$ and below the surface is imposed as a linear gradient from the surface temperature at $z = 0$ to the solidus temperature at z_h. Hence the temperature is continuous at z_h but its gradient is discontinuous there. In the lithosphere above z_h, the porosity is zero and the solid moves horizontally at the imposed half-spreading rate U_0. This imposed temperature field is inspired by the half-space cooling solution, in which vertical diffusion balances horizontal advection to give a

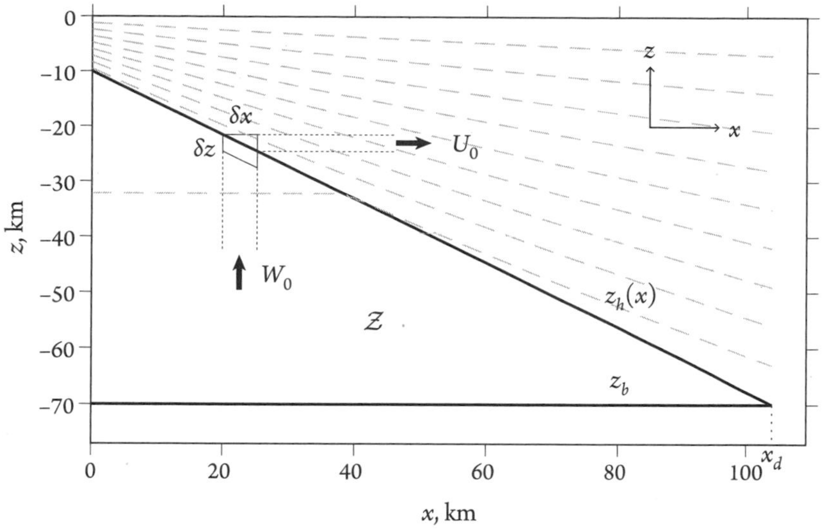

Figure 13.4. Schematic diagram of the model of mid-ocean ridge melt focusing. The ridge axis is located at $x=0, z=0$. The problem is assumed to have reflection symmetry about the z-axis. Dashed lines are temperature contours. The heavy solid line is the outline of the melting region $\mathcal{Z}$. A control volume along the top of the melting region is associated with a melting column beneath and a lithospheric strip to the right. The dip of the top of the melting regime is $\mathrm{d}z_h/\mathrm{d}x = -\tan\alpha$.

steady temperature field with isotherms that deepen with $t^{1/2}$. We simplify by assuming that isotherm depth is linear with time.

To treat the energetics of melting, we consider a one–chemical component system, as was done in section 11.2.2. Then the solidus temperature is a function of pressure only. We take this pressure to be lithostatic and hence write the solidus temperature as

$$T^{\mathcal{S}} = T_0^{\mathcal{S}} - \frac{\rho g z}{\vartheta}. \tag{13.23}$$

We further assume that $\mathcal{Z}$ can be divided into an array of melting columns in which two-phase flow is purely vertical. The upwelling rate W_0 is defined at z_b, the bottom of the melting region where the porosity departs from zero. $W_0(x)$ is obtained as a part of the solution. We assume that each column is independent of its neighbors and governed by the theory for a steady state, one-component mantle column in section 11.2.2. In particular,

$$F(z) = \frac{\rho g c_P}{L\vartheta}(z - z_b), \tag{13.24a}$$

$$\phi w = W_0 F, \tag{13.24b}$$

$$(1-\phi)W = W_0(1-F). \tag{13.24c}$$

These formulas are used to compute the flux from the melting region into the decompaction channel.

The decompaction channel is the key element of the focusing model. We assume that it lies along the line $z = z_h(x)$. We have seen in sections 11.4 and 13.3.1 that the decompaction channel is of finite thickness, but in the context of the tectonic-scale mid-ocean ridge domain, we assume that the decompaction channel is of negligible thickness. That is, we collapse all of the physics (i.e., freezing, lateral transport) onto a line through the two-dimensional domain. To derive equations governing the solid and liquid within this channel, we integrate statements of conservation of mass and energy over a control volume that spans z_h, as shown in figure 13.4. The control volume is of size δx by δz and these two are related by

$$\delta z = \frac{dz_h}{dx}\delta x. \tag{13.25}$$

We assume that the decompaction channel is at steady state and entirely contained within the control volumes. Volume integration of the mass conservation equations (4.11) and use of the divergence theorem gives

$$-\left.\phi w\right|_{z_h-\delta z}\delta x + \int_{z_h-\delta z}^{z_h}\left(\left.\phi u\right|_{x+\delta x} - \left.\phi u\right|_x\right) dz = \delta x \int_{z_h-\delta z}^{z_h}\frac{\Gamma}{\rho}\,dz, \tag{13.26a}$$

$$-(1-\phi)\left.W\right|_{z_h-\delta z}\delta x - U_0\frac{dz_h}{dx}\delta x = -\delta x\int_{z_h-\delta z}^{z_h}\frac{\Gamma}{\rho}\,dz. \tag{13.26b}$$

To simplify these expressions we define

$$q \equiv \int_{z_h-\delta z}^{z_h}\phi u\,dz, \tag{13.27}$$

$$G \equiv \int_{z_h-\delta z}^{z_h}\frac{\Gamma}{\rho}\,dz, \tag{13.28}$$

where q is the volumetric flow rate along the channel and G is the volumetric mass-transfer rate within the channel. Substituting these definitions into the system (13.26), taking the limit of $\delta z, \delta x \to 0$, Taylor expanding to first order the lateral flux at $x+\delta x$, and rearranging gives

$$\left.\phi w\right|_{z_h^-} + G = \frac{dq}{dx}, \tag{13.29a}$$

$$(1-\phi)\left.W\right|_{z_h^-} - G = -U_0\frac{dz_h}{dx}. \tag{13.29b}$$

The left-hand side is evaluated at $z = z_h^-$, the top of the melting column, infinitesimally beneath z_h. The first equation states that the divergence of the lateral channel flow is driven by the net of melt flux from below and phase change ($G < 0$, freezing); the second equation states that the solid added to the lithosphere is the net of the solid flux from below and phase change.

Summing equations (13.29) and using equations (13.24) to eliminate the column fluxes we obtain

$$W_0(x) = \frac{dq}{dx} - U_0\frac{dz_h}{dx}, \tag{13.30}$$

which is a bulk mass-conservation equation for the decompaction channel. Then combining (13.24b) with (13.29a) to eliminate W_0 from (13.30), we can write an equation for the divergence of q,

$$\frac{\mathrm{d}q}{\mathrm{d}x} = \frac{G - U_0 \frac{\mathrm{d}z_h}{\mathrm{d}x} F(z_h)}{1 - F(z_h)}. \tag{13.31}$$

Integration of this equation from the distal end of z_h to $x = 0$ would give us a predicted magma supply rate to the mid-ocean ridge, and hence a predicted crustal thickness. All of the quantities on the right-hand side are known except G, which quantifies the rate of freezing in the channel. Note that $\mathrm{d}z_h/\mathrm{d}x < 0$ so the second term in the numerator can potentially balance the first term.

The freezing rate is obtained through an energy balance over the control volume that straddles z_h. We consider the temperature equation (8.29) at steady state, without the isentropic gradient and neglecting any volumetric source terms. Volume integration and use of the divergence theorem gives

$$\kappa \left(\left.\frac{\mathrm{d}T}{\mathrm{d}z}\right|_{z_h} - \left.\frac{\mathrm{d}T}{\mathrm{d}z}\right|_{z_h - \delta z} \right) \delta x = \frac{L}{c_P} \delta x \int_{z_h - \delta z}^{z_h} \frac{\Gamma}{\rho} \, \mathrm{d}z. \tag{13.32}$$

Taking the limit $\delta z, \delta x \to 0$ and using the definition (13.28) of G, this becomes

$$G = \frac{\kappa c_P}{L} \left(\left.\frac{\mathrm{d}T}{\mathrm{d}z}\right|_{z_h^+} - \left.\frac{\mathrm{d}T}{\mathrm{d}z}\right|_{z_h^-} \right). \tag{13.33}$$

Recall that there is a discontinuity in the temperature gradient at z_h; we use z_h^+ to indicate that the gradient is evaluated on the lithospheric side and z_h^- to specify evaluation on the melting-region side. Hence this equation states that the latent heat of phase change balances the jump in heat flux across the decompaction channel. The temperature gradient in the melting region comes from the solidus relation (13.23) while that in the lithosphere comes from the linear temperature profile as $(T(z_h) - T(0))/z_h$. Both gradients are negative (recall that $z_h < 0$), but the lithospheric gradient is larger in absolute value. There is thus a divergence in the sensible heat flux. This drives freezing, which liberates latent heat.

Equation (13.31) must be evaluated numerically. A one-component model using empirical values for latent heat and the Clapeyron slope produces unrealistic results for F_{max} and z_b. Therefore, we instead impose $F_{\mathrm{max}} = 0.23$ and $z_b = 70$ km, and use these to calculate the L and ϑ appropriate for our simple model (see caption of fig. 13.5). The boundary condition is

$$q = 0 \quad \text{at} \quad x = x_d, \tag{13.34}$$

where x_d is the distal point of $\mathcal{Z}$, as shown in figure 13.4a. In some region adjacent to x_d, the divergence of q is negative, meaning that all melts that arrive in the decompaction channel are frozen into the lithosphere. This drives q below zero, which is unphysical. Hence, we shift the solution of (13.33) such that q departs from zero at exactly the point where $\mathrm{d}q/\mathrm{d}x$ becomes positive.

A solution to equation (13.31) is shown in figure 13.5. The half-spreading rate U_0 is 4 cm/yr and the top of the melting triangle dips at 30°; other parameters are given in the figure caption. Near the axis, z_h reaches shallow depths, associated with a sharp thermal gradient in the lithosphere above it (fig. 13.4). This leads to a large diffusive heat flux through the lithosphere near the axis, which causes rapid freezing. The rate of freezing, $-G$, is plotted in figure 13.5(a). It shows a gradual increase from x_d toward the axis, and

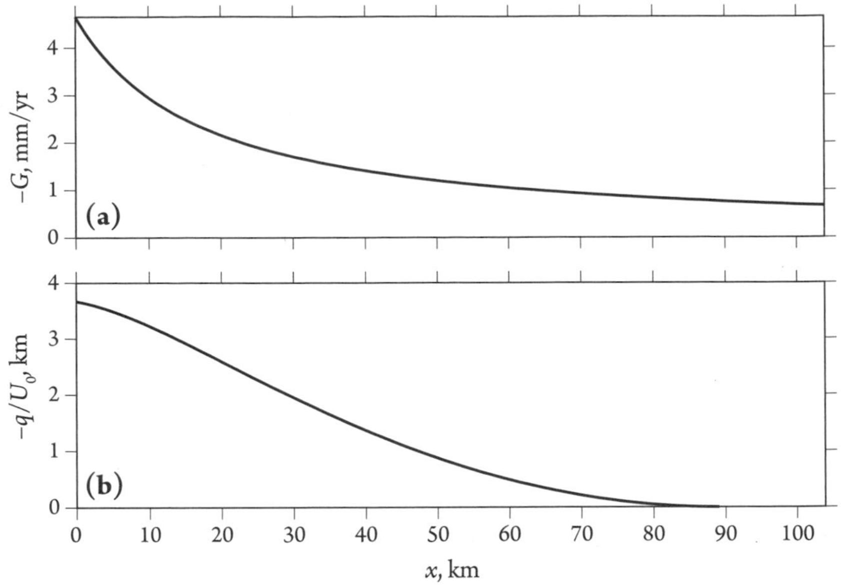

Figure 13.5. Results of the mid-ocean ridge focusing model with reference parameters. **(a)** The freezing rate $-G$. **(b)** The equivalent crustal thickness $-q/U_0$. Parameter values used here are $T_0^S = 1373$, $T_m = 1623$ K, $z_b = -70$ km, $z_{h0} = -10$ km, $z_h = z_{h0} - x\tan\alpha$, $\alpha = 30°$ (giving $x_d = 104$ km), $F_{\max} = 0.23$, $\Pi = F_{\max}/(z_h(0) - z_b)$, $\vartheta = z_b\rho g/(T_0^S - T_m)$, $L = c_P\rho g/\vartheta\Pi$, and other parameters as in table 8.1.

then a sharp increase within about 20 km of the axis. This rapid freezing diminishes the flow of melt through the decompaction channel, as shown in figure 13.5(b).

Figure 13.5(b) shows the magma flow rate in terms of an equivalent crustal thickness, $-q/U_0$ (neglecting the ratio of magma density to crustal density, which is approximately unity; the negative sign is present because flow is in the $-x$-direction). At distances greater than about 20 km from the ridge axis, the flux of magma into the decompaction channel from below dominates over freezing; the flow rate increases toward the axis. From 20 km to the axis, the increase in freezing rate shifts the balance such that the rate of flow through the decompaction channel slows and reaches a maximum.

Figure 13.6 shows how the solution to (13.31) varies with the dip of the focusing boundary $\alpha = \tan^{-1}(-\mathrm{d}z_h/\mathrm{d}x)$ and the spreading rate U_0. In (a), holding U_0 fixed, a larger dip leads to a thicker crust ($-q(0)/U_0$). In (b), for fixed α, larger spreading rate also leads to a thicker crust. In both cases, this is because W_0 is increased, causing more rapid melting and a greater melt flux from the melting columns to the decompaction channel. The effect is quantified by the second term in the numerator of (13.31), which becomes more positive. With increasing U_0, the effect on crustal thickness saturates as the freezing rate becomes negligible.

In the natural system, the spreading rate is a key control of the thermal profile, and hence of the depth to the melting region. This is demonstrated clearly in the half-space cooling model for the thermal structure of the oceanic lithosphere [Turcotte and Schubert, 2014]. The depth to any given isotherm scales with $\sqrt{\kappa x/U_0}$ and hence the slope

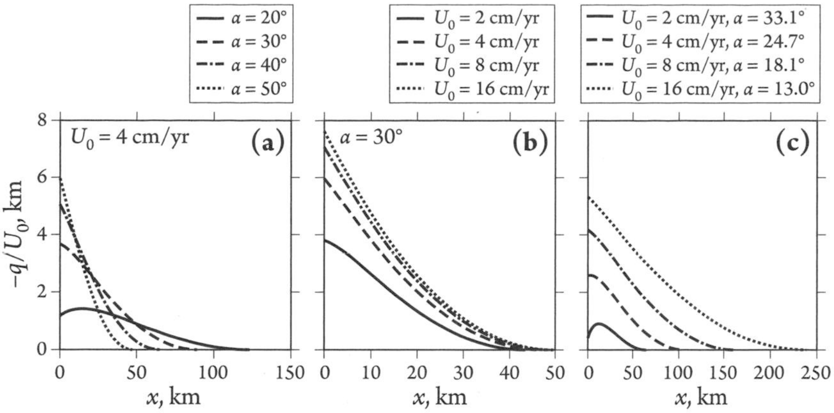

Figure 13.6. Focusing model sensitivity to lithospheric angle α and spreading rate U_0. **(a)** Variation in α for fixed U_0. **(b)** Variation in U_0 for fixed α. **(c)** Variation in U_0 that determines the angle α via the half-spacing cooling model [Turcotte and Schubert, 2014], evaluated at a distance of 10 km from the ridge axis.

of that isotherm also scales with $U_0^{-1/2}$. Figure 13.6(c) shows the behavior of the focusing model when the dip α of the top of the triangular melting region is computed using the slope of the half-space cooling model. The half-space model is evaluated at a a distance $x = 10$ km from the ridge axis and temperature that is 90% of the way to T_m. We hold the position z_{h0} of the apex of $\mathcal{Z}$ to be constant, independent of spreading rate U_0.

Figure 13.6(c) shows that faster spreading (with associated reduction in α) leads to a thicker crust. The models predicts that this effect does not saturate with increasing spreading rate. This is inconsistent with observations of crustal thickness as a function of spreading rate, which saturates at a half-rate of 2 cm/yr to a thickness of about 7 km.

Hence our simple model for sublithospheric focusing doesn't provide a good match with observations. One important source of error arises from the assumption of a single thermochemical component, and hence no mantle depletion with increased melting. The same issue also applies to freezing: there is no solidus depression as melts are distilled to higher concentrations of incompatible elements.

Another important source of error is the rough treatment of the energetics. We have imposed the structure of the melting region a priori and used a crude model of the temperature field (which controls the distribution of heat conduction). In a consistent model, the thermal and melting structure would be determined as part of the solution. In particular, our results are sensitive to the depth to the apex of the melting region. We hold this depth constant but in reality it varies sharply as a function of spreading rate.

These inconsistencies are readily corrected in a fully numerical model (see the Literature Notes at the end of this chapter). The model presented here, however, serves to illustrate the balance of melt flux, lateral transport, and freezing in a decompaction channel. More physically consistent versions of this approach might improve the match with observations. It is important to note, however, that sublithospheric focusing may be one of several mechanisms that contribute to melt focusing beneath mid-ocean ridges (see the Literature Notes at the end of this chapter).

13.4 Coupled Dynamics and Thermochemistry with the Enthalpy Method

Numerical simulations of magma generation and transport can avoid many of the assumptions and simplifications that were required above to attain solutions. Nonlinear numerical methods can, in principle, handle the coupling between the physical processes described by the PDEs. However, there are a variety of modeling choices to be made in establishing details of the coupling of fluid dynamics of magma/mantle flow with conservation of energy and composition, and with a thermodynamic model of phase change. Subtle issues and trade-offs accompany these choices and there is no broadly accepted solution. Here we develop a method that makes convenient simplifications and has some important advantages. This approach, called the enthalpy method, was pioneered in the metallurgy literature but has proven useful elsewhere as well (see the Literature Notes at the end of this chapter). We discuss it as simply as possible, assuming, for example, that the heat capacity is a constant, independent of temperature and material phase.

The most fundamental modeling choice required is whether to consider the thermochemical system to be in local equilibrium throughout the domain, or whether to model a disequilibrium system. Both of these approaches are discussed in chapter 10. In the case of disequilibrium, the rate of interphase mass transfer Γ must be computed according to thermochemical kinetics. This typically requires a calculation of the equilibrium state (phase fractions and compositions), and a comparison of the instantaneous state of the system to that equilibrium. For reaction rates that are large compared to the rate of magmatic segregation, the system should be close to equilibrium. It is therefore a reasonable approximation to assume that it is exactly in thermochemical equilibrium everywhere.

The enthalpy method is based on this powerful assumption. Thermochemical equilibrium means that the pressure P, bulk composition $\overline{c}_j$, and bulk specific enthalpy H at a point in space (really a representative/finite volume) uniquely determine the temperature, phase fractions (i.e., the porosity), and the phase compositions. This can be represented schematically, for a two-phase system, as

$$H, \overline{c}_j, P \longrightarrow \boldsymbol{\Theta} \longrightarrow T, \phi, c_j^s, c_j^\ell, \tag{13.35}$$

where $\boldsymbol{\Theta}$ represents some method of computing the equilibrium state. An example is given below.

Schematic equation (13.35) is a direct means of computing the porosity. Previously, the porosity was obtained by solving the mass-conservation equation for the solid phase (4.46c). But as discussed in section 10.3, the equilibrium calculation of ϕ does not provide an explicit expression for the melting rate Γ. In this context, the mass-conservation equation becomes the means for diagnosing the melting rate,

$$\Gamma = \frac{\partial \phi}{\partial t} - \boldsymbol{\nabla} \cdot (1-\phi)\boldsymbol{v}^s, \tag{13.36}$$

which states that the melting rate is the rate of change of porosity in excess of that arising from divergence of the solid flux. This can be evaluated to obtain Γ only after the variables on the right-hand side are known. And since this requires solutions for other, coupled equations, it is advantageous to formulate those equations to avoid the appearance of Γ. Returning to chapter 8 and considering the various forms the conservation of energy can take, it is evident that the equation for bulk specific enthalpy (8.14) is a suitable choice. Making the Boussinesq approximation and neglecting radiogenic and dissipative heating, this equation becomes

$$\frac{\partial H}{\partial t} + \rho c_P \nabla \cdot \overline{\boldsymbol{v}} T = \rho L \nabla \cdot (1-\phi)\boldsymbol{v}^s + k_T \nabla^2 T + \rho \alpha_\rho T \overline{\boldsymbol{v}} \cdot \mathbf{g}. \tag{13.37}$$

An expression relating temperature and enthalpy is then required to solve this equation (and as part of $\boldsymbol{\Theta}$, above). This is obtained by integrating the thermodynamic relationship for enthalpy, (8.15). In doing so, we retain our previous assumption that the thermodynamic pressure is the same as the lithostatic pressure. Hence, at a fixed depth, $\mathrm{d}P = 0$, and we can write the mass-specific enthalpy of phase i as $h^i = c_P^i (T - T_0) + h_0^i$, where a subscript 0 indicates a constant, reference quantity. Since the latent heat (per unit mass) is defined as $L \equiv h^\ell - h^s$, we can take $h_0^\ell = h_0^s + L$. And since we are interested only in changes of enthalpy, our choice of a reference state is arbitrary. Hence we can take $h_0^s = T_0 = 0$ without loss of generality. Then the phase enthalpies are

$$h^s = c_P T, \tag{13.38a}$$

$$h^\ell = c_P T + L. \tag{13.38b}$$

The definition of bulk, volumetric enthalpy in a two-phase system is $H \equiv \rho^\ell \phi h^\ell + \rho^s (1-\phi) h^s$ and hence, with the extended Boussinesq approximation, we can write

$$H = \rho c_P T + \rho L \phi. \tag{13.39}$$

This equation will be useful below.

In addition to the bulk enthalpy and its relation to temperature, we also require the bulk composition as an input to $\boldsymbol{\Theta}$. This set of quantities evolves according to the conservation equation for bulk species mass (9.4c), which is the sum of conservation equations for the two phases; conveniently, Γ drops out when taking this sum. The equation can be rewritten as

$$\frac{\partial \overline{c}_j}{\partial t} + \nabla \cdot \phi \boldsymbol{v}^\ell c_j^\ell + \nabla \cdot (1-\phi)\boldsymbol{v}^s c_j^s = \left(\mathcal{D}_j/\rho\right) \nabla \cdot \phi \nabla c_j^\ell, \tag{13.40}$$

using the Boussinesq approximation. Since, by the unity sum,

$$\overline{c}_k = 1 - \sum_{j=1, j\neq k}^{N} \overline{c}_j, \tag{13.41}$$

we need only solve $N-1$ versions of equation (13.40) for an N-component thermochemical system. To solve any of these, we need to relate the bulk composition to the phase compositions c_j^i and obtain the porosity. This must be done via determination of the equilibrium state. Hence we turn our attention to $\boldsymbol{\Theta}$.

Our approach for building a theory for $\boldsymbol{\Theta}$ follows the development of section 10.1.2; we use equilibrium partitioning coefficients $\check{K}_j (\equiv \check{c}_j^s / \check{c}_j^\ell)$. These could be computed as a function of pressure and temperature using ideal solution theory, for example. Simpler parameterizations or more detailed models for $\check{K}_j$ are also possible. We depart from section 10.1.2 here in that we take $\rho^s = \rho^\ell = \rho$, and hence we have $\phi = f$. We can then rewrite equation (10.20) as

$$\sum_{j=1}^{N} \frac{\bar{c}_j \left[1 - \check{K}_j\left(\frac{H-\rho L\phi}{\rho c_P}, P\right)\right]}{\phi + (1-\phi)\check{K}_j\left(\frac{H-\rho L\phi}{\rho c_P}, P\right)} = 0, \tag{13.42}$$

where the unknown temperature has been eliminated by rearranging equation (13.39), leaving the porosity ϕ as the only unknown. Hence, for a given set $\bar{c}_j$, H, and P and methods to compute $\check{K}_j(T,P)$, this is an implicit equation for ϕ; in general it must be solved numerically (by, e.g., Newton's method). Once ϕ has been computed, T is obtained from (13.39) and the phase compositions are given by

$$c_j^\ell = \bar{c}_j / \left[\phi + (1-\phi)\check{K}_j\right], \tag{13.43a}$$

$$c_j^s = \bar{c}_j / \left[\phi/\check{K}_j + (1-\phi)\right]. \tag{13.43b}$$

Hence, under the enthalpy method, the governing system of PDEs includes the modified Stokes equation (4.47b), the compaction equation (4.47a), which are both elliptic, and which model the instantaneous dynamics of the flow. And it includes the enthalpy equation (13.37) and conservation of bulk species mass (13.40), both hyperbolic, which model the evolution of porosity, temperature, and phase compositions. But what about conservation of mass equations, which are not included in the set above? Is conservation of mass preserved in a method where there is no explicit statement of continuity or porosity-evolution equation? If the dynamics are treated carefully, the answer is yes. If liquid pressure and solid velocity are obtained by solution of the compaction equation and the modified Stokes equation then the conservation of bulk mass is ensured by the use of equations (4.11) to derive (4.47a); continuity is contained within the compaction equation. Moreover, mass conservation equations (4.11) are used to derive PDEs for bulk enthalpy (13.37) and bulk composition (13.40). The thermochemical calculation of porosity simply partitions the bulk, conserved mass into liquid and solid parts.

Clearly, the enthalpy method lends itself to discretization and numerical solution, rather than to analytical methods. Without going into the details of such processes, we can make a few remarks. First, equations (13.37) and (13.40) are coupled, nonlinear PDEs because they depend on ϕ, T, c_j^s, and c_j^ℓ, which are themselves functions of the primary variables H and $\bar{c}_j$ via the thermodynamic calculation $\boldsymbol{\Theta}$. For some simple phase diagrams, this nonlinearity can be resolved analytically; more generally, it can be solved by applying Newton's method to the discretized, coupled, nonlinear algebraic system, or by a Picard iteration where the enthalpy method variables ϕ, T, c_j^s, and c_j^ℓ are lagged.

Compared to methods where the melting rate Γ is computed explicitly and used to evolve the porosity, the enthalpy method has the following advantages:

- The porosity is obtained by thermodynamic calculation in which it is inherently bounded by zero and unity. Methods that rely on the solution of a PDE for ϕ are not protected from violating these bounds.

- The number of equations to be solved is substantially less than with a disequilibrium method. For a system with N components in chemical disequilibrium, $2(N-1)$ equations are required to constrain the composition; only half that number is required if the system is in equilibrium.
- There is no need to prescribe poorly constrained kinetic parameters or laws for Γ.

Of course, the most general assumption about a natural system is that of disequilibrium, even if reaction rates are fast and the system remains close to equilibrium. Mathematically, equilibrium imposes rigid and generally nonlinear relationships between variables, making consistent numerical solutions difficult or inefficient to obtain. Use of nonconverged results may be worse than the uncertainties inherent in disequilibrium theory. The enthalpy method and related methods based on the assumption of equilibrium can be powerful tools for developing numerical models of tectonic-scale magma genesis and transport.

13.5 Literature Notes

The earliest models of two-phase dynamics in a plate-tectonic setting were by Spiegelman and McKenzie [1987] and Morgan [1987]. Both considered the consequences of mid-ocean ridge mantle corner flow on lateral melt transport (the material that is summarized above in section 13.2). Their models solve a version of the governing equations in which compaction and compaction pressure are neglected entirely, enabling analytical solutions but excluding some relevant behavior. One excluded behavior is decompaction boundary layers. Sparks and Parmentier [1991] and Spiegelman [1993c] showed that a consequence of compaction pressure is to create a high-permeability channel along the base of the lithosphere.

Around the same time, other investigators considered the potential for decompression melting to drive convection beneath the ridge axis. The component of the solid flow at mid-ocean ridges driven by local sources of buoyancy has been called *active flow*, in contrast to the passive, corner flow that is a response to tectonic plate separation. Two local sources of buoyancy were identified: melt-retention buoyancy is associated with the lower-density magma in the pores of the higher-density residual solid (right-hand side of equation (13.6a)); residual-depletion buoyancy is associated with the reduction in iron concentration in the residual solid after partial melting (which was not considered here). Work by Scott and Stevenson [1989], Buck and Su [1989], and Jha et al. [1994] used numerical solutions of the partially coupled governing equations to model the solid flow, melting, and porosity evolution. Melt segregation was included, but was driven exclusively by buoyancy.

Work by Spiegelman [1993c] and Spiegelman [1996] also considered active flow beneath mid-ocean ridges, but included a more complete coupling of solid and liquid flow. This coupling accounts for dynamic pressure and the influence of solid velocity, but excludes compaction and associated pressure gradients.

Three-dimensional simulations of mid-ocean ridges using partially coupled equations were considered by Barnouin-Jha et al. [1997], Magde and Sparks [1997], and Choblet and Parmentier [2001]. These were employed to investigate the distribution of melting, buoyancy, solid upwelling, and melt segregation along and across the ridge axis.

The fully coupled system of equations (4.46) was first solved by Ghods and Arkani-Hamed [2000], who considered the potential for active flow and for sublithospheric melt focusing. Their simulations were the first to resolve magmatic solitary waves within a tectonic-scale model and they provided support for the focusing hypothesis of Sparks and Parmentier [1991]. Their model for melt production and freezing lacked the rigor of their mechanical model. This imbalance was corrected by Katz [2008], who used the enthalpy method (section 13.4) and a two-component solid-solution phase diagram. Active convection driven by melt-retention buoyancy was investigated in the context of fully coupled, enthalpy-method-based models by Katz [2010]. The latter demonstrated the scaling of convective vigor with a dimensionless number analogous to the Rayleigh number. It also showed how active convection makes the ridge upwelling system prone to asymmetry. Tectonic-scale numerical models were used by Crowley et al. [2014] to model the impact of variations in sea level on mid-ocean ridge magmatism. They supported the argument by Huybers and Langmuir [2009] that glacial cycles can drive crustal thickness variations, though this idea has been controversial [e.g., Olive et al., 2015].

The use of multicomponent thermochemistry has allowed mid-ocean ridge models to begin to incorporate some of the complexity of the mantle petrological system, including simplified chemical heterogeneity [Katz and Weatherley, 2012] and volatile-enriched melting [Keller et al., 2017]. Chemical variability seems to promote channelization of flow, supporting a link between mantle heterogeneity and channelized melt transport. Incorporating more petrological complexity, as was done by Tirone et al. [2012], may represent a means to investigate this further. Grain size has also been considered recently in fully coupled numerical models of mid-ocean ridges [Turner et al., 2017].

Models of two-phase flow have been applied to other tectonic settings as well. Spiegelman and McKenzie [1987] considered the role of the dynamic pressure from corner flow in focusing liquid toward the wedge corner; Iwamori [1998] extended this model with melting and porosity evolution. Cagnioncle et al. [2007] excluded dynamic pressure and considered fluid segregation by buoyancy only; they also incorporated melting and investigated how solid flow affects fluid pathways. Wilson et al. [2014] developed two-phase models of water transport off the slab, neglecting melting, decoupling the solenoidal solid flow, and focusing instead on the role of compaction and the assocated pressure gradients. They demonstrated the potential role of a decompaction layer to focus melts toward the arc. Cerpa Gilvonio et al. [2017] incorporated grain size and associated permeability variations. The role of thermal transport by magma was considered by Rees Jones et al. [2018] using a melting-column model adapted to subduction zones and decoupled two-dimensional models.

The enthalpy method was developed in the metallurgy literature. It has been used to study solidification of mushy layers [Katz and Worster, 2008], partially molten ice in temperate glaciers [Aschwanden et al., 2012] and compacting, temperate firn layers [Meyer and Hewitt, 2017]. The same concepts were employed by Ribe [1985] in his mantle column models. Similar methods that couple equilibrium thermochemistry with fluid dynamics have been used in planetary science to investigate core segregation [Ricard et al., 2009; Šrámek et al., 2010], planetesimal differentiation [Šrámek et al., 2012], and mantle solidification [Boukaré and Ricard, 2017]. Application to magmatic emplacement and chemical evolution of the crust was made by Jackson et al. [2018], who directly parameterized the phase diagram in enthalpy–composition space.

13.6 Exercises

13.1 By integrating the steady state mass conservation equations (4.11) over the control volume shown in figure 13.4(b) and using the divergence theorem, derive (13.29).

13.2 Determine the marginal lithospheric dip angle α corresponding to $dq/dx = 0$. Express this angle as a function of depth z. Explain your result in terms of physical processes.

CHAPTER 14

Numerical Modeling of Two-Phase Flow

Most of the problems that we have considered in this book were analyzed with minimal recourse to numerical methods. But these were chosen for their pedagogical value; they are not representative of problems at the cutting edge of current research. In general, research problems involving magma/mantle dynamics benefit from (and frequently require) the use of numerical methods to obtain solutions. This doesn't mean that the analytical approach is less valuable; on the contrary, analytical solutions to simplified problems can be used to benchmark code and help to interpret results. And sometimes the simplifications that unlock an analytical approach are most directly recognized by studying the numerical solutions.

Given their importance, numerical methods and their application to magma/mantle dynamics merit much more space than the current chapter affords. However, for the PDEs considered above, the development of methods that are robust and efficient is ongoing. There is no consensus on the best approach, and the complexities of the two-phase physics pose significant challenges for commonly used numerical schemes. Hence this chapter provides only a short introduction that is intended to help the reader make informed choices about what techniques to pursue further.

Numerical methods leverage the power of computers to perform floating-point operations with great speed. These operations arise from discretization of the field variables in space and time. The PDEs that govern the continuous fields become become systems of linear or nonlinear algebraic equations that govern the discretized fields. By taking care in the construction of these systems, they can be put into a form that is generic and amenable to solution using established methods of linear algebra. Most of the relevant numerical algorithms are available through existing software. In the code provided in this chapter, we shall use MATLAB and its backslash operator to solve the system of the linear equation $\mathbf{A}\boldsymbol{x} = \boldsymbol{b}$, where $\mathbf{A}$ is a matrix and $\boldsymbol{x}$ and $\boldsymbol{b}$ are vectors. The vector $\boldsymbol{x}$ contains the set of unknown values representing the solution of the discretized PDE; its entries are known as the *degrees of freedom* (DOFs) of the problem. In MATLAB we write

$$x = A \backslash b; \tag{14.1}$$

to obtain the solution, $\boldsymbol{x}$, to the discrete problem. This is equivalent in its effect to taking $\boldsymbol{x} = \mathbf{A}^{-1}\boldsymbol{b}$. It is outside the scope of this book to consider how MATLAB implements the backslash operation. In software developed for research on problems with two-dimensional or three-dimensional domains, use of MATLAB's backslash operator is generally not recommended; instead, open-source software packages such as PETSc[1] provide a range of data structures and algorithms that are portable to high-performance computers. The online supplement of this book provides a Python implementation of the codes in this chapter.

Because we use MATLAB to illustrate the numerical computations in this book, we will use MATLAB's indexing convention: indexes start from a value of unity. For example, for a list of numbers q_i, the first entry in the list has index $i=1$. Some other languages would label the first entry with the index $i=0$.

Our focus will be on finite-difference methods, which are proven to work well for two-phase magma dynamics (see the Literature Notes at the end of this chapter). Finite-element methods are powerful and, in other fields, more widely used. Their use for two-phase magma dynamics is a topic of current research (again, see the Literature Notes). We consider finite elements only briefly in a one-dimensional problem, below.

The chapter uses the solitary-wave problem to illustrate instantaneous and time-dependent solutions in one dimension. Because it has an analytical solution, this problem (with suitable initial condition) is useful for validation of code. Moving into two dimensions, nontrivial analytical solutions are rare. Hence we introduce the *method of manufactured solutions* as a way to create analytical solutions for comparison with code output. Comparison with known solutions allows us to validate the code, but it is also valuable for checking the *convergence* of the numerical scheme (i.e., of the discretization and solution). A scheme is convergent if its solutions become arbitrarily close to the true solution as the discretization intervals become arbitrarily small (limited, of course, by the machine precision of the floating-point operations themselves).

14.1 The One-Dimensional Solitary Wave (Instantaneous)

Here we recall the magmatic solitary wave problem of section 6.4. The reader is encouraged to review that section before reading this one. A set of equations govern the problem: the compaction equation (6.31a) and the porosity evolution equation (6.31b). In this section we consider only a solution of the former, for the compaction rate at an instant in time. The equation reads

$$\frac{\mathrm{d}}{\mathrm{d}z}\varphi^n\frac{\mathrm{d}\mathcal{C}}{\mathrm{d}z}-\mathcal{C}=\frac{\mathrm{d}\varphi^n}{\mathrm{d}z}. \tag{14.2}$$

Recall that all symbols are nondimensional, lengths have been rescaled by the compaction length, and $\varphi=\phi/\phi_0$ is the porosity scaled by the background porosity ϕ_0. This equation is valid when the compaction viscosity is constant and uniform.

The boundary conditions that were used in the analysis of the solitary wave relied on a mirror symmetry of the problem about the center of the wave. Here we compute on a domain spanning the wave and hence impose boundary conditions requiring no disturbance to the background porosity in the far field:

$$\mathcal{C}=0 \text{ at } z\rightarrow\pm\infty. \tag{14.3}$$

This requires that any deviations of $\mathrm{d}\varphi/\mathrm{d}z$ from zero occur at finite z.

[1] The Portable, Extensible Toolkit for Scientific computation, http://www.mcs.anl.gov/petsc/

Given a distribution of porosity $\varphi(z)$, equation (14.2) allows us to solve for the associated compaction rate $\mathcal{C}(z)$. We showed in section 6.4 that if $\varphi(z)$ has the form of a solitary wave, the resulting compaction rate is described analytically. This will provide a useful benchmark for our numerical solutions below. Note, however, that equation (14.2) applies for any continuous, differentiable porosity below the dissagregation threshold. Hence, if $\varphi(z)$ satisfies these requirements and if its derivative tends to zero as $z \to \pm\infty$, it will be possible to obtain a numerical solution.

To enable comparison of numerical solutions with the analytical solutions of section 6.4, we take the porosity to be that of a solitary wave $\varphi = \Lambda$, with permeability exponent $n=3$ and peak value $\Lambda^* = 4$. Examination of the solitary wave profiles in figure 6.5 indicates that for these parameters, we require a domain extending to at least $z = \pm 30$ compaction lengths, such that we can safely impose the boundary conditions that formally hold at $z = \pm\infty$. Equations (6.46) and (6.49), respectively, are exact expressions of $\mathcal{C}$ and Λ.

To facilitate the development of the finite-difference and finite-element discretizations discussed below, it is helpful to rewrite the governing equation (14.2) in terms of an approximate solution $\tilde{\mathcal{C}}$ and its residual R,

$$\frac{\mathrm{d}}{\mathrm{d}z}\varphi^n \frac{\mathrm{d}\tilde{\mathcal{C}}}{\mathrm{d}z} - \tilde{\mathcal{C}} - \frac{\mathrm{d}\varphi^n}{\mathrm{d}z} \equiv R. \tag{14.4}$$

In a not-very-precise sense, finite-difference methods aim to find a $\tilde{\mathcal{C}}$ that minimizes R pointwise throughout the domain. They attempt to solve equation (14.2) directly, in its *strong form*. In contrast, finite-element methods seek a solution such that

$$\int_\Omega \mathcal{W} R \,\mathrm{d}\Omega = 0, \tag{14.5}$$

where Ω represents the domain and $\mathcal{W}$ is a *test function*—a function that weights the residual within the integral. (In this context, $\tilde{\mathcal{C}}$ is known as a *trial function*.) The integral in (14.5) is a functional of the residual. Applying this functional to the left-hand side of (14.4) results in the *weak form*, which is what we attempt to solve with finite elements.

14.1.1 FINITE-DIFFERENCE DISCRETIZATION

One classic approach to discretizing differential equations is to write the derivatives as finite differences. To this end, we define a set of N_z discrete values to represent the function $\tilde{\mathcal{C}}(z)$ over the domain. We assume that these are regularly spaced, such that

$$\lim_{N_z \to \infty} \tilde{\mathcal{C}}_i = \mathcal{C}(z_i) \text{ for } i = 1, 2, 3, \ldots, N_z. \tag{14.6}$$

This equation states that for any integer $1 \le i \le N_z$, the ith discrete entry of the numerical solution $\tilde{\mathcal{C}}_i$ should converge to the analytical solution $\mathcal{C}(z)$ evaluated at $z_i = (i-1)\Delta z$. The values of z_i are the set of discrete locations in space where discrete values of the compaction rate are sought. We shall refer to each location z_i as a *node*; the set of nodes will be termed the *grid*. Figure 14.1 shows a segment of the grid. The regular

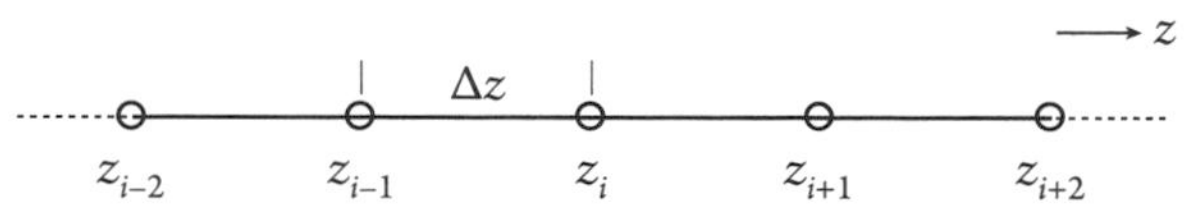

Figure 14.1. A segment of the grid for the finite-difference discretization. The spacing between nodes is Δz.

spacing between nodes is determined by

$$\Delta z = \frac{2z_m}{N_z - 1} \tag{14.7}$$

for a domain of $z \in [-z_m, z_m]$. On this finite domain, the boundary conditions (14.3) are approximated by the Dirichlet conditions

$$\tilde{\mathcal{C}}(z = \pm z_m) = 0. \tag{14.8}$$

Porosity disturbances must be sufficiently far from the boundaries that the influence of imposing this condition at $\pm z_m$ instead of $\pm\infty$ is negligible.

Like the compaction rate, porosity must also be specified on the nodes,

$$\varphi_i = \varphi(z_i) \text{ for } i = 1, 2, \ldots, N_z. \tag{14.9}$$

Here, however, we don't include the limit expressing convergence because we assume that $\varphi(z)$ is known a priori; equation (14.9) is understood as a sampling of that function at discrete points.

In the context of this discretization, derivatives are approximated as finite differences:

$$\lim_{N_z \to \infty} \frac{\tilde{\mathcal{C}}_{i+1} - \tilde{\mathcal{C}}_i}{\Delta z} = \left.\frac{\mathrm{d}\tilde{\mathcal{C}}}{\mathrm{d}z}\right|_{z_i + \Delta z/2}. \tag{14.10}$$

This statement is true only if the function being discretized and differentiated ($\mathcal{C}(z)$ in this case) is continuous and smooth. Note that the location at which the finite-difference approximation converges to the analytical derivative is halfway between the nodes z_i and z_{i+1}. Hence this is called a *central difference*. It is straightfoward to show that the standard finite-difference approximation for the second derivative is centered on the nodes themselves.

Examining equation (14.2), one should be concerned that discretized terms on the left-hand side are centered on nodes while the right-hand side term is offset by $\Delta z/2$. To avoid this offset, we define the set of half-nodes at $z_{i+1/2} \equiv (z_i + z_{i+1})/2$. Using linear interpolation, we define $\varphi_{i+1/2} \equiv (\varphi_i + \varphi_{i+1})/2$. This enables the central difference

$$\lim_{N_z \to \infty} \frac{\varphi^n_{i+1/2} - \varphi^n_{i-1/2}}{\Delta z} = \left.\frac{\mathrm{d}\varphi^n}{\mathrm{d}z}\right|_{z_i}. \tag{14.11}$$

We now have all the pieces required to write the finite-difference form of equation (14.2). It is

$$\frac{\varphi^n_{i+1/2}\left(\tilde{C}_{i+1}-\tilde{C}_i\right)-\varphi^n_{i-1/2}\left(\tilde{C}_i-\tilde{C}_{i-1}\right)}{\Delta z^2}-\tilde{C}_i=\frac{\varphi^n_{i+1/2}-\varphi^n_{i-1/2}}{\Delta z}. \tag{14.12}$$

Multiplying by Δz^2 and rearranging into stencil notation[2] gives

$$\left[\ \varphi^n_{i-1/2},\quad -\left(\Delta z^2+\varphi^n_{i-1/2}+\varphi^n_{i+1/2}\right),\quad \varphi^n_{i+1/2}\ \right]\tilde{C}_i=\Delta z\left(\varphi^n_{i+1/2}-\varphi^n_{i-1/2}\right). \tag{14.13}$$

Values in the stencil and on the right-hand side are known; the unknowns are the set $\tilde{C}_i$ for $i=1,2,\ldots,N_z$. The first and last entries are determined by the boundary conditions (14.8), $\tilde{C}_1=\tilde{C}_{N_z}=0$.

To make progress toward solving for $\tilde{C}_i$, we write (14.13) in the form $\mathbf{A}\boldsymbol{x}=\boldsymbol{b}$, where $\boldsymbol{b}$ is a column vector with $b_i=\Delta z\left(\varphi^n_{i+1/2}-\varphi^n_{i-1/2}\right)$. The unknowns go into column vector $\boldsymbol{x}$. The stencil is used to fill a tridiagonal band in the matrix $\mathbf{A}$. The first and last rows in $\mathbf{A}$ and $\boldsymbol{b}$ are constructed to satisfy the boundary conditions. The result is

$$\begin{pmatrix} 1 & 0 & & & & \\ [& S_2 &] & & & \\ & [& S_3 &] & & \\ & & & \ddots & & \\ & & & [& S_{N_z-1} &] \\ & & & & 0 & 1 \end{pmatrix}\begin{pmatrix}\tilde{C}_1\\ \tilde{C}_2\\ \tilde{C}_3\\ \vdots\\ \tilde{C}_{N_z-1}\\ \tilde{C}_{N_z}\end{pmatrix}=\begin{pmatrix}0\\ b_2\\ b_3\\ \vdots\\ b_{N_z-1}\\ 0\end{pmatrix}, \tag{14.14}$$

where $[\ S_i\ ]$ is the ith stencil, given in equation (14.13) above. Blank entries in the matrix are zeros. Note that most of the entries are zeros and hence this is referred to as a *sparse matrix*. In particular, there are only three diagonals of nonzero entries; the matrix is therefore considered *tridiagonal*. This specific sparsity pattern arises from the locality of the one-dimensional, differential operators: derivatives in z only depend on neighboring values in the grid.

Formation and solution of this matrix–vector equation is achieved in MATLAB by first defining a column array phi $=\varphi_i$ and a scalar dz $=\Delta z$ and then

```
K = ((phi(1:end-1) + phi(2:end))/2).^n;                % form permeability
b = [0; dz*(K(2:end) - K(1:end-1)); 0];                % form right-hand side
Dm = [K(1:end-1); 0];                                  % form subdiag of A
D  = [1; -(dz^2 + K(1:end-1) + K(2:end)); 1];          % form diag of A
Dp = [0; K(2:end)];                                    % form superdiag of A
Nz = length(phi);                                      % matrix size
A = spdiags([[Dm; 0], D, [0; Dp]], -1:1, Nz, Nz);      % create sparse matrix
cmp = A\b;                                             % solve
```

Numerical solutions using this finite-difference method can be compared with analytical solutions to assess their accuracy. To do so, we define the error of the numerical

[2]In stencil notation, multiplication of a stencil $[S_a,\ S_b,\ S_c]$ with a discrete variable q_i is computed as $S_aq_{i-1}+S_bq_i+S_cq_{i+1}$.

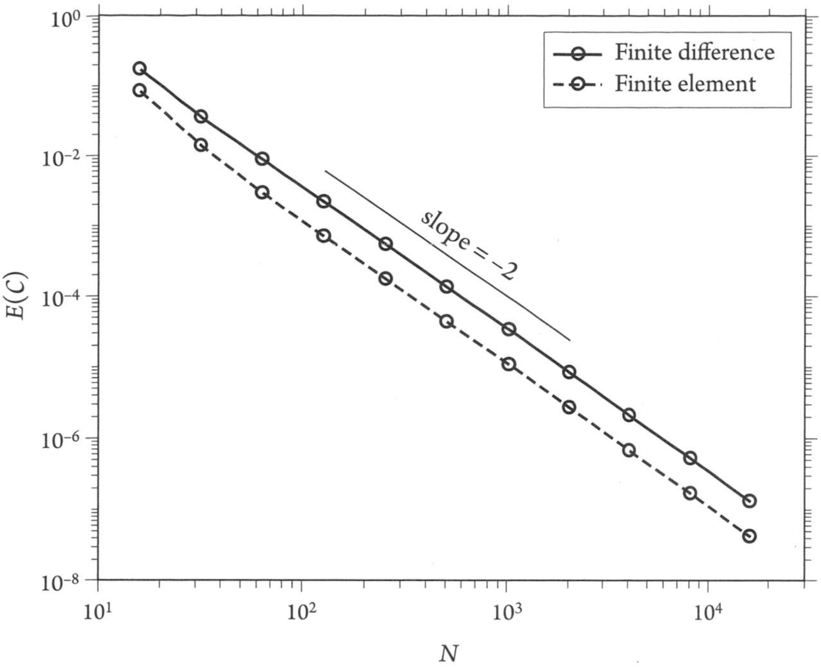

Figure 14.2. Error E versus number of nodes N_z for numerical solutions of the compaction equation (14.2). The solid line marks the error for the finite-difference method (14.13). The dashed line marks the error for the finite-element method. The analytical solution is shown in figure 6.5(b).

solution as

$$E(\mathcal{C}) \equiv \frac{\left|\left|\mathcal{C}(z_i) - \tilde{\mathcal{C}}_i\right|\right|_2}{||\mathcal{C}(z_i)||_2}, \tag{14.15}$$

where $||\cdot||_2$ indicates the $L2$ norm of the vector and $\mathcal{C}(z_i)$ is the analytical solution evaluated at the nodes. The denominator in this definition ensures that the size of the error is expressed relative to the size of the solution, rather than in absolute terms. A plot of E versus N_z is shown in figure 14.2. The slope is approximately -2, indicating that the error decreases by a factor of four when the number of nodes is increased by a factor of two. Hence we say that our discretization, equation (14.13), is *second order* in the grid spacing.

14.1.2 FINITE-ELEMENT DISCRETIZATION

The other classic approach to discretizing PDEs is the finite-element method, which solves the weak form of the PDE. Following the discussion surrounding equation (14.5), we obtain the weak form by multiplying the residual by a test function $\mathcal{W}$, integrating, and setting the result equal to zero

$$\int_\Omega \left(\mathcal{W}\frac{\mathrm{d}}{\mathrm{d}z}K\frac{\mathrm{d}\tilde{\mathcal{C}}}{\mathrm{d}z} - \mathcal{W}\tilde{\mathcal{C}} - \mathcal{W}\frac{\mathrm{d}K}{\mathrm{d}z}\right)\mathrm{d}\Omega = 0, \tag{14.16}$$

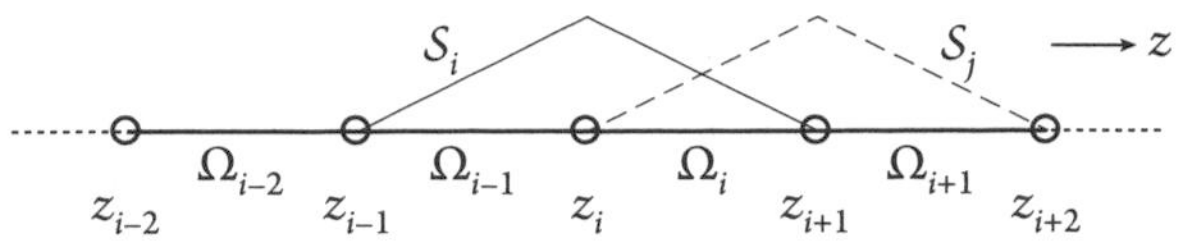

Figure 14.3. A segment of the grid for the finite-element discretization. The elements of the domain are labeled according to their index. The basis function is shown at two distinct values of *i*. Basis functions used here are linear "hat" functions with unit amplitude.

where, for concision, we have written $K \equiv \varphi^n$ and we refer to the domain as $\Omega \in [-z_m, z_m]$. The first and last terms in this equation can be integrated by parts:

$$\int_\Omega \mathcal{W}\frac{\mathrm{d}}{\mathrm{d}z}K\frac{\mathrm{d}\tilde{\mathcal{C}}}{\mathrm{d}z}\mathrm{d}\Omega = \left[\mathcal{W}K\frac{\mathrm{d}\tilde{\mathcal{C}}}{\mathrm{d}z}\right]_{-z_m}^{z_m} - \int_\Omega K\frac{\mathrm{d}\mathcal{W}}{\mathrm{d}z}\frac{\mathrm{d}\tilde{\mathcal{C}}}{\mathrm{d}z}\mathrm{d}\Omega, \tag{14.17a}$$

$$\int_\Omega \mathcal{W}\frac{\mathrm{d}K}{\mathrm{d}z}\mathrm{d}\Omega = [\mathcal{W}K]_{-z_m}^{z_m} - \int_\Omega K\frac{\mathrm{d}\mathcal{W}}{\mathrm{d}z}\mathrm{d}\Omega. \tag{14.17b}$$

Substituting these into (14.16), expressing derivatives with subscripts, and rearranging gives

$$\int_\Omega \left(K\mathcal{W}_z\tilde{\mathcal{C}}_z + \mathcal{W}\tilde{\mathcal{C}}\right)\mathrm{d}\Omega = \int_\Omega K\mathcal{W}_z\mathrm{d}\Omega - [\mathcal{W}K]_{-z_m}^{z_m} + \left[\mathcal{W}K\tilde{\mathcal{C}}_z\right]_{-z_m}^{z_m}. \tag{14.18}$$

This is the weak form of the PDE that we wish to solve numerically. The last term represents the boundary conditions on the gradient of the compaction rate; since we have Dirichlet boundary conditions, we will not need to evaluate this term (formally, we can take $\mathcal{W}(\pm z_m) = 0$)).

To discretize the weak form, we first need a grid of nodes. This is shown in figure 14.3. As we did for the finite-difference discretization, we choose to place the nodes at equally spaced intervals with $\Delta z = 2z_m/(N_z - 1)$. The spaces between nodes are the *elements* Ω_i; they are numbered according to their adjacent, smaller-z node. Instead of discretizing the derivatives as for the finite-difference method, in the finite-element method we discretize the unknown function by expressing it in terms of *basis functions*. There are many choices of basis function but here we use a simple one, the linear "hat" function,

$$\mathcal{S}_i(z) = \begin{cases} (z - z_{i-1})/\Delta z & \text{in } \Omega_{i-1}, \\ -(z - z_i)/\Delta z & \text{in } \Omega_i, \\ 0 & \text{otherwise.} \end{cases} \tag{14.19}$$

Two functions $\mathcal{S}_i$ and $\mathcal{S}_{i+1}$ are plotted in figure 14.3. This function is unity at the central node and zero at every other node. Hence it is orthogonal to all other $\mathcal{S}_{j\neq i}$ and provides a basis to uniquely approximate a function. This approximation is written

$$\tilde{\mathcal{C}}(z) \approx \sum_{i=1}^{N_z} c_i\mathcal{S}_i(z). \tag{14.20}$$

Here c_i is a set of constant, unknown coefficients that scale the basis functions to approximate $\tilde{\mathcal{C}}$. Finding a solution to the problem means finding the numerical values for c_i that minimize the integrated residual (14.5).

The test function $\mathcal{W}$ that weights the residual is an arbitrary function and hence we must choose a means to express it numerically. It is a convenient and canonical choice to approximate $\mathcal{W}$ using the same basis functions as were used for $\tilde{\mathcal{C}}$; this is known as *Galerkin finite elements*. Hence, we write

$$\mathcal{W}(z) \approx \sum_{j=1}^{N_z} \hat{c}_j \mathcal{S}_j(z). \tag{14.21}$$

Now, instead of an arbitrary function $\mathcal{W}$, we have an arbitrary set of coefficients $\hat{c}_j$.

Forging ahead, we substitute (14.20) and (14.21) into the weak form (14.18). Then, rearranging the order of integrals and sums, we have

$$\begin{aligned}\sum_{j=1}^{N_z} \hat{c}_j \sum_{i=1}^{N_z} c_i \int_\Omega \left(K\mathcal{S}_j'\mathcal{S}_i' + \mathcal{S}_j\mathcal{S}_i\right) \mathrm{d}\Omega = \sum_{j=1}^{N_z} \hat{c}_j \int_\Omega K\mathcal{S}_j' \mathrm{d}\Omega \\ - [\mathcal{W}K]_{-z_m}^{z_m} + \left[\mathcal{W}K\tilde{\mathcal{C}}_z\right]_{-z_m}^{z_m},\end{aligned} \tag{14.22}$$

where we use a prime to indicate z-derivatives of basis functions. Now recall that that the coefficients of the test-function expansion $\hat{c}_j$ are arbitrary. A convenient choice is $\hat{c}_j = \delta_{je}$, the Kronecker delta. Substituting this choice and dropping the now-obsolete sums over j gives

$$\sum_{i=1}^{N_z} c_i \int_{\Omega_e} \left(K\mathcal{S}_e'\mathcal{S}_i' + \mathcal{S}_e\mathcal{S}_i\right) \mathrm{d}\Omega = \int_{\Omega_e} K\mathcal{S}_e' \mathrm{d}\Omega. \tag{14.23}$$

The integrals, which previously were taken over the full domain Ω, are now taken over each element, indexed by e. Hence, by our choice of $\hat{c}_j$, we find that the weak form must be satisfied elementwise. We have dropped boundary terms evaluated at $e = 1, N_z$ ($\pm z_m$) because we have Dirichlet conditions (14.8), which immediately translate to $c_1 = c_{N_z} = 0$.

Equation (14.23) can be expressed in terms of a matrix–vector product $\mathbf{A}\boldsymbol{x} = \boldsymbol{b}$, where

$$\mathbf{A} = \mathscr{A} \int_{\Omega_e} \left(K\mathcal{S}_e'\mathcal{S}_i' + \mathcal{S}_e\mathcal{S}_i\right) \mathrm{d}\Omega \qquad \text{(the discrete \textit{bilinear form})}, \tag{14.24a}$$

$$\boldsymbol{b} = \mathscr{A} \int_{\Omega_e} K\mathcal{S}_e' \mathrm{d}\Omega \qquad \text{(the discrete \textit{linear form})}, \tag{14.24b}$$

and the vector of unknowns $\boldsymbol{x}$ represents the coefficients c_i. In (14.24), $\mathscr{A}$ is the assembly operator. To clarify the assembly operation, we write the bilinear and linear forms evaluated over one element Ω_j, which is written, respectively, as the submatrix and subvector

$$\mathbf{A}^{\Omega_j} = \int_{\Omega_j} \begin{pmatrix} K\mathcal{S}_j'\mathcal{S}_j' + \mathcal{S}_j\mathcal{S}_j & K\mathcal{S}_j'\mathcal{S}_{j+1}' + \mathcal{S}_j\mathcal{S}_{j+1} \\ K\mathcal{S}_{j+1}'\mathcal{S}_j' + \mathcal{S}_{j+1}\mathcal{S}_j & K\mathcal{S}_{j+1}'\mathcal{S}_{j+1}' + \mathcal{S}_{j+1}\mathcal{S}_{j+1} \end{pmatrix} \mathrm{d}\Omega, \tag{14.25a}$$

$$\boldsymbol{b}^{\Omega_j} = \int_{\Omega_j} \begin{pmatrix} K\mathcal{S}_j' \\ K\mathcal{S}_{j+1}' \end{pmatrix} \mathrm{d}\Omega. \tag{14.25b}$$

Then the assembly of the global matrix $\mathbf{A}$ and global vector $\boldsymbol{b}$ involve summing the entries of $\mathbf{A}^{\Omega_j}$ and $\boldsymbol{b}^{\Omega_j}$ into the correct locations (recalling that $\mathbf{A}$ is symmetrical).

Computing the entries in $\mathbf{A}^{\Omega_j}$ and $\boldsymbol{b}^{\Omega_j}$ requires integration; in general, this is performed by *numerical quadrature*. For our present purposes, however, the integrals can be approximated with the trapezoidal rule and formed analytically.

For the one-dimensional problem at hand, the bilinear form is a $N_z \times N_z$ tridiagonal matrix and the linear form is a $N_z \times 1$ vector. Matrix assembly is performed by looping over all of the $N_z - 1$ elements, performing the integral for all basis functions that overlap the element to form 2×2 matrix blocks, and adding those blocks into the global bilinear-form matrix. A similar procedure is followed for the linear form. A tabulated connectivity map facilitates matching of each element e with its bounding nodes i, $i+1$. In MATLAB, we set up our variables and arrays, allocate memory, and build the connectivity map with

```
Nz = length(phi);                        % number of nodes
Nz_elements = Nz-1;                      % number of elements
A = sparse([],[],[],Nz,Nz,3*Nz);         % allocate memory
b = zeros(Nz,1);                         % allocate memory
map = zeros(Nz_elements,2);              % allocate memory
for e=1:Nz_elements                      % loop over elements
   map(e,:) = [e e+1];                   % connectivity: elements & nodes
end
```

We then assemble the global matrix and right-hand-side vectors with

```
K = ((phi(1:end-1) + phi(2:end))/2).^n;  % permeability at element centers
for e=1:Nz_elements                      % loop over elements
   Ae = [1 -1; -1  1]*K(e)/dz + \ldots     %  element bilinear form
        [dz/3 dz/6; dz/6 dz/3];
   be  = [-1; 1]*K(e);                    %  element linear form
   map_e = map(e,:);                      %  map from element to nodes
   A(map_e,map_e) = A(map_e,map_e) + Ae;  %  assemble global matrix
   b(map_e) = b(map_e) + be;              %  assemble global right-hand side
end
```

This loop results in the $N_z \times N_z$ global matrix A =

$$\begin{pmatrix} A_{11}^{\Omega_1} & A_{12}^{\Omega_1} & & & & \\ A_{21}^{\Omega_1} & A_{22}^{\Omega_1} + A_{11}^{\Omega_2} & A_{12}^{\Omega_2} & & & \\ & A_{21}^{\Omega_2} & A_{22}^{\Omega_2} + A_{11}^{\Omega_3} & A_{12}^{\Omega_3} & & \\ & & & \ddots & & \\ & & & A_{21}^{\Omega_{N_z-2}} & A_{22}^{\Omega_{N_z-2}} + A_{11}^{\Omega_{N_z-1}} & A_{12}^{\Omega_{N_z-1}} \\ & & & & A_{21}^{\Omega_{N_z-1}} & A_{22}^{\Omega_{N_z-1}} \end{pmatrix}, \tag{14.26}$$

where the entries are given by (14.25a) with integration by the trapezoidal rule. Note that A is tridiagonal, as was the case for the finite-difference method.

Finally, we zero the rows corresponding to boundary nodes, impose the Dirichlet boundary conditions, and solve for the coefficients c_i with

```
A([1,Nz],:) = 0;                    % zero boundary rows
A(1,1) = 1; b(1) = 0;               % boundary condition
A(Nz,Nz) = 1; b(Nz) = 0;            % boundary condition
cmp = A\b;                          % solve
```

A plot of the error E (defined in equation (14.15)) as a function of the number of nodes N_z is shown in figure 14.2. Solid second-order convergence is evident in the slope of the line. For this test problem, the finite-element method is slightly more accurate than the finite-difference method.

14.2 The One-Dimensional Solitary Wave (Time-Dependent)

The solitary wave problem is time-dependent: the rate of change of the porosity is proportional to the compaction rate. Section 14.1 solved the instantaneous problem of computing the compaction rate given a distribution of porosity. In this section, we solve the coupled problem of obtaining the compaction rate and updating the porosity—marching forward through time in discrete steps. Indeed, time is discretized according to the time step

$$\Delta t = \frac{T}{N_t - 1}, \tag{14.27}$$

where the time domain is $t \in [0, T]$ and we discretize it into $N_t - 1$ intervals (steps) using N_t discrete values of time. We will refer to those values as t_j for $j = 1, 2, 3, \ldots, N_t$.

Stepping forward in time then involves updating the porosity from time t_j to time t_{j+1}. To do so we solve the dimensionless mass conservation equation for the solid phase, assuming small porosity $(1 - \phi \sim 1)$ and neglecting advection. Repeating equation (6.31b),

$$\frac{\partial \varphi}{\partial t} = \mathcal{C}, \tag{14.28}$$

which is coupled to equation (14.2). We previously obtained an analytical, traveling-wave solution for this system in section 6.4, so we aim to reproduce that solution here, including the wave speed.

Equation (14.28) can be discretized in a variety of ways, but one flexible and simple approach is to write

$$\frac{\varphi_i^{j+1} - \varphi_i^j}{\Delta t} = \theta \mathcal{C}_i^j + (1 - \theta)\mathcal{C}_i^{j+1}, \tag{14.29}$$

for some $0 \le \theta \le 1$. In this equation, subscripts indicate the position on the spatial grid whereas superscripts indicate the position in time. On the left-hand side is a finite-difference discretization of the time derivative. The porosity at the current time step φ_i^j is known for all $i \in [1, N_z]$, whereas the porosity at the future time step φ_i^{j+1} is unknown. On the right-hand side, we evaluate the compaction rate as a linear combination of the current and future steps.[3] Note that when we begin the time-step procedure, the

[3] Textbooks covering finite-difference methods should be consulted for a full discussion of these and other methods, their stability and their accuracy. See the Literature Notes at the end of this chapter.

compaction rate at the future time $\mathcal{C}_i^{j+1}$ is unknown; it depends on the porosity at the future time, which is also unknown.

Rearranging equation (14.29) to isolate the future porosity,

$$\varphi_i^{j+1} = \varphi_i^j + \Delta t\theta \mathcal{C}_i^j + \Delta t(1-\theta)\mathcal{C}_i^{j+1}. \tag{14.30}$$

This equation must be solved together with the discrete compaction equation (14.13) in stencil form, which we rewrite with superscripts to indicate the discrete time,

$$\left[\; K_{i-1/2}^{j+1}, \quad -\left(\Delta z^2 + K_{i-1/2}^{j+1} + K_{i+1/2}^{j+1}\right), \quad K_{i+1/2}^{j+1} \;\right] \tilde{\mathcal{C}}_i = \Delta z \left(K_{i+1/2}^{j+1} - K_{i-1/2}^{j+1}\right), \tag{14.31}$$

where we have again used K to concisely represent the dimensionless mobility.

Equations (14.30) and (14.31) are a coupled set of nonlinear equations. The nonlinearity comes from the porosity–permeability relationship. One approach to solving them would be to write down the full system of discrete, nonlinear algebraic equations for the porosity and the compaction rate and solve them together, using Newton's method at each time step. Here, to restrict ourselves to linear systems and solvers, we approach each time step using the following stages:

1. Solve equation (14.30) with $\theta = 1$ for $\left(\varphi_i^{j+1}\right)^*$, which is a first estimate of the future porosity.
2. Use $\left(\varphi_i^{j+1}\right)^*$ to solve equation (14.31) for $\left(\mathcal{C}_i^{j+1}\right)^*$, which is a first estimate of the future compaction rate.
3. Use $\left(\mathcal{C}_i^{j+1}\right)^*$ to solve equation (14.30) with $\theta = \frac{1}{2}$ for φ_i^{j+1}, the porosity at t_{j+1}.
4. Use φ_i^{j+1} to solve equation (14.31) for $\mathcal{C}_i^{j+1}$, the compaction rate at t_{j+1}.

Since we assume that the porosity (and permeability) are known in stages 2 and 4 of this algorithm, (14.31) is a linear equation for compaction rate (as in section 14.1, above). To gain additional accuracy, we can iterate stages 3 and 4; this is called a *fixed-point* or *Picard iteration*. MATLAB code implementing our algorithm includes

```
phi_j = phi_initial;                          % init porosity
cmp_j = SolveCmpRate(phi_initial,dz,n);       % init compaction rate
for i=1:Nt                                    % time-step loop
  phi_jp = phi_j + dt*cmp_j;                  % 1. update porosity
  cmp_jp = SolveCmpRate(phi_jp,dz,n);         % 2. solve cmp rate
  for k=1:picard_its                          % Picard iteration
    phi_jp = phi_j + 0.5*dt*(cmp_j + cmp_jp); % 3. update porosity
    cmp_jp = SolveCmpRate(phi_jp,dz,z);         % 4. solve cmp rate
  end
  t = t + dt;                                 % update time
  phi_j = phi_jp;                             % prepare for next step
  cmp_j = cmp_jp;                             % prepare for next step
end
```

where `SolveCmpRate()` is a function defined using the code from section 14.1.1, above. The initial condition is contained in `phi_initial`.

We consider the convergence of the time-dependent solution to the analytical solution for a solitary wave in exercise 14.3, below.

14.3 A Two-Dimensional Manufactured Solution (Instantaneous)

In this section we seek numerical solutions to the instantaneous flow problem for a given, two-dimensional porosity field. The system of governing equations, linear in the unknowns $\boldsymbol{v}^s$ and P^ℓ, are (4.47a) and (4.47b). Analytical solutions to these full equations can form the basis of a numerical benchmark (as in section 14.1.1 above), where they are available. For example, the benchmark described in section 14.5 is based on the simulated growth rate of porosity bands under simple shear of the solid phase. But that example illustrates how analytical solutions obtained under idealized conditions can be inconvenient as benchmarks of numerical models: the analytical growth rate of equation (14.48) is strictly valid for an infinite domain. An infinite domain is obviously difficult to realize in a code framework deployed on finite computing resources.

Here we use *the method of manufactured solutions* as an approach to check numerical results for convergence. This method allows us to create a benchmark problem without actually solving the PDEs analytically. In rough terms, it proceeds by choosing functions for each of the variable fields that are convenient and sufficiently differentiable, then introducing appropriate forcing terms into the equations such that the chosen functions are indeed a solution of the system. More detail is provided below.

For simplicity of exposition, we assume that $\eta_\phi \equiv \eta_0$, a constant. We further take $\zeta_\phi \equiv \frac{5}{3}\eta_0$, $k_\phi/\mu^\ell \equiv K_0(\phi/\phi_0)^n$ and rewrite the pressure as $\nabla P^\ell \equiv \nabla P + \rho^s\mathbf{g}$. We nondimensionalize the equations by choosing scales

$$\left[\boldsymbol{v}^s\right] = K_0\Delta\rho g,\ \ [P] = \Delta\rho g H,\ \ [\nabla] = 1/H, \tag{14.32}$$

where H is a length that is characteristic of the domain size. Substituting these into the governing equations (4.47a) and (4.47b) gives

$$-\nabla\cdot\boldsymbol{v}^s + \nabla\cdot\left[K_\phi\left(\nabla P + \hat{\boldsymbol{y}}\right)\right] = F_{\text{mfc}}, \tag{14.33a}$$

$$-\nabla P + \delta^2\nabla^2\boldsymbol{v}^s + 2\delta^2\nabla\left(\nabla\cdot\boldsymbol{v}^s\right) - \phi\hat{\boldsymbol{y}} = \boldsymbol{G}_{\text{mfc}}. \tag{14.33b}$$

where all symbols are now dimensionless and where $K_\phi = (\phi/\phi_0)^n$ is the mobility, $\hat{\boldsymbol{y}} = \mathbf{g}/g$ is the direction of gravitational acceleration, and $\delta = \delta_0/H$ is a nondimensional compaction length (δ_0 is the reference, dimensional compaction length). Here we have augmented the governing equations by introducing *forcing functions* F_{mfc} (scalar) and $\boldsymbol{G}_{\text{mfc}}$ (vector) on the right-hand side. For models of the natural system of partially molten rock, these are zero.

In the method of manufactured solutions, however, F_{mfc} and $\boldsymbol{G}_{\text{mfc}}$ are nonzero. They are obtained by substituting the analytical functions $\boldsymbol{v}^s_{\text{mfc}}$ and P_{mfc} into the augmented governing equations (14.33). The functions $\boldsymbol{v}^s_{\text{mfc}}$ and P_{mfc} are the "manufactured solution"–they are created, following some simple guidelines, to become a benchmark comparison for the numerical solution. Analytical expressions for F_{mfc} and $\boldsymbol{G}_{\text{mfc}}$ can be evaluated numerically, in the context of code designed to solve the governing system of equations (14.33). The numerical solution forced by F_{mfc} and $\boldsymbol{G}_{\text{mfc}}$ should converge to the manufactured solution. Hence we define the solution error as

$$E(P) \equiv \frac{||P_{\text{mfc}}(\boldsymbol{x}_i) - \tilde{P}_i||_2}{||P_{\text{mfc}}(\boldsymbol{x}_i)||_2}, \tag{14.34a}$$

$$E(\boldsymbol{v}^s) \equiv \frac{||\boldsymbol{v}^s_{\text{mfc}}(\boldsymbol{x}_i) - \tilde{\boldsymbol{v}}^s_i||_2}{||\boldsymbol{v}^s_{\text{mfc}}(\boldsymbol{x}_i)||_2}, \tag{14.34b}$$

and investigate the behavior of these error metrics as a function of grid spacing. Tildes above variables remind us that these are approximations of the true solution. Equations (14.34a) represent problems in any number of spatial dimensions if we allow i to index entries into vectors with a consistent ordering of unknowns (i.e., the ordering need not correspond to distance along a one-dimensional domain).

To proceed with a specific example, we must specify a manufactured solution and compute the associated forcing functions. It is convenient (though not necessary) to do so by using an analytically equivalent form of the governing equations, which we obtain with the vector identity $\nabla^2 \boldsymbol{V} = -\nabla\times\nabla\times \boldsymbol{V} + \nabla(\nabla\cdot \boldsymbol{V})$. Then equation (14.33b) becomes

$$-\nabla P - \delta^2\nabla\times\nabla\times \boldsymbol{v}^s + 3\delta^2\nabla\left(\nabla\cdot \boldsymbol{v}^s\right) - \phi\hat{\boldsymbol{y}} = \boldsymbol{G}_{\text{mfc}}. \tag{14.35}$$

This can be simplified further by making a Helmholtz decomposition of the velocity into two potentials (as in section 4.6.3),

$$\boldsymbol{v}^s \equiv \nabla\times\boldsymbol{\Psi} + \nabla\mathcal{U}, \tag{14.36}$$

where the vector potential can be expressed as $\boldsymbol{\Psi} \equiv \psi\hat{\boldsymbol{z}}$ for the purposes of our problem, in which we'll consider solutions that vary only in the x, y-plane. With this decomposition we can write $\nabla\cdot\boldsymbol{v}^s = \nabla^2\mathcal{U}$ and $-\nabla\times\nabla\times\boldsymbol{v}^s = \nabla\times\nabla^2(\psi\hat{\boldsymbol{z}})$.[4] Then the system (14.33) finally becomes

$$F_{\text{mfc}} = -\nabla^2\mathcal{U}_{\text{mfc}} + \nabla\cdot\left[\left(\frac{\phi_{\text{mfc}}}{\phi_0}\right)^n\left(\nabla P_{\text{mfc}} + \hat{\boldsymbol{y}}\right)\right], \tag{14.37a}$$

$$\boldsymbol{G}_{\text{mfc}} = -\nabla P_{\text{mfc}} + \delta^2\nabla\times\left(\nabla^2\psi_{\text{mfc}}\right)\hat{\boldsymbol{z}} + 3\delta^2\nabla\left(\nabla^2\mathcal{U}_{\text{mfc}}\right) - \phi_{\text{mfc}}\hat{\boldsymbol{y}}. \tag{14.37b}$$

To emphasize that this is a recipe for computing the forcing terms, we have rewritten the system with the forcing terms on the left-hand side and appended subscripts to the field variables.

Next we must choose forms for the manufactured solution $\phi_{\text{mfc}}, P_{\text{mfc}}, \psi_{\text{mfc}}, \mathcal{U}_{\text{mfc}}$, all of which are functions of x and y. In choosing these functions, it is helpful to follow some basic guidelines:

- Manufactured solutions should be expressed in terms of analytical functions that are smooth and conveniently differentiated;
- They should be sufficiently general that they make a nontrivial contribution to the forcing from every term in which they appear. This means that they should have more nonzero derivatives than the order of the terms in which they appear.

[4]Here we have used the vector identity $-\nabla\times\nabla\times\nabla\times \boldsymbol{V} = \nabla\times\nabla^2\boldsymbol{V}$. Note that since $\psi = \psi(x,y)$ we have

$$\nabla^2(\psi\hat{z}) = \left(\frac{\partial^2\psi}{\partial x^2} + \frac{\partial^2\psi}{\partial y^2}\right)\hat{z} = \left(\nabla^2\psi\right)\hat{z}.$$

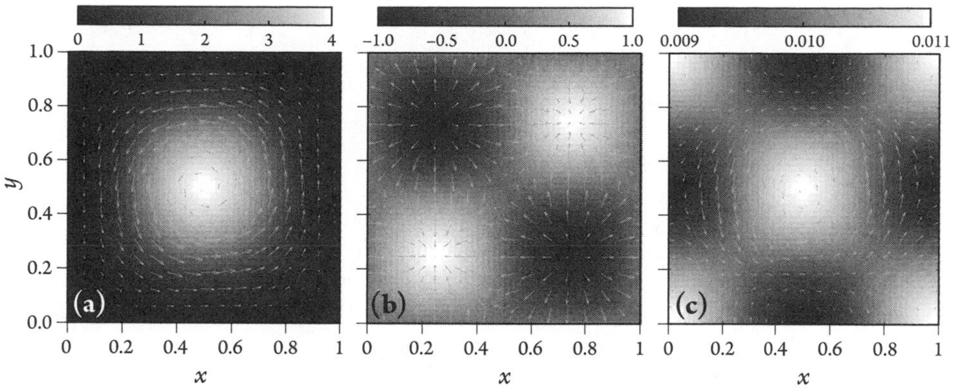

Figure 14.4. The manufactured solution, computed using equations (14.39), for $m=2$, $n=3$, $\delta=1$, $P^*=\psi^*=\mathcal{U}^*=1$, and $\phi^*=0.1$. All quantities are dimensionless. The pressure P_{mfc} is not shown. **(a)** The shear potential ψ_{mfc} is shown in gray scale; vectors illustrate $\nabla\times\psi_{\mathrm{mfc}}\hat{z}$, the incompressible part of the flow. **(b)** The compaction potential $\mathcal{U}_{\mathrm{mfc}}$ is shown in gray scale; vectors illustrate $\nabla\mathcal{U}_{\mathrm{mfc}}$, the compaction part of the flow. **(c)** The porosity ϕ_{mfc} is shown in gray scale; vectors illustrate the total solid-flow field $\boldsymbol{v}^s_{\mathrm{mfc}}$.

- The manufactured solution and relevant derivatives should be bounded by a small constant to ensure that they are feasibly resolved by floating-point numbers on a discrete grid with a reasonable number of nodes.
- They should be chosen so that the terms in the equation make physical sense (e.g., no negative porosities and permeabilities) and the numerical solution can converge.

These guidelines do not imply a requirement that the manufactured solution is physically realistic, which, in our case, might mean similar to the behavior of partially molten rocks.

For the unit-square domain

$$x\in[0,1],\ y\in[0,1], \tag{14.38}$$

a reasonable choice (among many) that satisfies the guidelines is

$$\phi_{\mathrm{mfc}}=\phi_0\left[1+\phi^*\cos\left(m\pi x\right)\cos\left(m\pi y\right)\right], \tag{14.39a}$$

$$P_{\mathrm{mfc}}=P^*\sin\left(m\pi x\right)\sin\left(m\pi y\right), \tag{14.39b}$$

$$\psi_{\mathrm{mfc}}=\psi^*\left[1-\cos\left(m\pi x\right)\right]\left[1-\cos\left(m\pi y\right)\right], \tag{14.39c}$$

$$\mathcal{U}_{\mathrm{mfc}}=\mathcal{U}^*\sin\left(m\pi x\right)\sin\left(m\pi y\right), \tag{14.39d}$$

where m is an even integer. Larger values of m introduce smaller length scales into the manufactured solution. The amplitudes $P^*, \psi^*, \mathcal{U}^*$ may be taken to be unity, but $|\phi^*|<1$ is required for positivity of the porosity; a reasonable choice is $\phi^*=0.1$.

Parts of the manufactured solution are shown in figure 14.4; (a) shows the incompressible part of the solid velocity while (b) shows the compaction part. In both cases, the associated velocity vectors are overlaid. Vectors in (c) show the full velocity field,

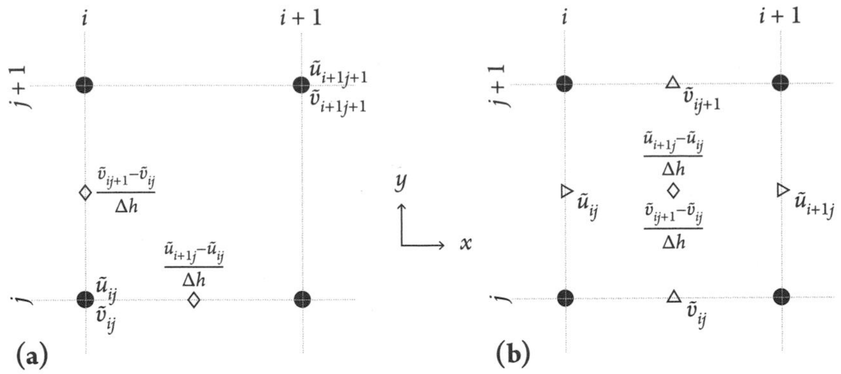

Figure 14.5. Patches of a finite-difference grids with grid spacing Δh and Δh and grid indices i (columns) and j (rows). Nodes are represented by filled black circles. **(a)** A node-centered grid with the velocity components located at the nodes. **(b)** A staggered, cell-centered grid with velocity components located at cell faces. Note that each velocity component is positioned on the face that has a normal vector pointing in the direction associated with that component.

with compaction and shear parts. Note that it is nonzero on the boundaries. The gray scale in (c) shows the porosity field.

Finite-difference discretization. We now discretize equations (14.33) to put them in a form that is amenable to numerical solution. For simplicity, we assume that the number of nodes in each direction, N_x, N_y is the same and equal to $\sqrt{N}$. Hence the grid spacing is the same in each direction, $\Delta x = \Delta y \equiv \Delta h$.

The approximation of derivatives by finite differences is conducted in the same manner as for the compaction problem, above. In the present case, however, we must deal with vectors and vector operators. The most important issue arises in handling the divergence of the velocity. Writing the two-dimensional velocity in terms of its components $\boldsymbol{v}^s = u\hat{\boldsymbol{x}} + v\hat{\boldsymbol{y}}$, we can approximate the divergence as

$$\nabla \cdot \tilde{\boldsymbol{v}}^s = \lim_{\sqrt{N}\to\infty} \left(\frac{\tilde{u}_{ji+1} - \tilde{u}_{ji}}{\Delta h} + \frac{\tilde{v}_{j+1i} - \tilde{v}_{ji}}{\Delta h} \right). \tag{14.40}$$

The location where $\nabla \cdot \tilde{\boldsymbol{v}}^s$ sits on the grid depends on the locations where $\tilde{u}$ and $\tilde{v}$ are stored; figure 14.5 illustrates two possibilities. In (a), velocity components are located at the grid nodes (shown as filled circles). This is known as a node-centered grid. Note that with this choice, the two terms of the discrete divergence (equation (14.40)) are located in different places (shown as open diamonds).

In figure 14.5(b), the velocity components are organized differently. They are located at the midpoint of the grid-cell faces. A *grid cell* is defined by the box connecting four nodes. The finite differences associated with (14.40) now place both terms in the same point, at the center of the cell. This arrangement is called a staggered, cell-centered grid. Velocity values are located at positions

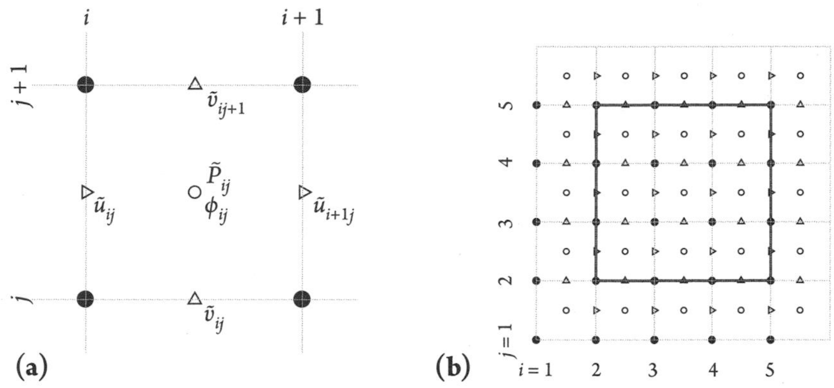

Figure 14.6. The staggered grid. $\hat{x}$ points in the direction of increasing grid-index i; $\hat{y}$ points in the direction of increasing j. **(a)** One cell within the staggered grid. Pressure and porosity are located at the cell centre; velocities are staggered on the faces. **(b)** A coarse staggered grid; symbols are as in (a). The boundaries of the domain (the unit square) are shown by the heavy black line. Variables outside the domain are used to impose boundary conditions.

$$\boldsymbol{x}\left(\tilde{u}_{ji}\right) = (i-2)\Delta h\hat{\boldsymbol{x}} + (j-3/2)\Delta h\hat{\boldsymbol{y}}, \tag{14.41a}$$

$$\boldsymbol{x}\left(\tilde{v}_{ji}\right) = (i-3/2)\Delta h\hat{\boldsymbol{x}} + (j-2)\Delta h\hat{\boldsymbol{y}}. \tag{14.41b}$$

This arrangement differs from that of the node-centered grid in three important ways. First, the locations of the horizontal and vertical velocity components are different. Second, neither of these components is colocated with a grid node; instead they are spaced halfway between grid nodes. Third, the offset between the node number (j, i) and the position of variables is such that $\boldsymbol{x}\left(\tilde{u}_{12}\right) = -\Delta h/2\hat{\boldsymbol{y}}$, i.e., the (1,2) entry is located outside the unit-square domain (similarly for $\tilde{v}_{ji}$). As we shall see below, this placement is used to enforce certain boundary conditions.

Figure 14.6(a) shows the position of all variables in the staggered grid. Pressure is located at the cell centers, as is porosity. The reason for this choice becomes evident when examining terms in the stress-balance equation (14.33b) for the partially molten aggregate. Considering the x-direction,

$$\hat{\boldsymbol{x}} \cdot (\nabla P) = \lim_{\sqrt{N}\to\infty} \frac{\tilde{P}_{ji} - \tilde{P}_{ji-1}}{\Delta h}, \tag{14.42a}$$

$$\hat{\boldsymbol{x}} \cdot \left(\nabla^2 \boldsymbol{v}^s\right) = \lim_{\sqrt{N}\to\infty} \left(\frac{\tilde{u}_{ji-1} - 2\tilde{u}_{ji} + \tilde{u}_{ji+1}}{\Delta h^2} + \frac{\tilde{u}_{j-1i} - 2\tilde{u}_{ji} + \tilde{u}_{j+1i}}{\Delta h^2}\right), \tag{14.42b}$$

$$\hat{\boldsymbol{x}} \cdot \nabla\left(\nabla \cdot \boldsymbol{v}^s\right) = \lim_{\sqrt{N}\to\infty} \left(\frac{\tilde{u}_{ji-1} - 2\tilde{u}_{ji} + \tilde{u}_{ji+1}}{\Delta h^2} + \frac{\tilde{v}_{j+1i} - \tilde{v}_{ji} - \tilde{v}_{j+1i-1} + \tilde{v}_{ji-1}}{\Delta h^2}\right). \tag{14.42c}$$

Notice that the result of these finite-difference approximations are located at the same place on the grid: the position of $\tilde{u}_{ji}$. In the y-direction, the pressure gradient and the viscous terms are centered on the position of $\tilde{v}_{ji}$. The forcing on the right-hand side, $\boldsymbol{G}_{\text{mfc}}$, can be evaluated on the relevant cell faces. Only the body-force term $\phi\hat{\boldsymbol{y}}$ requires interpolation because ϕ is located at cell centers.

A coarse but complete grid is shown in figure 14.6(b). The symbols correspond to discrete variables as specified in (a). The heavy black line marks the boundary of the domain. Although the different variables are located at different positions on the grid (none of which are the positions of the nodes), their logical distributions are identical: tabulated according to two integer indices. Hence their storage in arrays is unaffected by their staggering on the grid.

Note that on the bottom and top boundary of the domain, where $y=0,1$, there are $\tilde{v}_{ij}$ values but no $\tilde{u}_{ji}$. Similarly, on the boundaries where $x=0,1$, there are $\tilde{u}_{ji}$ values but no $\tilde{v}_{ji}$. In the case of a variable that is not located on the boundary, how can we enforce a Dirichlet boundary condition? We use the discrete value that is just outside the boundary and force the linear interpolant that crosses the boundary to satisfy the Dirichlet condition. For example, to enforce the boundary condition $\tilde{v}\big|_{x=0}=v_0$, we use the discrete equation

$$\left(\tilde{v}_{j1}+\tilde{v}_{j2}\right)/2=v_0. \tag{14.43}$$

In solving the discrete equations for the manufactured solution, we prescribe Dirichlet boundary conditions on all variables, using the manufactured solution itself. Some of these conditions need to be imposed as in (14.43).

Python code to solve the system (14.33) discretized by finite differences, including those given in (14.42), is provided in the online supplement to this book. The forcing terms $\boldsymbol{G}_{\text{mfc}}, F_{\text{mfc}}$ are computed using the manufactured solution (14.39), which is also used to obtain values for Dirichlet boundary conditions. The code is too long and involved to quote here but some discussion of its strategy is appropriate.

As in the examples above, the code puts the discrete problem in the form of a system of linear, algebraic equations. A block-structured version of this system (ignoring the boundary conditions) can be written

$$\begin{pmatrix} \delta^2\left(\nabla^2+2\frac{\partial^2}{\partial x^2}\right) & 2\delta^2\frac{\partial^2}{\partial x\partial y} & -\frac{\partial}{\partial x} \\ 2\delta^2\frac{\partial^2}{\partial y\partial x} & \delta^2\left(\nabla^2+2\frac{\partial^2}{\partial y^2}\right) & -\frac{\partial}{\partial y} \\ -\frac{\partial}{\partial x} & -\frac{\partial}{\partial y} & \nabla\cdot\tilde{K_\phi}\nabla \end{pmatrix}\begin{pmatrix}\tilde{u}\\ \tilde{v}\\ \tilde{P}\end{pmatrix}=\begin{pmatrix}\hat{\boldsymbol{x}}\cdot\boldsymbol{G}_{\text{mfc}}\\ \hat{\boldsymbol{y}}\cdot\boldsymbol{G}_{\text{mfc}}+\phi\\ F_{\text{mfc}}-\frac{\partial\tilde{K_\phi}}{\partial y}\end{pmatrix}. \tag{14.44}$$

In this equation, $\tilde{u}$ represents a subvector containing all of the degrees of freedom associated with the x-component of velocity; this is similarly so for $\tilde{v}$ and $\tilde{P}$. The blocks in the matrix are labeled with the continuous version of the operator that they represent. The gradient of the compaction stress term, $\boldsymbol{\nabla\nabla}\cdot\tilde{\boldsymbol{v}}^s$, has been separated into two parts, according to the velocity component on which they operate. On the right-hand side are the manufactured forcing terms plus any terms from the PDEs that are independent of the variables $\tilde{P}$ and $\tilde{\boldsymbol{v}}^s$. The mobility $K_\phi=(\phi/\phi_0)^n$ is computed using the discretized porosity ϕ; the latter must be interpolated onto the cell faces for consistency with the discrete operators (hence the tilde ["$\sim$"] on the mobility). This is clarified with regard to the $\boldsymbol{\nabla}\cdot\tilde{K_\phi}\boldsymbol{\nabla}$ operator in exercise 14.4, below.

Each row in the block-matrix equation (14.44) represents a PDE: the first two rows represent the x- and y-components of the force balance (14.33b) while the third row

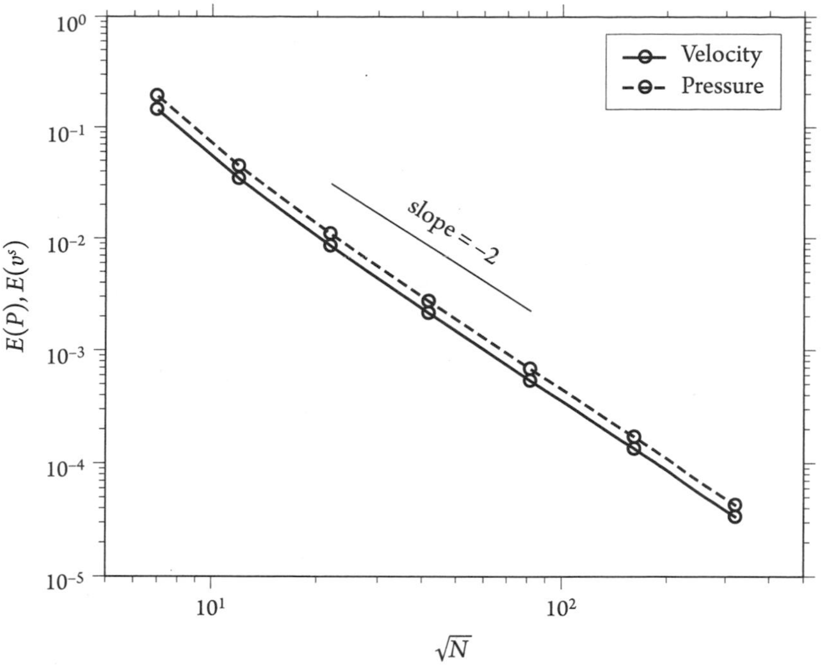

Figure 14.7. Error E versus number of nodes $\sqrt{N}$ along one direction. Solutions are obtained by finite-difference discretization of the Stokes/Darcy system (14.33). The solid line marks the velocity error; the dashed line marks the pressure error. The analytical (manufactured) solution is shown in figure 14.4.

represents the compaction equation (14.33a). Each row of the block matrix corresponds to N rows of the numerical $\mathbf{A}$ matrix; each column represents N columns. In the finite difference code, the nine sparse $N \times N$ blocks are formed separately and then assembled to form $\mathbf{A}$. Rows associated with boundary conditions are then appropriately reset. The right-hand-side vector is formed similarly, from three subvectors. MATLAB's backslash operator is used to obtain values for all the degrees of freedom. The solution fields $\tilde{P}_{ji}$ and $\tilde{\boldsymbol{v}}^s_{ji}$ are then extracted from the combined list. The error of the solution relative to the manufactured solution is computed according to equation (14.34a).

Figure 14.7 shows a plot of the error versus the number of grid points $\sqrt{N}$ in either Cartesian direction. As with the one-dimensional finite-difference calculations of section 14.1.1, we observe second-order convergence of the approximation. We are able to measure this convergence only because of our comparison with an analytical solution, which we manufactured without regard to any real, physical system. And yet the code is now verified to be correct and ready to be applied to a physical system.

14.4 Magmatic Solitary Waves as a Benchmark for Numerical Solutions

The analytical solution (6.49) for the form of a solitary wave provides the basis for benchmarking a one-dimensional solution of the equations governing time-dependent compaction. Because lower-dimensional solitary-wave solutions are unstable in higher dimensions, it is necessary to construct the solution of the dimension appropriate for the code to be tested.[5]

In benchmarking code, it is essential to recall that the analytical solutions for solitary waves are valid in the limit of small porosity. Free code that numerically solves a more general version of the equations will produce solutions that differ from the analytical results. These solutions should converge to the analytical solution for solitary porosity waves when $\phi_0 \ll 1$.

A sensible benchmark setup is therefore to initialize the porosity as $\phi_0 \ll 1$ and superimpose the analytical form of a solitary wave. The wave should have small amplitude and be of the appropriate dimension for the code. The domain should be large enough that the wave is initially distant from the boundaries. After running the code for a reasonable number of time steps, the output should be compared with the input. This can be done by recomputing the initial condition with the offset vt. Alternatively, one can impose a uniform background velocity on the solid phase of $\boldsymbol{v}^s = v\hat{\mathbf{g}}$—equal and opposite to that of the solitary wave. In this case, the numerically computed porosity field should be stationary and hence easily differenced with the initial condition. An advantage of this approach is that it tests the discretization of the porosity-advection term. With periodic boundary conditions in z, one can run the code such that the wave cycles once through the domain and returns to its starting position, where it can be differenced with the initial condition.

14.5 Porosity Bands as a Benchmark for Numerical Solutions

Results from the analysis of chapter 7 can be used as a benchmark for code that seeks to compute numerical solutions to the equations governing the mechanics. In fact, the analytical solutions of chapter 7 motivate two benchmarks: one for advection of porosity bands without growth or decay; the other for instantaneous growth of a porosity perturbation. For both cases, the calculation is initialized with a porosity field of

$$\phi(\boldsymbol{x},0) = \phi_0 + A\sin(\boldsymbol{k}_0 \cdot \boldsymbol{x}), \tag{14.45}$$

where $\phi^{(0)} = \phi_0$ corresponds to the leading-order, constant porosity in the stability analysis, and $A \ll \phi_0$ is the amplitude of perturbations. The initial wave vector is computed in terms of an initial angle θ_0 such that $\boldsymbol{k}_0 = k_0\left(\hat{\boldsymbol{x}}\sin\theta_0 + \hat{\boldsymbol{y}}\cos\theta_0\right)$. Setting the magnitude of the initial wavenumber $k_0 = 10\pi/H$ (where H is the height of the domain) provides a sufficient number of wavelengths within the domain. A good choice for θ_0 is $\pi/4$, although the benchmark should be conducted for a range of initial angles. Finally, it is advantageous to let the background compaction length δ_0 equal the domain height H.

Advection benchmark. If $\lambda = 0$ (and surface tension $\tau = 0$), the initial pattern of porosity variations is stable; it neither grows nor decays, but simply gets advected by the background, simple-shear flow, as was discussed in section 7.4.2. The flow rotates harmonic porosity perturbations, as described by equation (7.28). At any time $t > 0$, we can compare a numerically computed porosity field, which has evolved under advection

[5]Recipes and software for computing higher-dimensional solutions are provided by Simpson and Spiegelman [2011].

only, to the benchmark analytical solution,

$$\phi_B(\boldsymbol{x},t) = \phi_0 + A \sin\left[\boldsymbol{k}(t)\cdot\boldsymbol{x}\right] \tag{14.46}$$

with $\boldsymbol{k}(t) = k_0\left[\hat{\boldsymbol{x}}\sin\theta_0 + \hat{\boldsymbol{y}}(\cos\theta_0 - t\sin\theta_0)\right]$.

Compaction benchmark. If $\lambda > 0$ (and/or $\tau > 0$), the initial porosity pattern grows (or decays, depending on the angle θ_0 and wavenumber magnitude k) exponentially with time. The benchmark in this case is to compute the numerical compaction rate at $t=0$ and compare it with the analytical solution via equation (7.17). For the harmonic perturbation given in equation (14.45), this gives a benchmark compaction rate

$$\mathcal{C}_B = \frac{\dot{s}A}{1-\phi_0}\sin(\boldsymbol{k}_0\cdot\boldsymbol{x}), \tag{14.47}$$

where $\dot{s}$ is the appropriate growth-rate solution for the problem as formulated. If, for example, the viscosity is Newtonian, there is no surface tension and $\delta_0 k_0 = 10\pi$, $\dot{s}$ is given by (7.31) and the benchmark compaction rate is

$$\mathcal{C}_B = A\frac{\lambda\dot{\gamma}}{\nu_\phi}\cdot\frac{100\pi^2\sin 2\theta_0}{1+100\pi^2}\sin(\boldsymbol{k}_0\cdot\boldsymbol{x}). \tag{14.48}$$

Quantitative comparison can be made between this prediction and numerical results as a function of grid resolution and initial angle θ_0. It is important, however, to compare with a patch of the computational domain that is far from the influence of the boundaries; their effect on the solution is not captured by the benchmark analytical result.

14.6 Literature Notes

Numerical methods have been used in magma/mantle dynamics since the early days [e.g., Scott and Stevenson, 1984; Richter and McKenzie, 1984]. Until recently, however, there has been little analysis in the literature of the numerical schemes and their convergence. As an exception, Spiegelman [1993b] discussed the convergence of finite-difference solitary-wave solutions as a function of the nodes per compaction length. General methods for finite-difference and spectral discretization of differential equations (and analysis of their stability) are explained in Trefethen [1996].

One-dimensional problems are easily solved using methods from, for example, Numerical Recipes [Press et al., 2007] (now readily available in software such as MATLAB). Two- and three-dimensional problems can be more challenging to solve efficiently and accurately. Wiggins and Spiegelman [1995] adopted from Press et al. [2007] the use of a geometric multigrid for the compaction equation. A similar approach was used by Aharonov et al. [1997] for three-dimensional reactive flow calculations. In both of these examples, large-scale shear flows were neglected. Many authors have dealt with such flows by approximating them as incompressible and using Stokes solvers [e.g., Scott and Stevenson, 1989], or by writing the equations in potential form and using a multigrid method (although a description of the latter is difficult to find in the literature).

A solution of the fully-coupled, time-dependent equations in primitive-variable form (pressure and velocity) was first obtained by Ghods and Arkani-Hamed [2000] using pressure-correction methods developed by Patankar [1980] and Prakash and Patankar [1985]. In Katz et al. [2006, 2007], a different and more direct approach was developed. These studies discretized the primitive variables on a staggered grid using a finite-difference/finite-volume method. The coupled, nonlinear system of equations was solved using a Newton–Krylov scheme [e.g., Cai et al., 1997], obtained via an interface to PETSc [Balay et al., 2019, 2020]. This approach was extended by Katz and Worster [2008] and Katz [2008] to couple flow with the thermochemistry of phase change via the enthalpy method. Katz [2010] used staggered time stepping to address the separation of time scales between the liquid, which evolves rapidly, and the solid, which evolves slowly.

In thermochemical models, advective transport is closely coupled with melting and two-phase mechanics. Šrámek et al. [2010] made a careful comparison of advection schemes, including flux-limited total variation diminishing schemes, which were adopted by Katz and Weatherley [2012] for advecting mantle heterogeneity.

The finite-element method of discretizing equations and their solutions has found less-frequent use than have finite-difference methods; Keller et al. [2013] is an early example. It has the advantage of more flexible meshing, making the discretization of nonrectangular domains more straightforward. The convergence and efficient solution of finite-element discretizations for magma/mantle dynamics was analyzed by Rhebergen et al. [2014] and Rhebergen et al. [2015]. They were applied to model two-phase flow around a circular [Alisic et al., 2014] and spherical [Alisic et al., 2016] inclusion. [Wilson et al., 2014] used the finite-element method to model subduction. All of these applications leveraged FEniCS [Logg et al., 2012; Alnæs et al., 2015], an advanced numerical software library that offers automated code generation for finite-element problems.

One rather subtle issue in solving the two-phase dynamics is encountered when the domain contains patches where the porosity is zero. In such patches, the compaction equation drops out and certain variables (i.e., the liquid pressure and velocity) become undefined. Theory for handling this mathematical degeneracy was developed by Arbogast et al. [2017] and refined by Arbogast and Taicher [2017]. In the context of finite-difference and finite-volume methods, there are simpler (though less rigorous) means for dealing with this degeneracy [e.g., Katz et al., 2007; Dannberg et al., 2019].

For more information about the method of manufactured solutions, a convenient source is the Sandia technical report, Salari and Knupp [2000]. A more concise presentation is offered by Roache [2002].

14.7 Exercises

14.1 Show that the bilinear form (14.24a) for linear basis functions (14.19) is tridiagonal.

14.2 Explain how to write Neumann-type boundary conditions on a staggered grid. In particular, how could one enforce $\nabla P \cdot \hat{x} = 0$ on the $x = 0$ boundary in figure 14.6(b)?

14.3 Modify the code provided for the time-dependent problem of section 14.2 to create a periodic domain. Run your calculation to exactly the time within the model where the solitary wave returns to its initial position. Difference the final solution with the initial solution as a measure of error. How does the error change as a function of the number of Picard iterations taken? *Hint: Be very careful in your calculation of the grid spacing for the periodic domain.*

14.4 Write a finite-difference discretization of the term $\nabla \cdot \tilde{K_\phi} \nabla \tilde{P}$ for the staggered grid shown in figure 14.6. Note the implication for the grid locations of the discretized values of $\tilde{K_\phi}$. Explain three distinct ways to compute those values.

CHAPTER 15

Solutions to Exercises

15.1 Exercises from Chapter 3: One-Phase Mantle Dynamics

3.1 Combining $\boldsymbol{\nabla}\cdot\boldsymbol{v}=0$ with the definition of the streamfunction from equation (3.20) gives

$$\frac{\partial}{\partial x}\left(\frac{\partial\psi}{\partial y}\right)+\frac{\partial}{\partial y}\left(-\frac{\partial\psi}{\partial x}\right)=0.$$

Then, changing the order of differentiation of either term gives the expected result.

3.2 For an incompressible fluid, taking the divergence of equation (3.8) gives

$$\boldsymbol{\nabla}\cdot\boldsymbol{\sigma}=-\boldsymbol{\nabla}P+\eta\left[\nabla^2\boldsymbol{v}+\nabla\left(\nabla\cdot\boldsymbol{v}\right)\right]=-\boldsymbol{\nabla}P+\eta\nabla^2\boldsymbol{v}.$$

Substituting into (3.2) and noting that $\mathbf{g}=-g\hat{\boldsymbol{y}}$ yields (3.13).

3.3 (a) Substituting (3.23a) into (3.21) gives

$$\begin{aligned}\nabla^2\left(\nabla^2\psi\right)&=U_0\nabla^2\left(r^{-1}(f+f'')\right)\\&=U_0r^{-3}\left(f''''+2f''+f\right)\\&=0;\end{aligned}$$

therefore, $f(\theta)$ satisfies the ODE $f''''+2f''+f=0$.

(b) From (3.23b) we have

$$\begin{aligned}\nabla^2\boldsymbol{v}&=U_0\left(\frac{f'+f'''}{r^2}\hat{\boldsymbol{r}}+\frac{f+f''}{r^2}\hat{\boldsymbol{\theta}}\right)\\&=U_0\left(\frac{f'+f'''}{r^2}\hat{\boldsymbol{r}}-\frac{f''''+f''}{r^2}\hat{\boldsymbol{\theta}}\right)\\&=\frac{1}{\eta}\left(\frac{\partial P}{\partial r}\hat{\boldsymbol{r}}+\frac{1}{r}\frac{\partial P}{\partial\theta}\hat{\boldsymbol{\theta}}\right)\end{aligned}$$

using the solution to part (a) and (3.19a). We can then verify (3.23c) by substituting it into the result above.

Setting $f=e^{i\lambda\theta}$ in the ODE obtained in part (a) gives

$$\begin{aligned}\lambda^4-2\lambda^2+1&=\left(\lambda^2-1\right)^2\\&=0,\end{aligned}$$

giving double-root solutions $\lambda=\pm 1$. This corresponds to a general solution of the form

$$f(\theta)=(A_1+A_2\theta)e^{i\theta}+(A_3+A_4\theta)e^{-i\theta},$$

which is equivalent to (3.24) for appropriate constants A_i.

(c) For $\psi=r^nU_0g(\theta)$, (3.21) gives

$$\begin{aligned}\nabla^2\left(\nabla^2\psi\right)&=B_0\nabla^2\left(r^{n-2}(n^2g+g'')\right)\\&=B_0r^{n-4}\left((n-2)^2(n^2g+g'')+(n^2g''+g'''')\right)\\&=0,\end{aligned}$$

where setting $g(\theta)=e^{i\lambda\theta}$ gives

$$\begin{aligned}(n-2)^2(n^2-\lambda^2)-\lambda^2(n^2-\lambda^2)&=(n^2-\lambda^2)\left((n-2)^2-\lambda^2)\right)\\&=0,\end{aligned}$$

giving rise to roots $\lambda=\pm n$ and $\pm(n-2)$. For the cases $n=0$ and 2, we have $\lambda=\pm 2$ and a double root $\lambda=0$, corresponding to the general solution

$$g_0(\theta)=C_1\sin 2\theta+C_2\cos 2\theta+C_3\theta+C_4.$$

Otherwise (for $n\neq 1$) the general solution is given by

$$g_n(\theta)=C_1\sin(n-2)\theta+C_2\cos(n-2)\theta+C_3\sin n\theta+C_4\cos n\theta.$$

3.4 (a) From (3.24) we have

$$f(\theta)=C_1\sin\theta+C_2\theta\sin\theta+C_3\cos\theta+C_4\theta\cos\theta,$$

with boundary conditions $f(0)=f'(0)=f(-\theta_0)=0, f'(-\theta_0)=1$. Working through the algebra gives

$$C_1=\frac{\theta_0\sin(\theta_0)}{\theta_0^2-\sin^2\theta_0}=-C_4,\quad C_2=\frac{\sin\theta_0-\theta_0\cos\theta_0}{\theta_0^2-\sin^2\theta_0},\quad C_3=0,$$

from which we obtain the solution

$$f(\theta)=\frac{1}{\theta_0^2-\sin^2\theta_0}\left[(\theta+\theta_0)\sin\theta_0\sin\theta-\theta_0\theta\sin(\theta+\theta_0)\right].$$

(b) For $\psi = \frac{\tau}{\eta} r^2 g(\theta)$, we have

$$v_r = \frac{1}{r}\frac{\partial\psi}{\partial\theta} = \frac{\tau}{\eta} r g'(\theta), \qquad v_\theta = -\frac{\partial\psi}{\partial r} = \frac{-2\tau}{\eta} r g(\theta),$$

giving boundary conditions $g(0) = g'(0) = 0, g(-\theta_0) = 0, g'(-\theta_0) = 1$.
From exercise 3.3(c) we have that $g(\theta)$ is of the form

$$g(\theta) = C_1 \sin 2\theta + C_2 \cos 2\theta + C_3\theta + C_4,$$

where the boundary conditions give $C_3 = -2C_1, C_4 = -C_2$,

$$C_1 = \frac{1 - \cos 2\theta_0}{4(\sin 2\theta_0 - 2\theta_0 \cos 2\theta_0)}, \quad \text{and} \quad C_2 = \frac{2\theta_0 - \sin 2\theta_0}{4(\sin 2\theta_0 - 2\theta_0 \cos 2\theta_0)}.$$

From this we obtain the solution

$$g(\theta) = -\frac{\begin{array}{c}2(\theta + \theta_0) + \sin 2(\theta + \theta_0) - (\sin 2\theta + \sin 2\theta_0) \\ -(2\theta_0 \cos 2\theta + 2\theta \cos 2\theta_0)\end{array}}{\sin 2\theta_0 - 2\theta_0 \cos 2\theta_0},$$

which determines the velocity field, as required.

3.5 (a) Substituting the expansions into the governing equations and gathering $\mathcal{O}(\epsilon^0)$ terms, we obtain leading-order equations

$$\nabla^2 T^{(0)} = \frac{\mathrm{d}^2 T^{(0)}}{\mathrm{d}z^2} = 0,$$

$$\frac{\mathrm{d}P^{(0)}}{\mathrm{d}z} = -\rho^{(0)} g,$$

where $\rho^{(0)} = \rho_0[1 - \alpha(T^{(0)} - T_0)]$, according to equation (3.15). Therefore, by applying the temperature boundary conditions, we obtain solutions

$$T^{(0)} = T_0 + \delta T\left(1 - \frac{z}{h}\right),$$

$$P^{(0)} = P_0 + \rho^{(0)} g h \left(1 - \frac{z}{h}\right),$$

where P_0 is the pressure at $z = h$.

(b) Gathering $\mathcal{O}(\epsilon)$ terms, from the continuity equation we obtain

$$\nabla \cdot \boldsymbol{v}^{(1)} = 0,$$

and from the momentum equation we obtain

$$\nabla P^{(1)} = \eta \nabla^2 \boldsymbol{v}^{(1)} + \rho_0 \alpha g T^{(1)} \hat{\boldsymbol{z}},$$

and from the energy equation we obtain

$$\rho c_P \left(\frac{\partial T^{(1)}}{\partial t} + \boldsymbol{v}^{(1)} \cdot \boldsymbol{\nabla} T^{(0)} \right) = \boldsymbol{\nabla} \cdot k_T \boldsymbol{\nabla} T^{(1)},$$

$$\Rightarrow \frac{\partial T^{(1)}}{\partial t} - \frac{\delta T}{h} w^{(1)} = \kappa \nabla^2 T^{(1)},$$

$$\Rightarrow \left(\frac{\partial}{\partial t} - \kappa \nabla^2 \right) T^{(1)} = \frac{\delta T}{h} w^{(1)}.$$

where $w^{(1)} \equiv \boldsymbol{v}^{(1)} \cdot \hat{z}$ is the vertical component of the $\mathcal{O}(\epsilon)$ velocity.

To eliminate $P^{(1)}$ from the $\mathcal{O}(\epsilon)$ momentum equation we note that $\boldsymbol{\nabla} \times \boldsymbol{\nabla} \boldsymbol{A} = \boldsymbol{0}$ and $\boldsymbol{\nabla} \times (\boldsymbol{\nabla} \times \boldsymbol{A}) = \boldsymbol{\nabla} (\boldsymbol{\nabla} \cdot \boldsymbol{A}) - \nabla^2 \boldsymbol{A}$. Therefore taking the double curl of the $\mathcal{O}(\epsilon)$ momentum equation, using the continuity equation and, finally, dotting with $\hat{z}$, we obtain

$$\nabla^2 \nabla^2 w^{(1)} = -\frac{\rho_0 \alpha g}{\eta} \frac{\partial^2 T^{(1)}}{\partial x^2}.$$

(c) First we consider the no-normal-flux boundary condition. Since $\boldsymbol{v}(z=0,h) \cdot \hat{z} = \boldsymbol{0} + \boldsymbol{v}^{(1)}(z=0,h) \cdot \hat{z} = 0$, we have $w^{(1)}(z=0,h) = \hat{w}(z=0,1) = 0$. Next we consider the temperature boundary condition. Since the base-state satisfies the boundary conditions at $z=0,h$, the temperature perturbation must be zero at the boundaries. Hence we find that $T^{(1)}(z=0,h) = \hat{T}(z=0,1) = 0$. Finally we consider the no-shear-stress boundary condition, which can be expressed entirely in terms of the perturbation flow,

$$\tau_{xz}|_{z=0,h} = \frac{\eta}{2} \left(\frac{\partial u^{(1)}}{\partial z} + \frac{\partial w^{(1)}}{\partial x} \right) = 0. \tag{15.1}$$

The second term in the shear stress is zero by the no-normal-flux condition. Taking an x-derivative of the first term and using the continuity equation to eliminate $u^{(1)}$ gives the desired result.

(d) Substituting the solutions for $\hat{w}$ and $\hat{T}$ into the equations governing first-order perturbations gives

$$\left(n^2\pi^2 + k^2\right)^2 w_c = k^2 \mathrm{Ra} T_c$$

$$(n^2\pi^2 + k^2) T_c = w_c.$$

We then eliminate T_c, w_c to obtain the neutral stability curve

$$\mathrm{Ra} = \left(n^2\pi^2 + k^2\right)^3 / k^2.$$

Noting that Ra strictly increases with n, to minimize Ra one must take $n=1$. Then

$$\frac{\mathrm{dRa}}{\mathrm{d}k^2} = \frac{\left(\pi^2 + k^2\right)^2}{k^4} \left(2k^2 - \pi^2\right) = 0,$$

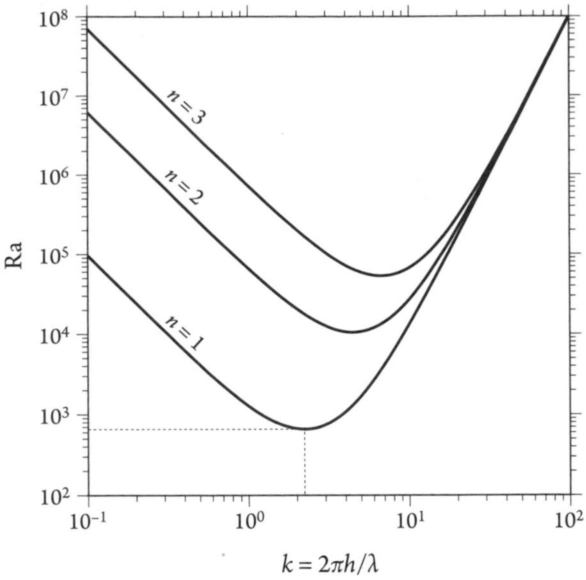

Figure 15.1. Rayleigh marginal stability curves for three values of n. These curves represent the Rayleigh number at which $s = 0$, as a function of the aspect ratio $2h/\lambda$ of the convection rolls, where λ is the dimensional horizontal wavelength.

$$\Rightarrow \quad k^2 = \frac{\pi^2}{2},$$

$$\Rightarrow \mathrm{Ra}_{\mathrm{crit}} = \frac{27\pi^4}{4}.$$

$\mathrm{Ra}_{\mathrm{crit}}$ is the minimal Rayleigh number that permits solutions of the given form in the case of marginal stability ($s = 0$). Therefore, assuming the relationship between s and Ra is continuous, larger Rayleigh numbers permit solutions corresponding to unstable perturbations. Therefore Ra_{crit} is the critical value of the Rayleigh number for the onset of mantle convection.

(e) The Rayleigh number provides a measure of the importance of buoyancy forces relative to viscous forces. By replacing stress-free boundaries with no-flow boundaries, we introduce a sink of kinetic energy through the viscous dissipation that occurs at rigid boundaries. This inhibits convection and so the critical Rayleigh number for onset of convection increases.

15.2 Exercises from Chapter 4: Conservation of Mass and Momentum

4.1 In an analogous way to (4.7), we have

$$\phi^s \boldsymbol{v}^s = \frac{1}{\delta V} \int_{\mathrm{RVE}} \Phi^s(\breve{\boldsymbol{x}})\, \breve{\boldsymbol{v}}(\breve{\boldsymbol{x}})\, \mathrm{d}^3\breve{\boldsymbol{x}}.$$

4.2 For steady and uniform porosity, we have $\frac{D_s\phi}{Dt} = 0$. Therefore, (4.15) gives

$$\begin{aligned} \mathcal{C} &= \frac{-\Gamma}{\rho^s(1-\phi)} \\ &= \frac{-1 \text{ kg/m}^3/\text{yr}}{0.98 \times 3300\text{kg m}^{-3}} \\ &\approx -3.09 \times 10^{-4}/\text{yr}. \end{aligned}$$

4.3 The extended Boussinesq approximation gives $\nabla\cdot\overline{\boldsymbol{v}} = 0$. In particular, we have

$$\begin{aligned} \nabla\cdot\left(\phi\boldsymbol{v}^\ell + (1-\phi)\boldsymbol{v}^s\right) &= \nabla\cdot\left(\boldsymbol{v}^s + \phi\left(\boldsymbol{v}^\ell - \boldsymbol{v}^s\right)\right) \\ &= \mathcal{C} + \nabla\cdot\left(\phi\left(\boldsymbol{v}^\ell - \boldsymbol{v}^s\right)\right) \\ &= 0, \end{aligned}$$

which gives the required result.

4.4 (a) Provided the choice of RVE is suitable, volume-averaged quantities do not depend on the choice of RVE. By construction of the RVE and the approximations we make when using it, small perturbations in the shape of the RVE domain cause variations in volume-averaged quantities that are sufficiently small so they can be neglected.

(b) By keeping the shape of the RVE fixed as $\boldsymbol{x}$ varies, the integral limits of $\phi^i(\boldsymbol{x})$ and δV are independent of $\boldsymbol{x}$. Moreover, the indicator function Φ^ℓ is constant away from solid–liquid interfaces, thus the gradient is only nonzero at those interfaces. Therefore, it follows that

$$\begin{aligned} \nabla\phi(\boldsymbol{x}) &= \nabla\left(\frac{1}{\delta V}\int_{\text{RVE}} \Phi^\ell(\boldsymbol{x}+\breve{\boldsymbol{x}})\,\mathrm{d}^3\breve{\boldsymbol{x}}\right) \\ &= \frac{1}{\delta V}\int_{\text{RVE}} \nabla\Phi^\ell(\boldsymbol{x}+\breve{\boldsymbol{x}})\,\mathrm{d}^3\breve{\boldsymbol{x}} \\ &= \frac{1}{\delta V}\int_{\mathcal{I}_{\text{RVE}}} \nabla\Phi^\ell(\boldsymbol{x}+\breve{\boldsymbol{x}})\,\mathrm{d}S_{\mathcal{I}}, \end{aligned}$$

as required.

4.5 We have $(1-\phi)\boldsymbol{\tau}^s = 2\eta_\phi\dot{\boldsymbol{\varepsilon}}^s$, where $\dot{\boldsymbol{\varepsilon}}^s$ is the deviatoric strain-rate tensor. Equation (4.41) gives

$$\begin{aligned} \text{trace}\,\dot{\boldsymbol{\varepsilon}}^s &= \tfrac{1}{2}\left[\nabla\cdot\boldsymbol{v}^s + \nabla\cdot\boldsymbol{v}^s - 2\mathcal{C}\right] \\ &= \nabla\cdot\boldsymbol{v}^s - \mathcal{C} \\ &= 0 \end{aligned}$$

by definition (4.16).

4.6 Substituting (4.32) and (4.35) into (4.19a) yields

$$-\nabla(\phi P^\ell)+\phi\rho^\ell \mathbf{g}=-\frac{\mu^\ell\phi^2}{k_\phi}\left(\boldsymbol{v}^s-\boldsymbol{v}^\ell\right)-P^\ell\nabla\phi,$$

hence dividing by ϕ and rearranging gives us (4.46a).

4.7 Eliminate liquid velocity by substituting (4.46d) into $\nabla\cdot$ (4.46a) to obtain

$$-\nabla\cdot\boldsymbol{v}^s+\nabla\cdot\frac{k_\phi}{\mu^\ell}\left(\nabla P^\ell-\rho^\ell\mathbf{g}\right)=0,$$

where $\mathcal{C}=\nabla\cdot\boldsymbol{v}^s$.

4.8 The nonextended Boussinesq approximation gives us the continuity equation (4.13) and hence the compaction rate is equal to

$$\mathcal{C}=-\nabla\cdot\left(\phi(\boldsymbol{v}^\ell-\boldsymbol{v}^s)\right)-\Gamma\Delta(1/\rho)$$

And hence taking the divergence of (4.46a) gives us

$$-\mathcal{C}-\Gamma\Delta(1/\rho)+\nabla\cdot\frac{k_\phi}{\mu^\ell}\left(\nabla P^\ell-\rho^\ell\mathbf{g}\right)=0$$

Noting that (4.46b) and (4.46c) will remain the same.

4.9 Assuming constant η_ϕ and ζ_ϕ, noting that $\mathcal{C}\equiv\nabla\cdot\boldsymbol{v}^s$ and that

$$\begin{aligned}\nabla\cdot\left(2\dot{\boldsymbol{\varepsilon}}^s\right)&=\nabla^2\boldsymbol{v}^s+\nabla\left(\nabla\cdot\boldsymbol{v}^s\right)-\tfrac{1}{3}\nabla\cdot\left(2\nabla\cdot\boldsymbol{v}^s\boldsymbol{I}\right)\\&=\nabla^2\boldsymbol{v}^s+\tfrac{1}{3}\nabla\left(\nabla\cdot\boldsymbol{v}^s\right),\end{aligned}$$

then using the vector identity $\nabla^2\boldsymbol{v}^s=\nabla(\nabla\cdot\boldsymbol{v}^s)-\nabla\times\nabla\times\boldsymbol{v}^s$ we obtain (4.50) by substituting the above results into (4.47b).

4.10 (a) Noting that $\boldsymbol{v}^s$ is of the form $f'(x)\hat{\boldsymbol{x}}+g'(y)\hat{\boldsymbol{y}}$, by observation it follows that

$$\boldsymbol{\Psi}=\mathbf{0}+c_1,\quad \mathcal{U}=\frac{\dot{\gamma}}{2}\left(x^2-y^2+c_2\right),$$

where c_j are arbitrary constants.

However, noting that $\boldsymbol{v}^s$ is also of the form $\partial_y\psi(x,y)\hat{\boldsymbol{x}}-\partial_x\psi(x,y)\hat{\boldsymbol{y}}$, it follows that

$$\boldsymbol{\Psi}=\dot{\gamma}xy\hat{\boldsymbol{z}}+c_1,\quad \mathcal{U}=c_2.$$

So linear combinations of the above can also give the correct velocity.

(b) We have

$$\begin{aligned}&\frac{\partial H}{\partial y}=x\\&\Rightarrow H(y)=xy+c\qquad\text{for } c \text{ constant,}\end{aligned}$$

which satisfies $\frac{\partial H}{\partial x}=y$ as required.

(c) We have

$$\begin{aligned}\nabla H &= y\hat{\boldsymbol{x}} + x\hat{\boldsymbol{y}} \\ \Rightarrow \boldsymbol{v}^s \cdot \nabla H &= \left(x\hat{\boldsymbol{x}} - y\hat{\boldsymbol{y}}\right) \cdot \left(y\hat{\boldsymbol{x}} + x\hat{\boldsymbol{y}}\right) \\ &= 0.\end{aligned}$$

As $\boldsymbol{v}^s$ is perpendicular to gradients in H at all points in the (x, y)-plane, it follows that trajectories coincide with contours of xy (i.e., lines on which $xy =$ const.).

15.3 Exercises from Chapter 5: Material Properties

5.1 (a) The shape of a pore with dihedral angle $\Theta = 60°$ is an equilateral triangle as illustrated below.

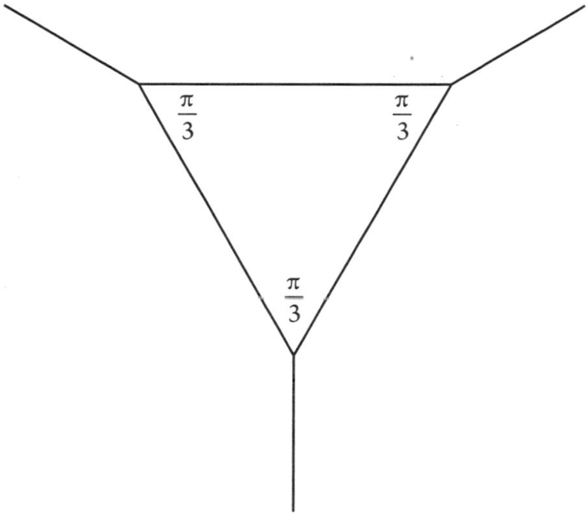

Figure 15.2. Two-dimensional pore with $\Theta = 60°$.

(b) For circular pores, at triple junctions the solid–liquid interfaces must be parallel. Therefore the required angle is $\Theta = 180°$.

5.2 Take the origin to be the lowest triple junction as illustrated in figure 5.2(b).

- In the same way as in figure 15.2, given the rotational symmetry of the pore we can draw an equilateral triangle by drawing straight lines between the junctions. As a result, it follows that

$$\tan\frac{\pi}{6} = \frac{y(x_J)}{x_J} \qquad \Rightarrow \qquad y(x_J) = x_J \tan\frac{\pi}{6}.$$

- As illustrated in figure 15.3, the gradient of $y(x)$ can be expressed as the tangent of the angle $y(x)$ makes with the horizontal. The solid–solid interface at $y = x_J$ makes an angle of $\pi/6$ with the horizontal axis. Also, by definition

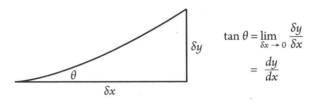

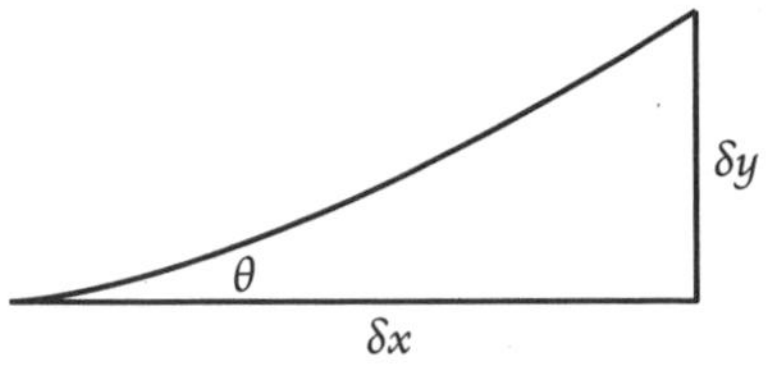

$$\tan\theta = \lim_{\delta x \to 0} \frac{\delta y}{\delta x} = \frac{dy}{dx}$$

Figure 15.3. Relationship between $y(x)$ and $\tan\theta$.

of the dihedral angle, the interface $y(x)$ makes an angle $\Theta/2$ with the solid–solid interface. Therefore at $x = x_J$ the interface $y(x)$ makes an angle of $\theta = \pi/6 - \Theta/2$ with the horizontal axis, giving the required boundary condition.

- By symmetry we require $y' = 0$ at $x = 0$.

5.3 (a) To relate x_Λ to ϕ and Θ it is sufficient to determine the porosity for a single, hexagonal grain because a hexagonal tessellation is space-filling. Noting that a regular hexagon with facet lengths d_f comprises six equilateral triangles of length d_f. Hence the area of one grain is given by

$$A = 6 \cdot \frac{\sqrt{3}}{4} d_f^2 = \frac{3\sqrt{3}}{2} d_f^2.$$

As illustrated in figure 5.1(b), each pore exists at a triple junction between three grains. Therefore a third of the area Δ of each pore is associated with each of the three neighboring grains. As a single grain has six vertices, the porosity of the matrix is given by

$$\phi = \frac{6}{A}\frac{\Delta}{3} = \frac{4}{3\sqrt{3}}\frac{\Delta}{d_f^2}.$$

It remains for us to determine $\Delta(x_\Lambda, \Theta)$. Assume that the pore shape can be described using (5.15). We consider an equilateral triangle circumscribed around the pore as shown in figure 15.4. The area of the pore is the area of this triangle less three times the area marked R. This circle has radius r_1. The area of R is the area of the circular sector (αr_1^2) minus the area of the triangle above it ($x_\Lambda r_1$). Hence,

$$\Delta(x_\Lambda, \Theta) = \sqrt{3}x_\Lambda^2 - \alpha r_1^2 + r_1 x_\Lambda, = x_\Lambda^2\left[\sqrt{3} - \alpha\left(1 + {y'_\Lambda}^{-2}\right) + \sqrt{1 + {y'_\Lambda}^{-2}}\right],$$

where $y'_\Lambda = \tan(\pi/6 - \Theta/2)$ and $\alpha = \arcsin(x_\Lambda/r_1) = \arcsin\left(1/\sqrt{1 + {y'_\Lambda}^{-2}}\right)$.

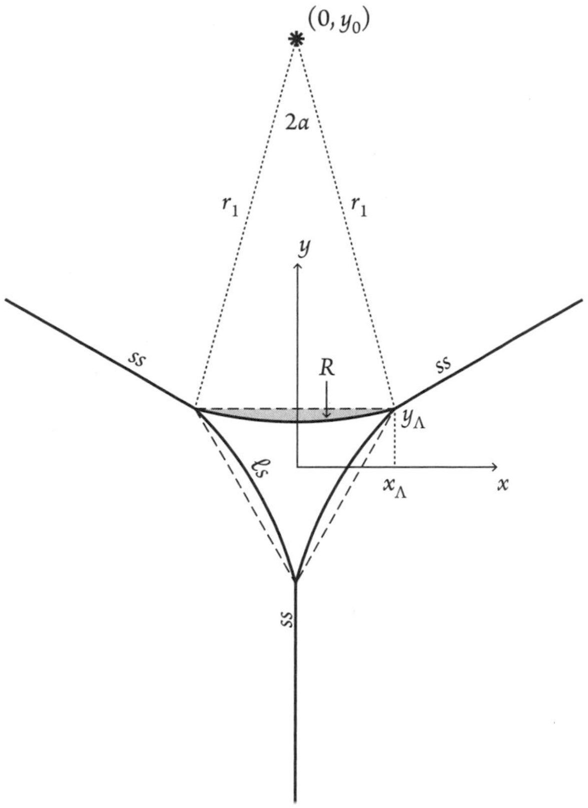

Figure 15.4.. Region R neighboring the pore space.

Substituting the pore area into the expression for the porosity we obtain the required relationship between ϕ, Θ, x_Λ, and d,

$$\phi = \frac{4}{3\sqrt{3}} \frac{x_\Lambda^2}{d_f^2} \left[\sqrt{3} - \alpha \left(1 + {y'_\Lambda}^{-2}\right) + \sqrt{1 + {y'_\Lambda}^{-2}} \right].$$

Rearranging this to solve for x_Λ gives

$$x_\Lambda = d_f \sqrt{\phi} \left[\frac{4}{3\sqrt{3}} \left(\sqrt{3} - \alpha \left(1 + {y'_\Lambda}^{-2}\right) + \sqrt{1 + {y'_\Lambda}^{-2}} \right) \right]^{-1/2}.$$

Since the diameter of the hexagonal grain D (connecting midpoints of opposing faces) is a factor of $\sqrt{3}$ larger than the facet length d_f,

$$x_\Lambda = D \sqrt{\phi} \left[\frac{4}{3} \left(\sqrt{3} - \alpha \left(1 + {y'_\Lambda}^{-2}\right) + \sqrt{1 + {y'_\Lambda}^{-2}} \right) \right]^{-1/2}.$$

(b) In the case $\Theta = 60°$, pores are equilateral triangles of side length $2x_\Lambda$ and area $\Delta = \sqrt{3}x_\Lambda^2$. Therefore as a pore area of 2Δ corresponds to each grain in the hexagonal tessellation, it follows that the porosity is given by

$$\phi = \frac{2\sqrt{3}x_\Lambda^2}{3\sqrt{3}d_f^2/2} = \frac{4}{3}\left(\frac{x_\Lambda}{d_f}\right)^2.$$

Geometrically, this means that $r_1 \to \infty$ and hence that the area of the region R is zero. This result can be shown to be consistent with the solution for ϕ in part (a) by taking ${y'_\Lambda}^{-2} \gg 1$ and $\alpha \ll 1$ such that $\sin\alpha \sim y'_\Lambda$.

5.4 (a) For radially symmetric, unidirectional flow, Stokes equations give

$$\mu \frac{1}{r}\frac{\partial}{\partial r}\left(r\frac{\partial w}{\partial r}\right) = -\rho g.$$

By imposing boundary conditions $w(R) = 0$ (no slip), and $\partial w/\partial r(0) = 0$ (symmetry about $r = 0$) we obtain the solution

$$w(r) = \frac{\rho g}{4\mu}\left(R^2 - r^2\right).$$

(b) The mean velocity through the tubule is given by

$$\begin{aligned} w_m &= \frac{1}{\pi R^2} \cdot 2\pi \int_0^R rw(r)\mathrm{d}r \\ &= \frac{\rho g}{2\mu R^2} \int_0^R r\left(R^2 - r^2\right)\mathrm{d}r \\ \Rightarrow \quad w_m &= \frac{\rho g R^2}{8\mu}. \end{aligned}$$

Therefore, as the solid lattice is immobile, the mean velocity per unit cross sectional area is given by

$$\begin{aligned} \overline{w} &= \frac{l\pi R^2}{l^3} w_m \\ &= \frac{\pi \rho g R^4}{8\mu l^2} \\ \Rightarrow \quad \overline{w} &= \frac{l^2\phi^2\rho g}{8\pi\mu}. \end{aligned}$$

(c) As there are no externally imposed pressure gradients, the Darcy velocity is related to the permeability by

$$\overline{w} = \frac{k_\phi \rho g}{\mu},$$

where comparing with the expression for $\overline{w}$ obtained in (b), we have

$$k_\phi = \frac{l^2}{8\pi}\phi^2.$$

Permeability associated with tubules at grain junctions generally gives rise to permeability–porosity relationships of the form ϕ^2. At higher porosities (and smaller dihedral angles), where grain faces are wetted, the relationship changes to a ϕ^3 form.

(d) The subtlety here lies in the fact that, in general, the permeability is tensor-valued, not scalar-valued. However, for isotropic media such as the cubic lattice model, the permeability is written as

$$\boldsymbol{k}_\phi = k_\phi \boldsymbol{I},$$

which permits expressing the permeability as being scalar-valued, a form used throughout this book. When the isotropic cubic lattice is tilted with respect to gravity, the permeability remains the same.

5.5 (a) We have

$$\begin{aligned}\dot{\boldsymbol{\varepsilon}}^{(0)} &= \frac{1}{2}\left[\nabla \boldsymbol{v}^{(0)} + \left(\nabla \boldsymbol{v}^{(0)}\right)^{\mathsf{T}}\right] - \frac{1}{3}\nabla\cdot \boldsymbol{v}^{(0)}\boldsymbol{I} \\ &= \frac{\dot{\gamma}}{2}\begin{pmatrix} 0 & 1 \\ 1 & 0 \end{pmatrix},\end{aligned}$$

therefore the tensor invariant $\dot{\varepsilon}_{II0}$ is given by

$$\begin{aligned}\dot{\varepsilon}_{II0} &= \sqrt{\dot{\boldsymbol{\varepsilon}}:\dot{\boldsymbol{\varepsilon}}/2} \\ &= \sqrt{\dot{\gamma}^2/4} \\ \Rightarrow \quad \dot{\varepsilon}_{II0} &= \frac{\dot{\gamma}}{2}.\end{aligned}$$

(b) Substituting $\boldsymbol{v} = \boldsymbol{v}^{(0)} + \epsilon \boldsymbol{v}^{(1)}$ into the strain rate gives

$$\dot{\varepsilon} = \frac{\dot{\gamma}}{2}\begin{pmatrix} \epsilon & 1 \\ 1 & \epsilon \end{pmatrix}.$$

(c) Substituting the expression for the strain rate obtained in (b) to the tensor invariant gives

$$\begin{aligned}\dot{\varepsilon}_{II} &= \sqrt{\frac{\dot{\gamma}^2}{4}\left(1+\epsilon^2\right)} \\ &= \frac{\dot{\gamma}}{2}\sqrt{1+\epsilon^2} \\ \Rightarrow \dot{\varepsilon}_{II} &= \frac{\dot{\gamma}}{2}\left(1 + \frac{1}{2}\epsilon^2 + \mathcal{O}(\epsilon^4)\right).\end{aligned}$$

Substituting the invariant into the shear viscosity given by (5.23), we obtain

$$\eta_\phi = \eta_0 \exp\left[-\lambda(\phi-\phi_0)/\mathfrak{n}\right]\left(1+\frac{1}{2}\epsilon^2+\mathcal{O}(\epsilon^4)\right)^{\frac{1-\mathfrak{n}}{\mathfrak{n}}}$$

$$= \eta_0 \exp\left[-\lambda(\phi-\phi_0)/\mathfrak{n}\right]\left(1+\frac{1-\mathfrak{n}}{2\mathfrak{n}}\epsilon^2+\mathcal{O}(\epsilon^4)\right);$$

therefore, the shear viscosity remains unchanged at first order, as required.

5.6 As each deformation mechanism responds to the same stress, for all k we have $\sigma = \eta_\phi^k \dot{\boldsymbol{\varepsilon}}^k$, where

$$\dot{\boldsymbol{\varepsilon}} = \sum_{k=1}^{n} \dot{\boldsymbol{\varepsilon}}^k$$

$$= \sum_{k=1}^{n} \frac{\sigma}{\eta_\phi^k}$$

$$= \sigma \sum_{k=1}^{N} \frac{1}{\eta_\phi^k}$$

$$\Rightarrow \sigma = \left(\sum_{k=1}^{N} \frac{1}{\eta_\phi^k}\right)^{-1} \dot{\boldsymbol{\varepsilon}}$$

$$= \eta_\phi \dot{\boldsymbol{\varepsilon}},$$

and so we recover (5.21), as required.

5.7 The melting rate Γ due to isentropic decompression of mantle that is upwelling at speed $W_0 > 0$ can be approximated by equation (5.39). The Lagrangian rate of change of pressure is $-\rho^s g W_0$; multiplying by the isentropic productivity $\mathrm{d}F/\mathrm{d}P|_s < 0$ gives Γ. The melting rate Γ_{sea} due to changes in sea level is associated with the same isentropic productivity because it is a property of the mantle that is melting. However, the rate of pressure change is given by $\rho^w g\dot{S}$, where ρ^w is the density of water and $\dot{S}$ is the rate of change of sea level. A measure of the relative size of Γ_{sea} is

$$\frac{\Gamma_{\mathrm{sea}}}{\Gamma} = \frac{\rho^w|\dot{S}|}{\rho^s W_0}.$$

Using values given in the problem, $\dot{S} \approx 10^{-2}$ m/yr and $W_0 = 3\times 10^{-2}$. The density ratio is approximately 1/3. Hence, the perturbations in melting rate due to sea level can be approximately 10% of the melting rate due to decompression.

15.4 Exercises from Chapter 6: Compaction and Its Inherent Length Scale

6.2 (a) The problem is equivalent to that outlined in section 6.1, with boundary conditions evaluated at $z=-\mathcal{H}$ instead of $z\to-\infty$. Integrating (6.2a) with respect to z and imposing $\frac{dP}{dz}=0$, $W=W_0/w_0$ on $z=-\mathcal{H}$ gives

$$\frac{dP}{dz}-W=\frac{d^2W}{dz^2}-W$$
$$=-\frac{W_0}{w_0}$$

using (6.2b). Imposing $W=W_0/w_0$ on $z=-\mathcal{H}$ and $W=0$ on $z=0$ gives

$$W=\frac{W_0}{w_0}\left(1-\cosh z-\coth\mathcal{H}\sinh z\right),$$

thus the dimensional compaction rate is

$$\mathcal{C}(z)=-\frac{W_0}{\delta}\left(\sinh\frac{z}{\delta}+\coth\frac{\mathcal{H}}{\delta}\cosh\frac{z}{\delta}\right).$$

(b) In the limit $\delta\gg\mathcal{H}$, as $z\sim\mathcal{H}$ we have (to leading-order, Taylor-expanding hyperbolic functions),

$$\mathcal{C}(z)\approx-\frac{W_0}{\delta}\left(\frac{z}{\delta}+\coth\frac{\mathcal{H}}{\delta}\right)$$
$$\approx-\frac{W_0}{\mathcal{H}},$$

therefore the compaction rate is uniform (to leading order) in the limit $\delta\gg\mathcal{H}$.

(c) In the limit $\delta/\mathcal{H}\to0$ we have $\coth(\mathcal{H}/\delta)\to1$, giving (for $z\le0$)

$$\mathcal{C}(z)\to-\frac{W_0}{\delta}\left(\sinh\frac{z}{\delta}+\cosh\frac{z}{\delta}\right)=-\frac{W_0}{\delta}e^{z/\delta}\quad\text{as}\quad\frac{\delta}{\mathcal{H}}\to0,$$

thus the half-space solution (6.9) is recovered, as required.

6.3 (a) The transformation $z\to-z$ gives $\frac{d}{dz}\to-\frac{d}{dz}$. Therefore (6.16) remains unchanged, so if $\mathcal{C}(z)$ is a solution of (6.16) then so is $\mathcal{C}(-z)$, and given the symmetric porosity profile it follows that (6.16) only admits solutions that can be written as a sum of symmetric and antisymmetric components. Moreover, given a solution $\mathcal{C}(z)$, (6.21) gives

$$\left[\delta_i^2\frac{d\mathcal{C}}{dz}\right]_{\mathcal{H}/2-\epsilon}^{\mathcal{H}/2+\epsilon}=\frac{k_0\Delta\rho g}{\mu^\ell}\left[c_i^n(1-c_i\phi_0)\right]_{\mathcal{H}/2-\epsilon}^{\mathcal{H}/2+\epsilon}$$
$$=-\frac{k_0\Delta\rho g}{\mu^\ell}\left[c_i^n(1-c_i\phi_0)\right]_{-(\mathcal{H}/2+\epsilon)}^{-(\mathcal{H}/2-\epsilon)}$$

$$= -\left[\delta_i^2 \frac{\mathrm{d}\mathcal{C}}{\mathrm{d}z}\right]_{-(\mathcal{H}/2+\epsilon)}^{-(\mathcal{H}/2-\epsilon)}$$

$$= \left[\delta_i^2 \frac{\mathrm{d}\mathcal{C}}{\mathrm{d}z}\right]_{-(\mathcal{H}/2-\epsilon)}^{-(\mathcal{H}/2+\epsilon)},$$

which can only hold if $\frac{\mathrm{d}\mathcal{C}}{\mathrm{d}z}(-z) = \frac{\mathrm{d}\mathcal{C}}{\mathrm{d}z}(z)$, i.e., if $\frac{\mathrm{d}\mathcal{C}}{\mathrm{d}z}$ is symmetric. Therefore $\mathcal{C}(z)$ must be antisymmetric.

(b) Similarly to section 6.2, we solve for the compaction rate in each layer separately using (6.16). As $\mathcal{C}(-z) = -\mathcal{C}(z)$, imposing far-field conditions and matching conditions at $z = \pm\mathcal{H}/2$ immediately gives

$$\mathcal{C}(z) = \mathcal{C}^* \begin{cases} e^{-\frac{z}{\delta_1}} & z \geq \frac{\mathcal{H}}{2}, \\ e^{-\frac{\mathcal{H}}{2\delta_1}} \dfrac{\sinh\left[\frac{z}{\delta_2}\right]}{\sinh\left[\frac{\mathcal{H}}{2\delta_2}\right]} & |z| \leq \frac{\mathcal{H}}{2} \\ -e^{\frac{z}{\delta_1}} & z \leq -\frac{\mathcal{H}}{2} \end{cases}.$$

To determine the constant $\mathcal{C}^*$ we use the jump condition at $z = \mathcal{H}/2$, employing (6.21). We find that

$$\mathcal{C}^* = \frac{k_0 \Delta\rho g}{\delta_0 \mu^\ell} \cdot \frac{\left[c_2^n(1 - c_2\phi_0) - c_1^n(1 - c_1\phi_0)\right] e^{\frac{\mathcal{H}}{2\delta_1}}}{c_1^{n/2} + c_2^{n/2} \coth\left[\frac{\mathcal{H}}{2\delta_2}\right]}.$$

(c) In the limit $\mathcal{H} \to \infty$ we have

$$\mathcal{C}^* = \frac{k_0 \Delta\rho g}{\delta_0 \mu^\ell} \cdot \frac{\left[c_2^n(1 - c_2\phi_0) - c_1^n(1 - c_1\phi_0)\right] e^{\frac{\mathcal{H}}{2\delta_1}}}{c_1^{n/2} + c_2^{n/2}} \equiv \frac{k_0 \Delta\rho g}{\delta_0 \mu^\ell} e^{\frac{\mathcal{H}}{2\delta_1}} c_{12}$$

thus making the substitution $\xi = z - \mathcal{H}/2$ gives

$$\mathcal{C}(\xi) = \frac{k_0 \Delta\rho g}{\delta_0 \mu^\ell} c_{12} \begin{cases} \exp\left(-\frac{\xi}{\delta_1}\right) & \text{for } \xi > 0 \\ \exp\left(\frac{\xi}{\delta_2}\right) & \text{for } \xi < 0 \end{cases},$$

thus the two-layer, permeability-step compaction rate is recovered.

6.4 (a) From (4.14) we have

$$\nabla \cdot \overline{\boldsymbol{v}} = \frac{\partial}{\partial z}\left(\phi w + (1-\phi) W\right)$$

$$= 0$$

$$\Rightarrow \phi \frac{\partial w}{\partial z} = -(1-\phi)\frac{\partial W}{\partial z}.$$

Taking the z-derivative of (6.1b) and using (6.1a) (for uniform porosity) gives

$$\frac{\partial^3 W}{\partial z^3} = \frac{1}{\delta_0^2}\frac{\partial W}{\partial z}.$$

Integrating (6.1a) and the Boussinesq condition using the boundary condition at $z=0$, and by eliminating the pressure in (6.1b), we obtain

$$\left(\zeta_\phi + \tfrac{4}{3}\eta_\phi\right)\frac{\partial^2 W}{\partial z^2} = -\frac{\mu\phi}{k_\phi}(w-W) + (1-\phi)\Delta\rho g.$$

(b) Using (6.56a) and boundary conditions $W=0$ at $z=0$, $\partial W/\partial z \to 0$ as $z \to \infty$ gives

$$W(z) = -w_0\phi\left(1 - e^{-\frac{z}{\delta_0}}\right)$$

where the constant w_0 is to be determined.

The extended Boussinesq velocity condition then gives

$$w(z) = w_0(1-\phi)\left(1 - e^{-\frac{z}{\delta_0}}\right)$$

$$\Rightarrow w - W = w_0\left(1 - e^{-\frac{z}{\delta_0}}\right);$$

therefore, (6.56b) gives

$$w_0 = \frac{k_\phi}{\mu}\frac{1-\phi}{\phi}\left(\rho^s - \rho^\ell\right)g.$$

Using (6.1c) (remembering to include the prefactor $(1-\phi)$), we have, at $t=0$,

$$\frac{\partial\phi}{\partial t} = -\frac{w_0}{\delta_0}\phi(1-\phi)e^{-\frac{z}{\delta_0}}.$$

6.5 (a) By employing the assumption $\nabla P^\ell = \nabla P + \rho^s \boldsymbol{g}$ and neglecting gravitational body forces, equation (4.47a) reduces to (6.57a).

As viscosities are uniform, (4.47b) can be written as

$$-\nabla P + (\zeta_\phi - \tfrac{2}{3}\eta_\phi)\nabla\left(\nabla\cdot\boldsymbol{v}^s\right) + \eta_\phi\nabla\cdot\left(\nabla\boldsymbol{v}^s + \left(\nabla\boldsymbol{v}^s\right)^{\mathsf{T}}\right) = \boldsymbol{0},$$

where using the vector identity $\nabla\times(\nabla\times\boldsymbol{v}^s) = \nabla(\nabla\cdot\boldsymbol{v}^s) - \nabla^2\boldsymbol{v}^s$ yields (6.57b).

(b) Using the length and velocity scalings given by (6.29) and the scaling

$$[P] = \frac{\mu\delta_\phi w_0\phi}{k_\phi},$$

equations (6.57) become

$$\nabla\cdot\boldsymbol{v}^s - \nabla^2 P = 0$$

$$-\nabla P + \nabla\left(\nabla\cdot\boldsymbol{v}^s\right) - \tfrac{1}{\nu_\phi}\nabla\times\left(\nabla\times\boldsymbol{v}^s\right) = \boldsymbol{0}.$$

(c) Taking the curl of (6.57b), we obtain

$$\nabla \times \left(\nabla \times \boldsymbol{\omega}^s\right) = \mathbf{0} \Rightarrow \nabla^2 \boldsymbol{\omega}^s = \mathbf{0},$$

as $\nabla \cdot \boldsymbol{\omega}^s = 0$.

Substituting $\nabla \cdot$ (6.57b) into (6.57a) to eliminate pressure, we obtain the required result:

$$-\nabla^2 C + C = 0.$$

6.6 (a) For spherically symmetric problems we have (noting $G = -r\partial G/\partial r$):

$$\begin{aligned}
\nabla^2 C &= \frac{1}{r^2}\frac{\partial}{\partial r}\left(r^2 \frac{\partial\left(G(r)e^{-r}\right)}{\partial r}\right) \\
&= \frac{1}{r^2}\frac{\partial}{\partial r}\left(r^2 e^{-r}\left(\frac{\partial G}{\partial r} - G\right)\right) \\
&= \frac{1}{r^2}\frac{\partial}{\partial r}\left((1+r)e^{-r} r^2 \frac{\partial G}{\partial r}\right) \\
&= (1+r)e^{-r}\nabla^2 G + \frac{1-(1+r)}{r^2} e^{-r} r^2 \frac{\partial G}{\partial r} \\
&= -(1+r)e^{-r}\delta^{(3)}(\boldsymbol{x}) + e^{-r} G.
\end{aligned}$$

Therefore, using the identity $f(\boldsymbol{x})\delta^{(3)}(\boldsymbol{x}) = f(\mathbf{0})\delta^{(3)}(\boldsymbol{x})$, we have

$$-\nabla^2 C + C = \delta^{(3)}(\boldsymbol{x}),$$

so $C(r) = G(r)e^{-r}$ satisfies (6.58) for $r > 0$, as required.

(b) We have

$$\begin{aligned}
\boldsymbol{v}^s &= \nabla\left(\frac{A}{r} + \frac{B}{r}e^{-r}\right) \\
&= \frac{\partial}{\partial r}\left(\frac{A}{r} + \frac{B}{r}e^{-r}\right)\nabla(r) \\
\Rightarrow \quad \boldsymbol{v}^s &= -\left(\frac{A}{r^2} + \frac{Be^{-r}}{r^2}(1+r)\right)\frac{\boldsymbol{x}}{r}.
\end{aligned}$$

Noting that $\boldsymbol{\omega}^s = \mathbf{0}$, (6.57b) gives

$$\begin{aligned}
\mathbf{0} &= \nabla\left(-P + \nabla \cdot \boldsymbol{v}^s\right) \\
&= \frac{\partial}{\partial r}\left(-P + \nabla \cdot \boldsymbol{v}^s\right)\frac{\boldsymbol{x}}{r} \\
\Rightarrow \quad 0 &= -P + \nabla \cdot \boldsymbol{v}^s,
\end{aligned}$$

which we obtain by assuming $P \to 0$ as $r \to \infty$. Therefore we have

$$\begin{aligned}
P &= \nabla \cdot \boldsymbol{v}^s \\
&= -\boldsymbol{x} \cdot \nabla \left(\frac{A}{r^3} + \frac{Be^{-r}}{r^3}(1+r) \right) - \left(\frac{A}{r^3} + \frac{Be^{-r}}{r^3}(1+r) \right) \nabla \cdot \boldsymbol{x} \\
&= -\frac{\boldsymbol{x} \cdot \boldsymbol{x}}{r} \frac{\partial}{\partial r} \left(\frac{A}{r^3} + \frac{Be^{-r}}{r^3}(1+r) \right) - \frac{3}{r^3} \left(A + Be^{-r}(1+r) \right) \\
&= -r \left[-\frac{3}{r^4} \left(A + Be^{-r}(1+r) \right) + \frac{Be^{-r}}{r^3}(1-(1+r)) \right] \\
&\quad - \frac{3}{r^3} \left(A + Be^{-r}(1+r) \right) \\
\Rightarrow \quad P &= \frac{Be^{-r}}{r} \quad .
\end{aligned}$$

(c) From (b) we have

$$\begin{aligned}
\nabla \boldsymbol{v}^s &= -\nabla \left[\left(\frac{A}{r^2} + \frac{Be^{-r}}{r^2}(1+r) \right) \frac{\boldsymbol{x}}{r} \right] \\
&= -\left[\left(\frac{A}{r^3} + \frac{Be^{-r}}{r^3}(1+r) \right) \boldsymbol{I} + \frac{\partial}{\partial r} \left(\frac{A}{r^3} + \frac{Be^{-r}}{r^3}(1+r) \right) \frac{\boldsymbol{x}\boldsymbol{x}}{r} \right],
\end{aligned}$$

where we note by symmetry that $\nabla \boldsymbol{v}^s = (\nabla \boldsymbol{v}^s)^\mathsf{T}$. Therefore it follows that

$$\begin{aligned}
\overline{\boldsymbol{\sigma}} \cdot \frac{\boldsymbol{x}}{r} &= \frac{2}{\nu_\phi} \left(-C\boldsymbol{I} + \nabla \boldsymbol{v}^s \right) \cdot \frac{\boldsymbol{x}}{r} \\
&= \frac{-2}{\nu_\phi} \left[\frac{Be^{-r}}{r} \frac{\boldsymbol{x}}{r} + \left(\frac{A}{r^3} + \frac{Be^{-r}}{r^3}(1+r) \right) \frac{\boldsymbol{x}}{r} \right. \\
&\qquad \left. + \frac{\partial}{\partial r} \left(\frac{A}{r^3} + \frac{Be^{-r}}{r^3}(1+r) \right) \boldsymbol{x} \right] \\
&= \frac{2}{\nu_\phi} \left[\frac{2A}{r^4} - \frac{Be^{-r}}{r^2} - \frac{Be^{-r}}{r^4}(1+r) - \frac{Be^{-r}}{r^3} \right. \\
&\qquad \left. + \frac{3Be^{-r}}{r^4}(1+r) + \frac{Be^{-r}}{r^3}(1+r) \right] \boldsymbol{x} \\
\Rightarrow \overline{\boldsymbol{\sigma}} \cdot \frac{\boldsymbol{x}}{r} &= \frac{4}{\nu_\phi} \left(\frac{A}{r^4} + \frac{Be^{-r}}{r^2}(1+r) \right) \frac{\boldsymbol{x}}{r^2},
\end{aligned}$$

as required.

(d) We have

$$\boldsymbol{v}^s - \nabla P = -\left(\frac{A}{r^2} + \frac{Be^{-r}}{r^2}(1+r) \right) \frac{\boldsymbol{x}}{r} + \frac{Be^{-r}}{r^2}(1+r) \frac{\boldsymbol{x}}{r}$$

$$= -\frac{A\boldsymbol{x}}{r^3}$$

$$= \frac{Q\boldsymbol{x}}{4\pi R^3} \qquad \text{at } r = R$$

$$\Rightarrow A = -\frac{Q}{4\pi} \quad .$$

The second boundary condition at $r = R$ gives

$$\frac{4}{\nu_\phi}\left[-\frac{Q}{4\pi R^2} + \frac{Be^{-R}}{R^2}(1+R)\right] = -Be^{-R}$$

$$\Rightarrow \frac{Q}{\pi R^2 \nu_\phi} = Be^{-R}\left(\frac{4}{\nu_\phi R^2}(1+R)+1\right)$$

$$\Rightarrow B = \frac{Qe^R}{\pi(4(1+R)+\nu_\phi R^2)} \quad .$$

(e) In the limit $R \to 0$ we have $A = -B = -Q/4\pi$. Therefore we have

$$\boldsymbol{v}^s = \frac{Q}{4\pi r^2}\left(1 - e^{-r}(1+r)\right)\frac{\boldsymbol{x}}{r},$$

$$P = \frac{Qe^{-r}}{4\pi r}.$$

6.7 Using (6.47) we have

$$C^2 = 2(2\Lambda^* + 1)\left(\frac{\Lambda - 1}{\Lambda}\right)^2 (\Lambda^* - \Lambda).$$

Substituting this into (6.49) gives

$$Z(\Lambda) = \frac{(2\Lambda^* + 1)}{[2(2\Lambda^* + 1)]^{1/2}} \int_{\Lambda^*}^{\Lambda} \frac{\Lambda \mathrm{d}\Lambda}{(\Lambda - 1)(\Lambda^* - \Lambda)^{1/2}}$$

$$= \left(\Lambda^* + \frac{1}{2}\right)^{1/2} \int_{\Lambda^*}^{\Lambda} \left[\frac{1}{(\Lambda^* - \Lambda)^{1/2}} + \frac{1}{(\Lambda - 1)(\Lambda^* - \Lambda)^{1/2}}\right] \mathrm{d}\Lambda$$

$$= \left(\Lambda^* + \frac{1}{2}\right)^{1/2} \left(\int_{\Lambda^*}^{\Lambda} \frac{(\Lambda^* - \Lambda)^{-1/2}}{\left((\Lambda^* - 1)^{1/2} - (\Lambda^* - \Lambda)^{1/2}\right)\left((\Lambda^* - 1)^{1/2} + (\Lambda^* - \Lambda)^{1/2}\right)} \mathrm{d}\Lambda \right.$$

$$\left. - 2(\Lambda^* - \Lambda)^{1/2}\right),$$

where the integral argument can be written as

$$\frac{1}{2}\frac{(\Lambda^*-\Lambda)^{-1/2}}{(\Lambda^*-1)^{1/2}}\left(\frac{1}{(\Lambda^*-1)^{1/2}-(\Lambda^*-\Lambda)^{1/2}}+\frac{1}{(\Lambda^*-1)^{1/2}+(\Lambda^*-\Lambda)^{1/2}}\right).$$

Therefore, using the identity $\frac{\mathrm{d}}{\mathrm{d}x}\ln(f(x))=f'(x)/f(x)$, it follows that

$$Z(\Lambda)=\left(\Lambda^*+\tfrac{1}{2}\right)^{\frac{1}{2}}\left[\frac{1}{(\Lambda^*-1)^{\frac{1}{2}}}\ln\left(\frac{(\Lambda^*-1)^{\frac{1}{2}}-(\Lambda^*-\Lambda)^{\frac{1}{2}}}{(\Lambda^*-1)^{\frac{1}{2}}+(\Lambda^*-\Lambda)^{\frac{1}{2}}}\right)\right.$$
$$\left.-2\left(\Lambda^*-\Lambda\right)^{\frac{1}{2}}\right].$$

6.8 Using the small-x approximation $\ln(1+x)=x$, (6.49) simplifies to

$$Z(\Lambda)=\left(\Lambda^*+\frac{1}{2}\right)^{1/2}\left[\frac{1}{(\Lambda^*-1)^{1/2}}\ln\left(1-\frac{2(\Lambda^*-\Lambda)^{1/2}}{(\Lambda^*-1)^{1/2}+(\Lambda^*-\Lambda)^{1/2}}\right)\right.$$
$$\left.-2(\Lambda^*-\Lambda)^{1/2}\right]$$
$$\approx -2(\Lambda^*-\Lambda)^{1/2}\left(\Lambda^*+\frac{1}{2}\right)^{1/2}\left[(\Lambda^*-1)^{-1/2}\right.$$
$$\left.\frac{1}{(\Lambda^*-1)^{1/2}+(\Lambda^*-\Lambda)^{1/2}}+1\right],$$

where noting that $(\Lambda^*-1)^{1/2}=\mathcal{O}(1)$ gives

$$Z(\Lambda)\approx-\frac{2\Lambda^*\left(\Lambda^*+\frac{1}{2}\right)^{1/2}}{\Lambda^*-1}(\Lambda^*-\Lambda)^{1/2}.$$

Rearranging the above expression for Z yields the first-order solution

$$\Lambda(Z)\approx\Lambda^*-\frac{(\Lambda^*-1)^2}{4\Lambda^{*2}(\Lambda^*+1/2)}Z^2;$$

therefore, we have

$$A=\frac{(\Lambda^*-1)^2}{4\Lambda^{*2}(\Lambda^*+1/2)},\qquad m=2.$$

15.5 Exercises from Chapter 7: Porosity-Band Emergence under Deformation

7.1 Recall that

$$\dot{\boldsymbol{\varepsilon}}^{(0)} = \frac{1}{2}\left[\nabla \boldsymbol{v}^{(1)} + (\nabla \boldsymbol{v}^{(1)})^{\intercal}\right] - \frac{1}{3}\nabla\cdot \boldsymbol{v}^{(1)}\boldsymbol{I}$$

Using this we find:
(a)

$$\dot{\boldsymbol{\varepsilon}}^{(0)} = \dot{\gamma}\begin{pmatrix} 1 & 0 \\ 0 & -1 \end{pmatrix}, \qquad \nabla\cdot\dot{\boldsymbol{\varepsilon}}^{(0)} = \boldsymbol{0}$$

(b)

$$\dot{\boldsymbol{\varepsilon}}^{(0)} = \frac{\dot{\gamma}}{2}\begin{pmatrix} 0 & 1 \\ 1 & 0 \end{pmatrix}, \qquad \nabla\cdot\dot{\boldsymbol{\varepsilon}}^{(0)} = \boldsymbol{0}$$

(c)

$$\dot{\boldsymbol{\varepsilon}}^{(0)} = \frac{G}{4\mu^{\ell}}\begin{pmatrix} 0 & h-2y \\ h-2y & 0 \end{pmatrix}, \qquad \nabla\cdot\dot{\boldsymbol{\varepsilon}}^{(0)} = -\frac{G}{2\mu^{\ell}}\hat{\boldsymbol{x}}$$

7.2 We have

$$\begin{aligned}
\dot{\boldsymbol{\varepsilon}}^{(1)} &= \frac{1}{2}\left[\nabla \boldsymbol{v}^{(1)} + (\nabla \boldsymbol{v}^{(1)})^{\intercal}\right] - \frac{1}{3}\nabla\cdot \boldsymbol{v}^{(1)}\boldsymbol{I} \\
\Rightarrow \nabla\cdot\dot{\boldsymbol{\varepsilon}}^{(1)} &= \frac{1}{2}\left[\nabla^2 \boldsymbol{v}^{(1)} + \nabla(\nabla\cdot \boldsymbol{v}^{(1)})\right] - \frac{1}{3}\nabla(\nabla\cdot \boldsymbol{v}^{(1)}) \\
\Rightarrow \nabla\times\left(\nabla\cdot\dot{\boldsymbol{\varepsilon}}^{(1)}\right) &= \frac{1}{2}\nabla^2\left(\nabla\times\boldsymbol{v}^{(1)}\right) \\
&= \frac{1}{2}\nabla^2\left(\nabla\times\nabla\times(\psi\hat{\boldsymbol{z}})\right) \\
&= \frac{1}{2}\nabla^2\left(\nabla(\nabla\cdot(\psi\hat{\boldsymbol{z}})) - \nabla^2\psi\hat{\boldsymbol{z}}\right) \\
\Rightarrow \nabla\times\left(\nabla\cdot\dot{\boldsymbol{\varepsilon}}^{(1)}\right) &= -\frac{1}{2}\nabla^4\psi\hat{\boldsymbol{z}} \quad ,
\end{aligned}$$

where substitution into (7.5b) immediately gives (7.8).

7.3 After substitution, we have

$$\eta_\phi = \eta_\phi^{(0)}\exp\left[-\tfrac{\lambda}{\mathrm{n}}\epsilon\phi^{(1)}\right]\left(1+\epsilon\dot{\varepsilon}_{II}^{(1)}/\dot{\varepsilon}_{II}^{(0)}\right)^{-\mathcal{N}}.$$

Taylor-expanding both terms gives

$$\eta_\phi = \eta_\phi^{(0)}\left(1-\epsilon\tfrac{\lambda}{\mathrm{n}}\phi^{(1)} + \mathcal{O}(\epsilon^2)\right)\left(1-\epsilon\mathcal{N}\dot{\varepsilon}_{II}^{(1)}/\dot{\varepsilon}_{II}^{(0)} + \mathcal{O}(\epsilon^2)\right).$$

Hence,

$$\eta_\phi = \eta_\phi^{(0)}\left(1-\epsilon\frac{\lambda\phi^{(1)}}{\mathrm{n}} - \epsilon\frac{\mathcal{N}\dot{\varepsilon}_{II}^{(1)}}{\dot{\varepsilon}_{II}^{(0)}}\right) + \mathcal{O}(\epsilon^2).$$

7.4 The strain-rate tensor is

$$\dot{\boldsymbol{\varepsilon}} = \begin{pmatrix} \epsilon \frac{\partial^2 \psi}{\partial x \partial y} & \frac{\dot{\gamma}}{2} + \frac{\epsilon}{2}\left(\frac{\partial^2 \psi}{\partial y^2} - \frac{\partial^2 \psi}{\partial x^2}\right) \\ \frac{\dot{\gamma}}{2} + \frac{\epsilon}{2}\left(\frac{\partial^2 \psi}{\partial y^2} - \frac{\partial^2 \psi}{\partial x^2}\right) & -\epsilon \frac{\partial^2 \psi}{\partial x \partial y} \end{pmatrix}.$$

Hence

$$\begin{aligned} \dot{\varepsilon}_{II} &= \sqrt{\left(\frac{\dot{\gamma}}{2} + \frac{\epsilon}{2}\left(\frac{\partial^2 \psi}{\partial y^2} - \frac{\partial^2 \psi}{\partial x^2}\right)\right)^2 + \left(\epsilon \frac{\partial^2 \psi}{\partial x \partial y}\right)^2} \\ &= \frac{\dot{\gamma}}{2}\sqrt{1 + \frac{2\epsilon}{\dot{\gamma}}\left(\frac{\partial^2 \psi}{\partial y^2} - \frac{\partial^2 \psi}{\partial x^2}\right) + \mathcal{O}(\epsilon^2)}. \end{aligned}$$

Finally, Taylor-expanding the square root gives

$$\dot{\varepsilon}_{II} = \frac{\dot{\gamma}}{2} + \frac{\epsilon}{2}\left(\frac{\partial^2 \psi}{\partial y^2} - \frac{\partial^2 \psi}{\partial x^2}\right) + \mathcal{O}(\epsilon^2).$$

7.5 (a) By setting the constants to the given values we obtain

$$\dot{s} \Big/ \left[\left(1 - \phi^{(0)}\right) \frac{\lambda \dot{\gamma}}{\mathrm{n}\nu_\phi^{(0)}}\right] = \frac{\left(\delta_{(0)} k\right)^2}{\left(\delta_{(0)} k\right)^2 + 1}\left[1 - \mathrm{D}_{\mathcal{I}}\left(\delta_{(0)} k\right)^2\right]$$

Substituting the scalings, we obtain (7.43).

(b) Differentiating $\dot{s}_*$ with respect to k_* gives

$$\frac{d\dot{s}_*}{dk_*} = \frac{2k_* - 2\mathrm{D}_{\mathcal{I}} k_*^5 - 4\mathrm{D}_{\mathcal{I}} k_*^3}{\left(k_*^2 + 1\right)^2}$$

To find the extremum, we set $\frac{d\dot{s}_*}{dk_*} = 0$, which we can rewrite as the condition

$$k_*^5 + 2k_*^3 - \frac{k_*}{\mathrm{D}_{\mathcal{I}}} = 0$$

Noting that $k_* = 0$ gives a minimum, we can divide by k_* and solve for k_*^2 since it is a quadratic.

$$(k_*^{max})^2 = -1 \pm \left(1 + \frac{1}{\mathrm{D}_{\mathcal{I}}}\right)^{\frac{1}{2}}$$

Taking the positive root, we finally obtain

$$k_*^{max} = \sqrt{\left(\frac{\mathrm{D}_{\mathcal{I}} + 1}{\mathrm{D}_{\mathcal{I}}}\right)^{\frac{1}{2}} - 1} = \sqrt{\frac{(\mathrm{D}_{\mathcal{I}} + 1)^{1/2} - \mathrm{D}_{\mathcal{I}}^{1/2}}{\mathrm{D}_{\mathcal{I}}^{1/2}}}$$

7.6 (a) Noting that (7.17) holds for general θ, substituting $\phi^{(1)} = Ae^{i\boldsymbol{k}\cdot\boldsymbol{x}+\dot{s}t}$ into (7.15a) gives

$$\dot{s}\left[-\delta_{(0)}^2\nabla^2\phi^{(1)} + \phi^{(1)}\right] = -2\lambda(1-\phi^{(0)})K_\phi^{(0)}\eta_\phi^{(0)}\dot{\gamma}\left(\frac{\partial^2\phi^{(1)}}{\partial x^2} - \frac{\partial^2\phi^{(1)}}{\partial y^2}\right)$$

$$\Rightarrow \dot{s}\left(1 + \left(2\pi\delta_{(0)}/l\right)^2\right) = -2\lambda(1-\phi^{(0)})K_\phi^{(0)}\eta_\phi^{(0)}\dot{\gamma}\left(\cos^2\theta - \sin^2\theta\right)$$

$$\Rightarrow \dot{s} = -2\left(1-\phi^{(0)}\right)\frac{\lambda\dot{\gamma}}{\nu_\phi^{(0)}}\cdot\frac{\left(2\pi\delta_{(0)}/l\right)^2}{1+\left(2\pi\delta_{(0)}/l\right)^2}\cos 2\theta.$$

(b) Setting $\theta = 0, \pi/2$ we immediately recover (7.18).

(c) Marginal stability is the case of zero growth rate. This occurs when the weaker, higher-porosity bands are under neither compression nor tension. This is only the case when the bands are oriented at 45° to the principal axes, i.e., when they are at $\theta = \pm\pi/4$.

7.7 (a) Using the chain rule we obtain

$$\frac{\partial\phi^{(1)}}{\partial t} = \frac{\partial\phi^{(1)}}{\partial r}\frac{\partial r}{\partial t} + \frac{\partial\phi^{(1)}}{\partial s}\frac{\partial s}{\partial t}$$
$$= -\frac{\partial\phi^{(1)}}{\partial r}x\dot{\gamma}\exp\left(-\dot{\gamma}t\right) + \frac{\partial\phi^{(1)}}{\partial s}y\dot{\gamma}\exp\left(\dot{\gamma}t\right).$$

Similarly

$$\frac{\partial\phi^{(1)}}{\partial x} = \frac{\partial\phi^{(1)}}{\partial r}\exp\left(-\dot{\gamma}t\right)$$
$$\frac{\partial\phi^{(1)}}{\partial y} = \frac{\partial\phi^{(1)}}{\partial s}\exp\left(\dot{\gamma}t\right)$$

Therefore

$$\dot{\phi}^{(1)} + \dot{\gamma}\left(x\frac{\partial\phi^{(1)}}{\partial x} - y\frac{\partial\phi^{(1)}}{\partial y}\right) = 0$$

(b) For the initial porosity perturbation of the form

$$\phi^{(1)}(\boldsymbol{x},0) = Ae^{i\boldsymbol{k}_0\cdot\boldsymbol{x}},$$

we have

$$\phi^{(1)}(\boldsymbol{x},t) = A\exp\left[i(k_{0x}x\exp\left(-\dot{\gamma}t\right) + k_{0y}\exp\left(\dot{\gamma}t\right))\right]$$

Hence

$$\boldsymbol{k}(t) = k_{0x}\exp\left(-\dot{\gamma}t\right) + k_{0y}\exp\left(\dot{\gamma}t\right),$$

and so the angle to the shear plane is

$$\theta(t) = \tan^{-1}\left[\frac{k_{0x}}{k_{0y}} \exp\left(-2\dot{\gamma} t\right)\right].$$

(c) Considering the growth rate of the porosity perturbations, the perturbation has the form

$$\phi^{(1)}(\boldsymbol{x}, t) = A \exp\left[i\boldsymbol{k}_0 \cdot \boldsymbol{x} + s(t)\right].$$

Substituting this into (7.45) (the homogeneous part cancels)

$$A\frac{\partial \exp\left(s(t)\right)}{\partial t} \exp\left[i\boldsymbol{k}_0 \cdot \boldsymbol{x}\right] = (1 - \phi^{(0)})\mathcal{C}^{(1)}$$

$$\Leftrightarrow \dot{s}\phi^{(1)} = (1 - \phi^{(0)})\mathcal{C}^{(1)}$$

Rearranging gives us (7.17).

7.8 As the leading-order porosity $\phi^{(0)}$ is uniform, the final term in (7.48) enters the analysis at $\mathcal{O}(\epsilon^2)$, so it is neglected in the linearized equation. With the modified form of the pressure gradient, (7.40) becomes

$$\dot{s}\frac{\left(\delta_{(0)}k\right)^2 + 1}{\left(1 - \phi^{(0)}\right)} = K_\phi^{(0)}\left[2\eta_\phi^{(0)} k_x k_y \left(\frac{\lambda\dot{\gamma}}{\mathfrak{n}} + \hat{\psi}\mathcal{N}\left(k_y^2 - k_x^2\right)\right) - \mathcal{S}^{(0)}\left(\mathcal{A}\mathcal{A}'' k^2 + k^4\right)\right],$$

from which we obtain

$$\dot{s} = \left(1 - \phi^{(0)}\right)\frac{\lambda\dot{\gamma}}{\mathfrak{n}\nu_\phi^{(0)}} \cdot \frac{\left(\delta_{(0)}k\right)^2}{\left(\delta_{(0)}k\right)^2 + 1}\left[\left(\frac{\sin 2\theta}{1 - \mathcal{N}\cos^2 2\theta} - \mathrm{D}_{\mathcal{I}}\mathcal{A}\mathcal{A}''\right) - \mathrm{D}_{\mathcal{I}}\left(\delta_{(0)}k\right)^2\right].$$

15.6 Exercises from Chapter 8: Conservation of Energy

8.1 Using summation notation, we have (writing $\partial/\partial x_i = \partial_i$)

$$\begin{aligned}
2\eta_\phi\left(\dot{\boldsymbol{\varepsilon}}^s\right)^2 &= \frac{1}{2}\eta_\phi\left(\partial_i v_j^s + \partial_j v_i^s - \frac{2}{3}\partial_k v_k^s \delta_{ij}\right)\left(\partial_i v_j^s + \partial_j v_i^s - \frac{2}{3}\partial_m v_m^s \delta_{ij}\right)\\
&= \frac{1}{2}\eta_\phi\left(2\partial_j v_i^s\left(\partial_i v_j^s + \partial_j v_i^s\right) - \frac{4}{3}\partial_i v_i^s\left(\partial_k v_k^s + \partial_m v_m^s\right)\right.\\
&\qquad\left. + 3\cdot\frac{4}{9}\partial_k v_k^s \partial_m v_m^s\right)\\
&= \eta_\phi\left(\partial_j v_i^s\left(\partial_i v_j^s + \partial_j v_i^s\right) - \frac{2}{3}\partial_i v_i^s \partial_k v_k^s\right)\\
&= \eta_\phi\left(\partial_i v_j^s + \partial_j v_i^s - \frac{2}{3}\partial_k v_k^s \delta_{ij}\right)\partial_j v_i^s
\end{aligned}$$

$$\Rightarrow 2\eta_\phi\left(\dot{\boldsymbol{\varepsilon}}^s\right)^2 = (1 - \phi)\boldsymbol{\tau}^s : \boldsymbol{\nabla}\boldsymbol{v}^s,$$

as required.

8.2 By the definition of the Lagrangian derivative, we have

$$\phi^i \rho^i \frac{\partial a^i}{\partial t} + \phi^i \rho^i \boldsymbol{v}^i \cdot \nabla a^i = \phi^i \rho^i \frac{\mathrm{D}_i a^i}{\mathrm{D}t}.$$

Multiplying (4.11) by a^i and adding the equation above we obtain the desired forms. Summing equations (8.55) gives (8.56).

8.3 For steady porosity, equation (8.45b) reduces to

$$\mathrm{e}^{-\lambda\phi} = \Pi(1-\phi)\phi.$$

Taylor-expanding and taking the linear-order term in ϕ, we obtain

$$1 - \lambda\phi = \Pi\phi$$
$$\Rightarrow \phi = \frac{1}{\lambda + \Pi}.$$

8.4 (a) Using the fact that the shear stress is constant we can rewrite (8.42) as

$$\dot{\mathcal{H}} = -\frac{\mathcal{F}}{\eta_0}\mathcal{H}\phi$$
$$\dot{\phi} = \frac{\mathcal{T}_0^2}{\rho^s L}\frac{e^{\lambda\phi}}{\eta_0} - \frac{\mathcal{F}\phi(1-\phi)}{\eta_0}.$$

Rescaling, as we did in chapter 8, we obtain

$$\dot{\mathcal{H}} = -\Pi\mathcal{H}\phi$$
$$\dot{\phi} = e^{\lambda\phi} - \Pi\phi(1-\phi).$$

(b) For the given assumptions, the porosity evolution equation becomes

$$\dot{\phi} = e^{\lambda\phi}.$$

Given the initial conditions we solve this and obtain

$$\phi = -\frac{\log(1-\lambda t)}{\lambda}.$$

Hence, we can solve for t^*

$$t^* = \frac{1 - e^{-0.3\lambda}}{\lambda}.$$

(c) Just by a slight modification of our code for the constant strain rate we have:

```
function S = CompactionDissipationSolutionStress(Alpha,lambda,phi0,
    tmax)

    if nargin<1; Alpha=1; end
```

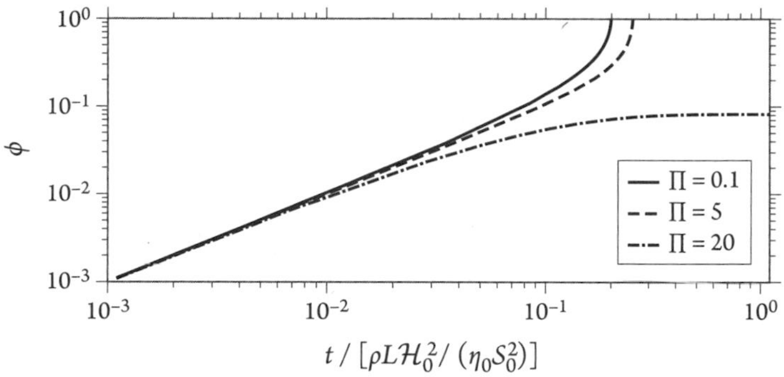

Figure 15.5. We can see that the biggest difference between the cases of constant shear stress and of constant strain rate is that the porosity diverges for Π below a certain value (more on this in exercise 8.6). At $\Pi = 20$ we observe that it resembles the behavior for constant strain rate.

```
    if nargin<2; lambda = 30; end
    if nargin<3; phi0 = 0; end
    if nargin<4; tmax = 0.1; end
    options = odeset('RelTol',1e-8,'Refine',10,'nonnegative',[1 1]);

    S = ode45(@derivatives,[0 0.0011 0.01 tmax],[1 phi0],options);

    function dfdt = derivatives(t,f)
        dfdt(1,1) = -Alpha*f(1)*f(2);
        dfdt(2,1) = exp(lambda*f(2)) - Alpha*(1-f(2))*f(2);
    end
end
```

From the data we find that for $\Pi = 0.1$, $t^* \approx 0.165$ and for $\Pi = 5$, $t^* \approx 0.205$. We can compare this to our approximate value from (b), which yields $t^* \approx 1.55$; this is consistent, as the increase in Π will cause the growth to slow down, hence it takes more time for disaggregation to occur.

8.5 (a) We have

$$
\begin{aligned}
\dot{\mathcal{H}} &= \frac{\mathrm{d}\mathcal{H}}{\mathrm{d}\phi}\dot{\phi} \\
&= \frac{\mathrm{d}\mathcal{H}}{\mathrm{d}\phi}\left(e^{-\lambda\phi} - \Pi(1-\phi)\phi\right) \\
&= -\Pi\mathcal{H}\phi
\end{aligned}
$$

$$
\Rightarrow \quad \frac{1}{\mathcal{H}}\frac{\mathrm{d}\mathcal{H}}{\mathrm{d}\phi} = \frac{-\Pi\phi}{e^{-\lambda\phi} - \Pi(1-\phi)\phi}
$$

$$
\Rightarrow \quad \frac{\mathrm{d}(\ln\mathcal{H})}{\mathrm{d}\phi} = \frac{-\Pi\phi}{e^{-\lambda\phi} - \Pi(1-\phi)\phi},
$$

as required.

(b) Using the given approximations, we have

$$\frac{\mathrm{d}(\ln \mathcal{H})}{\mathrm{d}\phi} \approx \frac{-\Pi\phi}{(1-\lambda\phi)-\Pi\phi}$$

$$= \frac{\Pi}{\lambda+\Pi}\left(1-\frac{1}{1-(\lambda+\Pi)\phi}\right)$$

$$\Rightarrow \ln \mathcal{H} = \frac{\Pi}{\lambda+\Pi}\left(\phi+\frac{1}{\lambda+\Pi}\ln|1-(\lambda+\Pi)\phi)|\right) + \text{constant},$$

therefore using the initial conditions (for dimensionless variables) $\mathcal{H}=1, \phi=\phi_0$ we obtain

$$\mathcal{H}(\phi) = \left[\frac{1-(\lambda+\Pi)\phi}{1-(\lambda+\Pi)\phi_0}\right]^{\frac{\Pi}{(\lambda+\Pi)^2}} e^{\frac{\Pi}{\lambda+\Pi}(\phi-\phi_0)}.$$

(c) For $\Pi\phi \ll 1$, (8.45b) gives

$$\dot{\phi} = e^{-\lambda\phi}$$

$$\Rightarrow \quad e^{\lambda\phi}\dot{\phi} = 1$$

$$= \frac{1}{\lambda}\frac{\mathrm{d}}{\mathrm{d}t}e^{\lambda\phi};$$

Therefore, using the initial condition $\phi=\phi_0$ at $t=0$, gives

$$\phi(t) = \frac{1}{\lambda}\ln\left(\lambda t + e^{\lambda\phi_0}\right).$$

For $\Pi\phi \gg 1$, using the same approximations made in (b), (8.45b) gives

$$\dot{\phi} = -\Pi\phi$$

$$\Rightarrow \quad \phi(t) = \phi_0 e^{-\Pi t}.$$

8.6 (a) In the evolution equation for the porosity

$$\dot{\phi} = \mathrm{e}^{-\lambda\phi} - \Pi(1-\phi)\phi$$

we can see that the right-hand side is positive for $\phi=0,\ 1$. For small enough Π the right-hand side is positive, hence the porosity keeps increasing. This behavior changes when Π is increased to a value Π_c at which the right-hand side has a root, i.e., there is a unique steady state. Therefore, we require that $\dot{\phi}(\phi_c) = \ddot{\phi}(\phi_c) = 0$:

$$\mathrm{e}^{-\lambda\phi_c} - \Pi(1-\phi_c)\phi_c = 0$$

$$-\lambda\mathrm{e}^{-\lambda\phi_c} - \Pi(1-2\phi_c) = 0$$

First we express ϕ_c by eliminating the exponential

$$-\lambda(1-\phi_c)\phi_c = (1-2\phi_c).$$

Solving this quadratic and noting that the porosity must be positive, hence taking the positive root, gives us

$$\phi_c = \frac{\lambda - 2 + \sqrt{\lambda^2+4}}{2\lambda}.$$

Hence, the steady-state condition gives us

$$\Pi_c^{strain} = \frac{e^{-\lambda\phi_c}}{\phi_c(1-\phi_c)} = \frac{\lambda^2 e^{-\frac{\lambda-2+\sqrt{\lambda^2+4}}{2}}}{\sqrt{\lambda^2+4}-2}.$$

(b) The evolution equation for constant-shear stress is

$$\dot{\phi} = \mathrm{e}^{\lambda\phi} - \Pi(1-\phi)\phi;$$

hence all the algebra before works with the only difference in the justification for the choice of root ϕ_c. In this case both roots are positive, but we choose the smaller one as the larger root is greater than 1 and we know that $0 \leq \phi \leq 1$. Therefore,

$$\Pi_c^{shear} = \frac{\lambda^2 e^{-\frac{-\lambda-2+\sqrt{\lambda^2+4}}{2}}}{\sqrt{\lambda^2+4}-2}.$$

(c) Notice that

$$\Pi_c^{shear} = \Pi_c^{strain} e^{\lambda} > \Pi_c^{strain}.$$

The dissipation is

- $\Psi = \eta_0 \left(\frac{\mathcal{S}_0}{\mathcal{H}_0}\right)^2 e^{-\lambda\phi}$ for constant strain rate,
- $\Psi = \frac{1}{\eta_0}\left(\frac{\mathcal{S}_0}{\mathcal{H}_0}\right)^2 e^{\lambda\phi}$ for constant shear stress.

Hence the melting rate is

- $\Gamma = L\eta_0 \left(\frac{\mathcal{S}_0}{\mathcal{H}_0}\right)^2 e^{-\lambda\phi}$ for constant strain rate,
- $\Gamma = L\frac{1}{\eta_0}\left(\frac{\mathcal{S}_0}{\mathcal{H}_0}\right)^2 e^{\lambda\phi}$ for constant shear stress.

This tells us that for constant strain rate, the melting rate decreases with porosity, whereas for constant shear stress it increases with porosity. Recalling that the compaction dissipation number represents the ratio of compaction to shear-driven dissipative heating, we can see why the compaction dissipation number is bigger for the case of constant shear stress. Since with increasing porosity the melting increases, which in turn drives the porosity growth even further, we need compaction to counteract this effect. In the case of constant shear stress the melting rate decreases with porosity, hence the effect of compaction does not have to be as big.

15.7 Exercises from Chapter 9: Conservation of Chemical-Species Mass

9.1 We start off by solving the second equation. Rearranging the terms gives us

$$\frac{1}{c_j^s}\frac{\mathrm{d}c_j^s}{\mathrm{d}t} = -\frac{\left(\frac{1}{D_j}-1\right)\mathcal{G}}{1-\mathcal{G}t}.$$

This can be easily integrated:

$$\log(c_j^s) = \left(\frac{1}{D_j}-1\right)\log(1-\mathcal{G}t) + \text{const.}$$

Hence

$$c_j^s = c_0(1-\mathcal{G}t)^{\frac{1}{D_j}-1}.$$

For the liquid concentration, we rewrite the first equation as

$$t\frac{\mathrm{d}c_j^\ell}{\mathrm{d}t} + c_j^\ell = \frac{1}{D_j}c_j^s$$

$$\Leftrightarrow \qquad \frac{\mathrm{d}\left(tc_j^\ell\right)}{\mathrm{d}t} = \frac{c_0}{D_j}(1-\mathcal{G}t)^{\frac{1}{D_j}-1}$$

$$\Leftrightarrow \qquad tc_j^\ell = -\frac{c_0}{\mathcal{G}}(1-\mathcal{G}t)^{\frac{1}{D_j}} + \text{const.}$$

Since c_j^ℓ is not singular, we require this equation to hold at $t=0$, hence const. $=\frac{c_0}{\mathcal{G}}$. Therefore,

$$c_j^\ell = c_0\frac{1-(1-\mathcal{G}t)^{\frac{1}{D_j}-1}}{\mathcal{G}t}.$$

Recalling that $F=\mathcal{G}t$, we see that it is indeed equivalent to the canonical fractional melting model.

9.2 The phase-specific evolution equations become

$$\phi^\ell\rho^\ell\frac{\mathrm{D}_\ell c_j^\ell}{\mathrm{D}t} = \left(c_j^\Gamma - c_j^\ell\right)\Gamma + \phi^\ell\rho^\ell\left(\lambda_{j-1}c_{j-1}^\ell - \lambda_j c_j^\ell\right) + \boldsymbol{\nabla}\cdot\phi^\ell\rho^\ell\mathcal{D}\boldsymbol{\nabla}c_j^\ell,$$

$$\phi^s\rho^s\frac{\mathrm{D}_s c_j^s}{\mathrm{D}t} = -\left(c_j^\Gamma - c_j^s\right)\Gamma + \phi^s\rho^s\left(\lambda_{j-1}c_{j-1}^s - \lambda_j c_j^s\right).$$

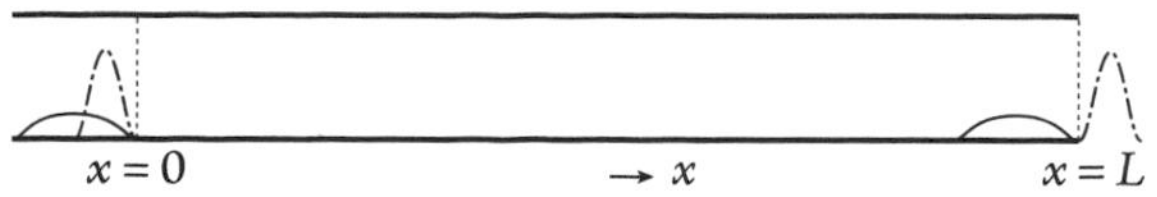

Figure 15.6. The two species enter the channel at $x = 0$. We want the length of the channel to be sufficient such that when the faster species exits the channel, the slower is still inside.

Neglecting diffusion, making the Boussinesq approximation and assuming that the trace elements are in equilibrium, we obtain

$$\phi^\ell \rho \left(\frac{\partial c_j^\ell}{\partial t} + \boldsymbol{v}^\ell \cdot \nabla c_j^\ell \right) = \left(c_j^\Gamma - c_j^\ell \right) \Gamma + \phi^\ell \rho \left(\lambda_{j-1} c_{j-1}^\ell - \lambda_j c_j^\ell \right),$$

$$\phi^s \rho \left(\frac{\partial D_j c_j^\ell}{\partial t} + \boldsymbol{v}^s \cdot \nabla D_j c_j^\ell \right) = - \left(c_j^\Gamma - D_j c_j^\ell \right) \Gamma + \phi^s \rho \left(\lambda_{j-1} D_{j-1} c_{j-1}^\ell - \lambda_j D_j c_j^\ell \right).$$

By expanding the derivative of the product in the second equation, summing these equations, and rearranging, we obtain (9.32).

9.3 We first need to express the transport velocities of each species. Recalling the definition, we have, in this case,

$$v^{D_j} = \frac{\phi v^\ell}{\phi + D_j(1-\phi)}.$$

Since we have $D_1 > D_2$, we know that $v^{D_1} < v^{D_2}$. We want for both species to exit the channel completely while not overlapping. Hence, at the time of exit of the faster species, we want the front end of the slower species to be still within the channel (see fig. 15.6).

The time for the slower species to reach the end is given by

$$t_{out} = \frac{L}{v^{D_1}}.$$

We require that

$$t_{out} v^{D_2} \geq L + d_2$$

This gives us the condition

$$v^{D_2} t_{out} = (L) \frac{v^{D_2}}{v^{D_1}} \geq L + d_2.$$

Expressing L and substituting expressions for transport velocities we get

$$L \geq \frac{d_2}{\frac{\phi + D_1(1-\phi)}{\phi + D_2(1-\phi)} - 1}.$$

15.8 Exercises from Chapter 10: Petrological Thermodynamics of Liquid and Solid Phases

10.1 Algebraic rearrangement gives

$$\phi = \frac{\rho^s}{\rho^\ell(1/f - 1) + \rho^s}.$$

10.2 See the online supplementary materials.

15.9 Exercises from Chapter 11: Melting-Column Models

11.1 (a) Assuming no melting, mass conservation equations simply become

$$\frac{\mathrm{d}(\phi w)}{\mathrm{d}z} = 0$$
$$\frac{\mathrm{d}((1-\phi)W)}{\mathrm{d}z} = 0.$$

If ϕ, w, W are constants, then these equations are satisfied. Combined with the boundary conditions we get $\phi(z) = \phi_0$ for $z \geq 0$.

(b) If $\Gamma = 0$, (11.4) gives us $F(z) = 0$ for $z \geq 0$. Therefore from (11.12) we obtain

$$\mathcal{Q}\phi^n + \phi = 0.$$

This has two solutions, $\phi = 0$ and $\phi = (-1/\mathcal{Q})^{1/n}$; however, as the porosity is always real and nonnegative, the only physically relevant solution is $\phi = 0$. This is consistent with part (a) of this exercise, as the derivation of (11.12) relied on the boundary condition $\phi = 0$ at $z = 0$.

11.2 First recall the vertically integrated mass conservation equations (11.3). With these, we can rewrite the equations as

$$F\frac{\mathrm{d}c_j^\ell}{\mathrm{d}z} = +\left(c_j^s/D_j - c_j^\ell\right) \left.\tfrac{\mathrm{d}F}{\mathrm{d}z}\right|_{\mathrm{s}},$$
$$(1-F)\frac{\mathrm{d}c_j^s}{\mathrm{d}z} = -\left(c_j^s/D_j - c_j^s\right) \left.\tfrac{\mathrm{d}F}{\mathrm{d}z}\right|_{\mathrm{s}}.$$

Next, from the fact that we have constant, uniform melting rate Γ, we can obtain the degree of melting from (11.4), $F = \left.\frac{\mathrm{d}F}{\mathrm{d}z}\right|_{\mathrm{s}}(z - z_0)$. Substituting this expression into our equations, we can directly integrate the second differential equation to obtain

$$\ln(c_j^s) = \left(\frac{1}{D_j} - 1\right)\ln\left(1 - \left.\tfrac{\mathrm{d}F}{\mathrm{d}z}\right|_{\mathrm{s}}(z - z_0)\right) + \text{const.},$$

and hence

$$c_j^s = c_0\left[1 - \left.\tfrac{\mathrm{d}F}{\mathrm{d}z}\right|_{\mathrm{s}}(z - z_0)\right]^{\left(\frac{1}{D_j} - 1\right)}.$$

Substituting this into the first equation, we can integrate the equation by writing

$$\left[c_j^\ell + (z - z_0)\frac{\mathrm{d}c_j^\ell}{\mathrm{d}z}\right] = c_j^s / D_j,$$

where we notice that the left-hand side is a total derivative. Thus, imposing the same regularity condition as in exercise 9.1, we get

$$c_j^\ell = c_0 \frac{1 - \left[1 - \left.\frac{\mathrm{d}F}{\mathrm{d}z}\right|_s (z - z_0)\right]^{\frac{1}{D_j} - 1}}{\left.\frac{\mathrm{d}F}{\mathrm{d}z}\right|_s (z - z_0)}.$$

Since $F = \left.\frac{\mathrm{d}F}{\mathrm{d}z}\right|_s (z - z_0)$, this gives us exactly the canonical fractional melting model:

$$c_j^s = c_0\,(1 - F)^{\left(\frac{1}{D_j} - 1\right)},$$

$$c_j^\ell = c_0 \frac{1 - (1 - F)^{\frac{1}{D_j} - 1}}{F}.$$

11.3 Rearranging equation (11.22), we can express

$$z_0 = \frac{L F_{\max}^{1c}}{g c_P} \cdot \frac{1}{\rho/\vartheta - \alpha_\rho T_0^S / c_P}.$$

Using the values in the table we can then estimate $z_0 \approx 60$ km.

11.4 Again, we start by expressing

$$z_0 = \frac{F_{\max}^{2c}}{g} \cdot \frac{L/c_P + M^S \Delta c}{\rho/\vartheta - \alpha_\rho T_0^S / c_P}.$$

Here we again obtain $z_0 \approx 60$ km.

11.5 Recalling that $C \equiv \frac{\mathrm{d}W}{\mathrm{d}z}$, we can sum the mass conservation equations to obtain

$$-\frac{\mathrm{d}^2(\phi w)}{\mathrm{d}z^2} = \frac{\mathrm{d}C}{\mathrm{d}z}.$$

Therefore the momentum conservation can be rewritten as

$$\phi w = \left(\frac{\phi}{\phi_0}\right)^n \left[-\delta_0^2 \frac{\mathrm{d}^2(\phi w)}{\mathrm{d}z^2} + q_\infty\right].$$

Substituting the expression from the definition of the shape function gives us

$$q_\infty \left[1 - G(z/\delta_f)\right] = \left(\frac{\phi}{\phi_0}\right)^n \left[-\frac{\delta_0^2}{\delta_f^2} q_\infty G''(z/\delta_f) + q_\infty\right].$$

Rearranging gives us

$$\frac{\phi}{\phi_0} = \left[\frac{1 - G(z/\delta_f)}{1 - R^2 G''(z/\delta_f)}\right]^{1/n}.$$

11.6 (a) Noticing that $G(z/\delta_f) = G'(z/\delta_f) = G''(z/\delta_f) = \exp(z/\delta_f)$, we see that $G'' > 0$ for all z and hence the pressure gradient is always positive. Furthermore, $G(-\infty) = 0$ and $G(0) = 1$ thereby satisfying boundary conditions (11.54).

(b) Substituting the shape function into (11.57) gives us

$$\frac{\phi}{\phi_0} = \left[\frac{1 - \exp(z/\delta_f)}{1 - R_\delta^2 \exp(z/\delta_f)}\right]^{1/n}.$$

For there to be a local extremum we require the first derivative to be zero at some point for some value R_δ. Differentiating this equation yields

$$\frac{1}{\phi_0}\frac{\mathrm{d}\phi}{\mathrm{d}z} = \frac{1}{n}\left(\frac{\phi}{\phi_0}\right)^{1-n} \times \left(\frac{R_\delta^2 \frac{\exp(z/\delta_f)}{\delta_f}\left(1 - \exp(z/\delta_f)\right) - \frac{\exp(z/\delta_f)}{\delta_f}\left(1 - R_\delta^2 \exp(z/\delta_f)\right)}{\left(1 - R_\delta^2 \exp(z/\delta_f)\right)^2}\right).$$

This is zero only if the numerator of the last bracket is zero; hence we require

$$R_\delta^2 \frac{\exp(z/\delta_f)}{\delta_f}\left(1 - \exp(z/\delta_f)\right) - \frac{\exp(z/\delta_f)}{\delta_f}\left(1 - R_\delta^2 \exp(z/\delta_f)\right) = 0.$$

After simplification we obtain $R_\delta^2 = 1$. However, if $R_\delta^2 = 1$, the porosity is uniform:

$$\frac{\phi}{\phi_0} = \left[\frac{1 - \exp(z/\delta_f)}{1 - \exp(z/\delta_f)}\right]^{1/n} = 1.$$

Hence there is never a extremum for this shape function.

11.7 First we calculate the z-derivative of (11.65):

$$\frac{\mathrm{d}a_j^l}{\mathrm{d}z} = \left(\frac{(1 - D_j) F_{max} a_j^s|_0}{\left(D_j + (1 - D_j) F_{max} z\right)^2}\right) \tilde{a}^\ell + \left(\frac{a_j^s|_0}{D_j + (1 - D_j) F_{\mathrm{max}} z}\right) \frac{\mathrm{d}\tilde{a}_j^\ell}{\mathrm{d}z}.$$

Therefore we have

$$\frac{1}{a_j^\ell}\frac{\mathrm{d}a_j^l}{\mathrm{d}z} = \left(\frac{(1-D_j)F_{max}}{D_j+(1-D_j)F_{max}z}\right) + \frac{1}{\tilde{a}_j^\ell}\frac{\mathrm{d}\tilde{a}_j^\ell}{\mathrm{d}z}.$$

Substituting this into (11.64) and multiplying by $\tilde{a}_j^\ell$ gives us

$$\frac{\mathrm{d}\tilde{a}_j^\ell}{\mathrm{d}z} = \frac{\lambda_j z_0}{w^{D_j}}\left[\left(\frac{\phi+(1-\phi)D_{j-1}}{\phi+(1-\phi)D_j}\right)a_{j-1}^\ell - a_j^\ell\right].$$

Finally, we express the ratio of activities of elements in terms of their respective ingrowth factors

$$\frac{a_{j-1}^\ell}{a_j^\ell} = \left(\frac{D_j+(1-D_j)F_{max}}{D_{j-1}+(1-D_{j-1}F_{max})}\right)\frac{\tilde{a}_{j-1}^\ell}{\tilde{a}_j^\ell},$$

where we have used the fact that the unmolten solid is in secular equilibrium. Recalling the expression for effective transport speeds, (11.63), we finally get

$$\frac{\mathrm{d}\tilde{a}_j^\ell}{\mathrm{d}z} = \lambda_j z_0\left[\frac{\tilde{a}_{j-1}^\ell}{w^{D_{j-1}}} - \frac{\tilde{a}_j^\ell}{w^{D_j}}\right].$$

11.8 (a) The degree of melting is defined as $F = \frac{\int_0^z \Gamma dz}{\rho^s W_0}$; therefore, we can express the melting rate as $\Gamma = \rho^s W_0 \frac{\mathrm{d}F}{\mathrm{d}z}$. Substituting this and the equation for the solid velocity into (11.79b), we obtain

$$W_0(1-F)\frac{\mathrm{d}c_j^s}{\mathrm{d}z} = -c_j^s(1/D_j-1)W_0\frac{\mathrm{d}F}{\mathrm{d}z}.$$

Rearranging this gives us

$$\frac{1}{c_j^s}\frac{\mathrm{d}c_j^s}{\mathrm{d}z} = -(1/D_j-1)\frac{1}{1-F}\frac{\mathrm{d}F}{\mathrm{d}z},$$

which can be integrated and solved as

$$c_j^s = c_0^s\,(1-F)^{(1/D_j-1)}.$$

Now we need to express F. The temperature equation can be rewritten as

$$\frac{\mathrm{d}T}{\mathrm{d}z} = -\frac{L}{c_e}\frac{\mathrm{d}F}{\mathrm{d}z} - \frac{\alpha_\rho g T_0^S}{c_e},$$

which, after integrating from 0 to z and assuming that the temperature is the solidus temperature, becomes

$$\left(\frac{\alpha_\rho g T_0^S}{c_e} - \frac{g\rho}{\vartheta}\right) z + M^S\left(c_j^s - c_0^s\right) = -\frac{L}{c_e} F.$$

Thus, the implicit solution is given by

$$c_j^s = c_0^s \left\{1 + \frac{c_e}{L}\left[\left(\frac{\alpha_\rho g T_0^S}{c_e} - \frac{g\rho}{\vartheta}\right) z + M^S\left(c_j^s - c_0^s\right)\right]\right\}^{(1/D_j - 1)}.$$

(b) The approximation of the solid velocity reduces (11.79b) to

$$\rho^s W_0 \frac{\mathrm{d}c_j^s}{\mathrm{d}z} = -c_j^s (1/D_j - 1)\Gamma,$$

but this time we express Γ using the temperature and the solidus temperature equations:

$$\Gamma = -\frac{W_0 c_e \rho^s}{L}\left(M^S \frac{\mathrm{d}c^s}{\mathrm{d}z} - g\rho/\vartheta + \frac{\alpha_\rho g T_0^S}{c_e}\right).$$

Therefore, we can rewrite (11.79b) as

$$\frac{\mathrm{d}c_j^s}{\mathrm{d}z} = c_j^s (1/D_j - 1)\frac{c_e}{L}\left(M^S \frac{\mathrm{d}c^s}{\mathrm{d}z} - g\rho/\vartheta + \frac{\alpha_\rho g T_0^S}{c_e}\right).$$

Define $\alpha = (1/D_j - 1)\frac{c_e M^S}{L}$ and $\beta = -\frac{g\rho}{M^S \vartheta} + \frac{\alpha g T_0^S}{c_e M^S}$ to rewrite the equation in a more convenient form

$$\frac{\mathrm{d}c_j^s}{\mathrm{d}z} = \alpha c_j^s \left(\frac{\mathrm{d}c_j^s}{\mathrm{d}z} + \beta\right).$$

Rearranging this then gives us

$$(1 - \alpha c_j^s)\frac{\mathrm{d}c_j^s}{\mathrm{d}z} = \alpha\beta c_j^s,$$

which after premultiplying by $-\alpha$ and substituting $C = -\alpha c_j^s$ becomes

$$(1 + C)\frac{\mathrm{d}C}{\mathrm{d}z} = \alpha\beta C.$$

Finally, by defining $y \equiv e^{\alpha\beta z}$ we can rewrite the differential operator as $\frac{\mathrm{d}}{\mathrm{d}z} = \frac{\mathrm{d}y}{\mathrm{d}z}\frac{\mathrm{d}}{\mathrm{d}y} = \frac{\mathrm{d}}{\mathrm{d}y}\alpha\beta e^{\alpha\beta z} = \alpha\beta y \frac{\mathrm{d}}{\mathrm{d}y}$, and hence the equation is equivalent to

$$y(1+C)\frac{\mathrm{d}C}{\mathrm{d}y} = C,$$

which means that $C(y) = \mathcal{W}(\gamma_0 y)$. Thus, we can express the solid concentration

$$c_j^s = -\frac{1}{\alpha}\mathcal{W}(\gamma_0 e^{\alpha\beta z}).$$

To find the value of the constant γ, we use the boundary condition $c_j^s = c_0$ at $z=0$. We require that

$$c_0 = -\frac{1}{\alpha}\mathcal{W}(\gamma_0).$$

Using the property of the Lambert W function $\mathcal{W}(ze^z) = z$, we deduce that $\gamma_0 = -\alpha c_0 e^{-\alpha c_0}$.

Note that since $e^{\alpha\beta z}$ is always positive, this gives a single-valued function (as long as $D_j > 1$), as the Lambert W function is single-valued on the positive real axis.

15.10 Exercises from Chapter 12: Reactive Flow and the Emergence of Melt Channels

12.1 Given the approximations we know that

$$a \sim \frac{1}{2}\left(\frac{n\mathcal{K}/\mathcal{S} + \sigma k^2/\mathrm{Da}}{\sigma\mathcal{K}}\right);$$

hence

$$a \sim \frac{1}{2}\left(\frac{1}{(1-\epsilon)\mathcal{S}} + \frac{k^2}{\mathrm{Da}\mathcal{K}}\right).$$

Taylor-expanding this gives us

$$(1-\epsilon)^{-1} \sim 1 + \epsilon + \mathcal{O}(\epsilon^2) \quad \text{and}$$

$$(1 + k^2/(\mathrm{Da\,Pe}))^{-1} \sim 1 - \frac{k^2}{\mathrm{Da\,Pe}} + \mathcal{O}(\epsilon^2),$$

and we get that at the lowest order,

$$a \sim \frac{1}{2\mathcal{S}} + \frac{k^2}{2\mathrm{Da}}.$$

For the ϵ estimate we get from (12.47b) that

$$\frac{a^2+b^2}{k^2} \sim \frac{n}{\sigma\mathcal{K}} - 1,$$

which can be rewritten as

$$\frac{1}{(1-\epsilon)(1+\frac{k^2}{\mathrm{Da\,Pe}})} \sim \frac{a^2+b^2}{k^2}+1.$$

Using the same Taylor expansions we arrive at

$$1+\epsilon-\frac{k^2}{\mathrm{Da\,Pe}}+\mathcal{O}(\epsilon^2) \sim \frac{a^2+b^2}{k^2}+1,$$

which after rearranging becomes

$$\epsilon \sim \frac{a^2+b^2}{k^2}+\frac{k^2}{\mathrm{Da\,Pe}}.$$

12.2 The expression for the growth rate simplifies to

$$\sigma=\pm\frac{n}{\mathcal{K}}\left\{\left[k^4-k^2\frac{\mathcal{K}^2}{\mathcal{S}^2}\right]^{1/2}\pm k^2\right\}\left[2k^2\right]^{-1}.$$

Noting that in the given limits $\mathcal{K}\approx 1$ we can rearrange this to obtain

$$\frac{\sigma}{n}\approx\frac{1\pm\left[1-\frac{1}{\mathcal{S}^2k^2}\right]}{2}.$$

Thus we can see that as $k\to\infty$, $\frac{\sigma}{n}$ goes to 0 or 1, hence there is no high wavenumber cutoff.

12.3 (a) The only term which will change due to the nonconstant compaction viscosity in the simplified system (12.22) is the pressure term in the mass conservation equation. Therefore the full system after changing the compaction viscosity term becomes (after nondimensionalizing)

$$\frac{\partial\phi}{\partial t}=\mathcal{P}\phi+\mathcal{G}+\chi, \tag{15.2a}$$

$$\mathcal{M}\frac{\partial\phi}{\partial t}+\nabla\cdot\phi\boldsymbol{v}^\ell=\mathcal{M}\left(\mathcal{G}+\chi\right), \tag{15.2b}$$

$$\phi\boldsymbol{v}^\ell\cdot\left(\nabla\chi/\mathrm{Da}-\hat{\boldsymbol{z}}\right)=\frac{1}{\mathrm{Da\,Pe}}\frac{\partial}{\partial x}\phi\frac{\partial\chi}{\partial x}-\chi, \tag{15.2c}$$

$$\phi\boldsymbol{v}^\ell=\phi^n\left(\hat{\boldsymbol{z}}-\mathcal{S}\nabla\mathcal{P}\right). \tag{15.2d}$$

Perturbing around the base state (12.23) we find that the base state equations don't change and are exactly the same as (12.24), hence we will assume the approximate solution (12.30). Gathering the first-order perturbation terms we obtain the system

$$\phi_t^{(1)} = \mathcal{P}^{(1)}\phi^{(0)} + \phi^{(1)}\mathcal{P}^{(0)} + \chi^{(1)}, \tag{15.3a}$$

$$\phi_z^{(1)} + \nabla\cdot \boldsymbol{v}^{(1)} = 0, \tag{15.3b}$$

$$w^{(1)} + \phi^{(1)} = \frac{\chi_z^{(1)}}{\text{Da}} + \chi^{(1)} - \frac{1}{\text{Da Pe}}\chi_{xx}^{(1)} \tag{15.3c}$$

$$\boldsymbol{v}^{(1)} = (n-1)\,\phi^{(1)}\hat{z} - \mathcal{S}\nabla\mathcal{P}^{(1)}. \tag{15.3d}$$

(b) Focusing only on the conservation of mass equation, we can rewrite it as

$$\left(\frac{\partial}{\partial t} + 1\right)\phi^{(1)} = \mathcal{P}^{(1)} + \chi^{(1)},$$

where we have used the values of our base state. If we call the operator $D_t \equiv \left(\frac{\partial}{\partial t} + 1\right)$, the system becomes of the same form as (12.31) previously in the chapter.

Therefore we can directly deduce from the relation between the respective operators that the new growth rate is related to the old growth rate by

$$\sigma_{old} = \sigma_{new} + 1.$$

Hence $\sigma_{new} = \sigma_{old} - 1$.

(c) As mentioned earlier in the chapter, if the growth rate is positive, we get an instability. Therefore, we require $\sigma < 0$ for stability.

(d) From the second part of this question we know that $\sigma_{new} = \sigma_{old} - 1 < \sigma_{old}$. Thus, the system with variable compaction viscosity suppresses instabilities.

(e) If $\zeta_\phi = -\zeta_0\left(1 - \log\frac{\phi}{\phi_0}\right)$, then the mass conservation equation (dimensionless) instead becomes

$$\frac{\partial\phi}{\partial t} = \mathcal{P}\frac{1}{1-\log\phi} + \mathcal{G} + \chi.$$

We now look at the effect of perturbation on the pressure term.

$$\frac{1}{1-\log\phi} \approx \frac{1}{1-\log\left(1+\epsilon\phi^{(1)} + \mathcal{O}(\epsilon)\right)}.$$

Taking the Taylor expansion of the right-hand side we obtain

$$\frac{1}{1-\log\phi} \approx 1 + \epsilon\phi^{(1)} + \mathcal{O}(\epsilon).$$

Noting that the first-order perturbation is the same as for compaction viscosity $\zeta_\phi = \zeta_0\frac{\phi_0}{\phi}$ we can conclude that to the first order, both systems behave identically.

15.11 Exercises from Chapter 13: Tectonic-Scale Models

13.1 Taking the volume integral of the steady-state mass conservation equations gives us

$$\int_V \nabla\cdot\left[\phi \boldsymbol{v}^\ell\right] dV = \int_V \frac{\Gamma}{\rho} dV,$$

$$\int_V \nabla\cdot\left[(1-\phi)\boldsymbol{v}^s\right] dV = -\int_V \frac{\Gamma}{\rho} dV.$$

First recall that $\boldsymbol{v}^\ell = u\hat{\boldsymbol{x}} + w\hat{\boldsymbol{z}}$ and that we assume translational invariance in the x-direction for the melting rate, hence $\frac{\partial\Gamma}{\partial x} = 0$. Using the divergence theorem we arrive at

$$\int_{\delta V} \left[\phi \boldsymbol{v}^\ell\right]\cdot \boldsymbol{n}\, dl = \delta x \int_{z_h-\delta z}^{z_h} \frac{\Gamma}{\rho} dz,$$

$$\int_{\delta V} \left[(1-\phi)\boldsymbol{v}^s\right]\cdot \boldsymbol{n}\, dl = -\delta x \int_{z_h-\delta z}^{z_h} \frac{\Gamma}{\rho} dz,$$

By expanding the dot product we obtain

$$-\phi w|_{z_h-\delta z}\,\delta x + \int_{z_h-\delta z}^{z_h} \left(\phi u|_{x+\delta x} - \phi u|_x\right) \mathrm{d}z = \delta x \int_{z_h-\delta z}^{z_h} \frac{\Gamma}{\rho}\,\mathrm{d}z,$$

$$-(1-\phi) W|_{z_h-\delta z}\,\delta x - U_0 \frac{\mathrm{d}z_h}{\mathrm{d}x}\delta x = -\delta x \int_{z_h-\delta z}^{z_h} \frac{\Gamma}{\rho}\,\mathrm{d}z.$$

By defining the volumeric flow rate q and the volumeric mass-transfer rate G we can rewrite these by noting the following

$$\int_{z_h-\delta z}^{z_h} \left(\phi u|_{x+\delta x} - \phi u|_x\right) \mathrm{d}z = q(x+\delta x) - q(x) \approx \frac{\mathrm{d}q}{\mathrm{d}x}\delta x$$

Therefore, if we divide both equations by δx and take the limit $\delta x \to 0$ we obtain

$$\phi w|_{z_h^-} + G = \frac{\mathrm{d}q}{\mathrm{d}x},$$

$$(1-\phi) W|_{z_h^-} - G = -U_0 \frac{\mathrm{d}z_h}{\mathrm{d}x}.$$

13.2 Identifying $\tan\alpha = -\mathrm{d}z_h/\mathrm{d}x$, rearranging equation (13.31), and setting $\mathrm{d}q/\mathrm{d}x = 0$, we obtain

$$\alpha_{\text{crit}} = \tan^{-1} \frac{-G(1-F(z))}{U_0 F(z)},$$

for $z = z_h$. Evaluating this result using the reference parameter values used to produce figure 13.5 gives the result shown in figure 15.7. The shape of the curve is controlled by two factors, evident if we consider that for small α,

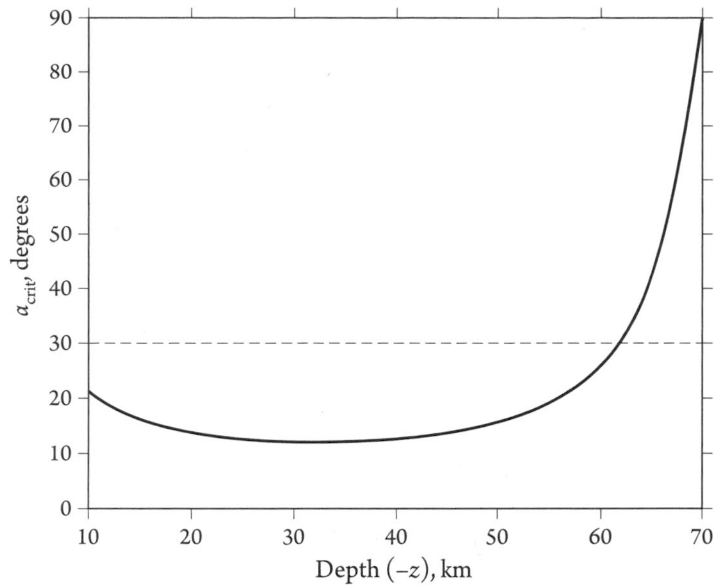

Figure 15.7. Minimum, critical angle α for melt focusing, as a function of depth along the base of the lithosphere. The dashed line is shows the value of α used to in figure 13.5.

$U_0\alpha_{crit} \sim -G(1/F - 1)$. Starting from the minimum of the curve and moving left (toward the ridge axis along z_h), the critical angle increases because freezing $(-G)$ is more rapid at shallow depths. A steeper slope is required to enable focusing. Starting from the minimum and moving right (away from the ridge along z_h), the critical angle increases because $1/F$ becomes large. Freezing is slow, but the melt supply is small and so a steep slope of the base of the lithosphere is required for focusing.

15.12 Exercises from Chapter 14: Numerical Modeling of Two-Phase Flow

14.1 To show that the bilinear form (14.24a) generates a tridiagonal matrix, we consider two submatrix entries in the assembly sum, $\mathbf{A}^{\Omega_n}$ and $\mathbf{A}^{\Omega_{n-1}}$, as given in equation (14.25a). Each of these is a 2×2 matrix. $\mathbf{A}^{\Omega_n}$ contributes to the diagonal entries A_{nn} and $A_{n+1\,n+1}$ and to the off-diagonal entries $A_{n+1\,n}$ and $A_{n\,n+1}$, whereas $\mathbf{A}^{\Omega_n-1}$ contributes to the diagonal entries $A_{n-1\,n-1}$ and A_{nn}, and to the off-diagonal entries $A_{n\,n-1}$ and $A_{n-1\,n}$.

After summing these entries into the bilinear form matrix A, we consider row n. It contains contributions to entries A_{nn} and $A_{n\,n+1}$ from $\mathbf{A}^{\Omega_n}$ and contributions to A_{nn} and $A_{n\,n-1}$ from $\mathbf{A}^{\Omega_{n-1}}$. Because the submatrices are 2×2, no other submatrices contribute to row n. Hence row n is tridiagonal.

14.2 Neumann boundary conditions are imposed by setting the value of the buffer point (the variable located *outside* the physical domain) to enforce the condition. To enforce $\hat{\boldsymbol{x}} \cdot \boldsymbol{\nabla} P$ on the boundary at $x = 0$ we write

$$\frac{\tilde{P}_{j2} - \tilde{P}_{j1}}{\Delta h} = 0 \quad \Rightarrow \quad \tilde{P}_{j1} = \tilde{P}_{j2}.$$

We impose this condition for $j \in [2, N_y - 1]$. The same approach can be used to set a Neumann condition on the component of the velocity that is tangential to a boundary. This is useful in imposing a condition on the shear traction, for example.

14.4 For simplicity we consider only one dimension. A second-order accurate discretization is

$$\frac{1}{\Delta h^2}\left[\tilde{K_\phi}_{i+1/2}\left(\tilde{P}_{i+1} - \tilde{P}_i\right) - \tilde{K_\phi}_{i-1/2}\left(\tilde{P}_i - \tilde{P}_{i-1}\right)\right].$$

We note that here the mobility is evaluated on the cell faces, rather than its native location at the cell centers.

Recall that the dimensionless mobility is written as $\tilde{K_\phi} = (\phi/\phi_0)^n$. The different means for computing the mobility on the boundary are associated with the different methods of averaging:

$$\tilde{K_\phi}_{i+1/2} = \begin{cases} \frac{\tilde{\phi}_i^n + \tilde{\phi}_{i+1}^n}{2\tilde{\phi}_0^n} & \text{arithmetic mean,} \\ \frac{1}{2\tilde{\phi}_0^n}\left(\frac{1}{\tilde{\phi}_i^n} + \frac{1}{\tilde{\phi}_{i+1}^n}\right)^{-1} & \text{harmonic mean,} \\ \frac{1}{\tilde{\phi}_0^n}\sqrt{\tilde{\phi}_i^n \tilde{\phi}_{i+1}^n} & \text{geometric mean.} \end{cases}$$

For the latter two, the mobility is zero if one of the two porosities is zero.

Bibliography

C. B. Agee and D. Walker. Olivine flotation in mantle melt. *Earth and Planetary Science Letters*, 114(2-3):315–324, 1993.

E. Aharonov, J. Whitehead, P. Kelemen, and M. Spiegelman. Channeling instability of upwelling melt in the mantle. *Journal of Geophysical Research*, 1995.

E. Aharonov, M. Spiegelman, and P. Kelemen. Three-dimensional flow and reaction in porous media: Implications for the earth's mantle and sedimentary basins. *Journal of Geophysical Research: Solid Earth*, 102(B7):14821–14833, 1997.

J. Ahern and D. Turcotte. Magma migration beneath an ocean ridge. *Earth and Planetary Science Letters*, 45:115–122, 1979.

L. Alisic, J. F. Rudge, Katz, Richard F, G. N. Wells, and S. Rhebergen. Compaction around a rigid, circular inclusion in partially molten rock. *Journal of Geophysical Research: Solid Earth*, 119(7): 5903–5920, 2014.

L. Alisic, S. Rhebergen, J. F. Rudge, R. F. Katz, and G. N. Wells. Torsion of a cylinder of partially molten rock with a spherical inclusion: Theory and simulation. *Geochemistry, Geophysics, Geosystems*, 17(1):143–161, 2016.

J. Allwright and R. Katz. Pipe Poiseuille flow of viscously anisotropic, partially molten rock. *Geophysical Journal International*, 199(3):1608–1624, 2014. doi: 10.1093/gji/ggu345.

M. Alnæs, J. Blechta, J. Hake, A. Johansson, B. Kehlet, A. Logg, C. Richardson, J. Ring, M. E. Rognes, and G. N. Wells. The FEniCS Project version 1.5. *Archive of Numerical Software*, 3(100):9–23, 2015.

G. M. Anderson. *Thermodynamics of natural systems*. Cambridge University Press, 3rd edition, 2017.

T. Arbogast and A. L. Taicher. A cell-centered finite difference method for a degenerate elliptic equation arising from two-phase mixtures. *Computational Geosciences*, 21(4):701–712, 2017.

T. Arbogast, M. A. Hesse, and A. L. Taicher. Mixed methods for two-phase Darcy–Stokes mixtures of partially melted materials with regions of zero porosity. *SIAM Journal on Scientific Computing*, 39 (2):B375–B402, 2017.

A. A. Arzi. Critical phenomena in the rheology of partially melted rocks. *Tectonophysics*, 44(1-4): 173–184, 1978.

A. Aschwanden, E. Bueler, C. Khroulev, and H. Blatter. An enthalpy formulation for glaciers and ice sheets. *Journal of Glaciology*, 58(209), 2012. doi: 10.3189/2012JoG11J088.

P. Asimow and E. Stolper. Steady-state mantle-melt interactions in one dimension: I. Equilibrium transport and melt focusing. *Journal of Petrology*, 40, 1999.

P. Asimow, M. Hirschmann, and E. Stolper. An analysis of variations in isentropic melt productivity. *Philosophical Transactions of the Royal Society London A*, 355, 1997.

H. Atkinson. Overview no. 65: Theories of normal grain growth in pure single phase systems. *Acta Metallurgica*, 36(3):469–491, 1988.

N. J. Austin and B. Evans. Paleowattmeters: A scaling relation for dynamically recrystallized grain size. *Geology*, 35(4):343, 2007.

D. Bailey and J. Tarney, editors. *Evidence for Chemical Heterogeneity in the Earth's Mantle. A Royal Society Discussion*, volume 297 of *Series A*, 1980. The Royal Society.

L. Baker Hebert, P. Antoshechkina, P. Asimow, and M. Gurnis. Emergence of a low-viscosity channel in subduction zones through the coupling of mantle flow and thermodynamics. *Earth and Planetary Science Letters*, 278(3-4):243–256, 2009. doi: 10.1016/j.epsl.2008.12.013.

S. Balay, W. D. Gropp, L. C. McInnes, and B. F. Smith. Efficient management of parallelism in object oriented numerical software libraries. In E. Arge, A. M. Bruaset, and H. P. Langtangen, editors, *Modern Software Tools in Scientific Computing*, pages 163–202. Birkhäuser Press, 1997.

S. Balay, S. Abhyankar, M. F. Adams, J. Brown, P. Brune, K. Buschelman, L. Dalcin, A. Dener, V. Eijkhout, W. D. Gropp, D. Karpeyev, D. Kaushik, M. G. Knepley, D. A. May, L. C. McInnes, R. T. Mills, T. Munson, K. Rupp, P. Sanan, B. F. Smith, S. Zampini, H. Zhang, and H. Zhang. PETSc Web page. https://www.mcs.anl.gov/petsc, 2019.

S. Balay, S. Abhyankar, M. F. Adams, J. Brown, P. Brune, K. Buschelman, L. Dalcin, A. Dener, V. Eijkhout, W. D. Gropp, D. Karpeyev, D. Kaushik, M. G. Knepley, D. A. May, L. C. McInnes, R. T. Mills, T. Munson, K. Rupp, P. Sanan, B. F. Smith, S. Zampini, H. Zhang, and H. Zhang. PETSc users manual. Technical Report ANL-95/11 - Revision 3.14, Argonne National Laboratory, 2020. https://www.mcs.anl.gov/petsc.

V. Barcilon and O. Lovera. Solitary waves in magma dynamics. *Journal of Fluid Mechanics*, 204:121–133, 1989.

V. Barcilon and F. Richter. Nonlinear waves in compacting media. *Journal of Fluid Mechanics*, 164: 429–448, 1986.

K. Barnouin-Jha, E. Parmentier, and D. Sparks. Buoyant mantle upwelling and crustal production at oceanic spreading centers: On-axis segmentation and off-axis melting. *Journal of Geophysical Research*, 102(B6):11979–11989, 1997.

G. Batchelor. *An Introduction to Fluid Mechanics*. Cambridge University Press, 1967.

M. D. Behn and T. L. Grove. Melting systematics in mid-ocean ridge basalts: Application of a plagioclase-spinel melting model to global variations in major element chemistry and crustal thickness. *Journal of Geophysical Research: Solid Earth*, 120(7):4863–4886, 2015.

M. D. Behn, G. Hirth, and J. R. Elsenbeck. Implications of grain size evolution on the seismic structure of the oceanic upper mantle. *Earth and Planetary Science Letters*, 282(1-4):178–189, 2009.

D. Bercovici. The generation of plate tectonics from mantle convection. *Earth and Planetary Science Letters*, 205(3-4):107–121, 2003.

D. Bercovici and C. Michaut. Two-phase dynamics of volcanic eruptions: compaction, compression and the conditions for choking. *Geophysical Journal International*, 182(2):843–864, 2010.

D. Bercovici and Y. Ricard. Energetics of a two-phase model of lithospheric damage, shear localization and plate-boundary formation. *Geophysical Journal International*, 152, 2003.

D. Bercovici and Y. Ricard. Tectonic plate generation and two-phase damage: Void growth versus grain size reduction. *Journal of Geophysical Research: Solid Earth*, 110(B3), 2005.

D. Bercovici and J. F. Rudge. A mechanism for mode selection in melt band instabilities. *Earth and Planetary Science Letters*, 433:139–145, 2016. doi: 10.1016/j.epsl.2015.10.051.

D. Bercovici, Y. Ricard, and G. Schubert. A two-phase model for compaction and damage 1. General theory. *Journal of Geophysical Research*, 106, 2001a.

D. Bercovici, Y. Ricard, and G. Schubert. A two-phase model for compaction and damage 3. Applications to shear localization and plate boundary formation. *Journal of Geophysical Research*, 106, 2001b.

M. Beuthe. Spatial patterns of tidal heating. *Icarus*, 223(1):308–329, 2013.

C. Bierson and F. Nimmo. A test for Io's magma ocean: Modeling tidal dissipation with a partially molten mantle. *Journal of Geophysical Research: Planets*, 121(11):2211–2224, 2016.

M. A. Biot. General theory of three-dimensional consolidation. *Journal of Applied Physics*, 12(2):155–164, 1941.

T. Bo, R. F. Katz, O. Shorttle, and J. F. Rudge. The melting column as a filter of mantle trace-element heterogeneity. *Geochemistry, Geophysics, Geosystems*, 19(12):4694–4721, 2018. doi: 10.1029/2018GC007880.

Y. Bottinga and D. F. Weill. The viscosity of magmatic silicate liquids; a model calculation. *American Journal of Science*, 272(5):438–475, 1972.

C.-E. Boukaré and Y. Ricard. Modeling phase separation and phase change for magma ocean solidification dynamics. *Geochemistry, Geophysics, Geosystems*, 18(9):3385–3404, 2017.

C.-E. Boukaré, Y. Ricard, and G. Fiquet. Thermodynamics of the MgO-FeO-SiO_2 system up to 140 GPa: Application to the crystallization of earth's magma ocean. *Journal of Geophysical Research: Solid Earth*, 120(9):6085–6101, 2015.

D. J. Bower, P. Sanan, and A. S. Wolf. Numerical solution of a non-linear conservation law applicable to the interior dynamics of partially molten planets. *Physics of the Earth and Planetary Interiors*, 274:49–62, 2018.

R. S. Bradley. Thermodynamic calculations on phase equilibria involving fused salts; part I, General theory and application to equilibria involving calcium carbonate at high pressure. *American Journal of Science*, 260(5):374–382, 1962a.

R. S. Bradley. Thermodynamic calculations on phase equilibria involving fused salts; Part 2, Solid solutions and application to the olivines. *American Journal of Science*, 260(7):550–554, 1962b.

M. Braun and P. Kelemen. Dunite distribution in the Oman ophiolite: Implications for melt flux through porous dunite conduits. *Geochemistry, Geophysics, Geosystems*, 3:8603, 2002. doi: 10.1029/2001GC000289.

T. Breithaupt, L. N. Hansen, S. Toppaladoddi, and R. F. Katz. The role of grain-environment heterogeneity in normal grain growth: A stochastic approach. *Acta Materialia* 209:116699, 2021 doi: 10.1016/j.actamat.2021.116699.

W. Buck and W. Su. Focused mantle upwelling below mid-ocean ridges due to feedback between viscosity and melting. *Geophysical Research Letters*, 16(7):641–644, 1989.

J. Bulau, H. Waff, and J. Tyburczy. Mechanical and Thermodynamic Constraints on Fluid Distribution in Partial Melts. *Journal of Geophysical Research*, 84:6102–6108, 1979.

J. Burke and D. Turnbull. Recrystallization and grain growth. *Progress in Metal Physics*, 3:220–292, 1952.

R. Buscall and L. R. White. The consolidation of concentrated suspensions. Part 1.—The theory of sedimentation. *Journal of the Chemical Society, Faraday Transactions 1: Physical Chemistry in Condensed Phases*, 83(3):873–891, 1987.

S. Butler. The effects of buoyancy on shear-induced melt bands in a compacting porous medium. *Physics of the Earth and Planetary Interiors*, 173(1-2):51–59, 2009. doi: 10.1016/j.pepi.2008.10.022.

S. Butler. Numerical Models of Shear-Induced Melt Band Formation with Anisotropic Matrix Viscosity. *Physics of the Earth and Planetary Interiors*, 200-201:28–36, 2012. doi: 10.1016/j.pepi.2012.03.011.

A.-M. Cagnioncle, E. Parmentier, and L. Elkins-Tanton. Effect of solid flow above a subducting slab on water distribution and melting at convergent plate boundaries. *Journal of Geophysical Research*, 112, 2007. doi: 10.1029/2007JB004934.

X. Cai, D. Keyes, and V. Venkatakrishnan. Newton-Krylov-Schwarz: An implicit solver for CFD. In *Proceedings of the Eighth International Conference on Domain Decomposition Methods*, pages 387–400, 1997.

Z. Cai and D. Bercovici. Two-phase damage models of magma-fracturing. *Earth and Planetary Science Letters*, 368:1–8, 2013.

Z. Cai and D. Bercovici. Two-dimensional magmons with damage and the transition to magma-fracturing. *Physics of the Earth and Planetary Interiors*, 256:13–25, 2016.

N. G. Cerpa, D. W. Rees Jones, and R. F. Katz. Consequences of glacial cycles for magmatism and carbon transport at mid-ocean ridges. 528, 2019. doi: 10.1016/j.epsl.2019.115845.

N. Cerpa Gilvonio, I. Wada, and C. R. Wilson. Fluid migration in the mantle wedge: Influence of mineral grain size and mantle compaction. *Journal of Geophysical Research: Solid Earth*, 122(8): 6247–6268, 2017. doi: 10.1002/2017JB014046.

J. Chadam, D. Hoff, E. Merino, P. Ortoleva, and A. SEN. Reactive infiltration instabilities. *IMA Journal Of Applied Mathematics*, 36(3):207–221, 1986.

M. Cheadle, M. Elliott, and D. M[c]Kenzie. Percolation threshold and permeability of crystallizing igneous rocks: The importance of textural equilibrium. *Geology*, 32:757–760, 2004.

G. Choblet and E. Parmentier. Mantle upwelling and melting beneath slow spreading centers: effects of variable rheology and melt productivity. *Earth and Planetary Science Letters*, 184:589–604, 2001.

C. Claude-Ivanaj, B. Bourdon, and C. J. Allègre. Ra–Th–Sr isotope systematics in Grande Comore Island: a case study of plume–lithosphere interaction. *Earth and Planetary Science Letters*, 164(1-2): 99–117, 1998.

J. A. D. Connolly. The geodynamic equation of state: What and how. *Geochemistry, Geophysics, Geosystems*, 10(10), 2009.

J. A. D. Connolly and Y. Y. Podladchikov. Compaction-driven fluid flow in viscoelastic rock. *Geodinamica Acta*, 11(2-3):55–84, 1998.

J. A. D. Connolly and Y. Y. Podladchikov. Decompaction weakening and channeling instability in ductile porous media: Implications for asthenospheric melt segregation. *Journal of Geophysical Research*, 112:B10205, 2007.

J. A. D. Connolly, M. W. Schmidt, G. Solferino, and N. Bagdassarov. Permeability of asthenospheric mantle and melt extraction rates at mid-ocean ridges. *Nature*, 462(7270):209–212, 2009. doi: 10.1038/nature08517.

R. F. Cooper and D. L. Kohlstedt. *Interfacial energies in the olivine-basalt system. Advances in Earth and Planetary Sciences*, 12:217–228, 1982.

R. F. Cooper and D. L. Kohlstedt. Solution-precipitation enhanced diffusional creep of partially molten olivine-basalt aggregates during hot-pressing. *Tectonophysics*, 107(3-4):207–233, 1984.

J. W. Crowley, R. F. Katz, P. Huybers, C. H. Langmuir, and S.-H. Park. Glacial cycles drive variations in the production of oceanic crust. *Science*, 347(6227):1237–1240, 2014. doi: 10.1126/science.1261508.

M. Daines and D. Kohlstedt. Influence of deformation on melt topology in peridotites. *Journal of Geophysical Research*, 102:10257–10271, 1997.

M. Daines and D. L. Kohlstedt. The transition from porous to channelized flow due to melt/rock reaction during melt migration. *Geophysical Research Letters*, 21(2):145–148, 1994.

J. Dannberg, Z. Eilon, U. Faul, R. Gassmöller, P. Moulik, and R. Myhill. The importance of grain size to mantle dynamics and seismological observations. *Geochemistry, Geophysics, Geosystems*, 18(8): 3034–3061, 2017.

J. Dannberg, R. Gassmöller, R. Grove, and T. Heister. A new formulation for coupled magma/mantle dynamics. *Geophysical Journal International*, 219(1):94–107, 2019.

G. F. Davies. Dynamic Earth: Plates, plumes and mantle convection, 2000.

S. De Groot and P. Mazur. *Non-Equilibrium Thermodynamics*. Dover Publications, 1984.

K. Denbigh. *The Principles of Chemical Equilibrium*. Cambridge University Press, 4th edition, 1981.

D. B. Dingwell, P. Courtial, D. Giordano, and A. R. L. Nichols. Viscosity of peridotite liquid. *Earth and Planetary Science Letters*, 226(1-2):127–138, 2004.

D. Drew. Mathematical modeling of two-phase flow. *Annual Review of Fluid Mechanics*, 15:261–291, 1983. doi: 10.1146/annurev.fl.15.010183.001401.

D. A. Drew. Averaged Field Equations for Two-Phase Media. *Studies in Applied Mathematics*, 50(2): 133–166, 1971.

D. A. Drew and L. A. Segel. Averaged Equations for Two-Phase Flows. *Studies in Applied Mathematics*, 50(3):205–231, 1971.

P. Duval. The role of the water content on the creep rate of polycrystalline ice. *IAHS Publications*, 118: 29–33, 1977.

I. Eksinchol, J. F. Rudge, and J. Maclennan. Rate of melt ascent beneath Iceland from the magmatic response to deglaciation. *Geochemistry, Geophysics, Geosystems*, 20(6):2585–2605, 2019.

L. Elkins, G. Gaetani, and K. Sims. Partitioning of U and TH during garnet pyroxenite partial melting: Constraints on the source of alkaline ocean island basalts. *Earth and Planetary Science Letters*, 265 (1-2):270–286, 2008.

L. J. Elkins, B. Bourdon, and S. Lambart. Testing pyroxenite versus peridotite sources for marine basalts using U-series isotopes. *Lithos*, 332:226–244, 2019.

T. Elliott and M. W. Spiegelman. Melt Migration in Oceanic Crustal Production: A U-series Perspective. In K. K. Turekian and H. D. Holland, editors, *Treatise on Geochemistry*, pages 465–510. Elsevier, 2003. doi: 10.1016/B0-08-043751-6/03031-0.

O. Evans, M. Spiegelman, and P. B. Kelemen. A poroelastic model of serpentinization: Exploring the interplay between rheology, surface energy, reaction, and fluid flow. *Journal of Geophysical Research: Solid Earth*, 123(10):8653–8675, 2018.

O. Evans, M. Spiegelman, and P. B. Kelemen. Phase-field modeling of reaction-driven cracking: Determining conditions for extensive olivine serpentinization. *Journal of Geophysical Research: Solid Earth*, 125(1):e2019JB018614, 2020.

D. F. Faizullaev. *Laminar motion of multiphase media in conduits*. Consultants Bureau, 1969.

W. S. Farren and G. I. Taylor. The heat developed during plastic extension of metals. *Proceedings of the Royal Society of London. Series A, Containing Papers of a Mathematical and Physical Character*, 107(743):422–451, 1925.

U. Faul. Permeability of partially molten upper mantle rocks from experiments and percolation theory. *Journal of Geophysical Research*, 102:10299–10311, 1997.

U. Faul, D. Toomey, and H. Waff. Intergranular basaltic melt is distributed in thin, elongated inclusions. *Geophysical Research Letters*, 21:29–32, 1994.

G. Fiquet, A. Auzende, J. Siebert, A. Corgne, H. Bureau, H. Ozawa, and G. Garbarino. Melting of peridotite to 140 gigapascals. *Science*, 329(5998):1516–1518, 2010.

A. Fowler. A mathematical model of magma transport in the asthenosphere. *Geophysical and Astrophysical Fluid Dynamics*, 33, 1985.

A. Fowler. *Mathematical Geoscience*. Springer Science & Business Media, 2011.

A. C. Fowler. On the transport of moisture in polythermal glaciers. *Geophysical and Astrophysical Fluid Dynamics*, 28:99–140, 1984.

A. C. Fowler. Generation and Creep of Magma in the Earth. *SIAM Journal on Applied Mathematics*, 49(1):231–245, 1989.

F. C. Frank. Two-component flow model for convection in the Earth's upper mantle. *Nature*, 220(5165): 350–352, 1968.

W. S. Fyfe. *Fluids in the Earth's crust: Their significance in metamorphic, tectonic and chemical transport process*. Elsevier, 2012.

J. Ganguly. Adiabatic decompression and melting of mantle rocks: An irreversible thermodynamic analysis. *Geophysical Research Letters*, 32(6), 2005.

J. Ganguly. *Thermodynamics in Earth and planetary sciences*. Springer, 2008.

G. Garapic, U. H. Faul, and E. Brisson. High-resolution imaging of the melt distribution in partially molten upper mantle rocks: evidence for wetted two-grain boundaries. *Geochemistry, Geophysics, Geosystems*, 14(3):556–566, 2013.

D. Gebhardt and S. Butler. Linear analysis of melt band formation in a mid-ocean ridge corner flow. *Geophysical Research Letters*, 43(8):3700–3707, 2016.

S. Ghanbarzadeh, M. A. Hesse, M. Prodanović, and J. E. Gardner. Deformation-assisted fluid percolation in rock salt. *Science*, 350(6264):1069–1072, 2015.

M. S. Ghiorso, M. M. Hirschmann, P. W. Reiners, and V. C. Kress III. The pMELTS: A revision of MELTS for improved calculation of phase relations and major element partitioning related to partial melting of the mantle to 3 GPa. *Geochemistry, Geophysics, Geosystems*, 3(5):1–35, 2002.

A. Ghods and J. Arkani-Hamed. Melt migration beneath mid-ocean ridges. *Geophysical Journal International*, 140:687–697, 2000.

D. Giordano, J. K. Russell, and D. B. Dingwell. Viscosity of magmatic liquids: A model. *Earth and Planetary Science Letters*, 271(1-4):123–134, 2008.

E. Glueckauf. Theory of chromatography. Part 10.¯Formulæ for diffusion into spheres and their application to chromatography. *Journal of the Chemical Society, Faraday Transactions*, 51(0):1540–1551, 1955.

C. Hall and E. Parmentier. Spontaneous melt localization in a deforming solid with viscosity variations due to water weakening. *Geophysical Research Letters*, 27(1):9–12, 2000.

C. E. Hall and E. M. Parmentier. Influence of grain size evolution on convective instability. *Geochemistry, Geophysics, Geosystems*, 4(3), 2003.

L. N. Hansen, M. E. Zimmerman, and D. L. Kohlstedt. Laboratory measurements of the viscous anisotropy of olivine aggregates. *Nature*, 492(7429):415–418, 2012. doi: 10.1038/nature11671.

M. Haseloff, I. Hewitt, and R. Katz. Englacial pore water localizes shear in temperate ice stream margins. *Journal of Geophysical Research: Earth Surface*, 124(11):2521–2541, 2019.

C. Havlin, E. Parmentier, and G. Hirth. Dike propagation driven by melt accumulation at the lithosphere–asthenosphere boundary. *Earth and Planetary Science Letters*, 376:20–28, 2013.

L. B. Hebert and L. G. J. Montési. Generation of permeability barriers during melt extraction at mid-ocean ridges. *Geochemistry, Geophysics, Geosystems*, 11(12):Q12008, 2010.

K. R. Helfrich and J. A. Whitehead. Solitary waves on conduits of buoyant fluid in a more viscous fluid. *Geophysical and Astrophysical Fluid Dynamics*, 51(1-4):35–52, 2006.

M. A. Hesse, A. R. Schiemenz, Y. Liang, and E. M. Parmentier. Compaction-dissolution waves in an upwelling mantle column. *Geophysical Journal International*, 187(3):1057–1075, 2011. doi: 10.1111/j.1365-246X.2011.05177.x.

D. R. Hewitt, D. T. Paterson, N. J. Balmforth, and D. M. Martinez. Dewatering of fibre suspensions by pressure filtration. *Physics of Fluids*, 28(6):063304, 2016.

I. Hewitt and A. Fowler. Partial melting in an upwelling mantle column. *Philosophical Transactions of the Royal Society London A*, 2008. doi: 10.1098/rspa.2008.0045.

I. J. Hewitt. Modelling melting rates in upwelling mantle. *Earth and Planetary Science Letters*, 300: 264–274, 2010. doi: 10.1016/j.epsl.2010.10.010.

I. J. Hewitt and C. Schoof. Models for polythermal ice sheets and glaciers. *The Cryosphere*, 11(1): 541–551, 2017.

S. Hier-Majumder. Development of anisotropic mobility during two-phase flow. *Geophysical Journal International*, 186(1):59–68, 2011.

S. Hier-Majumder, Y. Ricard, and D. Bercovici. Role of grain boundaries in magma migration and storage. *Earth and Planetary Science Letters*, 248:735–749, 2006. doi: 10.1016/j.epsl.2006.06.015.

M. Hillert. On the theory of normal and abnormal grain growth. *Acta Metallurgica*, 13(3):227–238, 1965.

E. Hinch and B. Bhatt. Stability of an acid front moving through porous rock. *Journal of Fluid Mechanics*, 212:279–288, 1990.

M. M. Hirschmann, P. D. Asimow, M. Ghiorso, and E. Stolper. Calculation of peridotite partial melting from thermodynamic models of minerals and melts. iii. Controls on isobaric melt production and the effect of water on melt production. *Journal of Petrology*, 40(5):831–851, 1999.

G. Hirth and D. Kohlstedt. Rheology of the upper mantle and the mantle wedge: A view from the experimentalists. In *Inside the Subduction Factory*, volume 138 of *AGU Geophysical Monograph*. American Geophysical Union, 2003.

G. Hirth and D. L. Kohlstedt. Experimental constraints on the dynamics of the partially molten upper mantle: Deformation in the diffusion creep regime. *Journal of Geophysical Research*, 100(B2):1981–2001, 1995a.

G. Hirth and D. L. Kohlstedt. Experimental constraints on the dynamics of the partially molten upper mantle: 2. Deformation in the dislocation creep regime. *Journal of Geophysical Research*, 100(B8): 15441–15449, 1995b.

M. Hoefner and H. S. Fogler. Pore evolution and channel formation during flow and reaction in porous media. *AIChE Journal*, 34(1):45–54, 1988.

M. B. Holness, Z. Vukmanovic, and E. Mariani. Assessing the role of compaction in the formation of adcumulates: a microstructural perspective. *Journal of Petrology*, 58(4):643–673, 2017.

B. Holtzman and D. Kohlstedt. Stress-driven melt segregation and strain partitioning in partially molten rocks: Effects of stress and strain. *Journal of Petrology*, 48:2379–2406, 2007. doi: 10.1093/petrology/egm065.

B. Holtzman, N. Groebner, M. Zimmerman, S. Ginsberg, and D. Kohlstedt. Stress-driven melt segregation in partially molten rocks. *Geochemistry, Geophysics, Geosystems*, 4, 2003. doi: 10.1029/2001GC000258.

P. Huybers and C. Langmuir. Feedback between deglaciation and volcanic emissions of CO_2. *Earth and Planetary Science Letters*, 2009. doi: 10.1016/j.epsl.2009.07.014.

G. Ito and S. Martel. Focusing of magma in the upper mantle through dike interaction. *Journal of Geophysical Research*, 107:2223, 2002. doi: 10.1029/2001JB000251.

H. Iwamori. Melt-solid flow with diffusion controlled chemical-reaction. *Geophysical Research Letters*, 19, 1992.

H. Iwamori. Dynamic disequilibrium melting model with porous flow and diffusion-controlled chemical equilibration. *Earth and Planetary Science Letters*, 114(2-3):301–313, 1993a.

H. Iwamori. A model for disequilibrium mantle melting incorporating melt transport by porous and channel flows. *Nature*, 366(6457):734–737, 1993b.

H. Iwamori. ^{238}U–^{230}Th–^{226}Ra and ^{235}U–^{231}Pa disequilibria produced by mantle melting with porous and channel flows. *Earth and Planetary Science Letters*, 125(1-4):1–16, 1994.

H. Iwamori. Transportation of H_2O and melting in subduction zones. *Earth and Planetary Science Letters*, 160:65–80, 1998.

H. Iwamori, D. McKenzie, and E. Takahashi. Melt generation by isentropic mantle upwelling. *Earth and Planetary Science Letters*, 134(3-4):253–266, 1995.

M. Jackson, J. Blundy, and R. Sparks. Chemical differentiation, cold storage and remobilization of magma in the earth's crust. *Nature*, 564(7736):405–409, 2018.

H. Jacobson and C. Raymond. Thermal effects on the location of ice stream margins. *Journal of Geophysical Research: Solid Earth*, 103(B6):12111–12122, 1998.

G. T. Jarvis and D. P. McKenzie. Convection in a compressible fluid with infinite Prandtl number. *Journal of Fluid Mechanics*, 96(3):515–583, 1980.

K. Jha, E. Parmentier, and J. Morgan. The role of mantle-depletion and melt-retention buoyancy in spreading-center segmentation. *Earth and Planetary Science Letters*, 125(1-4):221–234, 1994.

Z. M. Jin, H. W. Green, and Y. Zhou. Melt topology in partially molten mantle peridotite during ductile deformation. *Nature*, 372(6502):164–167, 1994.

J. S. Jordan and M. A. Hesse. Reactive transport in a partially molten system with binary solid solution. *Geochemistry, Geophysics, Geosystems*, 16(12):4153–4177, 2015.

M. Jull and D. McKenzie. The effect of deglaciation on mantle melting beneath Iceland. *Journal of Geophysical Research*, 101:21815–21828, 1996.

M. Jull, P. Kelemen, and K. Sims. Consequences of diffuse and channelled porous melt migration on uranium series disequilibria. *Geochimica et Cosmochimica Acta*, 66, 2002.

H. Jung and H. S. Waff. Olivine crystallographic control and anisotropic melt distribution in ultramafic partial melts. *Geophysical Research Letters*, 25(15):2901–2904, 1998.

S. Karato. *Deformation of earth materials: An introduction to the rheology of solid Earth*. Cambridge University Press, 2008.

S. Karato and P. Wu. Rheology of the upper mantle: A synthesis. *Science*, 260, 1993.

R. Katz. Magma dynamics with the enthalpy method: Benchmark solutions and magmatic focusing at mid-ocean ridges. *Journal of Petrology*, 2008. doi: 10.1093/petrology/egn058.

R. Katz. Porosity-driven convection and asymmetry beneath mid-ocean ridges. *Geochemistry, Geophysics, Geosystems*, 10(Q0AC07), 2010. doi: 10.1029/2010GC003282.

R. Katz and Y. Takei. Consequences of viscous anisotropy in a deforming, two-phase aggregate: 2. Numerical solutions of the full equations. *Journal of Fluid Mechanics*, 734:456–485, 2013. doi: 10.1017/jfm.2013.483.

R. Katz and S. Weatherley. Consequences of mantle heterogeneity for melt extraction at mid-ocean ridges. *Earth and Planetary Science Letters*, 335-336:226–237, 2012. doi: 10.1016/j.epsl.2012.04.042.

R. Katz and M. Worster. Simulation of directional solidification, thermochemical convection, and chimney formation in a Hele-Shaw cell. *Journal of Computational Physics*, 2008. doi: 10.1016/j.jcp.2008.06.039.

R. Katz, M. Spiegelman, and B. Holtzman. The dynamics of melt and shear localization in partially molten aggregates. *Nature*, 442, 2006. doi: 10.1038/nature05039.

R. Katz, M. Knepley, B. Smith, M. Spiegelman, and E. Coon. Numerical simulation of geodynamic processes with the Portable Extensible Toolkit for Scientific Computation. *Physics of the Earth and Planetary Interiors*, 163:52–68, 2007. doi: 10.1016/j.pepi.2007.04.016.

P. Kelemen, N. Shimizu, and V. Salters. Extraction of mid-ocean-ridge basalt from the upwelling mantle by focused flow of melt in dunnite channels. *Nature*, 375(6534):747–753, 1995a.

P. Kelemen, G. Hirth, N. Shimizu, M. Spiegelman, and H. Dick. A review of melt migration processes in the adiabatically upwelling mantle beneath oceanic spreading ridges. *Philosophical Transactions of the Royal Society London A*, 355(1723):283–318, 1997.

P. B. Kelemen. Reaction between ultramafic rock and fractionating basaltic magma i. phase relations, the origin of calc-alkaline magma series, and the formation of discordant dunite. *Journal of Petrology*, 31(1):51–98, 1990.

P. B. Kelemen, H. Dick, and J. Quick. Formation of hartzburgite by pervasive melt rock reaction in the upper mantle. *Nature*, 358(6388):635–641, 1992.

P. B. Kelemen, J. A. Whitehead, E. Aharonov, and K. Jordahl. Experiments on flow focusing in soluble porous-media, with applications to melt extraction from the mantle. *Journal of Geophysical Research*, 100:475–496, 1995b.

P. B. Kelemen, M. Braun, and G. Hirth. Spatial distribution of melt conduits in the mantle beneath oceanic spreading ridges: Observations from the Ingalls and Oman ophiolites. *Geochemistry, Geophysics, Geosystems*, 1(7), 2000.

T. Keller and R. F. Katz. The role of volatiles in reactive melt transport in the asthenosphere. *Journal of Petrology*, 57(6):1073–1108, 2016. doi: 10.1093/petrology/egw030.

T. Keller and J. Suckale. A continuum model of multi-phase reactive transport in igneous systems. *Geophysical Journal International*, 219(1):185–222, 2019.

T. Keller, D. A. May, and B. J. P. Kaus. Numerical modelling of magma dynamics coupled to tectonic deformation of lithosphere and crust. *Geophysical Journal International*, 195(3):1406–1442, 2013. doi: 10.1093/gji/ggt306.

T. Keller, R. F. Katz, and M. M. Hirschmann. Volatiles beneath mid-ocean ridges: Deep melting, channelised transport, focusing, and metasomatism. *Earth and Planetary Science Letters*, 464:55–68, 2017. doi: 10.1016/j.epsl.2017.02.006.

P. M. Kenyon. Trace element and isotopic effects arising from magma migration beneath mid-ocean ridges. *Earth and Planetary Science Letters*, 101(2-4):367–378, 1990.

K. Key, S. Constable, L. Liu, and A. Pommier. Electrical image of passive mantle upwelling beneath the northern east pacific rise. *Nature*, 495(7442):499–502, 2013.

D. King, M. Zimmerman, and D. Kohlstedt. Stress-driven melt segregation in partially molten olivine-rich rocks deformed in torsion. *Journal of Petrology*, 51:21–42, 2010. doi: 10.1093/petrology/egp062.

D. S. H. King, S. Hier-Majumder, and D. L. Kohlstedt. An experimental study of the effects of surface tension in homogenizing perturbations in melt fraction. *Earth and Planetary Science Letters*, 307 (3-4):349–360, 2011a.

D. S. H. King, B. K. Holtzman, and D. L. Kohlstedt. An experimental investigation of the interactions between reaction-driven and stress-driven melt segregation: 1. Application to mantle melt extraction. *Geochemistry, Geophysics, Geosystems*, 12, 2011b. doi: 10.1029/2011GC003684.

S. Labrosse, J. Hernlund, and N. Coltice. A crystallizing dense magma ocean at the base of the earth's mantle. *Nature*, 450(7171):866–869, 2007.

W. Landuyt, D. Bercovici, and Y. Ricard. Plate generation and two-phase damage theory in a model of mantle convection. *Geophysical Journal International*, 174(3):1065–1080, 2008.

M. Le Bars and M. G. Worster. Interfacial conditions between a pure fluid and a porous medium: implications for binary alloy solidification. *Journal of Fluid Mechanics*, 550:149–173, 2006.

Y. Liang. Kinetics of crystal-melt reaction in partially molten silicates: 1. Grain scale processes. *Geochemistry, Geophysics, Geosystems*, 4(5), 2003.

Y. Liang. Trace element fractionation and isotope ratio variation during melting of a spatially distributed and lithologically heterogeneous mantle. *Earth and Planetary Science Letters*, 552:116594, 2020.

Y. Liang and B. Liu. Simple models for disequilibrium fractional melting and batch melting with application to REE fractionation in abyssal peridotites. *Geochimica Et Cosmochimica Acta*, 173:181–197, 2016.

Y. Liang and B. Liu. Stretching chemical heterogeneities by melt migration in an upwelling mantle: An analysis based on time-dependent batch and fractional melting models. *Earth and Planetary Science Letters*, 498:275–287, 2018.

Y. Liang, J. Price, D. Wark, and E. Watson. Nonlinear pressure diffusion in a porous medium: Approximate solutions with applications to permeability measurements using transient pulse decay method. *Journal of Geophysical Research*, 106:529–535, 2001.

Y. Liang, A. Schiemenz, M. Hesse, E. Parmentier, and J. Hesthaven. High-porosity channels for melt migration in the mantle: Top is the dunite and bottom is the harzburgite and lherzolite. *Geophysical Research Letters*, 2010. doi: 10.1029/2010GL044162.

Y. Liao, F. Nimmo, and J. A. Neufeld. Heat production and tidally driven fluid flow in the permeable core of enceladus. *Journal of Geophysical Research: Planets*, 2020. doi: 10.1029/2019JE006209.

B. Liu and Y. Liang. The prevalence of kilometer-scale heterogeneity in the source region of MORB upper mantle. *Science Advances*, 3(11):e1701872, 2017.

A. Logg, K.-A. Mardal, and G. Wells. *Automated solution of differential equations by the finite element method: The FEniCS book*. Springer Science & Business Media, 2012.

J. Longhi. Some phase equilibrium systematics of lherzolite melting: I. *Geochemistry, Geophysics, Geosystems*, 3(3):1–33, 2002.

S. Maaløe and Å. Scheie. The permeability controlled accumulation of primary magma. *Contributions To Mineralogy And Petrology*, 81(4):350–357, 1982.

J. Maclennan, M. Jull, D. McKenzie, L. Slater, and K. Grönvold. The link between volcanism and deglaciation in Iceland. *Geochemistry, Geophysics, Geosystems*, 2002. doi: 10.1029/2001GC000282.

C. W. MacMinn, E. R. Dufresne, and J. S. Wettlaufer. Large deformations of a soft porous material. *Physical Review Applied*, 5(4):044020, 2016.

L. Magde and D. Sparks. Three-dimensional mantle upwelling, melt generation, and melt migration beneath segment slow spreading ridges. *Journal of Geophysical Research*, 102:20571–20583, 1997.

A. Malthe-Sørenssen, B. Jamtveit, and P. Meakin. Fracture patterns generated by diffusion controlled volume changing reactions. *Physical Review Letters*, 96(24):245501, 2006.

T. R. Marchant and N. F. Smyth. Approximate solutions for magmon propagation from a reservoir. *IMA Journal of Applied Mathematics*, 70:796–813, 2005.

D. McKenzie. The generation and compaction of partially molten rock. *Journal of Petrology*, 25, 1984.

D. McKenzie. ^{230}Th–^{238}U disequilibrium and the melting processes beneath ridge axes. *Earth and Planetary Science Letters*, 1985. doi: 10.1016/0012-821X(85)90001-9.

D. McKenzie. Compaction and crystallization in magma chambers: towards a model of the skaergaard intrusion. *Journal of Petrology*, 52(5):905–930, 2011.

D. P. McKenzie. Speculations on the consequences and causes of plate motions. *Geophysical Journal International*, 18(1):1–32, 1969.

A. K. McNamara. A review of large low shear velocity provinces and ultra low velocity zones. *Tectonophysics*, 760:199–220, 2019.

S. Mei, W. Bai, T. Hiraga, and D. Kohlstedt. Influence of melt on the creep behavior of olivine-basalt aggregates under hydrous conditions. *Earth and Planetary Science Letters*, 201:491–507, 2002.

C. R. Meyer and I. J. Hewitt. A continuum model for meltwater flow through compacting snow. *Cryosphere*, 11, 2017.

C. Michaut, D. Bercovici, and R. S. J. Sparks. Ascent and compaction of gas rich magma and the effects of hysteretic permeability. *Earth and Planetary Science Letters*, 282:258–267, 2009.

C. Michaut, Y. Ricard, D. Bercovici, and R. S. J. Sparks. Eruption cyclicity at silicic volcanoes potentially caused by magmatic gas waves. *Nature Geoscience*, 6(10):856, 2013.

K. J. Miller, W.-l. Zhu, L. G. J. Montési, and G. A. Gaetani. Experimental quantification of permeability of partially molten mantle rock. *Earth and Planetary Science Letters*, 388:273–282, 2014. doi: 10.1016/j.epsl.2013.12.003.

K. J. Miller, L. G. J. Montési, and W.-L. Zhu. Estimates of olivine–basaltic melt electrical conductivity using a digital rock physics approach. *Earth and Planetary Science Letters*, 432:332–341, 2015. doi: 10.1016/j.epsl.2015.10.004.

K. J. Miller, W.-l. Zhu, L. G. Montési, G. A. Gaetani, V. Le Roux, and X. Xiao. Experimental evidence for melt partitioning between olivine and orthopyroxene in partially molten harzburgite. *Journal of Geophysical Research: Solid Earth*, 121(8):5776–5793, 2016.

L. G. Montési and G. Hirth. Grain size evolution and the rheology of ductile shear zones: from laboratory experiments to postseismic creep. *Earth and Planetary Science Letters*, 211(1-2):97–110, 2003.

W. Moore. Tidal heating and convection in Io. *Journal of Geophysical Research: Planets*, 108(E8), 2003.

J. P. Morgan. Melt migration beneath mid-ocean spreading centers. *Geophysical Research Letters*, 14 (12):1238—1241, 1987.

O. Navon and E. Stolper. Geochemical consequences of melt percolation: The upper mantle as a chromatographic column. *Journal of Geology*, 95(3):285–307, 1987.

A. Nicolas. A melt extraction model based on structural studies in mantle peridotites. *Journal of Petrology*, 27(4):999–1022, 1986.

A. Nicolas and M. Jackson. High temperature dikes in peridotites: Origin by hydraulic fracturing. *Journal of Petrology*, 23(4):568–582, 1982. doi: 10.1093/petrology/23.4.568.

J. Nye and F. Frank. Hydrology of the intergranular veins in a temperate glacier. In *Symposium on the Hydrology of Glaciers*, pages 157–161, 1973.

E. Ohtani, Y. Nagata, A. Suzuki, and T. Kato. Melting relations of peridotite and the density crossover in planetary mantles. *Chemical Geology*, 120(3-4):207–221, 1995.

J. A. Olive, M. D. Behn, G. Ito, W. R. Buck, J. Escartßn, and S. Howell. Sensitivity of seafloor bathymetry to climate-driven fluctuations in mid-ocean ridge magma supply. *Science*, 350(6258):310–313, 2015.

B. Oliveira, J. C. Afonso, S. Zlotnik, and P. Diez. Numerical modelling of multiphase multicomponent reactive transport in the earth's interior. *Geophysical Journal International*, 212(1):345–388, 2018.

B. Oliveira, J. C. Afonso, and R. Tilhac. A disequilibrium reactive transport model for mantle magmatism. *Journal of Petrology*, 2020.

J. Oliver and B. Isacks. Deep earthquake zones, anomalous structures in the upper mantle, and the lithosphere. *Journal of Geophysical Research*, 72(16):4259–4275, 1967.

P. Olson and U. Christensen. Solitary wave propagation in a fluid conduit within a viscous matrix. *Journal of Geophysical Research: Solid Earth*, 91(B6):6367–6374, 1986.

T. C. O'Reilly and G. F. Davies. Magma transport of heat on Io: A mechanism allowing a thick lithosphere. *Geophysical Research Letters*, 8(4):313–316, 1981.

P. J. Ortoleva. *Geochemical self-organization*. Oxford University Press: Clarendon Press, 1994.

R. A. Parsons, F. Nimmo, J. W. Hustoft, B. K. Holtzman, and D. L. Kohlstedt. An experimental and numerical study of surface tension-driven melt flow. *Earth and Planetary Science Letters*, 267(3-4): 548–557, 2008.

S. Patankar. *Numerical heat transfer and fluid flow*. CRC Press, 1980.

V. Pawan, X. Wu, and V. Berdichevsky. Entropy decay during grain growth. *Scientific Reports (Nature Research)*, 10(1), 2020.

S. J. Peale, P. Cassen, and R. T. Reynolds. Melting of Io by tidal dissipation. *Science*, 203(4383):892–894, 1979.

M. Pec, B. K. Holtzman, M. Zimmerman, and D. L. Kohlstedt. Reaction infiltration instabilities in experiments on partially molten mantle rocks. *Geology*, 43(7):575–578, 2015.

M. Pec, B. K. Holtzman, M. E. Zimmerman, and D. L. Kohlstedt. Reaction infiltration instabilities in mantle rocks: An experimental investigation. *Journal of Petrology*, 58(5):979–1003, 2017. doi: 10.1093/petrology/egx043.

M. Pec, B. Holtzman, M. Zimmerman, and D. L. Kohlstedt. Influence of lithology on reactive melt flow channelization. *Geochemistry, Geophysics, Geosystems*, 21(8):e2020GC008937, 2020.

A. Philpotts and J. Ague. *Principles of igneous and metamorphic petrology*. Cambridge University Press, 2009.

J.-P. Poirier. *Introduction to the Physics of the Earth's Interior*. Cambridge University Press, 2000.

C. Prakash and S. Patankar. A control volume-based finite-element method for solving the Navier-Stokes equations using equal-order velocity-pressure interpolation. *Numerical Heat Transfer*, 8(3): 259–280, 1985.

W. H. Press, S. A. Teukolsky, W. T. Vetterling, and B. P. Flannery. *Numerical recipes: The art of scientific computing*, 3rd edition. Cambridge University Press, 2007.

C. Qi, Y.-H. Zhao, and D. L. Kohlstedt. An experimental study of pressure shadows in partially molten rocks. *Earth and Planetary Science Letters*, 382:77–84, 2013.

C. Qi, D. Kohlstedt, R. Katz, and Y. Takei. An experimental test of the viscous anisotropy hypothesis for partially molten rocks. *Proceedings of the National Academy of Sciences of the USA*, 2015. doi: 10.1073/pnas.1513790112.

Z. Qin. Disequilibrium partial melting model and its implications for trace element fractionations during mantle melting. *Earth and Planetary Science Letters*, 112(1-4):75–90, 1992.

J. E. Quick. The origin and significance of large, tabular dunite bodies in the Trinity peridotite, northern California. *Contributions to Mineralogy and Petrology*, 78(4):413–422, 1982.

H. Ramberg. Temperature changes associated with adiabatic decompression in geological processes. *Nature*, 234(5331):539–540, 1971.

D. W. Rees Jones and R. F. Katz. Reaction-infiltration instability in a compacting porous medium. *Journal of Fluid Mechanics*, 852:5–36, 2018. doi: 10.1017/jfm.2018.524.

D. W. Rees Jones, R. F. Katz, M. Tian, and J. F. Rudge. Thermal impact of magmatism in subduction zones. *Earth and Planetary Science Letters*, 481:73–79, 2018. doi: 10.1016/j.epsl.2017.10.015.

D. W. Rees Jones and J. F. Rudge. Fast magma ascent, revised estimates from the deglaciation of Iceland. *Earth and Planetary Science Letters*, 542:116324, 2020. doi: 10.1016/j.epsl.2020.116324.

D. W. Rees Jones, H. Zhang, and R. F. Katz. Magmatic channelization by reactive and shear-driven instabilities at mid-ocean ridges: A combined analysis. *Geophysical Journal International*, 226(1):582–609, 2021. https://doi.org/10.1093/gji/ggab112

J. P. Renaud and W. G. Henning. Increased tidal dissipation using advanced rheological models: Implications for Io and tidally active exoplanets. *The Astrophysical Journal*, 857(2):98, 2018.

J. Renner, K. Viskupic, G. Hirth, and B. Evans. Melt extraction from partially molten peridotites. *Geochemistry, Geophysics, Geosystems*, 4, 2003. doi: 10.1029/2002GC000369.

S. Rhebergen, G. N. Wells, R. F. Katz, and A. J. Wathen. Analysis of Block Preconditioners for Models of Coupled Magma/Mantle Dynamics. *SIAM Journal on Scientific Computing*, 36(4):A1960–A1977, 2014.

S. Rhebergen, G. N. Wells, A. J. Wathen, and R. F. Katz. Three-field block preconditioners for models of coupled magma/mantle dynamics. *SIAM Journal on Scientific Computing*, 37(5):A2270–A2294, 2015. doi: 10.1137/14099718X.

N. Ribe. The generation and composition of partial melts in the earth's mantle. *Earth and Planetary Science Letters*, 73, 1985.

N. M. Ribe. Theory of melt segregation: A review. *Journal Of Volcanology And Geothermal Research*, 33(4):241–253, 1987.

N. M. Ribe. *Theoretical Mantle Dynamics*. Cambridge University Press, 2018.

Y. Ricard. Treatise on Geophysics, In *Mantle Physics*, volume 7 of *Physics of Mantle Convection*. Elsevier, 2007.

Y. Ricard and D. Bercovici. A continuum theory of grain size evolution and damage. *Journal of Geophysical Research: Solid Earth*, 114(B1), 2009.

Y. Ricard, D. Bercovici, and G. Schubert. A two-phase model for compaction and damage: 2. Applications to compaction, deformation, and the role of interfacial surface tension. *Journal of Geophysical Research*, 106:8907–8924, 2001.

Y. Ricard, O. Šrámek, and F. Dubuffet. A multi-phase model of runaway core–mantle segregation in planetary embryos. *Earth and Planetary Science Letters*, 284(1–2):144–150, 2009.

G. C. Richard, S. Kanjilal, and H. Schmeling. Solitary-waves in geophysical two-phase viscous media: a semi-analytical solution. *Physics of the Earth and Planetary Interiors*, 198:61–66, 2012.

C. Richardson. Melt flow in a variable viscosity matrix. *Geophysical Research Letters*, 25(7):1099–1102, 1998.

F. M. Richter. Simple models for trace element fractionation during melt segregation. *Earth and Planetary Science Letters*, 77(3-4):333–344, 1986.

F. M. Richter and S. F. Daly. Dynamical and chemical effects of melting a heterogeneous source. *Journal of Geophysical Research: Solid Earth*, 94(B9):12499–12510, 1989.

F. M. Richter and D. McKenzie. Dynamical models for melt segregation from a deformable matrix. *The Journal of Geology*, 92(6):729–740, 1984.

G. N. Riley and D. L. Kohlstedt. Kinetics of melt migration in upper mantle-type rocks. *Earth and Planetary Science Letters*, 105:500–521, 1991.

P. Rios and D. Zöllner. Critical assessment 30: Grain growth–unresolved issues. *Materials Science and Technology*, 34(6):629–638, 2018.

E. Rivalta, B. Taisne, A. Bunger, and R. Katz. A review of mechanical models of dike propagation: Schools of thought, results and future directions. *Tectonophysics*, 638:1–42, 2015.

P. J. Roache. Code verification by the method of manufactured solutions. *Journal of Fluids Engineering*, 124(1):4–10, 2002.

J. H. Roberts and F. Nimmo. Tidal heating and the long-term stability of a subsurface ocean on enceladus. *Icarus*, 194(2):675–689, 2008.

A. Røyne, B. Jamtveit, J. Mathiesen, and A. Malthe-Sørenssen. Controls on rock weathering rates by reaction-induced hierarchical fracturing. *Earth and Planetary Science Letters*, 275(3-4):364–369, 2008.

A. Rozel. Impact of grain size on the convection of terrestrial planets. *Geochemistry, Geophysics, Geosystems*, 13(10), 2012.

A. Rozel, Y. Ricard, and D. Bercovici. A thermodynamically self-consistent damage equation for grain size evolution during dynamic recrystallization. *Geophysical Journal International*, 184(2): 719–728, 2011.

A. M. Rubin. Dike ascent in partially molten rock. *Journal of Geophysical Research: Solid Earth*, 103 (B9):20901–20919, 1998.

J. Rudge. Textural equilibrium melt geometries around tetrakaidecahedral grains. *Philosophical Transactions of the Royal Society London A*, 2018a. doi: 10.1098/rspa.2017.0639.

J. Rudge and D. Bercovici. Melt-band instabilities with two-phase damage. *Geophysical Journal International*, 201(2):640–651, 2015. doi: 10.1093/gji/ggv040.

J. F. Rudge. The viscosities of partially molten materials undergoing diffusion creep. *Journal of Geophysical Research: Solid Earth*, 123(12):10–534, 2018b.

J. F. Rudge, P. B. Kelemen, and M. Spiegelman. A simple model of reaction-induced cracking applied to serpentinization and carbonation of peridotite. *Earth and Planetary Science Letters*, 291(1-4): 215–227, 2010.

J. F. Rudge, D. Bercovici, and M. Spiegelman. Disequilibrium melting of a two phase multicomponent mantle. *Geophysical Journal International*, 184(2):699–718, 2011.

K. Salari and P. Knupp. Code verification by the method of manufactured solutions. Technical report, Sandia National Laboratories, Albuquerque, NM (US); Sandia National Laboratories, Livermore, CA (US), 2000.

A. Schiemenz, Y. Liang, and E. Parmentier. A high-order numerical study of reactive dissolution in an upwelling heterogeneous mantle—I. Channelization, channel lithology and channel geometry. *Geophysical Journal International*, 186(2):641–664, 2011.

H. Schmeling, J. P. Kruse, and G. Richard. Effective shear and bulk viscosity of partially molten rock based on elastic moduli theory of a fluid filled poroelastic medium. *Geophysical Journal International*, 190(3):1571–1578, 2012. doi: 10.1111/j.1365-246X.2012.05596.x.

C. Schoof and I. J. Hewitt. A model for polythermal ice incorporating gravity-driven moisture transport. *Journal of Fluid Mechanics*, 797:504–535, 2016. doi: 10.1017/jfm.2016.251.

G. Schubert, D. Turcotte, and P. Olsen. *Mantle convection in the Earth and Planets*. Cambridge University Press, 2001.

D. Scott and D. Stevenson. Magma ascent by porous flow. *Journal of Geophysical Research*, 91, 1986.

D. Scott and D. Stevenson. A self-consistent model of melting, magma migration and buoyancy-driven circulation beneath mid-ocean ridges. *Journal of Geophysical Research*, 94:2973–2988, 1989.

D. R. Scott. The competition between percolation and circulation in a deformable porous medium. *Journal of Geophysical Research*, 93(B6):6451–6462, 1988.

D. R. Scott and D. J. Stevenson. Magma solitons. *Geophysical Research Letters*, 11(11):1161–1164, 1984.

D. R. Scott, D. J. Stevenson, and J. A. Whitehead. Observations of solitary waves in a viscously deformante pipe. *Nature*, 319:759–761, 1986.

S. J. Sim, M. Spiegelman, D. R. Stegman, and C. Wilson. The influence of spreading rate and permeability on melt focusing beneath mid-ocean ridges. *Physics of the Earth and Planetary Interiors*, page 106486, 2020.

G. Simpson and M. Spiegelman. Solitary wave benchmarks in magma dynamics. *Journal of Scientific Computing*, 49:268–290, 2011.

G. Simpson, M. Spiegelman, and M. I. Weinstein. Degenerate dispersive equations arising in the study of magma dynamics. *Nonlinearity*, 20(1):21–49, 2007.

G. Simpson, M. Spiegelman, and M. Weinstein. A multiscale model of partial melts: 1. Effective equations. *Journal of Geophysical Research*, 115, 2010a. doi: 10.1029/2009JB006375.

G. Simpson, M. Spiegelman, and M. Weinstein. A multiscale model of partial melts: 2. Numerical results. *Journal of Geophysical Research*, 115, 2010b. doi: 10.1029/2009JB006376.

N. Sleep. Tapping of melt by veins and dikes. *Journal of Geophysical Research*, 93, 1988.

N. H. Sleep. Segregation of magma from a mostly crystalline mush. *Geological Society of America Bulletin*, 85:1225–1232, 1974.

N. H. Sleep. Tapping of magmas from ubiquitous mantle heterogeneities: an alternative to mantle plumes? *Journal of Geophysical Research: Solid Earth*, 89(B12):10029–10041, 1984.

C. S. Smith. Grains, phases, and interphases: an interpretation of microstructure. *Transactions of the Metallurgical Society of AIME*, 175:15–51, 1948.

D. Sparks and E. Parmentier. Melt extraction from the mantle beneath spreading centers. *Earth and Planetary Science Letters*, 105, 1991.

D. C. Spencer, R. F. Katz, and I. J. Hewitt. Magmatic intrusions control Io's crustal thickness. *Journal of Geophysical Research*, 125(6), 2020a. doi: 10.1029/2020je006443.

D. C. Spencer, R. F. Katz, I. J. Hewitt, D. A. May, and L. P. Keszthelyi. Compositional layering in Io driven by magmatic segregation and volcanism. *Journal of Geophysical Research: Planets*, 125(9): e2020JE006604, 2020b.

M. Spiegelman. Flow in deformable porous media. Part 1: Simple analysis. *Journal of Fluid Mechanics*, 247, 1993a.

M. Spiegelman. Flow in deformable porous media. Part 2: Numerical analysis—The relationship between shock waves and solitary waves. *Journal of Fluid Mechanics*, 247, 1993b.

M. Spiegelman. Physics of melt extraction: theory, implications, and applications. *Philosophical Transactions of the Royal Society London A*, 342, 1993c.

M. Spiegelman. Geochemical consequences of melt transport in 2-d: The sensitivity of trace elements to mantle dynamics. *Earth and Planetary Science Letters*, 139, 1996.

M. Spiegelman. UserCalc: A Web-based uranium series calculator for magma migration problems. *Geochemistry, Geophysics, Geosystems*, 1(8), 2000. doi: 10.1029/1999GC000030.

M. Spiegelman. Linear analysis of melt band formation by simple shear. *Geochemistry, Geophysics, Geosystems*, 2003. doi: 10.1029/2002GC000499.

M. Spiegelman and T. Elliott. Consequences of melt transport for uranium series disequilibrium in young lavas. *Earth and Planetary Science Letters*, 118, 1993.

M. Spiegelman and P. Kelemen. Extreme chemical variability as a consequence of channelized melt transport. *Geochemistry, Geophysics, Geosystems*, 4(7), 2003. doi: 10.1029/2002GC000336.

M. Spiegelman and P. Kenyon. The requirements for chemical disequilibrium during magma migration. *Earth and Planetary Science Letters*, 1992. doi: 10.1016/0012-821X(92)90119-G.

M. Spiegelman and D. McKenzie. Simple 2-D models for melt extraction at mid-ocean ridges and island arcs. *Earth and Planetary Science Letters*, 83, 1987.

M. Spiegelman and J. Reynolds. Combined dynamic and geochemical evidence for convergent melt flow beneath the east pacific rise. *Nature*, 402, 1999.

M. Spiegelman, P. Kelemen, and E. Aharonov. Causes and consequences of flow organization during melt transport: the reaction infiltration instability in compactible media. *Journal of Geophysical Research*, 106, 2001.

O. Šrámek, Y. Ricard, and F. Dubuffet. A multiphase model of core formation. *Geophysical Journal International*, 181(1):198–220, 2010.

O. Šrámek, L. Milelli, Y. Ricard, and S. Labrosse. Thermal evolution and differentiation of planetesimals and planetary embryos. *Icarus*, 217(1):339–354, 2012.

D. Stevenson. Self-regulation and melt migration (can magma oceans exist?). *Eos, Transactions, American Geophysical Union*, 61:1021, 1980.

D. Stevenson. Spontaneous small-scale melt segregation in partial melts undergoing deformation. *Geophysical Research Letters*, 16, 1989.

D. Stevenson and D. Scott. Mechanics of fluid-rock systems. *Annual Review of Earth and Planetary Sciences*, 23, 1991.

L. Stixrude and B. Karki. Structure and freezing of $MgSiO_3$ liquid in Earth's lower mantle. *Science*, 310 (5746):297–299, 2005.

R. L. Stocker and R. B. Gordon. Velocity and internal friction in partial melts. *Journal of Geophysical Research*, 80(35):4828–4836, 1975.

E. Stolper, D. Walker, B. H. Hager, and J. F. Hays. Melt segregation from partially molten source regions: The importance of melt density and source region size. *Journal of Geophysical Research*, 86(B7): 6261–6271, 1981.

A. Stracke. Earth's heterogeneous mantle: A product of convection-driven interaction between crust and mantle. *Chemical Geology*, 330:274–299, 2012.

A. Stracke and B. Bourdon. The importance of melt extraction for tracing mantle heterogeneity. *Geochimica et Cosmochimica Acta*, 73(1):218–238, 2009.

A. Stracke, B. Bourdon, and D. McKenzie. Melt extraction in the Earth's mantle: Constraints from U-Th-Pa-Ra studies in oceanic basalts. *Earth and Planetary Science Letters*, 244:97–112, 2006. doi: 10.1016/j.epsl.2006.01.057.

P. Szymczak and A. J. C. Ladd. Interacting length scales in the reactive-infiltration instability. *Geophysical Research Letters*, 40(12):3036–3041, 2013.

P. Szymczak and A. J. C. Ladd. Reactive-infiltration instabilities in rocks. Part 2. Dissolution of a porous matrix. *Journal of Fluid Mechanics*, 738:591–630, 2014.

P. Tackley. Mantle convection and plate tectonics: Toward an integrated physical and chemical theory. *Science*, 288:2002–2007, 2000a.

P. J. Tackley. Self-consistent generation of tectonic plates in time-dependent, three-dimensional mantle convection simulations. *Geochemistry, Geophysics, Geosystems*, 1(8), 2000b.

Y. Takei. Stress-induced anisotropy of partially molten rock analogue deformed under quasi-static loading test. *Journal of Geophysical Research*, 115:B03204, 2010. doi: 10.1029/2009JB006568.

Y. Takei and S. Hier-Majumder. A generalized formulation of interfacial tension driven fluid migration with dissolution/precipitation. *Earth and Planetary Science Letters*, 288:138–148, 2009.

Y. Takei and B. Holtzman. Viscous constitutive relations of solid-liquid composites in terms of grain boundary contiguity: 1. Grain boundary diffusion control model. *Journal of Geophysical Research*, 2009a. doi: 10.1029/2008JB005850.

Y. Takei and B. Holtzman. Viscous constitutive relations of solid-liquid composites in terms of grain boundary contiguity: 2. Compositional model for small melt fractions. *Journal of Geophysical Research*, 2009b. doi: 10.1029/2008JB005851.

Y. Takei and B. Holtzman. Viscous constitutive relations of solid-liquid composites in terms of grain boundary contiguity: 3. Causes and consequences of viscous anisotropy. *Journal of Geophysical Research*, 2009c. doi: 10.1029/2008JB005852.

Y. Takei and R. Katz. Consequences of viscous anisotropy in a deforming, two-phase aggregate: 1. Governing equations and linearised analysis. *Journal of Fluid Mechanics*, 734:424–455, 2013. doi: 10.1017/jfm.2013.482.

Y. Takei and R. F. Katz. Consequences of viscous anisotropy in a deforming, two-phase aggregate. Why is porosity-band angle lowered by viscous anisotropy? *Journal of Fluid Mechanics*, 784:199–224, 2015. doi: 10.1017/jfm.2015.592.

G. I. Taylor and H. Quinney. The latent energy remaining in a metal after cold working. *Proceedings of the Royal Society of London. Series A, Containing Papers of a Mathematical and Physical Character*, 143(849):307–326, 1934.

J. Taylor-West and R. F. Katz. Melt-preferred orientation, anisotropic permeability and melt-band formation in a deforming, partially molten aggregate. *Geophysical Journal International*, 203(2): 1253–1262, 2015. doi: 10.1093/gji/ggv372.

M. Tirone, J. Ganguly, and J. P. Morgan. Modeling petrological geodynamics in the Earth's mantle. *Geochemistry, Geophysics, Geosystems*, 10(4), 2009.

M. Tirone, G. Sen, and J. Phipps Morgan. Petrological geodynamic modeling of mid-ocean ridges. *Physics of the Earth and Planetary Interiors*, 190:51–70, 2012.

A. Toramaru and N. Fujii. Connectivity of melt phase in a partially molten peridotite. *Journal of Geophysical Research*, 91:9239–9252, 1986.

L. N. Trefethen. Finite difference and spectral methods for ordinary and partial differential equations. http://people.maths.ox.ac.uk/trefethen/pdetext.html, 1996.

D. L. Turcotte. Magma Migration. *Annual Review of Earth and Planetary Sciences*, 10(1):397–408, 1982.

D. L. Turcotte and J. L. Ahern. A Porous Flow Model for Magma Migration in the Asthenosphere. *Journal of Geophysical Research*, 83(B2):767–772, 1978.

D. L. Turcotte and E. R. Oxburgh. Mantle convection and the new global tectonics. *Annual Review of Fluid Mechanics*, 4(1):33–66, 1972.

D. L. Turcotte and G. Schubert. *Geodynamics*. Cambridge University Press, 2014.

A. J. Turner, R. F. Katz, and M. D. Behn. Grain-size dynamics beneath mid-ocean ridges: Implications for permeability and melt extraction. *Geochemistry, Geophysics, Geosystems*, 16(3):925–946, 2015.

A. J. Turner, R. F. Katz, M. D. Behn, and T. Keller. Magmatic focusing to mid-ocean ridges: The role of grain-size variability and non-Newtonian viscosity. *Geochemistry, Geophysics, Geosystems*, 23(1): 15–14, 2017. doi: 10.1002/2017GC007048.

S. Turner, S. Black, and K. Berlo. ^{210}Pb–^{226}Ra and ^{228}Ra–^{232}Th systematics in young arc lavas: Implications for magma degassing and ascent rates. *Earth and Planetary Science Letters*, 227(1-2):1–16, 2004.

R. J. Twiss. Theory and applicability of a recrystallized grain size paleopiezometer. In *Stress in the Earth*, pages 227–244. 1977.

J. A. van Orman, T. L. Grove, and N. Shimizu. Rare earth element diffusion in diopside: influence of temperature, pressure, and ionic radius, and an elastic model for diffusion in silicates. *Contributions to Mineralogy and Petrology*, 141(6):687–703, 2001.

P. J. Vaughan, D. L. Kohlstedt, and H. Waff. Distribution of the glass phase in hot-pressed, olivine-basalt aggregates: An electron microscopy study. *Contributions to Mineralogy and Petrology*, 81(4): 253–261, 1982.

Z. Vestrum and S. Butler. Effects of ongoing melting and buoyancy on melt band evolution in a compacting porous layer. *Physics of the Earth and Planetary Interiors*, page 106485, 2020.

N. von Bargen and H. Waff. Permeabilities, interfacial-areas and curvatures of partially molten systems: Results of numerical computation of equilibrium microstructures. *Journal of Geophysical Research*, 91, 1986.

O. Šrámek, Y. Ricard, and D. Bercovici. Simultaneous melting and compaction in deformable two-phase media. *Geophysical Journal International*, 2007. doi: 10.1111/j.1365-246X.2006.03269.x.

I. Wada, M. D. Behn, and J. He. Grain-size distribution in the mantle wedge of subduction zones. *Journal of Geophysical Research: Solid Earth*, 116(B10), 2011.

H. Waff and J. Bulau. Equilibrium fluid distribution in an ultramafic partial melt under hydrostatic stress conditions. *Journal of Geophysical Research*, 84:6109–6114, 1979.

H. S. Waff. Effects of the gravitational field on liquid distribution in partial melts within the upper mantle. *Journal of Geophysical Research*, 85(B4):1815–1825, 1980.

D. Waldbaum. Temperature changes associated with adiabatic decompression in geological processes. *Nature*, 232(5312):545–547, 1971.

D. Walker, E. M. Stolper, and J. F. Hays. A numerical treatment of melt/solid segregation: Size of the eucrite parent body and stability of the terrestrial low-velocity zone. *Journal of Geophysical Research*, 83(B12):6005–6013, 1978.

S. Wang, S. Constable, C. A. Rychert, and N. Harmon. A lithosphere-asthenosphere boundary and partial melt estimated using marine magnetotelluric data at the central middle atlantic ridge. *Geochemistry, Geophysics, Geosystems*, 21(9):e2020GC009177, 2020.

D. Wark and E. Watson. Grain-scale permeabilities of texturally equilibrated, monomineralic rocks. *Earth and Planetary Science Letters*, 164, 1998.

D. Wark, C. Williams, E. Watson, and J. Price. Reassessment of pore shapes in microstructurally equilibrated rocks, with implications for permeability of the upper mantle. *Journal of Geophysical Research*, 108, 2003. doi: 10.1029/2001JB001575.

S. M. Weatherley and R. F. Katz. Melting and channelized magmatic flow in chemically heterogeneous, upwelling mantle. *Geochemistry, Geophysics, Geosystems*, 13(1), 2012. doi: 10.1029/2011GC003989.

S. M. Weatherley and R. F. Katz. Melt transport rates in heterogeneous mantle beneath mid-ocean ridges. *Geochimica et Cosmochimica Acta*, 172:39–54, 2016. doi: 10.1016/j.gca.2015.09.029.

L. T. White, N. Rawlinson, G. S. Lister, F. Waldhauser, B. Hejrani, D. A. Thompson, D. Tanner, C. G. Macpherson, H. Tkalčić, and J. P. Morgan. Earth's deepest earthquake swarms track fluid ascent beneath nascent arc volcanoes. *Earth and Planetary Science Letters*, 521:25–36, 2019.

R. S. White, D. McKenzie, and R. K. O'Nions. Oceanic crustal thickness from seismic measurements and rare earth element inversions. *Journal of Geophysical Research: Solid Earth*, 97(B13):19683–19715, 1992.

C. Wiggins and M. Spiegelman. Magma migration and magmatic solitary waves in 3-D. *Geophysical Research Letters*, 22:1289–1292, 1995.

C. R. Wilson, M. Spiegelman, P. E. van Keken, and B. R. Hacker. Fluid flow in subduction zones: The role of solid rheology and compaction pressure. *Earth and Planetary Science Letters*, 401:261–274, 2014.

M. Wilson. *Igneous Petrologensis: A Global Tectonic Approach*. Springer, 1989.

X. Xiao, B. Evans, and Y. Bernabé. Permeability evolution during non-linear viscous creep of calcite rocks. *Pure and Applied Geophysics*, 163(10):2071–2102, 2006.

B. Yakobson. Morphology and rate of fracture in chemical decomposition of solids. *Physical Review Letters*, 67(12):1590, 1991.

V. Yarushina, Y. Podladchikov, and L. Wang. Model for (de)compaction and porosity waves in porous rocks under shear stresses. *Journal of Geophysical Research: Solid Earth*, 125(8):e2020JB019683, 2020.

V. M. Yarushina, Y. Y. Podladchikov, and J. A. D. Connolly. (De)compaction of porous viscoelastoplastic media: Solitary porosity waves. *Journal of Geophysical Research: Solid Earth*, 120(7):4843–4862, 2015.

T. Yoshino, Y. Nishihara, and S.-i. Karato. Complete wetting of olivine grain boundaries by a hydrous melt near the mantle transition zone. *Earth and Planetary Science Letters*, 256(3-4):466–472, 2007.

W. Zhu and G. Hirth. A network model for permeability in partially molten rocks. *Earth and Planetary Science Letters*, 212:407–416, 2003.

Index

activity, 151–157, 164, 203–205, 207–209; ratio, 152, 153, 156, 157, 204, 205, 208–210
adiabatic, 74, 79, 127, 128, 138, 180
advection, 41, 62, 107, 109–111, 113, 120, 123, 126, 143, 153, 154, 185, 204, 230–234, 246, 267, 276, 278
affinity, 175, 176, 216
algorithm, 180, 181, 268
ammonium chloride, 170, 180
anisotropy, 6, 17, 26, 76, 77, 117, 118
ansatz, 38, 224
assembly operator, 265
asthenosphere, 3, 6–10, 12, 14, 24, 26, 49, 55, 63, 74, 94–96, 118, 119, 122, 124, 129, 130, 141, 161, 182, 189, 199, 214, 216, 226
asymmetry, 113, 256
asymptotic, 108, 115, 140, 185, 186, 198, 210, 223, 226, 230, 235

basalt, 5, 13, 25, 40, 63, 64, 66–68, 73, 76, 77, 94, 95, 101, 130, 158, 161, 167–171, 174, 185, 236
base state, 37, 38, 103–105, 109, 111, 113, 115, 117, 118, 218, 221–225, 231, 232, 233, 234, 237, 283, 316, 317
basis function, 264–266, 278
benchmark, 14, 210, 258, 260, 269, 275–277
Bercovici, David, 26
biharmonic, 33, 34, 36, 54, 111
bilinear, 265, 266, 278, 319
binary, 161, 170, 181, 210
binary-loop, 210
boundary layer, xix, 10, 14, 24, 83, 140, 182, 185–187, 196–200, 209, 212, 218, 222, 226, 243
Boussinesq, 3, 31, 36, 43, 55
brittle-ductile transition, 10
Bulau, J.R., 19
buoyancy, 1, 6–8, 10, 13, 14, 20, 23, 33, 73, 83, 85–87, 90, 93, 94, 100, 102, 116, 118, 183–187, 190, 194, 197, 199, 202, 210, 238, 240–242, 244, 245, 255, 256, 284

capillarity, 114, 115
carbon, 1, 5, 8, 10, 11, 141, 145, 161, 194
central difference, 261
channelization, 8, 9, 25, 28, 95, 158, 214, 232, 235, 236, 256
channels, 8, 25
chemical potential, xix, 60, 64, 121, 122, 138, 161, 163–165, 175
chemical transport, 5, 9, 11, 25
chromatographic, 153, 158, 206–209
Clausius Clapeyron, xviii, 19, 135, 136, 168, 188, 189, 193–196, 249
Coble creep, 72, 117
column, xix, 8, 13, 14, 19, 20, 83, 97, 137, 158, 180, 182–199, 201–212, 214, 222, 234, 242, 243, 247, 248, 256, 310
compaction, xviii, 4, 8, 13, 14, 19–25, 44, 46, 50–56, 68–73, 77, 80–88, 90–98, 100, 102–104, 107–109, 111, 120–122, 124, 128, 130–132, 134, 139, 140, 178, 183–188, 190, 196–199, 202, 215, 217–219, 222, 224–226, 228, 230–237, 239–241, 246, 254–256, 259–261, 263, 264, 267, 268, 271, 272, 274–278, 286, 293, 294, 307, 316, 317; decompaction, xviii, 14, 23, 107, 199–201, 211, 212, 218, 243–251, 255; decompaction layer, 202, 203, 211, 243, 245, 256; equation, 52, 107; length, xviii, 13, 24, 82–84, 86–88, 93–96, 104, 108, 112–114, 116, 130, 131, 184, 187, 197, 200, 202, 218–220, 233, 234, 238, 244–246, 259, 269, 276, 277; potential, 111, 271
compatible, 74, 147
compressibility, 3, 4, 44, 50, 73, 74, 78, 97
concentration, xix, 7, 16, 63, 71, 125, 141–154, 159, 161–164, 168–170, 172–174, 176, 177, 179, 181, 191, 192, 194–196, 203, 204, 212, 213, 215, 216, 218–221, 255, 308, 315
conductivity, xviii, 31, 37, 76, 122, 128
contiguity, 3, 117, 118
continuity, 31, 36, 43, 70, 91, 254, 282, 283, 286
convection, xiii, xiv, 1, 2, 9, 12, 15, 20, 22, 27, 29, 31–33, 35–37, 39, 45, 54, 76, 100, 103, 129, 138, 151, 240, 255, 256, 284
convention, 30
corner flow, 8, 24, 33, 34, 35–37, 118, 183, 185, 203, 235, 241, 242, 255, 256
corrosive, 232

crystal zoning, 141
cylindrical, 70, 72, 78, 241

damage, 26, 55, 75, 97, 137
Damköhler, 220, 225–227, 229, 231, 232
Darcy, 3, 19, 20, 22, 24, 46, 49, 51, 52, 78, 85, 94, 95, 124, 130, 197, 198, 200, 217, 231, 232, 275, 290; drag, xviii, 13, 24, 48, 49, 66, 86, 93, 107, 184–187, 190, 194, 197; region, 197, 210; solution, 186, 193, 196, 197, 199, 203, 205, 208; speed, 94
decay, xix, 7, 31, 82–84, 86, 103, 105, 141, 143, 151–156, 159, 160, 203, 204, 206, 207, 209, 211, 218, 242, 276
decompaction, 24
decompression, 74
deep mantle, 11
deglaciation, 7, 185, 211
degree of melting, xviii, 8, 74, 121, 135–138, 153, 159, 173, 174, 183–187, 189–196, 203–205, 207, 211, 213, 310, 313
degrees of freedom, 258, 274, 275
Δ, 15
depleted, 168, 190, 195
diagnostic, 26, 178
diffuse interface, 114, 115
diffusion, 3, 9, 25, 41, 60–62, 64, 68, 70, 71, 77, 96, 117, 118, 122, 123, 126, 138, 140–143, 148–150, 152, 158, 182, 187, 191, 212, 214, 220, 221, 230–233, 246, 309
diffusion creep, 71, 77, 96, 117
dihedral angle, xviii, 11, 19, 21, 47, 63–66, 76, 78, 114, 287, 288, 291
dike, 6, 9, 10, 26
dilution, 157, 158, 204, 209
Dirichlet, 261, 264, 265, 267, 274
disaggregation, xviii, 51, 52, 55, 66, 71, 98, 139, 305
discretization, 26, 27, 129, 254, 258–261, 263, 264, 267, 272, 275–279, 320
disequilibrium, 7, 9, 13, 14, 17, 25, 26, 74, 76, 122, 127, 138, 144, 147–150, 152, 153, 157–160, 174, 175, 177, 179–181, 196, 203, 207, 209, 211, 212, 214, 220, 236, 252, 255
dislocations, 3, 59, 61, 75
dislocation creep, 68, 112, 117, 118
dispersion, 15, 24, 66, 88, 90, 109, 115, 143, 225–230, 235
dispersivity, 143
dissipation, xviii, 11, 12, 31, 36, 75, 122, 124, 126–130, 132, 134, 138–140, 175, 176, 178, 212, 253, 284, 307
dissolve, 63, 77, 169, 216, 217
Drew, Donald A., 21
dunite, 5, 25, 214, 216, 234, 235

effective component, 162
effective pressure, 50
effective stress, 49
efficiency, 150, 244, 245
eigenfunction, 112, 227, 229
eigenvalue, 155, 224
eigenvector, 156
energy minimum, 62, 63, 161
enstatite, 4, 144–146
enthalpy, xix, 13, 121, 124–129, 161, 164, 165, 252–254, 256; method, 14, 178, 179, 238, 252, 254–256, 278
entropy, xix, 13, 75, 121, 122, 126, 127, 135, 137, 161, 164, 165, 175, 176
equilibrium, xix, 7, 9, 13, 18–20, 22, 23, 25, 46, 47, 64, 70, 74, 121, 123, 134, 147–150, 152, 153, 161, 162, 164–167, 169, 171, 173–181, 188, 196, 216–220, 235, 252, 253, 255, 256, 309; chemical, 121, 154, 161, 163, 165, 174, 191, 192, 204, 220, 226, 235, 238, 252; secular, 152, 156–158, 203–208, 313; textural, 62, 63, 66, 68, 76, 100; thermodynamic, 74, 121, 122, 124, 127, 163, 177, 188
error, 6, 24, 166, 251, 262, 263, 267, 269, 270, 275, 276, 279
Eulerian, 128, 139
eutectic, 162, 170–172, 178, 180
expansivity, xviii, 122, 126, 128, 188
experiments, 100

Faizullaev, Dzharulla F., 19
fayalite, 4, 167, 181
feedback, xiv, 11, 62, 76, 107
FEniCS, 27
fertile, 5, 145, 173, 191, 193, 196
Fick's law, 142
filter-press, 185
finite difference, 14, 179, 198, 259–261, 263, 264, 267, 272, 274, 275, 277, 278
finite element, 14, 27, 259, 260, 263–265, 267, 278
finite strain, 113
finite volume, 27, 129, 252, 278
first law of thermodynamics, 13, 122
fluid dynamics, 13, 14, 17, 35, 39, 44, 69, 80, 121, 124, 129, 177, 179, 215, 252, 256
flux melting, 10, 214
focusing, 8, 9, 14, 25, 46, 158, 203, 211, 238, 242–248, 250–252, 256, 317, 319
forsterite, 4, 143–146, 167, 181
Fourier, 123, 127, 158
Fowler, Andrew C., 21
fractional crystallization, 150
fractional melting, 150, 159, 212, 308, 311
fractionate, xiii, 9, 27, 147, 148, 153, 157, 159, 203, 208
Frank, F. Charles, 18
freezing, xiii, xix, 14, 44, 97, 101, 126, 136, 144, 146, 150, 153, 190, 199–203, 211, 217, 244–246, 248–251, 256, 319
functional, 260
fusible, 3, 168, 190, 191

Galerkin, 265
garnet, 4, 152, 153, 161, 209

geochemical, 8–10, 17, 27, 193, 203, 214, 234, 236
geochemistry, 9, 25, 27
geodynamic, xiii, xiv, 2, 4, 35, 36, 60–62, 75, 124, 125, 181
geological, 1, 8, 25, 59, 62, 79, 158, 214, 226, 227, 234
geophysical, 17, 68, 193
geotherm, 11, 129, 134, 138, 199
Gibbs free energy, 138, 161, 163–165, 180
glacial cycles, 256
governing equations, 12, 13, 20, 22, 23, 26, 29, 32, 37, 38, 52, 53, 55, 58, 73, 77, 80, 87, 88, 97, 100–104, 114, 131, 146, 182, 189, 191, 199, 215, 217, 221, 239, 240, 255, 269, 270, 282
grain scale, 6
gravity, xvii, 8, 19, 20, 24, 29, 32, 35, 45, 52, 58, 74, 78, 80, 83, 92, 94, 98, 122, 123, 128, 129, 134, 138, 162, 182, 199, 202, 203, 210, 243, 244, 269, 291, 295

half-life, 156, 157, 205–207, 209
half-space cooling model, 250, 251
harzburgite, 25
heat capacity, xviii, 31, 122, 126, 128, 164, 188, 189, 252
Helmholtz decomposition, 53–55, 57, 98, 104, 239, 270
heterogeneity, xiii, 9, 10, 26–28, 158, 190, 191, 236, 256, 278
Hillert model, 60, 75
homogenization, 54
hydrology, 11

ice, 2, 3, 7, 11, 18, 19, 21, 22, 122, 130, 138, 256
Iceland, 7, 8, 95, 185, 211
ideal solution, 13, 161, 163, 164, 167, 168, 170, 173, 175, 178, 180, 254
immiscible, 41, 46, 161, 162, 168
impermeable, 4, 14, 20, 63, 97, 98, 130, 131, 202, 241, 244, 246
incompatible, 5, 147, 148, 179, 194, 196, 251
incompressible, 4, 12, 31, 33, 36, 43, 44, 48–50, 53, 62, 69, 70, 78, 104, 107, 238–240, 271, 277, 280
incongruent, 144–146, 216, 217, 219
indicator function, xviii, 41, 42, 46, 285; gradient, 42, 46
ingrowth, 153–158, 204–209, 212, 313
instability, 13, 25, 33, 55, 75, 96, 101, 105, 107, 108, 115, 116, 118, 224, 226, 230, 233–237, 317; reaction-infiltration, 118, 233–235
interface, xviii, 23, 26, 42, 43, 45–47, 58, 63–66, 70, 114, 115, 117, 121–125, 142, 143, 148, 216, 278, 285, 287, 288; and melting, 43
internal energy, xviii, 13, 121–125, 127, 128
interphase force, 45, 46
interpolate, 274
invariant, xvii, 27, 68, 78, 105, 111, 143, 291, 292
isentropic, 3, 36, 74, 122, 127, 129, 134, 135, 137, 138, 182, 187, 210, 214, 215, 249
isotope, xiii, 7, 17, 26, 27, 151–153, 156, 157, 204–206

kinetics, 75, 175, 216, 252
Kohlstedt, David L., 25, 116
Kozeny-Carman, 66

laboratory, xiv, 5, 13, 16, 17, 25, 60, 68, 77, 95, 96, 100, 101, 108, 113, 116, 118, 122, 130, 193, 235
Lagrangian, 31, 44, 52, 62, 71, 80, 125, 126, 129, 137, 139, 143, 178, 304
Lagrangian derivative, 44
latent heat, xviii, 125, 126, 128, 129, 132, 137, 166, 181, 188, 189, 236, 249, 253
lever rule, 163, 173, 193
limestone, 235
linear algebra, 254, 258, 268
linearize, 71, 87, 88, 90, 97, 102, 103, 105, 113, 117, 134, 173, 174, 177, 180, 191–195, 210, 211, 214, 221, 303
linearized stability analysis, 103
liquidus, xix, 136, 163, 167–169, 171–174, 178, 179, 190–194
lithological, 7, 12, 216
lithosphere, xix, 1, 3, 8, 10, 12, 14, 24–26, 33, 34, 37, 95, 96, 119, 187, 199, 203, 217, 241, 246, 248–250, 255, 319
lithostatic, xviii, 20, 23, 69, 85, 122, 124, 127–129, 136, 148, 179, 189, 192, 212, 239, 240, 247, 253
localize, xiv, 14, 24, 25, 60, 73, 90, 113, 138, 214, 218, 241
low-productivity tail, 196

magmon, 24, 88
major element, 141, 143, 147, 148
mantle plume, 3, 238
manufactured solutions, method of, 259, 269–271, 274, 275, 278
matched asymptotic, 186, 210
MATLAB, 258, 259, 262, 266, 268, 275, 277
Maxwell, 121
McKenzie, Dan, xv, 22–24, 27, 28, 77
melting interval, 193
melting rate, 7, 13, 24, 43, 52, 53, 58, 74, 126, 129, 132, 137, 138, 140, 145, 148, 150, 153, 156, 157, 159, 174–179, 183, 186–190, 192, 194, 195, 200, 203, 211, 212, 214–216, 232, 233, 235, 252, 254, 307, 310, 313, 318
melting region, 3, 24, 137, 185, 233, 234, 238, 246, 247, 249–251
microscopic, 6, 39, 47
microstructure, 65
mid-ocean ridge, xiii, 1–3, 5, 8–10, 12, 14, 18, 24–26, 33, 34, 75, 79, 94, 95, 118, 122, 146, 151, 158, 169, 170, 182, 183, 185, 193, 203, 210, 213, 233–236, 238, 241, 243, 246–249, 250, 252, 255, 256
minerals, 4
mixing, xiii, 9, 27, 56, 161, 164, 175
mixture theory, 27, 54
mobility, xviii, 102, 183, 200, 215, 218, 243, 268, 269, 274, 320

model complexity, 5
molar mass, xix, 145, 146, 165, 181
molar partition coefficient, xix, 165
momentum, 44
multi-component, 21, 26, 138, 180, 190, 215, 256

Nabarro–Herring creep, 71
Newton's method, 91, 167, 172, 254, 268
Newton–Krylov, 278
nonlinear, 10, 24, 50, 88, 89, 102, 105, 115, 195, 234, 239, 252, 254, 268
normal force, 202
normal grain growth, 59
notation, 14
nucleide, 206
numerical, 8, 10, 14, 16, 17, 26, 55, 77, 92–94, 96, 117, 118, 124, 128, 133, 134, 148, 167, 172, 179, 187, 198, 199, 201, 205, 207, 208, 210, 211, 225, 227, 232, 235, 236, 238–240, 251, 252, 254–256, 258–260, 262, 263, 265, 266, 269, 271, 272, 275–278
numerical methods, 26
Nye, John F., 18

olivine, 4, 5, 13, 25, 40, 63, 64, 66–68, 74, 76, 77, 95, 101, 113, 130, 143, 145, 161, 162, 164, 167, 168, 170, 171, 191, 214, 216, 217, 234, 236
Oman, 234
one-component, 138, 188–190, 192, 193, 196, 212, 247, 249
Onsager, 176
ophiolites, 25, 214, 234
orthoenstatite, 144
orthopyroxene, 145, 146

paleo-wattmeter, 75
paleopiezometer, 62, 75
parent–daughter, 156, 157, 205, 208, 209
partition coefficient, xix, 147, 153, 157, 163, 166, 178, 207, 209
passive advection trajectories, 110, 111, 113, 114
pattern formation, 100
Péclet, 220, 230–233
peridotite, 5, 77, 158, 236
permeability, xviii, 3, 7, 8, 11, 13, 19, 20, 22, 24, 25, 48, 52, 53, 55, 58, 62, 63, 66–69, 76–78, 83, 84, 86–88, 93–97, 100, 102, 107, 108, 118, 131, 134, 140, 183, 186, 199, 208, 209, 211, 225, 230–233, 240, 241, 245, 255, 256, 260, 268, 290, 291, 294; step, 83, 84, 86, 93, 97, 185; threshold, 76
perturbation, 92, 93, 103–110, 112, 116–120, 221–225, 228, 229, 231, 237, 276, 277, 283, 302, 303, 316, 317
petrology, xix, 4, 5, 13, 17, 74, 129, 158, 162, 167, 179, 180, 234
PETSc, 26, 259, 278
phase diagram, 168, 170, 172, 173, 180, 188, 192–194, 210, 254, 256
phase fractions, 42
Picard, 254, 268, 279
plagioclase, 4, 161
planetary, xiv, 5, 12, 138, 256
planets, 2, 12; Enceladus, 139; Io, 2, 12, 138, 180; Jupiter, 2, 12, 138
plasticity, 26, 30, 55, 97
plate tectonics, xiii, xiv, 1, 2, 5, 9, 15, 18, 22, 24, 30, 33, 36, 255; environments, 10
Poiseuille, 22, 118, 119
polybaric, 146
polycrystalline, 3, 11, 18, 21, 130, 171
polynomial, 168, 184, 224, 226–228
post-glacial rebound, 3
postglacial, 68
potential energy, xiii, 58, 59, 122, 129, 189
potential temperature, xviii, 129, 130
principle axes, 100, 113
productivity, 74, 136, 186, 189, 193, 195, 196, 198, 215, 233
pseudo-component, 167–169
pure shear, 57, 105, 108, 110, 111, 117, 119, 120
pyroxene, 4, 25, 73, 76, 145, 161, 191, 216, 217, 234–236
Python, xiv, xv, 160, 259, 274

quartz, 145

radiogenic, xiii, xviii, xix, 13, 17, 123, 126, 127, 129, 134, 140, 141, 143, 151, 156, 158, 178, 182, 204, 253
radium, 156, 158, 209
Rational Thermodynamics, 181
Rayleigh number, 33
reaction, xix, 1, 10, 11, 118, 144–147, 150, 158, 175–177, 214–217, 219, 220, 233, 234, 236, 252
reactive flow, 8, 9, 11, 14, 25, 75, 95, 117, 118, 158, 177, 180, 190, 214–220, 231–236, 277, 315
reactive melting, 190, 214–217, 219, 220, 222, 231, 232, 234
reactive potential, 220
recrystallization, 61, 117
refractory, 5, 145, 162, 190, 191, 193, 196
regularize, 114
replacive, 25, 214
representative volume element, 39
rescaling, 16
residual, 1, 22, 74, 174, 190, 216, 255, 260, 263, 265
reversible, 127, 130
Reynolds number, 29, 44
rheology, 26, 35, 48, 49, 55, 97
Ricard, Yanick, xv, 26
rigid, 53
rotation, 108, 110, 113, 215, 287
Rudge, John F., xv, 28, 211
RVE, xviii, 39–42, 44–47, 49, 56, 121, 123, 141–143, 149, 285

sea level, 79, 211
second law of thermodynamics, 127, 176
second order, 263
second invariant, 105

segregation, xiii, 1–3, 12–14, 18–22, 24, 25, 27, 44, 47, 52, 53, 55, 58, 72–76, 81–86, 92–94, 97, 102, 107, 116, 118, 124, 127, 132, 134–138, 143, 153, 158, 182, 184–186, 190, 193, 199, 203, 204, 210, 235, 241, 242, 252, 255, 256; channelized, 8; rate, 7
sensible heat, 126, 129, 137, 249
separation of scales, 6, 39, 118, 238
series expansion, 103
shape function, xix, 200, 201, 212, 244, 246, 311, 312
silicate, 2, 5, 12, 15, 73, 77, 100, 125, 138, 164, 169, 175, 178–180, 196, 211, 216, 220, 236
Simon–Glatzel, 168, 181
simple shear, 78, 105, 107–111, 115, 117–119, 131, 132, 269, 276
singularity, 72, 73, 77, 96, 186, 187
Sleep, Norman H., 19
small-porosity, 88, 186, 197, 239, 241
solenoidal, 53, 54, 240, 256
solid solution, 4, 162, 167, 170, 172, 180, 181, 256
solidus, xix, 10, 78, 134–138, 140, 163, 167–174, 178, 179, 188, 190–196, 212, 246, 247, 249, 251, 314
solitary wave, 13, 24, 88, 90–94, 96, 97, 122, 256, 259, 260, 267, 268, 275–277, 279
solubility, 216, 218, 219, 232, 233
sparse matrix, 262
Spiegelman, Marc W., 24, 25, 28
spinel, 4, 5, 161
spontaneous, 59, 62, 93, 114, 161, 164, 176
spreading rate, 183, 185, 246, 249–251
staggered grid, 27, 273, 278, 279
stencil, 262, 268
Stevenson, David J., 25
stiffness, 220, 228, 231–233
Stokes, 3, 12, 14, 19, 29, 33, 36, 47, 52, 53, 78, 107, 238, 240, 254, 275, 277, 290
subduction, 1, 3, 7, 10, 15, 24, 33, 37, 75, 169, 210, 238, 256, 278
subscript, 16
supercritical, 161
superscript, 15
surface tension, 23, 26, 45, 46, 55, 58, 62, 63, 97, 114, 116–118, 120, 276, 277
symmetry, 26, 34, 55, 63, 70, 72, 113, 140, 247, 259, 287, 288, 290, 297

tabular, 214, 234, 235
Taylor series, 88, 105, 119, 130, 139, 177, 207, 222, 248, 293, 300, 301, 304, 315–317
Taylor-Quinney fraction, 61
temperature, 31
tension, 30, 46
Terzaghi's principle, 49
test function, 260, 263, 265
texture, 62, 76
thermochemical, 5, 33, 121, 143, 144, 146, 162, 163, 166, 168–170, 174, 176, 178–180, 183, 210, 215, 251–254, 278
thermodynamic, xiv, 4, 5, 13, 15, 19, 23, 26, 64, 73–75, 78, 121, 122, 124–127, 132, 135, 138, 141, 143, 144, 161, 163, 164, 167, 170, 172, 173, 175, 176, 180, 181, 186–190, 193, 210, 214, 216, 252–254, 310; flow, 175, 176; force, 175, 176; pressure, 23, 121, 124, 125, 127, 129, 137, 138, 162, 192, 253
thorium, 156, 158, 209
three-component, 169–171, 196, 210
tomography, xiii, 8, 67, 95, 185
torsion, 100–102, 108, 117, 118
tortuosity, 47, 66, 143
trace element, 9, 13, 17, 25, 26, 141, 143, 146-152, 154, 158, 159, 163, 204, 309
trial function, 260
tridiagonal, 262, 266, 267, 278, 319
Turcotte, D.L., 19
two-component, 167–170, 173, 174, 180, 187, 190–196, 210–212, 215, 256

U-series, 7, 26
underpressure, 86, 187, 232, 235
undersaturated, 25, 216, 217, 220–222, 231, 232, 236
univariate, 165
uranium, 7, 26, 154, 158, 209, 211; series, 7, 8, 14, 17, 26, 95, 151, 152, 155, 156, 158, 203, 204, 206, 208, 209, 211, 236

validation, 16, 101, 259
vapor, 161, 162
vein, 6, 9, 21
viscogravitational, 185, 186, 196, 197, 199
viscoelastic, 23, 26, 55, 97, 138
viscosity, 48; aggregate, 51, 68, 102, 105, 113; anisotropic, 26, 51, 77, 117, 118; bulk, 50; compaction, 4, 23, 50, 70, 77; constant, 53; exponent, 115; magma, 3, 73, 77; mantle, 30, 77, 95; Newtonian, 106; non-Newtonian, 21, 30, 51, 69, 108, 111–114, 117; shear, 105
volatiles, xiii, 1, 8, 10, 26, 73, 141, 143, 145, 169, 180, 193–196, 210–212, 236, 256
volcanic plumbing, 211
volcano, xiii, 5, 8, 10, 12, 55, 151, 238
volcanology, 55
vorticity, xvii, 32, 33, 54, 98, 239

Waff, Harve S., 19
water, 1, 4, 5, 8–11, 15, 18, 21, 22, 49, 73, 83, 116, 122, 130, 138, 139, 141, 145, 161, 169–171, 194, 235, 256
wave, 3, 12, 13, 15, 24, 83, 87–94, 96, 97, 105, 110, 118, 120, 122, 159, 182, 235, 256, 259, 260, 267, 268, 275–277, 279; length, 88, 90, 108–110, 112, 114–118, 221, 225, 226, 230–232, 234, 235, 284; number, 88, 112, 114, 115, 221, 224, 225, 227, 228, 232, 236, 276, 277, 316; solitary, 88, 96; train, 93; vector, 106, 109, 110, 120
wavevector, 110
weak form, 260, 263, 264, 265

Zener pinning, 60
zero-compaction-length, 93, 97, 185, 203; approximation, 93, 184, 187, 193, 197, 199